# The Indian Nitrogen Assessment

## Related titles

*Nitrogen in the Environment, 2nd Edition*
(ISBN 978-0-12-374347-3)

*Global Biogeochemical Cycles in the Climate System*
(ISBN 978-0-12-631260-7)

*Environmental Biotechnology, 2nd Edition*
(ISBN 978-0-12-407776-8)

*Environmental Geochemistry*
(ISBN 978-0-444-53159-9)

# The Indian Nitrogen Assessment

## Sources of Reactive Nitrogen, Environmental and Climate Effects, Management Options, and Policies

**Edited by**
Yash P. Abrol
Tapan K. Adhya
Viney P. Aneja
Nandula Raghuram
Himanshu Pathak
Umesh Kulshrestha
Chhemendra Sharma
Bijay Singh

ELSEVIER

Woodhead Publishing is an imprint of Elsevier
The Officers' Mess Business Centre, Royston Road, Duxford, CB22 4QH, United Kingdom
50 Hampshire Street, 5th Floor, Cambridge, MA 02139, United States
The Boulevard, Langford Lane, Kidlington, OX5 1GB, United Kingdom



**Library of Congress Cataloging-in-Publication Data**
A catalog record for this book is available from the Library of Congress

**British Library Cataloguing-in-Publication Data**
A catalogue record for this book is available from the British Library

ISBN: 978-0-12-811836-8 (print)
ISBN: 978-0-12-811904-4 (online)

For information on all Woodhead Publishing publications visit our website at https://www.elsevier.com/books-and-journals



*Publisher:* Cathleen Sether
*Acquisitions Editor:* Laura S. Kelleher
*Editorial Project Manager:* Jill Cetel
*Production Project Manager:* Omer Mukthar
*Designer:* Greg Harris

Typeset by TNQ Books and Journals

# Contents

# List of Contributors

**Y.P. Abrol** ING-SCON, DPS Marg, New Delhi, India

**S. Adhikari** ICAR-Central Institute of Freshwater Aquaculture, Bhubaneswar, Odisha, India

**T.K. Adhya** South Asia Nitrogen Centre, DPS Marg, New Delhi, India; KIIT University, Bhubaneswar, Odisha, India

**S. Agrawal** Public Health Foundation of India (New Delhi NCR), Gurgaon, India

**A. Ahmed** Aligarh Muslim University (AMU), Aligarh, India

**S. Anandan** ICAR-NIANP, Bangalore, India

**V.P. Aneja** North Carolina State University, Raleigh, NC, United States

**K.S. Anil** ICAR-National Bureau of Soil Survey and Land Use Planning Regional Station, Bangaluru

**J.S. Arunkumar** University of Agricultural Sciences, Bengaluru, India

**D. Balachandar** Tamil Nadu Agricultural University, Coimbatore, Tamilnadu, India

**K. Batabyal** Bidhan Chandra Krishi Viswavidyalaya, Mohanpur, India

**G. Beig** Indian Institute of Tropical Meteorology, Pune, India

**D.K. Benbi** Punjab Agricultural University, Ludhiana, India

**S.K. Bhanja** ICAR-Central Avian Research Institute, Izatnagar, India

**A. Bhatia** ICAR-Indian Agricultural Research Institute (IARI), New Delhi, India

**S. Bhattacharya** IORA Ecological Solutions Pvt. Ltd., New Delhi, India

**R. Bhattacharyya** CESCRA, ICAR-Indian Agricultural Research Institute, Pusa, India

**Bijay-Singh** Punjab Agricultural University, Ludhiana, Punjab, India

**A.K. Biswas** ICAR-Indian Institute of Soil Science, Bhopal, India

**W. Brownlie** NERC Centre for Ecology & Hydrology (CEH), Edinburgh Research Station, Midlothian, United Kingdom; University of Edinburgh, Edinburgh, United Kingdom

**T.K. Chanda** The Fertiliser Association of India, New Delhi, India

**G.N. Chattopadhyay** Ashirbad, Sevapalli, Santiniketan, West Bengal, India

**O.P. Chaturvedi** ICAR-Central Agroforestry Research Institute, Jhansi, India

**S. Datta** ICAR-Central Institute of Fisheries Education, Kolkata, West Bengal, India

**I. Dev** ICAR-Central Agroforestry Research Institute, Jhansi, India

**M.L. Dotaniya** ICAR-Indian Institute of Soil Science, Bhopal, India

**U. Dragosits** NERC Centre for Ecology & Hydrology (CEH), Edinburgh Research Station, Midlothian, United Kingdom

**J. Drewer** NERC Centre for Ecology & Hydrology (CEH), Edinburgh Research Station, Midlothian, United Kingdom

**B.S. Dwivedi** ICAR-Indian Agricultural Research Institute, New Delhi, India

**A.N. Ganeshamurthy** ICAR-Indian Institute of Horticultural Research, Bengaluru, India

**D. Ganguly** National Centre for Sustainable Coastal Management, Ministry of Environment, Forests and Climate Change, Anna University Campus, Chennai, India

**B.N. Ghosh** ICAR-National Bureau of Soil Survey and Land Use Planning Regional Station, Kolkata, India

**P.K. Ghosh** ICAR-Indian Grassland and Fodder Research Institute, Jhansi, India

**S.D. Ghude** Indian Institute of Tropical Meteorology, Pune, India

**N.K.S. Gowda** ICAR-NIANP, Bangalore, India

**A.B. Gupta** Malaviya National Institute of Technology, Jaipur, Rajasthan, India

**R. Gupta** Government Post Graduate College, Dausa, India

**S.K. Gupta** Consultant Pediatrician & Neonatologist and Scientist of Environmental Medicine, Jaipur, Rajasthan, India

**B.R. Gurjar** Indian Institute of Technology (I.I.T.) – Roorkee, Roorkee, India

**J. Hillier** University of Aberdeen, Aberdeen, United Kingdom

**S. Hooda** Guru Gobind Singh Indraprastha University (IPU), Dwarka, India

**C.M. Howard** NERC Centre for Ecology & Hydrology (CEH), Edinburgh Research Station, Midlothian, United Kingdom; University of Edinburgh, Edinburgh, United Kingdom

**N. Jain** ICAR-Indian Agricultural Research Institute (IARI), New Delhi, India

**V. Jain** Indian Council of Agricultural Research, New Delhi, India

**R. Jeyabaskaran** Central Marine Fisheries Research Institute, Kochi, India

**Pramod Jha** ICAR-Indian Institute of Soil Science, Bhopal, India

**E.J.M. Joy** London School of Hygiene & Tropical Medicine, London, United Kingdom; Leverhulme Centre for Integrative Research on Agriculture and Health (LCIRAH), London, United Kingdom

**D. Kalaivanan** ICAR-Indian Institute of Horticultural Research, Bengaluru, India

**V. Kripa** Central Marine Fisheries Research Institute, Kochi, India

**U. Kulshrestha** Jawaharlal Nehru University, New Delhi, India

**Dhiraj Kumar** ICAR-Central Agroforestry Research Institute, Jhansi, India

**Dinesh Kumar** ICAR-Indian Agricultural Research Institute (IARI), New Delhi, India

**N. Kumar** ICAR-Central Agroforestry Research Institute, Jhansi, India

**R.M. Kumar** ICAR-Indian Institute for Rice Research (IIRR), Hyderabad, India

**S. Kundu** ICAR-Indian Institute of Soil Science, Bhopal, India

**Brij Lal Lakaria** ICAR-Indian Institute of Soil Science, Bhopal, India

**M. Lalitha** ICAR-National Bureau of Soil Survey and Land Use Planning Regional Station, Bangaluru

**S.K. Mahanta** ICAR-Indian Grassland and Fodder Research Institute, Jhansi, India

**M. Maheswari** ICAR-Central Research Institute for Dryland Agriculture, Hyderabad, Telangana, India

**S. Maji** Indian Institute of Tropical Meteorology, Pune, India

**A.B. Mandal** ICAR-Central Avian Research Institute, Izatnagar, India

**B. Mandal** Bidhan Chandra Krishi Viswavidyalaya, Mohanpur, India

**B.P. Meena** ICAR-Indian Institute of Soil Science, Bhopal, India

**M.C. Meena** ICAR-Indian Agricultural Research Institute, New Delhi, India

**M. Mohini** ICAR-NDRI, Karnal, India

**A. Moring** NERC Centre for Ecology & Hydrology (CEH), Edinburgh Research Station, Midlothian, United Kingdom; University of Edinburgh, Edinburgh, United Kingdom

**M. Muralidhar** ICAR-Central Institute of Brackishwater Aquaculture, Chennai, India

**A.N.G. Murthy** ICAR-Indian Institute of Horticultural Research, Bengaluru, Karnataka, India

**A. Nagpure** University of Minnesota, United States

**D.R. Nayak** University of Aberdeen, Aberdeen, United Kingdom

**C.N. Neeraja** ICAR-Indian Institute for Rice Research (IIRR), Hyderabad, India

**A. Paneer Selvam** National Centre for Sustainable Coastal Management, Ministry of Environment, Forests and Climate Change, Anna University Campus, Chennai, India

**A.S. Panicker** Indian Institute of Tropical Meteorology, Pune, India

**H. Pathak** ICAR-Indian Agricultural Research Institute (IARI), New Delhi, India; ICAR-National Rice Research Institute, Cuttack, Odisha, India; National Rice Research Institute (NRRI), Cuttack, India

**Ashok K. Patra** ICAR-Indian Institute of Soil Science, Bhopal, India

**E.V.S. Prakasa Rao** CSIR-Centre for Mathematical Modelling and Computer Simulation, Bengaluru, India

**C.S. Prasad** ICAR-NIANP, Bangalore, India

**R. Prasanna** ICAR-Indian Agricultural Research Institute (IARI), New Delhi, India

**D. Prema** Central Marine Fisheries Research Institute, Kochi, India

**A. Price** University of Aberdeen, Aberdeen, United Kingdom

**N. Priya** Jawaharlal Nehru University, New Delhi, India

**R. Purvaja** National Centre for Sustainable Coastal Management, Ministry of Environment, Forests and Climate Change, Anna University Campus, Chennai, India

**K. Puttanna** Central Institute of Medicinal and Aromatic Plants, Bengaluru, India

**T.K. Radha** ICAR-Indian Institute of Horticultural Research, Bengaluru, India

**N. Raghuram** Guru Gobind Singh Indraprastha University (IPU), New Delhi, India

**S. Rajendiran** ICAR-Indian Institute of Soil Science, Bhopal, India

**A. Ram** ICAR-Central Agroforestry Research Institute, Jhansi, India

**S.N. Ram** ICAR-Indian Grassland and Fodder Research Institute, Jhansi, India

**B. Ramakrishnan** ICAR-Indian Agricultural Research Institute (IARI), New Delhi, India

**AL. Ramanathan** Jawaharlal Nehru University, New Delhi, India

**R. Ramesh** National Centre for Sustainable Coastal Management, Ministry of Environment, Forests and Climate Change, Anna University Campus, Chennai, India

**P. Ranjan** Jawaharlal Nehru University, New Delhi, India

**D.L.N. Rao** Indian Institute of Soil Science, Bhopal, Madhya Pradesh, India

**D.S. Reay** University of Edinburgh, Edinburgh, United Kingdom

**R.S. Robin** National Centre for Sustainable Coastal Management, Ministry of Environment, Forests and Climate Change, Anna University Campus, Chennai, India

**J. Rudek** Environmental Defense Fund, New York, NY, United States

**T.R. Rupa** ICAR-Indian Institute of Horticultural Research, Bengaluru, India

**S.K. Sahu** Utkal University, Bhubaneswar, India

**V. Sahu** Indian Institute of Technology (I.I.T.) – Roorkee, Roorkee, India

**S. Samanta** ICAR-Central Inland Fisheries Research Institute, Barrackpore, West Bengal, India

**S.M. Sappal** Jawaharlal Nehru University, New Delhi, India

**R. Saraswathy** ICAR-Central Institute of Brackishwater Aquaculture, Chennai, India

**D. Sarkar** Bidhan Chandra Krishi Viswavidyalaya, Mohanpur, India

**A.K. Shanker** ICAR-Central Research Institute for Dryland Agriculture, Hyderabad, Telangana, India

**C. Sharma** CSIR-National Physical Laboratory, New Delhi, India

**K. Sharma** ICAR, New Delhi, India

**A. Singh** CSIR-Central Road Research Institute, New Delhi, India

**G. Singh** National Centre for Sustainable Coastal Management, Ministry of Environment, Forests and Climate Change, Anna University Campus, Chennai, India

**Renu Singh** ICAR-Indian Agricultural Research Institute (IARI), New Delhi, India

**Richa Singh** CSIR-National Physical Laboratory, New Delhi, India

**V.K. Singh** ICAR-Indian Agricultural Research Institute, New Delhi, India

**V.V. Singh** Central Marine Fisheries Research Institute, Kochi, India

**U. Skiba** NERC Centre for Ecology & Hydrology (CEH), Edinburgh Research Station, Midlothian, United Kingdom

**J.U. Smith** University of Aberdeen, Aberdeen, United Kingdom

**S. Sohi** University of Edinburgh, Edinburgh, United Kingdom

**K.R. Sooryanarayana** Central Ground Water Board, Ministry of Water Resources, River Development & Ganga Rejuvenation, Bengaluru, India

**A. Subba Rao** ICAR-Indian Institute of Soil Science, Bhopal, India

**D. Subrahmanyan** ICAR-Indian Institute for Rice Research (IIRR), Hyderabad, India

**K. Surekha** ICAR-Indian Institute for Rice Research (IIRR), Hyderabad, India

**M.A. Sutton** NERC Centre for Ecology & Hydrology (CEH), Edinburgh Research Station, Midlothian, United Kingdom

**J. Syama Dayal** ICAR-Central Institute of Brackishwater Aquaculture, Chennai, India

**M. Tak** Leverhulme Centre for Integrative Research on Agriculture and Health (LCIRAH), London, United Kingdom; School of African and Oriental Studies, University of London, London, United Kingdom

**R.K. Tewatia** The Fertiliser Association of India, New Delhi, India

**H.S. Thind** Punjab Agricultural University, Ludhiana, India

**A.R. Uthappa** ICAR-Central Agroforestry Research Institute, Jhansi, India

**H.J.M. van Grinsven** PBL Netherlands Environmental Assessment Agency, The Hague, The Netherlands

**K.K. Vass** National Academy of Agricultural Sciences, New Delhi, India

**M. Vieno** NERC Centre for Ecology & Hydrology (CEH), Edinburgh Research Station, Midlothian, United Kingdom

**S.R. Voleti** ICAR-Indian Institute for Rice Research (IIRR), Hyderabad, India

# Preface

Nitrogen is the most commonly yield-limiting nutrient in agriculture production. But the scenario changed completely across the world following the use of fertilizer-N. In India too, the scenario changed from "ship to mouth" in 1960s to "farm to fork" in the 1990s following extensive use of fertilizer-N. This rising dependence on fertilizer-N, which represents the largest human interference in the biospheric N-cycle, coupled with the notoriously low N-use efficiency, started adversely affecting the ecosystem and environment. Improvement in the lifestyle of the people resulting into increased demand for energy and transport further complicated the scenario through interference in global N-cycle through ever-increasing value of "Reactive Nitrogen" (Nr).

The Indian aspect of the "N-cycle" is not understood well. For this we need a precise understanding of the scale of nitrogen use/misuse/release through various agricultural, industrial, vehicular, and other activities and their respective contribution to pollution of water and air with special reference to various point/nonpoint sources, and the biogeochemical N-cycle. In this respect, one of the major challenges before the scientific community is to keep policymakers posted with reliable estimates of Nr transfers to different ecosystems and to explain to them balanced, cost-effective, and feasible strategies to reduce Nr where it is detrimental. For this we need to develop an integrative approach to research and policy regarding Nr in Indian agriculture, industry, and environment.

This necessity was formally recognized in 2004, when the Society for Conservation of Nature (SCON), a voluntary body of scientists, brought together some concerned Indian experts from diverse backgrounds to discuss the Nr issues. This was followed by a series of nationwide consultations in association with National Academy of Agricultural Sciences (NAAS), India in 2005, and with the Department of Biotechnology (DBT), Govt. of India and Indian National Science Academy (INSA) in 2006 while drawing active support from other agencies such as the Ministry of Environment and Forests, Govt. of India, and Council of Scientific and Industrial research (CSIR). A set of research bulletins were written and published including a special issue of the journal *Current Science* to sensitize the Indian scientific community at large on the importance of achieving a better understanding of N-cycle. These efforts by SCON lead to setting up of a network of Nr researchers and experts named "Indian Nitrogen Group" (ING) under the aegis of SCON as an outcome of an INSA workshop in 2006. Since then the major task of ING is to undertake interdisciplinary dialogue to generate awareness and also pursue advocacy for an integrated approach toward research and policy on Nr. Identification of all significant sources of Nr flows and their contribution to regional/national/global N-cycle are important for making informed decisions on the policy of sustainable N management. SCON-ING now intends to prepare a countrywide assessment report for India on the lines of the exercise conducted by USA through North American Nitrogen Center (www.nitrogennorthamerica.org) and by Europe through the European Nitrogen Assessment (http://www.nine.esf.org/sites/nine.esf.org/files/ena.doc/.../ENA.pre.pdf).

The present compilation (exercise) on "Indian Nitrogen Assessment" is an outcome of a series of meetings spread over a period of more than a year. The first of these was that SCON organized a scoping workshop ("Group Discussion") with 25 leading scientists, from a range of disciplines engaged in research on various aspects of N-Use (N-cycle), to choose topics to be covered in N-assessment and identifying Indian scientists to be involved in relation to N-use in different ecological systems and N-flows in soil, water, and air. This scoping exercise was followed by a workshop on "Reactive Nitrogen Assessment" supported primarily by the Ministry of Earth Sciences (MoES), Govt. of India, and then a third workshop on "Reactive Nitrogen in South Asia" was organized with support from MoES and the UNEP International Nutrient Management System (INMS) funds, managed by INI.

Subsequent to the above exercises, more than 30 scientists working on various aspects of nitrogen, i.e., nitrogen input in cropping systems, grasslands, horticulture, forestry, followed by N-flows in animal systems including livestock and poultry, flows into river systems, coastal regions, and oceanic systems including fisheries (freshwater, brackish and marine) are invited to contribute chapters/assessment reports in these critical areas. Also, reports from energy, industry, and transport sectors are included. The broad themes (categories) in which these contributions on Nr assessment are organized are:

1. Nitrogen in India in a global perspective
2. Nitrogen processes in the biosphere
3. Nitrogen flows in air, soil, and water
4. N and environmental and climate impacts
5. Managing N in relation to key societal effects
6. N policies and impacts

**Editors**

# Foreword

Introduction of nitrogen responsive Mexican Wheat Germplasm/cultivars set the tone for enhancement in cereal production, particularly wheat, in India followed by other crops in the 1960s ushering in what has now become popularly known as "green Revolution." This was very closely linked to a spurt in the utilization of nitrogenous fertilizers and also ensuring irrigation water to the extent possible. It was a very timely step to feed the teeming millions who were living from "ship to mouth." With chronic deficiency of nitrogen in Indian soils and ever-increasing population from around 250 million to around 1.3 billion, consumption of fertilizer-N increased from 0.6 Mt in 1965–66 to 17.4 Mt in 2015–16. The country has now emerged as the second largest producer and consumer of fertilizer in the world. It is estimated that demand of N in India by 2030 would be 23.45 Mt. Further, it is forecast that by 2050 the fertilizer consumption shall increase twofold as compared to the present figures (FAI, personal communication). In addition, with the improvement of lifestyle and greater disposable income, increased demands on energy and fossil-fuel consumption also contribute significantly to the release of reactive-N.

However, an enigma is that while nitrogen provided food security to the millions, its use efficiency is abysmally low in the range of 32%–35% or may be even less in some ecosystems. With the lowering response of crop varieties to fertilizer-N, farmers tend to apply more and more fertilizer. This is resulting in an uncontrolled accumulation of reactive forms of nitrogen in soil, water, and air which not only causes pollution, ill health, and adversely affects biodiversity and other ecosystem services, but also exacerbates climate change and poses associative challenges to our national planning and international negotiations. Hence alternative strategies need to be worked out that could minimize both loss of this valuable nutrient and reduce pollution from fertilizer overuse. It is required to explore possibilities of sharing and exchanging data on a regular basis through a network of institutions working in the area of N-use. It is also necessary to develop awareness programs through social media for judicious use of fertilizer-N to contribute to the well-being of society.

It is imperative that identification of all major sources of reactive-N flows and their significant contribution to our understanding of the national/regional/global N-cycle are most important for informed decisions on domestic sustainable management for food security, energy, industry, health, and environment. This would also strengthen our negotiating positions on international platforms (at present, there is no integrated mechanism) to understand and assess the nitrogen cycle that encompasses agriculture, urbanization, industry, transport and environment involving as many ministries, and was not sufficiently clear from our own narrow disciplinary boundaries till a few years ago.

Initiative by the Indian Nitrogen Group (ING-SCON) to conduct an exercise on identification, quantification of input of Nr into the cropping systems, N flows in soil, water and air is laudable and timely. I congratulate the scientists. This study will contribute to better understanding of the regional/global N-cycle.

**M S Swaminathan**

# 1

# Technical Summary

Y.P. Abrol[1], T.K. Adhya[2,3]

[1]ING-SCON, DPS MARG, NEW DELHI, INDIA; [2]SOUTH ASIA NITROGEN CENTRE, DPS MARG, NEW DELHI, INDIA; [3]KIIT UNIVERSITY, BHUBANESWAR, ODISHA, INDIA

**CHAPTER OUTLINE**



Being an integral component of global food and energy security, nitrogen cycling has undergone massive transformations in terms of its structure and functions at various scales during the last six decades, and anthropologically remains the most perturbed biogeochemical cycle. Changes in N-cycle are visible in the form of altered transport, conversion, and exchange processes of various fractions of combined or reactive-N ($N_R$). It has resulted in increased transfer between different ecological compartments and substantial loss of $N_R$ invoking disastrous consequences to the environment, such as eutrophication and atmospheric deposition, biodiversity loss, and negative effects on ecosystem services. Globally, there has been an increasing awareness on the issue where "nutrients are turning into pollutants," and policies are being actively framed to control the release of reactive N ($N_R$) to the environment so as to ebb the damages done (UNEP and WHRC, 2007).

Nitrogen, in the form of chemical fertilizers, is the mainstay of increased agricultural productivity for feeding the growing population of India. During 1970 to 2010, fertilizer-N consumption in India increased by about 11 times whereas the uptake by crop increased 3 times and loss of reactive N ($N_R$) increased 4 times. Apart from fertilizer-N, cultivation of crops also adds N inputs through biological N-fixation (BNF). A conservative estimate of BNF in Indian agriculture suggests a contribution of 5.20–5.76 Tg N amounting approximately to 9.5%–10.6% of the global agricultural BNF with cereals contributing 32% and grain legumes contributing 43%. Although environmentally benign, as far as their production is concerned, BNF also leads to the production of $N_R$ albeit to a lesser extent. Further, with economic development, societal demands have increased on energy and transport, adding to the skewed balance of $N_R$. Necessity for considering the usefulness of the $N_R$ challenge for India is apparently more of economic nature considering the huge subsidy burden to the country's exchequer for fertilizer-N and additionally the environmental burden including adverse climate and associated health impacts, converted into absolute economic values. Thus, an in-depth analysis and policy interventions to stem the tide should be a win–win option.

The Indian Nitrogen Assessment. http://dx.doi.org/10.1016/B978-0-12-811836-8.00001-X

Drawing a more focused approach to understand and manage the N-cycle in the Indian context, it was felt necessary to begin an "Indian Nitrogen Assessment." The aim of this exercise was to analyze the N scenario in India in a global perspective (Section A), to analyze the N processes in the biosphere (Section B), N flows in air, soil, and water (Section C), environmental and climate impacts (Section D), managing N in relation to key societal impacts (Section E), and finally enlisting and evaluation of current policies of the state for managing N and predicting possible future scenarios for the country (Section F). The project used an interactive approach wherein leaders in different disciplines were requested to organize teams of experts based on peer recognition and scientific contributions, including consultations on outlines during workshops held on the subject.

Consuming around 15.5% of the global production of fertilizer-N in 2015–16, dependence of India on the use of fertilizer to drive the crop productivity remains undisputed. Considering that India needs to double its food production by 2050, growth trend of fertilizer-N use currently growing at a rate of ~1.9%, almost at par with the population growth rate, is likely to continue. Based on a tentative selling price of US$ 78.8 per ton, cash subsidy of around US$ 7 billion puts a heavy burden on the country's exchequer. In addition to this, N-use efficiency (NUE) in India is abysmally low. It has been estimated that average use efficiency of N by crops in 2008 was 22% and full-chain NUE (including livestock) was 20%, suggesting that a major amount of applied $N_R$ is wasted, making it a nonperforming subsidy. Hence, the alternative is to increase the NUE (Andrews and Lea, 2013) so that considerable savings could be made on fertilizer-N use and subsidy without adversely affecting the crop yield.

Agroecosystems constitute the most predominant terrestrial ecosystems in India accounting for the largest quantum of fertilizer-N use. Although the largest proportion of fertilizer-N is applied for growing cereals, unfortunately, the use efficiency is also the lowest causing ~65% of applied N being subject to loss including to the environment. Among the other terrestrial ecosystems, horticultural crops have a lower N content than the agricultural crops, but modern high intensity horticultural production system causes leakage of considerable amount of $N_R$ into the environment. The scenario is further complicated by the livestock sector where cattle and buffaloes are the largest contributors of ammonia accounting for 56.1% and 23.6%, respectively. The poultry industry, on the other hand, with a projected annual growth rate of 6% is estimated to excrete $N_R$ to the tune of 0.415 million tons in 2016 that is anticipated to increase to 1.089 million tons by 2030. The entire scenario, therefore, becomes quite alarming from the viewpoint of total environmental impact of $N_R$.

Enhanced N mobilization through cultivation, animal husbandry, and industrial and domestic wastewater discharges harms the aquatic environment. In states such as Punjab, Haryana, Delhi, Maharashtra, Andhra Pradesh, and Karnataka with extensive economic and agricultural activities, nitrate concentrations are increasing in surface and groundwater. Nitrate concentrations in river systems tend to be higher in region of extensive economic activity and high population pressure as experienced in the major river basins of the country. Overexploitation of groundwater accompanied by urban development and intense agricultural activities are resulting in the deterioration of the groundwater availability and quality thereby influencing human and ecosystem health. Large-scale inputs of $N_R$ from anthropogenic activities have also led to ecological damage in large part of coastal areas along Indian coastline. Nitrogen assessment in the surface waters, as concentration of $NO_2$ and $NO_3$ carried out in the Indian exclusive economic zones and the adjoining areas during the years 1998–2007 revealed that the annual concentrations (μM) of $NO_2$ ranged from 0 to 0.4, 0 to 0.6, and 0 to 0.7 in the Arabian Sea, Bay of Bengal, and the Andaman Sea, respectively. The corresponding values for $NO_3$ (μM) were 0–2.5, 0 to 3 and 0 to 3.5, respectively. Comprehensive monitoring of $N_R$ concentration in the aquatic systems would greatly help in quantifying the degree of $N_R$ leakage to the coastal waters and its impact on coastal ecosystem.

In addition to several negative impacts on human health and ecosystem pristineness, emission of $NO_x$ and ammonia serve as an indirect source of $N_2O$, an important greenhouse gas, which also contributes to fine particulate matter and regional haze concentration in the atmosphere. $NO_x$ influences the oxidation capacity of the atmosphere through −OH and nitrate, and influences the radiation budget of the atmosphere through $O_3$ formation. Total $N_2O$ emission has increased during 2000–2010 from 264.16 Gg to 370.38 Gg with a compounded annual growth rate of 3.44%, and had a direct relationship with the increased fertilizer-N use. Indian $N_R$ emissions are estimated to be 6.24 Tg per year whereas annual $N_R$ depositions are estimated to be 3.61 Tg per year. Wet deposition (1.97 Tg) of $N_R$ species is almost equal to the dry deposition (1.67 Tg) and the Indo-Gangetic plain is reported as the region of high deposition of $NH_4$ and $NO_3$ (Clarisse et al., 2009).

A major negative impact of $N_R$ originating from inorganic fertilizer and atmospheric deposition on soil quality is acidification. Indian soils mostly have low total N concentration and crop plants respond, at least in terms of growth, to application of fertilizer-N. Hence, application of excessive amounts of N fertilizers may impair soil health through acidification, long-term soil carbon degradation and adverse impacts on the structure and function of the soil biological community. As enough data are not available to understand these complex interactions, there is an urgent need to conduct intensive studies on novel crop management practices including balanced application of organic and inorganic fertilizers that are being promoted for maintenance of soil health and crop productivity.

A rapidly growing economy (at >7% per year) and increasing disposable income of the people in India have the potential to influence societal demand for higher energy consumption, increased personalized transport, and changes in dietary choices. While Indian energy sector was estimated to emit 12.06 Gg $N_2O$ in the year 2010, demand for energy is expected to drive this figure northward. Increasing urbanization and push for personalized transport has made the air quality of big urban conglomerations from bad to worse. As per World Health Organization survey in 2015, of the 1600 world cities, Delhi had the worst air quality of any major city in the world. In a recent episode of smog that continued for the entire first week of November, 2016 in Delhi, $NO_2$ concentrations in the air in excess of 500 μg/m$^3$ was measured which is six times more than the ambient standard prescribed by the Central Pollution Control Board of India (80 μg m$^{-3}$ for residential, industrial, and eco-sensitive areas in a 24 h period). Also, there exist distinct and diverse dietary patterns in India; animal protein consumption is low across the population. Interestingly, in urban areas diverse dietary choices are emerging and these are expected to affect the agricultural production as dietary choices are the drivers of demand for food production.

With free movement of $N_R$ between different environmental compartments, strategies to reduce leakage of $N_R$ to the environment need to be highly interactive and integrated. As major contribution to the $N_R$ budget in India comes from agriculture, the first and foremost strategy is to increase the use efficiency of fertilizer-N. Research evidences suggest that the NUE can be enhanced substantially though new fertilizer formulations, use of urease and nitrification inhibitors, site-specific N management, integrating mineral and organic sources of N, and management of applied N following principles of right source, time, rate, and method of N application. Several policy initiatives both at the government and at the industry level are aiming to harness the actual benefits through useful implementation of different measures of improving NUE. In addition, scientific management of water and wastewater could help in retrieving some fractions of $N_R$. As far as energy and transportation sectors are concerned, concerted efforts should be made to replace the use of coal and popularize the alternate renewable energy sources.

United Nations Environment Program (UNEP) in preparation of the 3rd Inter-Governmental Review of the Global program of action for the Protection of the Marine Environment from Land-based Activities (GPA), proposed an aspirational goal of improving relative NUE by 20% in a 5-year span (UNEP, 2012). Monetization of such projections indicates huge savings. It is now suggested that ensuring the efficient use of fertilizers and associated crop management practices through effective policy implementation in India can lead to substantial savings and at least 30% reduction in the use of N. There is a huge opportunity for India to seize the N challenge by taking along all the stakeholders and increasing the public and institutional awareness of the many benefits and threat perceptions of increased $N_R$ in the environment.

Controlling unwanted $N_R$ release through policy initiatives is difficult because much of the $N_R$ release in India, like other countries with agricultural production base, is from food and energy production and reactive-N species can be transported to long distance in the atmosphere and aquatic systems. Hence, management strategies to reduce $N_R$ emissions to the environment require an integrated approach to take advantage of increasing adoptions of NUE indicators. It is assumed that with the support of favorable government policies around 30% of the emission of $N_2O$ or 0.08 million tons of emission can be avoided as against business as usual (BAU) by 2030 from the agriculture sector itself. In addition, interaction of the nitrogen cycle with other nutrients and the water cycle can provide higher use efficiency as well as provide strategy to counter $N_R$-related water pollution. It is presumed that with about 50% of the urban population wastewater be treated through modern technologies, emission reduction of 50% can be achieved by 2050 with respect to BAU scenario. Given the level of environmental challenges in the country, nitrogen loading into the atmosphere arising out of various sources, often leads to underproductivity and even deaths. While it is clear that technologies exist, they need to be implemented as policies backed by sufficient funds.

Improving nitrogen management including reduction of $N_R$ losses from various sectors including agriculture, industry, transport and energy, and waste through increased public awareness and appropriate policy interventions has the potential to improve the efficiency of nitrogen policies in an integrated manner. Developing and unifying various net anthropogenic nitrogen inputs (Swaney et al., 2105) as indicators to assess the links between $N_R$ flows in different sectors will provide useful guidance for an overall assessment at regional level that could eventually be connected to global level (Reis et al., 2016).

## References

Andrews, M., Lea, P.J., 2013. Our nitrogen "footprint": the need for increased crop nitrogen use efficiency. Annals of Applied Biology 163, 165–169.

Clarisse, L., Clerbaux, C., Dentener, F., Hurtmans, D., Coheur, P.-F., 2009. Global ammonia distribution derived from infrared satellite observations. Nature Geoscience 2, 479–483.

Reis, S., Bekunda, M., Howard, C.M., Karanja, N., Winiwarter, W., Yan, X., Bleeker, A., Sutton, M.A., 2016. Synthesis and review: tackling the nitrogen management challenge: from global to local scales. Environment Research Letters 11, 120205. http://dx.doi.org/10.1088/1748-9326/11/12/120205.

Swaney, D.P., Hong, B., Paneer Selvam, A., Howarth, R.W., Ramesh, R., Purvaja, R., 2015. Net anthropogenic nitrogen inputs and nitrogen fluxes from Indian watersheds: an initial assessment. Journal of Marine Systems 141, 45–58.

UNEP [United Nations Environment Program], 2012. Report of the third session of the Intergovernmental review meeting on the implementation of the global programme of action for the protection of the marine environment from land-based activities. UNEP/GPA/IGR.3/6. www.unep.org/gpa/who-we-are/governance/third-intergovernmental-review-meeting-implementation-gpa.

UNEP [United Nations Environment Program] and WHRC [Woods Hole Research Centre], 2007. Reactive Nitrogen in the Environment: Too Much or Too Little of a Good Thing. United Nations Environment Programme, Paris, pp. 1–56.

SECTION A

# Nitrogen in India in a Global Perspective

2

# The Indian Nitrogen Challenge in a Global Perspective

M.A. Sutton[1], J. Drewer[1], A. Moring[1,2], T.K. Adhya[3,4], A. Ahmed[5], A. Bhatia[6], W. Brownlie[1,2], U. Dragosits[1], S.D. Ghude[7], J. Hillier[8], S. Hooda[9], C.M. Howard[1,2], N. Jain[6], Dinesh Kumar[6], R.M. Kumar[10], D.R. Nayak[8], C.N. Neeraja[10], R. Prasanna[6], A. Price[8], B. Ramakrishnan[6], D.S. Reay[2], Renu Singh[6], U. Skiba[1], J.U. Smith[8], S. Sohi[2], D. Subrahmanyan[10], K. Surekha[10], H.J.M. van Grinsven[11], M. Vieno[1], S.R. Voleti[10], H. Pathak[12], N. Raghuram[9]

[1]*NERC CENTRE FOR ECOLOGY & HYDROLOGY (CEH), EDINBURGH RESEARCH STATION, MIDLOTHIAN, UNITED KINGDOM;* [2]*UNIVERSITY OF EDINBURGH, EDINBURGH, UNITED KINGDOM;* [3]*SOUTH ASIA NITROGEN CENTRE, DPS MARG, NEW DELHI, INDIA;* [4]*KIIT UNIVERSITY, BHUBANESWAR, ODISHA, INDIA;* [5]*ALIGARH MUSLIM UNIVERSITY (AMU), ALIGARH, INDIA;* [6]*ICAR-INDIAN AGRICULTURAL RESEARCH INSTITUTE (IARI), NEW DELHI, INDIA;* [7]*INDIAN INSTITUTE OF TROPICAL METEOROLOGY, PUNE, INDIA;* [8]*UNIVERSITY OF ABERDEEN, ABERDEEN, UNITED KINGDOM;* [9]*GURU GOBIND SINGH INDRAPRASTHA UNIVERSITY (IPU), NEW DELHI, INDIA;* [10]*ICAR-INDIAN INSTITUTE FOR RICE RESEARCH (IIRR), HYDERABAD, INDIA;* [11]*PBL NETHERLANDS ENVIRONMENTAL ASSESSMENT AGENCY, THE HAGUE, THE NETHERLANDS;* [12]*NATIONAL RICE RESEARCH INSTITUTE (NRRI), CUTTACK, INDIA*

## CHAPTER OUTLINE



**The Indian Nitrogen Assessment. http://dx.doi.org/10.1016/B978-0-12-811836-8.00002-1**

Human activities have massively altered the global nitrogen (N) cycle, doubling annual production of reactive N ($N_r$) compounds from atmospheric dinitrogen ($N_2$). The use of 120 Mt year$^{-1}$ fertilizer N, with a global terrestrial/atmospheric N fixation of 285 Mt year$^{-1}$, has provided huge benefits for global food production. However, nitrogen use efficiency (NUE) of the world food system is only ~15%. The lost $N_r$ creates a cascade of air and water pollution and greenhouse gas emissions, until it is eventually denitrified back to $N_2$.

India clearly illustrates a dual N challenge for food and environment, consuming 17 Mt of N fertilizer annually (14% of the global total), which has increased since 1970 at 6% year$^{-1}$ approximately. Emissions of nitrogen oxides ($NO_x$) from combustion sources are also increasing rapidly at 6.5% year$^{-1}$ currently. By comparison, population growth rate is lower (2% year$^{-1}$), while ammonia ($NH_3$) emission increase is even less (1%), pertaining to smaller changes in livestock numbers. At current rate, Indian $NO_x$ emissions will exceed $NH_3$ emissions by 2055. India currently loses $N_r$ worth US$10 billion year$^{-1}$ as fertilizer value, while costs of $N_r$ to health, ecosystems, and climate are estimated at US$75 (38–151) billion year$^{-1}$.

Only a small fraction of the Indian population consumes animal products, hence per capita $N_r$ use and pollution is much less than in many developed countries. However, rates of meat consumption are increasing. While published projections from the UN Food and Agriculture Organization anticipate a doubling of South Asian fertilizer consumption from 2006 to 2050 (equivalent to 1.9% year$^{-1}$ increase), these projections lack transparency and require reevaluation. In practice, the future nitrogen cycle for India will depend on scientific advances in agronomy, genetics and environment, and the extent to which government and society grasp the emerging opportunities for optimizing N management.

## The Global Nitrogen Challenge

The global nitrogen challenge can be summed up as the goal to provide enough reactive nitrogen ($N_r$) to meet societal needs for food and energy production, while avoiding damaging threats of residual $N_r$ flows to the environment, climate and human health. With current global nitrogen use efficiency (NUE) in the food production system being only around 15%,[1] this challenge may seem almost impossible, as 85% of $N_r$ inputs are lost in many forms to the environment.

It is likely that there will always be some adverse impacts associated with the societal use of $N_r$, such as water and air pollution, climate change, and threats to biodiversity (Galloway et al., 2008a,b; Sutton et al., 2011; Fowler et al., 2013). This reflects the diverse web of human and biological processes involved in N transformations. On the other hand, there is huge potential to improve societal use of $N_r$ in a sustainable manner, which would reduce these threats substantially. This would simultaneously increase the

[1]Based on annual agricultural $N_r$ input of 180 Mt leading to 28 Mt $N_r$ in food (Sutton et al., 2013).

benefits of N-use for the farmers, industrialists, and citizens across the world. For example, it has been estimated that even a *relative* improvement in current rates of NUE by 20% (e.g., from 15% to 18% or from 20% to 24% NUE) would have net economic benefits for the green economy worth around US $170 billion year$^{-1}$. This includes both the value of saved fertilizer and the reduction in societal costs to health, ecosystems, and climate. The value of fertilizer savings alone would be worth around US $23 billion year$^{-1}$, of which 3, 3, and 6 billion US $ would be saved by India, the United States, and China, respectively (Sutton et al., 2013). This relative 20% improvement provides a simple benchmark, and it is clear from the differences between the countries that much more ambitious goals and savings could be envisaged.

In order to understand the global nitrogen challenge, it is necessary to recognize first that there are many different compounds that contribute to the global nitrogen cycle. By far the largest N pool is dinitrogen gas ($N_2$), which is highly unreactive. It is the major constituent of the Earth's atmosphere, making up to 78% of the air we all breathe. By contrast, other more reactive N compounds exist in relatively modest amounts. This is because of the stable, low-energy state of $N_2$, which means that other N compounds are eventually broken down or "denitrified" back to $N_2$.

Traditionally, these flows have been viewed as the "nitrogen cycle." This idea of a cycle emphasizes that the pool of atmospheric $N_2$ is converted to form N compounds—often called "fixed nitrogen" or "reactive nitrogen" ($N_r$)—which are, in turn, eventually converted back to form $N_2$. This view offers a neutral perspective that applies to both natural and human influenced systems but tends to underplay the importance of N losses and their impacts. As can be seen from Fig. 2.1, however, it is hard to make the numbers add up, as losses, such as those to seminatural terrestrial ecosystems or aquatic systems, are both uncertain and often omitted from such diagrams.

Nevertheless, by including the magnitude of the N flows, this perspective can be used to illustrate the inefficiencies of N-use in the system. Fig. 2.1 shows this by including numbers and line thickness in proportion to N flows. Global annual terrestrial N fixation comprises around 285 million tons per year (Mt year$^{-1}$), including from fertilizers (120 Mt year$^{-1}$), biological nitrogen fixation (natural c. 58; agriculture c. 60 Mt year$^{-1}$), and 47 Mt year of unintended fixation as $NO_x$ (combustion sources 40; lightning 7). The figure shows how this ends up delivering only 28 Mt year$^{-1}$ of $N_r$ in proteins and other constituents of human food. If all these sources are included, the contribution of total N fixation to human nutrition amounts to a full-system NUE of only 10% (excluding oceans, as the largest $N_r$ flows are decoupled).

Another way to consider the numbers is to compare the natural and anthropogenic flows. Natural N fixation from terrestrial biological nitrogen fixation (BNF) and lightning amounts to approximately 65 Mt year$^{-1}$, while anthropogenic N fixation from BNF, fertilizers, and combustion adds up to 220 Mt year$^{-1}$. Although human activities have reduced the natural rates of BNF (by converting natural land to agricultural land), even if the natural area has been halved (implying contribution of natural BNF amounting to approximately to 116 Mt year$^{-1}$), then this still suggests that humans have more than

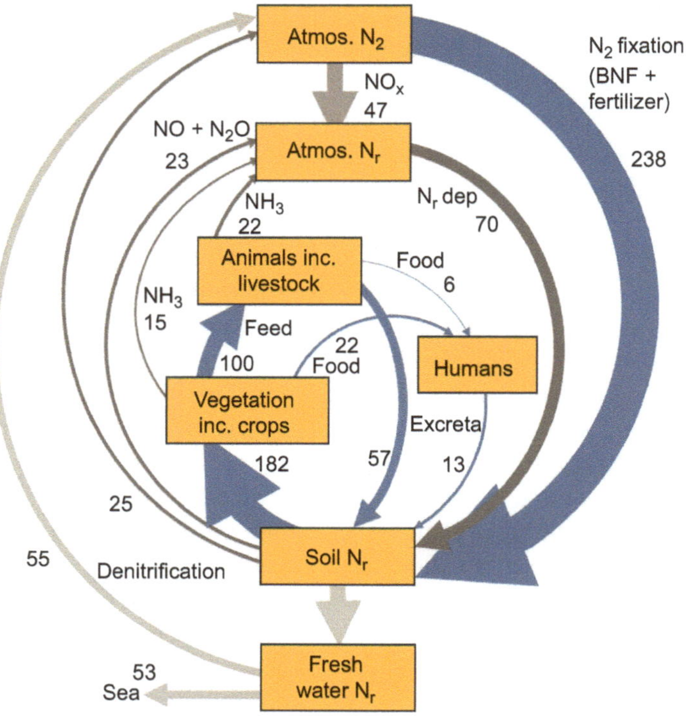


**FIGURE 2.1** Simplified summary of the global nitrogen cycle. *Blue lines* indicate intended flows. *Light gray lines* are often omitted from such a diagram. The numbers are million tons N per year (Mt $year^{-1}$) for around the year 2000. *Based on Sutton, M.A., Bleeker, A., Howard, C.M., Bekunda, M., Grizzetti, B., de Vries, W., van Grinsven, H.J.M., Abrol, Y.P., Adhya, T.K., Billen, G., Davidson, E.A., Datta, A., Diaz, R., Erisman, J.W., Liu, X.J., Oenema, O., Palm, C., Raghuram, N., Reis, S., Scholz, R.W., Sims, T., Westhoek, H., Zhang, F.S., with contributions from Ayyappan, S., Bouwman, A.F., Bustamante, M., Fowler, D., Galloway, J.N., Gavito, M.E., Garnier, J., Greenwood, S., Hellums, D.T., Holland, M., Hoysall, C., Jaramillo, V.J., Klimont, Z., Ometto, J.P., Pathak, H., Plocq Fichelet, V., Powlson, D., Ramakrishna, K., Roy, A., Sanders, K., Sharma, C., Singh, B., Singh, U., Yan, X.Y., Zhang, Y., 2013. Our Nutrient World: The Challenge to Produce More Food and Energy With Less Pollution. Global Overview of Nutrient Management. Centre for Ecology and Hydrology, Edinburgh on Behalf of the Global Partnership on Nutrient Management and the International Nitrogen Initiative. pp. 114.*

doubled global rates of terrestrial N fixation (factor 2.5). This means that the environmental aspect of the nitrogen challenge must address a massive perturbation to the global system.

An alternative perspective to the "nitrogen cycle" was introduced by Galloway et al. (2003), termed the "nitrogen cascade." This emphasizes the raised flows of N and their multiple impacts under anthropogenic influences. The idea of a cascade represents various forms of $N_r$ produced as anthropogenic manufactured $N_r$ is introduced to the system, with each of these forms being associated to several different impacts.

This cascade or "waterfall" idea also reflects the dissipation of a substantial amount of energy. Around 2% of the world energy is used to produce the annual 120 Mt $N_r$ for fertilizer and other uses, mainly as ammonia ($NH_3$) through the Haber-Bosch process. In

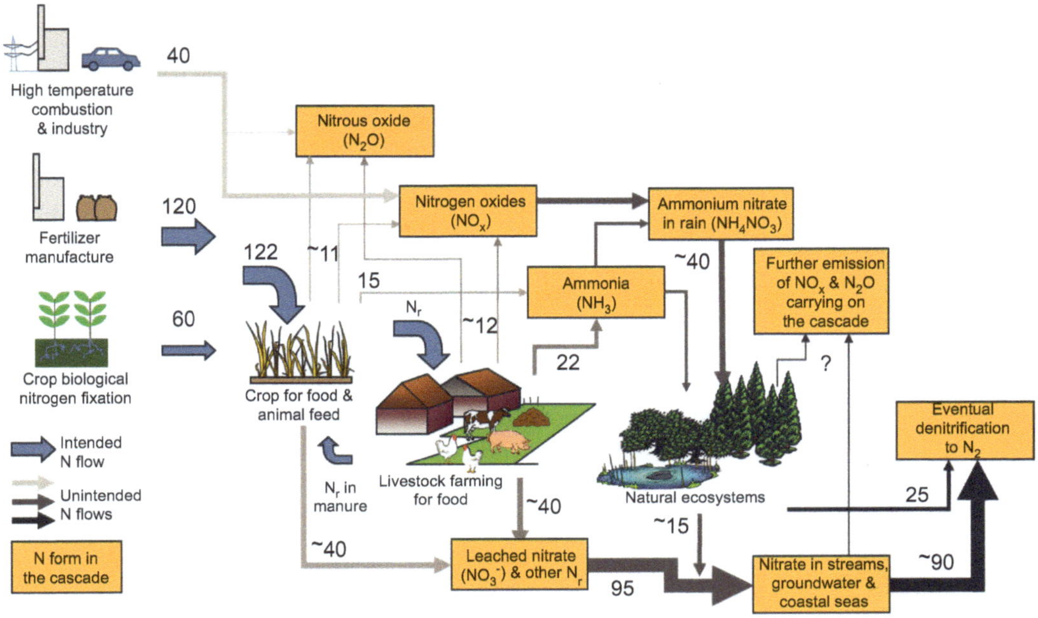


**FIGURE 2.2** Simplified summary of the global nitrogen cascade. The view emphasizes anthropogenic fixation of $N_2$ to $N_r$, the multiple forms of $N_r$ losses, and the dissipation of energy as $N_r$ is eventually denitrified back to $N_2$. Numbers are Mt N year$^{-1}$. *Summarized from Sutton, M.A., Bleeker, A., Howard, C.M., Bekunda, M., Grizzetti, B., de Vries, W., van Grinsven, H.J.M., Abrol, Y.P., Adhya, T.K., Billen, G., Davidson, E.A., Datta, A., Diaz, R., Erisman, J.W., Liu, X.J., Oenema, O., Palm, C., Raghuram, N., Reis, S., Scholz, R.W., Sims, T., Westhoek, H., Zhang, F.S., with contributions from Ayyappan, S., Bouwman, A.F., Bustamante, M., Fowler, D., Galloway, J.N., Gavito, M.E., Garnier, J., Greenwood, S., Hellums, D.T., Holland, M., Hoysall, C., Jaramillo, V.J., Klimont, Z., Ometto, J.P., Pathak, H., Plocq Fichelet, V., Powlson, D., Ramakrishna, K., Roy, A., Sanders, K., Sharma, C., Singh, B., Singh, U., Yan, X.Y., Zhang, Y., 2013. Our Nutrient World: The Challenge to Produce More Food and Energy With Less Pollution. Global Overview of Nutrient Management. Centre for Ecology and Hydrology, Edinburgh on Behalf of the Global Partnership on Nutrient Management and the International Nitrogen Initiative. pp. 114.*

addition, substantial energy is associated with the formation of $N_r$ by BNF, and even with nitrogen oxides ($NO_x$), which are inadvertently formed during combustion processes. These primary forms of $N_r$ gradually dissipate energy as they react to produce different $N_r$ compounds, such as ammonium containing ($NH_4^+$) salts, organic N compounds such as amines ($R\text{-}NH_2$) and proteins, oxidized N compounds such as nitrates ($NO_3^-$) and nitrites ($NO_2^-$), and the greenhouse gas nitrous oxide ($N_2O$). Eventually, the energy is completely dissipated as $N_r$ compounds are finally denitrified back to $N_2$, reaching the bottom of the cascade (Fig. 2.2).

Both these views of nitrogen transformations—the nitrogen cycle and the nitrogen cascade—illustrate the complexity of the global nitrogen challenge. Even though they are sketched out here very simply, it is obvious that $N_r$ flows are vital for life on Earth, and that optimizing these flows, while meeting societal needs for food and energy, requires strategies that takes multiple interactions between various forms of $N_r$ into account.

## Nitrogen in Relation to Other Nutrient Cycles

It is worth comparing the nitrogen challenge with the challenges for other major nutrients such as phosphorus (P) and potassium (K), as there are key similarities and differences among them.

In the case of N and P, the challenges include several forms of each element with different effects on their respective cycles. For N, the prime distinction is between $N_2$ and various usable $N_r$ compounds that are naturally in short supply. For P, the key distinction is between bound P compounds that are hard for plants to access (such as in geological deposits or in soil complexes) compared to more easily available soluble forms of P, such as dissolved and reversibly adsorbed phosphate ($PO_4^{3-}$) in soils. Both $N_r$ and $PO_4^{3-}$ are used in agriculture and tend to become unavailable—$N_r$ as it cascades back to $N_2$, and $PO_4^{3-}$ as it gets tightly bound to soil complexes or is lost as pollutant to the wider environment. In the case of P, however, a buildup of soil P stocks represents a "phosphorus capital" that can ultimately be recycled into useful products. By contrast, once $N_r$ is denitrified to $N_2$, substantial energy is needed to convert it back to useful $N_r$ compounds.

Other major differences between N and P concern the range of impacts and the nature of the ultimate resource availability. Although both the elements have volatile and particulate phases, only a tiny fraction of P enters the atmosphere, for example, as particulate $PO_4^{3-}$ or phosphine gas ($PH_3$). This means that in terms of environmental threats, P is primarily a challenge for water quality in freshwater and marine systems. By contrast, many forms of $N_r$ are extremely volatile, such as $NO_x$, which include a combination of nitric oxide (NO) and nitrogen dioxide ($NO_2$), as well as their reaction products including nitric acid ($HNO_3$) and nitrous acid (HONO). Similarly, $NH_3$, volatile organic compounds (VOCs, such as R-$NH_2$), and $N_2O$ are all readily emitted into the atmosphere. With this plethora of volatile forms being emitted, the global nitrogen challenge is clearly multidimensional. It spans from water pollution issues and atmospheric challenges from urban air pollution ($NO_x$ and particulate matter), to transboundary pollution (acid rain and nitrogen deposition) to hemispheric and global-scale effects on stratospheric ozone and climate.

Concerning timescales, there has been much discussion about "peak phosphorus" and eventual depletion of available phosphate-rock reserves and resources (Cordell and White, 2014). The term "reserves" refers to currently minable deposits which are opened up and accessible according to current technologies and economics. The term "resources" refers to the total known deposits that could be opened up and potentially accessible in future. Phosphate reserves have recently been estimated to be around 370 years, with total minable phosphate resources around 1500 years based on the current rates. Morocco and the Western Sahara are estimated to hold 70% of the global phosphate-rock reserves, while China has the highest share of current production at 38% (figures for 2012; USGS, 2012a,b; Sutton et al., 2013).

Although there has been much less discussion about the topic, a situation similar to that for P also applies for K (often termed potash and expressed as $K_2O$), where global reserves have a lifetime of around 260 years, and resources a lifetime of 7000 years. However, 80% of the reserves are in only two countries, Canada and Russia (Sutton et al., 2013). By contrast, since K is highly mobile, it is not typically limiting in aquatic ecosystems and has therefore not attracted attention from an environmental perspective.

This means that there is a strong case to develop more sustainable nutrient practices in future, and to investigate future options for each of $N_r$, P, and K recovery from "waste" pools. By contrast, the total available N pool is much larger. At current rates of natural plus anthropogenic N fixation, it would take around 8 million years to consume all the available $N_2$ in the atmosphere—assuming that there was no denitrification (Sutton et al., 2013). In reality with denitrification, this means that there is an effectively infinite amount of the $N_2$ resource, with the only limiting factor being the substantial energy requirement needed to fix nitrogen from $N_2$ into $N_r$ compounds.

## Emerging Priorities for the Indian Nitrogen Challenge

There has been a growing recognition of the environmental accumulation of $N_r$ in the Indian scientific community for over a decade (Abrol et al., 2007, 2008; Galloway et al., 2008a) and the challenge it poses for sustainable development (Raghuram et al., 2011; Abrol et al., 2012). It is useful to consider the nitrogen challenge for India in relation to three main headings: the economic case for improving resource efficiency, the societal challenges associated with a rapidly growing economy, and the implications for science and technological development.

### The Economic Case for Improving NUE

The first reason for improving NUE in India must be part of a general drive to improve resource use efficiency, offering the potential for substantial cost savings. India is estimated to import around 30% of its N and P fertilizer, while almost all of its potash is imported (2014–15, FAI, 2016). It is also notable that the US Geological Survey (2012a) does not list any significant phosphate-rock reserves for India, while identified Indian potash reserves are estimated to have a lifetime of just 5 years.

Based on a nominal price of US \$0.60 $kg^{-1}$ N[2], with total import of 4.8 million tons N in 2014–2015, these imports amount to an annual cost of nearly US\$3 billion to India, even without considering the cost of P and K import. In order to support historical food security goals, the Government of India also provides a substantial subsidy on fertilizers use for farmers, equivalent to approximately 70% of the fertilizer cost (Ministry of Chemicals and Fertilizers (2016)). With total fertilizer N consumption in India for 2014–2015 at 16.9 Mt $year^{-1}$, this equates to a cash subsidy of around

[2]www.indexmundi.com urea monthly price, based on World Bank GEM Commodities.

US$7 billion year. [3] Considering the cost of fertilizer imports, the cost of the fertilizer subsidy, and the limited mineral P and K stocks of India, there is a substantial case for the Government of India to improve NUE and wider nutrient use efficiency.

In preparation for the 3rd Intergovernmental Review (IGR-3, January 2012, Manila) of the Global Programme of Action for the Protection of the Marine Environment from Land-based Activities (GPA), it was proposed by the United Nations Environment Programme (UNEP) that an aspirational goal should be set to improve NUE by a relative amount of 20% by 2016–2017 (UNEP, 2011; para 48). In the end, this goal was not agreed by the countries signing the resulting Manila Declaration, which focused on general rather than quantitative objectives (UNEP, 2012). The relevant numbers were, however, recorded as part of the *Our Nutrient World* report for UNEP. It was estimated there that crop NUE for India in 2008 was 22%, while a simple estimate of full-chain NUE (including livestock) was 20%. Achieving a goal to increase NUE relatively by 20% (i.e., to 26% and 24%, respectively), would (at US$1 $kg^{-1}$ N) lead to savings worth US$2.9 billion $year^{-1}$ (Sutton et al., 2013).

One of the conclusions of *Our Nutrient World* was that the societal costs of N pollution are even larger than the direct economic costs as wasted fertilizer. Using a willingness-to-pay approach for Europe, Brink et al. (2011) estimated total societal costs of European N pollution at 70 to 320 billion Euro/y for around the year 2000. Van Grinsven et al. (2013) subsequently updated these numbers to give estimates of 75–485 billion Euro/y for Europe.

It is extremely challenging to apply such an approach at a global scale, or even to transfer to other continents. While recognizing the substantial uncertainties, the approach can still be informative. In scaling up to a global level, the *Our Nutrient World* report took account of differences in gross national income (GNI) for the year 2000 and estimated the global societal costs of nitrogen pollution to be 200 to 2000 billion US$ $year^{-1}$. In Table 2.1 we apply the same approach for India, in addition to comparing estimates for the years 2000 and 2015. Overall, this indicates societal costs of Indian $N_r$ pollution (for health, ecosystems, and climate) to be 17 (8.5–34) billion US$ for 2000 and 75 (38–151) billion US$ for 2015. The values in brackets indicate ranges of uncertainty based on the willingness-to-pay approach used.

It is relevant to compare the direct fertilizer costs with the estimated societal costs for health, ecosystems, and climate. For example, the earlier studies found that the societal costs greatly exceed the direct fertilizer costs of $N_r$ losses by a factor of 20 in Europe (based on Brink et al., 2011) and by a factor of 7 globally (Sutton et al., 2013). Even though there is substantial uncertainty in the exact value of the total societal cost, it is evident that it dominates compared with the fertilizer loss.

[3]For comparison, FAI (2016) report the central subsidy on urea fertilizers for India at 12,300 Rs crore (imports) plus 38,200 Rs crore (indigenous manufactured fertilizer) for 2015–16, equivalent to a total of 7.6 billion USD per year. An additional subsidy equivalent to US$ 3.4 billion is provided on P and K fertilizers.

**Table 2.1** Summary of Reactive Nitrogen Emissions in India for 2000 and 2015 With a First Indication of the Societal Costs of $N_r$ Pollution per unit N and Scaled up to the Country

| Year: 2000 | Emission | Health costs[a] | Ecosystem costs[a] | Climate costs[a] | Total costs[a] | Health costs | Ecosystem costs | Climate costs | Total costs |
|---|---|---|---|---|---|---|---|---|---|
| N losses | Mt N/year$^{-1}$ | USD kg$^{-1}$ N | | | | Billion USD year$^{-1}$ | | | |
| $N_r$ to water | 5.6[b] | 0.1 | 1.1 | – | 1.2 | 0.5 | 6.4 | – | 6.9 |
| $NH_3$ to air | 5.0[c] | 1.1 | 0.2 | – | 1.3 | 5.7 | 0.9 | – | 6.6 |
| $NO_x$ to air | 1.6[c] | 1.7 | 0.2 | – | 1.9 | 2.7 | 0.3 | – | 3.0 |
| $N_2O$ to air | 0.4[c] | 0.2 | – | 0.9 | 1.0 | 0.1 | – | 0.4 | 0.5 |
| Total | | | | | | | | | 17.1 |
| **Year: 2015** | **Emission** | **Health** | **Ecosystem** | **Climate** | **Total** | **Health** | **Ecosystem** | **Climate** | **Total** |
| **N losses** | **Mt N year$^{-1}$** | **USD kg$^{-1}$ N** | | | | **Billion USD year$^{-1}$** | | | |
| $N_r$ to water | 7.5 | 0.3 | 3.6 | 0.0 | 3.9 | 2.3 | 27.1 | - | 29.3 |
| $NH_3$ to air | 6 | 3.6 | 0.6 | 0.0 | 4.2 | 21.6 | 3.6 | - | 25.2 |
| $NO_x$ to air | 3.2 | 5.4 | 0.6 | 0.0 | 6.0 | 17.3 | 1.9 | - | 19.2 |
| $N_2O$ to air | 0.5 | 0.6 | 0.0 | 2.7 | 3.3 | 0.3 | – | 1.4 | 1.7 |
| Total | | | | | | | | | 75.4 |

Overall uncertainties in the societal costs are estimated to be a factor of ±2, i.e., billion US$ 75 (38–151) per year for 2015.

[a]Unit societal costs are based on the European Nitrogen Assessment (Brink et al., 2011) corrected for differences in per capita gross national income PPP (purchasing power parity, from World Bank, 2016) between the EU and India and converted to US$.

[b]Noting that full-chain NUE is similar in Europe and India (Sutton et al., 2013; Annex), a similar ratio of $N_r$ loss (water) to N fertilizer input (0.45) is here assumed to be the estimates of Brink et al. (2011).

[c]From EDGAR (2016), version 4.3.1 ($NO_x$ and $NH_3$) and version 4.2 ($N_2O$), which for $NO_x$ are consistent with the satellite-based estimation of Ghude et al. (2013).

In the case of India, let us assume a fertilizer commodity price of US$ 0.22 kg$^{-1}$ N for the year 2000, as compared with US$ 0.6 kg$^{-1}$ N for 2015. For 2000, the fertilizer value of $N_r$ losses (excluding substantial denitrification to $N_2$) was around US$2.5 billion, with the comparable value for 2015 being US$10.3 billion. Both the $N_r$ price and amount of $N_r$ losses increased over the period, leading to an overall rise in the fertilizer value of $N_r$ losses of 310%. If denitrification to $N_2$ were included, these values could easily increase by an extra third. The estimated overall societal costs have also increased substantially by 340%. This reflects both larger $N_r$ losses and GNI in India, which is assumed to be proportional to willingness-to-pay to prevent impacts of N pollution. The ratio of societal costs to fertilizer costs in India is estimated at 6.8 for 2000 and 7.3 for 2015, broadly similar to the global ratio reported in *Our Nutrient World*.

Overall, these values demonstrate a strong case for improving NUE in India, both to reduce the health, ecosystem, and climate costs of $N_r$ pollution and to save farmers and the Government of India trillions of Rupees per year.

## Societal Challenge of a Rapidly Growing Economy

The consequences of the rapidly growing Indian economy are directly reflected in the increased $N_r$ use and associated losses/emissions. This poses an even greater challenge if India has to balance greater economic activity with the maintenance of environmental quality, while simultaneously meeting goals associated with food and energy security. Not only have $N_r$ losses been increasing, but there is clear evidence that NUE has also decreased substantially over recent decades. For example, rice NUE decreased from 23% in 1970 to around 8% in 2005 (Adhya et al., 2010).

Majority of the important changes over the past 45 years are illustrated in Fig. 2.3. The primary driver has been the rapid increase in fertilizer N input, which was the main feedstock for the Green Revolution, allowing increased yields of the newly introduced genetic varieties for rice, wheat, and maize. Overall, since 1970, fertilizer N input has increased by an average of 6% year$^{-1}$. Emissions of $NO_x$ (mainly from combustion sources such as industries, vehicles, cooking, and residue burning) represent another source of new $N_r$ for India, the deposition of which could in principle have benefits for crop yields. However, any such fertilizing effect would have to be offset against crop

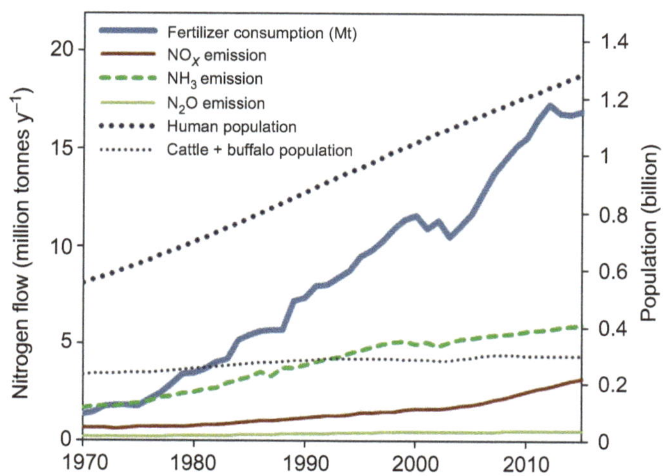


**FIGURE 2.3** Trends in nitrogen flows in India through fertilizer consumption, emissions of nitrogen oxides ($NO_x$), ammonia ($NH_3$) and nitrous oxide ($N_2O$), as compared with trends in the population of humans and major livestock (cattle plus buffalo). $NO_x$ and $NH_3$ values from 2011 and $N_2O$ values from 2009 are linear projections based on the preceding six years. *Based on FAI 2016. Fertiliser Association of India. All India Consumption of Fertiliser Nutrients, and Central Subsidy on Fertilisers. http://www.faidelhi.org/statistical-database.htm; EDGAR, 2016 Global Emissions Data Version 4.3.1. January 2016.http://edgar.jrc.ec.europa.eu/overview.php?v=431; FAO, 2016. United Nations Food and Agriculture Organization, Statistics. Faostat3.fao.org.*

losses due to increased tropospheric ozone concentrations, which are produced as a reaction product of $NO_x$ and VOCs. Over the last decade, the rate of $NO_x$ emissions have increased in line with economic development, with the most recent estimates indicating an increase of 6.5% $year^{-1}$.

These rates of increase are much faster than the annual growth rate of the human population of India at 2% $year^{-1}$, while the livestock numbers have increased at a much slower rate. Although the numbers of poultry increased rapidly from the mid-1990s (up to 7% $year^{-1}$), this has slowed since 2007, with the current annual increase being 2%. In terms of emissions, however, poultry makes a relatively minor contribution in India, with the largest emissions from livestock associated to cattle and buffalo. The total population of these animals reached a maximum in 2007 and has since been decreasing at 0.3% per year. The net result is that $NH_3$ emissions, which result from both livestock and urea fertilizer application, are estimated to have been increasing modestly at 1% $year^{-1}$.

Expressed as nitrogen, $NH_3$ emissions in India are still nearly double of $NO_x$ emissions (Table 2.1), indicating the priority to reduce $NH_3$ emissions, especially as these represent a direct loss of fertilizer value for crops. However, the rapid increase in $NO_x$ emissions should also be seen as of particular concern. If the current trends continue, $NO_x$ emissions will exceed $NH_3$ emissions after 2055, illustrating the challenge to manage these emissions in future. Although not as extreme, this represents a similar situation to China where $NO_x$ emissions are increasing even more rapidly and estimated to exceed $NH_3$ emissions around 2025 (Liu et al., 2013; Sutton and Bleeker, 2013). At current rates of increase, Indian $NO_x$ emissions will be 8.8 Mt by 2055, equivalent to the total fertilizer input to India in the mid-1990s. Avoiding such an increase will require substantial effort in utilizing improved technology for $NO_x$ removal, such as catalytic and non-catalytic reduction to $N_2$. However, it also poses an opportunity for innovation in developing future $NO_x$ recapture and utilization technology, where $NO_x$ is scrubbed from combustion sources to be collected as $HNO_3$, providing a valuable fertilizer feedstock (Yasuda et al., 2011; Sutton et al., 2013 under Chapter 6, Key Action 5) worth billions of dollars per year.

With rapidly growing GNI (at >7% $year^{-1}$), increasing wealth also has the potential to affect the consumption and dietary patterns of Indian citizens. It has been reported that only around 4% of people in India currently eat animal products (Sutton et al., 2013, Section 7.8). In fact, this figure may substantially underestimate the fraction of people eating animal products in India. For example, the Hindu-CNN-IBN State of the Nation Survey found that the popular image of vegetarian India is off the mark, as only 31% of individuals and 21% of families (with all the members) were recorded to be vegetarian (Yadav and Kumar, 2006). However, there may be an issue concerning definitions, as eating eggs is considered nonvegetarian in India, while eggs are typically associated with dairy products in many parts of Europe. Nevertheless, the per capita consumption of meat in India at 4 kg per annum is certainly among the lowest in the world, compared with 86 kg $year^{-1}$ in the United States, 76 kg $year^{-1}$ in Europe, and 108 kg $year^{-1}$ in Oceania (Ali and Pappa, 2015 based on FAO statistics for 2014).

If the per capita meat consumption in India were to increase significantly, it would require substantial additional $N_r$ flows, with extra pollution associated with livestock feed and animal production. In the case of Europe, Leip et al. (2015) estimated that the livestock typically accounted for 70–80% of the contribution of agriculture to a range of $N_r$ environmental indicators including air quality, water quality, biodiversity loss, and greenhouse gas emissions. With increasing human population, combined with a higher fraction of livestock products in Indian diets, a massive increase in $N_r$ fertilizer inputs and $N_r$ pollution losses would be expected.

The opposite situation has been described by Westhoek et al. (2015) in scenarios for Europe. They found that decreasing European meat and dairy intake (including eggs) by 50% would reduce $N_r$ pollution levels by around 40%, while freeing up substantial areas of agricultural land for other production (such as bioenergy). While it is obvious that the poorest in India need better diets, there is the risk that rapid development leads to an overshoot in the already well-fed sections of the population. These points indicate the need to identify and converge toward *combined dietary guidelines* for both health and environment.

The scale of the potential change has been illustrated by the UN Food and Agriculture Organization (FAO; Alexandratos and Bruinsma, 2012), as summarized by Sutton and Bleeker (2013) who compared total fertilizer use (N, P, and K) per ha of agricultural land in 1962, 2006, and 2050 (see Fig. 2.4). If the FAO projections for 2050 are realistic, then fertilizer use in South Asia will more than double between 2006 and 2060, with a similar increase in Latin America. Such a change may seem extreme, but is equivalent to an annual growth rate of fertilizer consumption of just 1.9%. This is much less than the

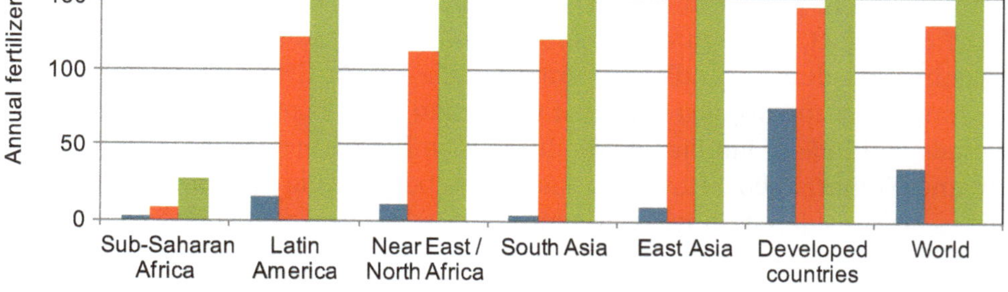


**FIGURE 2.4** Estimated changes in total annual fertilizer input to agricultural land, representing the sum of N, phosphorus (as $P_2O_5$) and potassium (as $K_2O$). *Based on Alexandratos, N., Bruinsma, J., 2012. World Agriculture Towards 2030/2050, The 2012 Revision. Rome, FAO.*

growth witnessed in recent decades, and is similar to the population growth rate.[4] The problem is that these FAO projections lack transparency in the main assumptions. They therefore require reevaluation in the light of the different main drivers: reductions in $N_r$ losses, improvements in field NUE, livestock NUE, population change, dietary change, and the extent of anticipated net imports.

Such a future increase is easily feasible and would lead to a huge increase in the scale of the "Indian Nitrogen Challenge." In particular, a growing livestock sector would increase the $N_r$ threat to air quality, water quality, greenhouse gas emissions, and loss of biodiversity unless substantial technical improvements in N management are made.

## Science and Technological Needs

The economic case for reducing $N_r$ losses, combined with the risk of a substantially worsening situation for Indian $N_r$ pollution, points to an urgent need for better scientific knowledge and major technological advances. Already India has the highest levels of $NH_3$ pollution in the world (see, e.g., Clarisse et al., 2009), contributing to particulate matter pollution and threatening human health. It is hard to imagine the scale of these problems if losses double again within the next 40 years.

It should be obvious from the multiple sources and impacts of $N_r$ and the transboundary nature of the $N_r$ pollution that a broad scientific approach is needed to address the "Indian Nitrogen Challenge." In particular, it is necessary to focus not just on the traditional challenges of crop genetic improvement, although these remain important, but also to extend more widely to develop improved agronomic practices, to consider the opportunities and barriers at farm and village scales, and to consider the regional and national situations. These last levels of upscaling are essential so as to better quantify the drivers of change, their environmental impacts, and the possible barriers to change. National and international policies also have a key role to play in improving nitrogen management in India, and it will be vital that each of the major opportunities for change is recognized and addressed.

This vision for a broad perspective is clearly outlined in the objectives of the newly established "Virtual Joint Centre on Nitrogen Efficiency of Whole-cropping Systems for improved performance and resilience in agriculture"(*NEWS India–UK*). The vision is summarized in Fig. 2.5, indicating how the path to increasing NUE includes research at different scales: from genetics to agronomy, to the farm scale, and to the national scale.

*NEWS India–UK* is developing a transdisciplinary perspective as a basis for building durable cooperation, while developing a platform for strengthening stakeholder engagement. It includes five main scientific objectives (Sutton et al., 2016):

1. **To characterize and explain plant NUE differences**, including the establishment of a phenotypic ranking of Indian rice varieties for NUE supported by improved

[4]It is not necessary to assume that diets change to reproduce the FAO projections, or even assume that decreasing system efficiency associated with increased animal intake might be offset by NUE gains through better genetics, agronomy and livestock management.

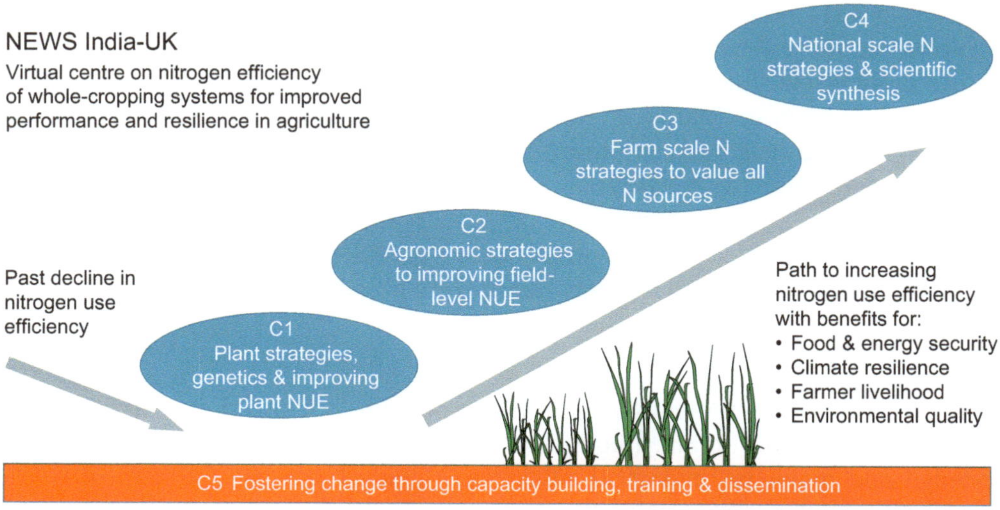


**FIGURE 2.5** Main elements in the research strategy of the NEWS India–UK Virtual Joint Center on Nitrogen Efficiency of Whole-cropping Systems for improved performance and resilience in agriculture.

functional understanding of traits, comparing this with what can be achieved by a known genetic modification for improved NUE.

2. **To compare the effectiveness of agronomic and biological approaches to improving NUE**, and to assess by how much the combined approaches maximize the overall crop performance.
3. **To quantify the relationships between NUE improvement and measured reduction in N losses**, considering the different N forms and N loss pathways.
4. **To test how a better use of all available N resources can improve Farm NUE** in rice–wheat rotations, including from organic manures, atmospheric N deposition and BNF, and to explore options for overcoming the barriers to better manure recycling.
5. **To establish a national N agricultural budget for India** to examine scenarios of improved N management and resilience, and the associated environmental benefits.

In particular, it should be emphasized that the research agenda established in this new initiative considers all available sources of $N_r$, including the potential for improved recycling of manures and other organic wastes, aiming to strengthen the value chain associated with these resources. This kind of thinking is essential as solution to the Indian Nitrogen Challenge align with the developing Circular Economy.

There are also specific challenges associated with many of the scientific elements. For example, the behavior of soil nitrogen under wet soil conditions of lowland rice is markedly different from its behavior under dry soil conditions. Under flooded conditions, most N to be taken up by rice is in $NH_4^+$ form. This means that specific practices need to be evaluated in relation to overall functioning of the nitrogen cycle. An example would be evaluating the practice of Alternate Wetting and Drying (AWD). This practice results in periodic aerobic soil conditions, stimulating sequential nitrification and denitrification losses (Buresh and Haefele, 2010). Growing rice under AWD could therefore consequently lead to a greater loss of applied fertilizer and soil nitrogen compared with that under submergence conditions (Peng et al., 2006). Cabangon et al. (2004) reported that the nitrogen recovery of rice under AWD (about 20%) was significantly lower than that under submergence (about 40%). Increasing NUE in AWD and other practices will help farmers to reduce the amount of N fertilizer, increase their income, and facilitate the adoption of AWD to cope with water scarcity. A combination of AWD and other high NUE management practices and efficient cultivars will also benefit the environment by reducing the amount of water used and chemical pollution entering the surroundings.

In addition to the core science objectives of *NEWS India–UK*, there are even wider challenges to address. For example, measurement techniques and monitoring networks for $N_r$ components in the air and water in India would benefit from much greater development. Here a program of *NEWS India–UK* Junior and Senior Fellowships and a Massive Online Open Course (MOOC) are building the foundations to enable future action.

In relation to scientific synthesis and support for international policy development, India is also leading the regional contribution of South Asia to the emerging International Nitrogen Management System (INMS). This new program is currently being established jointly between the International Nitrogen Initiative (INI) and the UNEP, with financial support through the Global Environment Facility. In particular, it will provide a mechanism for delivering regional demonstrations that assess the major flows of N in each major world region (including South Asia), identifying the N threats, and assessing the opportunities and barriers to better nitrogen management, underpinning global assessment of the nitrogen challenge.

In this regard, it is vital that the international scientific community works even more closely with actors across society, with government representatives, business, civil society, and farmers. This is needed to give perspective to refresh the research priorities, while at the same time helping to mobilize the necessary actors to stimulate change. The global nitrogen cycle crosses all our interests and only by grasping the bigger picture will we be able to identify the win–win opportunities and highlight the importance of each part of the puzzle.

## Conclusion: India as an Emerging "Nitrogen Champion"

In practice, the future nitrogen cycle for India will depend not only on scientific advances and interventions through agronomy, genetics, and environmental science, but also on the extent to which government and society grasp the opportunities for better N management. Until now, despite the growing recognition of the importance of $N_r$, it has had only a limited public profile and policy attention. The result is that there has so far been little international agreement by countries to improve nitrogen management. The INMS is only now beginning to be established, while the international "Nitrogen Policy Arena" requires much more attention to improving coordination and developing synergies. To address this gap requires champion countries. It should therefore be asked: Could India become such a champion?

The answer must surely be yes. There is a huge opportunity for India in grasping the nitrogen challenge. The financial value of fertilizer and other $N_r$ losses in India are huge. Expressed simply as fertilizer value, India's annual N fertilizer subsidy is equivalent to US$ 7 billion year$^{-1}$ while $N_r$ pollution losses represent a lost fertilizer value of US$ 10 billion year$^{-1}$. This includes losses from agriculture as $NH_3$ and $N_2O$ to air and $N_r$ to water, as well as $NO_x$ from combustion sources to air. If $N_r$ denitrification to form $N_2$ were included, the figures would be even bigger. These numbers point to a major financial opportunity for India to lead science, innovation, and policy actions on better nitrogen management.

The societal costs of nitrogen pollution are even larger, estimated at US$75 (40–150) billion year$^{-1}$. Although these are based on a willingness-to-pay approach and extremely hard to value, they clearly indicate the threats of $N_r$ losses to air pollution (especially particulate matter levels and ozone) affecting human health, of air and water pollution threatening ecosystems and livelihoods (e.g., through threats to fish stocks and crop yields) and of climate change. Whatever the exact value of the total societal costs, the general conclusion is that these substantially exceed the value of lost fertilizer, giving an even stronger case for India to lead the way internationally.

Perhaps the most pressing reason for India to develop as a nitrogen champion concerns the risks for the future, as growing wealth and population point potentially to a future with double the amounts of $N_r$ losses and pollution levels by 2050. Although such projections are highly uncertain, the equivalent annual growth rate of 1.8% is much less than current rates, demonstrating that it is a realistic possible future.

There are already welcome signs that India is starting to grasp the opportunity. For example, the desire to reduce its $N_r$ fertilizer subsidy has led to India requiring (since 2015) all urea fertilizer be coated with neem oil, a natural nitrification inhibitor. Aside from avoiding financial leakage of fertilizer subsidy to nonagricultural uses, this holds the prospect to increase NUE by reducing nitrification and denitrification.[5] India has also accepted chairmanship (in 2016) of UNEP's Global Partnership on Nutrient Management (GPNM), a multi-actor network that stimulates cooperation in improving nutrient cycling.

There are, however, still major challenges ahead. The next Intergovernmental Review of the GPA on marine pollution (IGR-4) is scheduled for Indonesia in 2017. Following 5 years after IGR-3 in Manila, will the countries choose this time to agree on any

[5]It remains a scientific task under the *NEWS India–UK* center to show whether neem poses a significant risk of increasing $NH_3$ emissions and the techniques by which that could be avoided (e.g., urease inhibitors or urea deep placement).

aspirational goals related to nitrogen? For example, substantial progress has been made on consensus development about NUE since IGR-3 (e.g., Sutton et al., 2013; Norton et al., 2015; Oenema et al., 2015), so will the countries be ready this time to agree to a goal to increase NUE, for example, by 30% by 2030? This would clearly complement the goal to reduce nutrient pollution to the marine environment by 20% (by 2025) under the "Our Ocean Action Plan" (US Department of State (2014)).

A challenge and specific global opportunity for India to increase NUE and to slow down the increasing $N_r$ pollution is to use its strong vegetarian tradition to control the growth of consumption of animal products. Education and marketing by food industry and retail are important instruments for this, which could provide major opportunities to promote the international profile of Indian cuisine.

Perhaps the biggest challenge will be to bring the different benefits of $N_r$ together internationally. The UNECE Air Convention agreed its Gothenburg Protocol revision in 2012 to reduce $NO_x$ and $NH_3$ emissions (Reis et al., 2012)—yet there still is no global convention for air pollution. The Paris Agreement on climate will have implications for reducing $N_2O$ emissions, while also holding significant opportunities to reduce stratospheric ozone depletion due to $N_2O$ (UNEP, 2013). The Convention on Biological Diversity has also included nitrogen deposition as part of its Aichi Targets (CBD, 2012). Yet, with such fragmentation across the "Nitrogen Policy Arena," it is currently hard for each process to see all the benefits on its own as a basis to mobilize action.

This poses a major technical challenge for the scientific community, in each of INI, INMS, and *NEWS India–UK*, to work with national champions, such as India, and the wider policy community. Together the goal must be to develop a shared vision of simultaneous benefits for food, energy, and environment if the global nitrogen challenge is ultimately to be addressed.

## Acknowledgments

The authors gratefully acknowledge funding through the UK Biological and Biotechnological Research Council (BBSRC), the UK Natural Environment Research Council (NERC) and the Indian Department of Biotechnology (DBT) for support of the Newton-Bhaba NEWS India–UK Virtual Joint Research Centre. MAS, CMH, DR, and WB thank NERC (International Opportunities Programme) for support of the International Nitrogen Management System—Pump Priming (INMSpp) project and the United Nations Environment Programme, and the Global Environment Facility for financing through the Toward INMS project. This chapter is a contribution to the work of the International Nitrogen Initiative.

## References

Abrol, Y.P., Raghuram, N., Sachdev, M.S., 2007. Agricultural Nitrogen Use & Its Environmental Implications. IK International, New Delhi, 552 pp.

Abrol, Y.P., Raghuram, N., Hoysall, C., 2008. Special issue on reactive nitrogen in agriculture, environment and health. Current Science 94 (11), 1343–1344.

Abrol, Y.P., Pandey, R., Raghuram, N., Ahmad, A., 2012. Nitrogen cycle sustainability and sustainable technologies for nitrogen fertilizer and energy management. Journal Indian Institute of Science 92 (1), 17–36.

Adhya, T.K., Shukla, A.K., Panda, D., 2010. Nitrogen losses, N-use efficiency and N management in rice and rice- based cropping system, 51p. In: ING Bulletins on Regional Assessment of Reactive Nitrogen, Bulletin No. 10, (Ed. Bijay Singh), SCON-ING, New Delhi, pp i-iv & 1–51. See Also FAI (2009) Fertilizer Statistics 2008–09. Fertilizer Association of India, New Delhi.

Alexandratos, N., Bruinsma, J., 2012. World Agriculture Towards 2030/2050: The 2012 Revision. FAO, Rome.

Ali, J., Pappa, E., 2015. Global meat market: structural change across geographic regions. South Asia Research 35 (2), 143–157.

Brink, C., van Grinsven, H., Jacobsen, B.H., Rabl, A., Gren I.-, M., Holland, M., Klimont, Z., Hicks, K., Brouwer, R., Dickens, R., Willems, J., Termansen, M., Velthof, G., Alkemade, R., van Oorschot, M., Webb, J., 2011. Costs and benefits of nitrogen in the environment. In: Sutton, M.A., Howard, C.M., Erisman, J.W., Billen, G., Bleeker, A., Grennfelt, P., van Grinsven, H.J.M., Grizzetti, B. (Eds.), The European Nitrogen Assessment. Cambridge University Press, Cambridge, pp. 513–540.

Buresh, R.J., Haefele, S.M., 2010. Changes in paddy soils under transition to water saving and diversified cropping systems. In: Paper to be Presented at World Congress of Soil Science, Brisbane, Australia, 1–6 August 2010.

Cabangon, R.J., Tuong, T.P., Castillo, E.G., Bao, L.X., Lu, G., Wang, G., Cui, Y., Bouman, B.A.M., Li, Y., Chen, C.D., Wang, J.Z., 2004. Effect of irrigation method and N fertilizer management on rice yield, water productivity and nutrient use efficiencies in typical lowland rice conditions in China. Paddy Water Environment 2, 195–206.

CBD, 2012. Decisions Adopted by the Conference of the Parties to the Convention on Biological Diversity at its Eleventh Meeting (Hyderabad, India, 8–19 October 2012). UNEP/CBD/COP/11/35. (see p. 102 for nitrogen indicator under Aichi Biodiversity Target 8). http://www.cbd.int/doc/decisions/cop-11/full/cop-11-dec-en.pdf.

Clarisse, L., Clerbaux, C., Dentener, F., Hurtmans, D., Coheur, P.-F., 2009. Global ammonia distribution derived from infrared satellite observations. Nature Geoscience 2, 479–483.

Cordell, D., White, S., 2014. Insert title annual review of Environ. Resources 39, 161–181.

EDGAR, 2016. Global Emissions Data Version 4.3.1 (January 2016). http://edgar.jrc.ec.europa.eu/overview.php?v=431.

FAI, 2016. Fertiliser Association of India. All India Consumption of Fertiliser Nutrients, and Central Subsidy on Fertilisers. http://www.faidelhi.org/statistical-database.htm.

FAO, 2016. United Nations Food and Agriculture Organization, Statistics. Faostat3.fao.org.

Fowler, D., Pyle, J.A., Raven, J.A., Sutton, M.A., 2013. The global nitrogen cycle of the twenty-first century. Introduction. Philosophical Transactions of the Royal Society, B 368, 2130165. http://dx.doi.org/10.1098/rstb.2013.0165.

Galloway, J.N., Aber, J.D., Erisman, J.W., Seitzinger, S.P., Howarth, R.W., Cowling, E.B., Cosby, B.J., 2003. The nitrogen cascade. Bioscience 53, 341–356.

Galloway, J.N., Raghuram, N., Abrol, Y.P., 2008a. A perspective on reactive nitrogen in global, Asian and Indian context. Current Science 94 (11), 1375–1381.

Galloway, J.N., Townsend, A.R., Erisman, J.W., Bekunda, M., Cai, Z., Freney, J.R., Martinelli, L.A., Seitzinger, S.P., Sutton, M.A., 2008b. Transformation of the nitrogen cycle: recent trends, questions and potential solutions. Science 320, 889–892. http://dx.doi.org/10.1126/science.1136674.

Ghude, S.D., Pfister, G.G., Jena, C., van der, A.R.J., Emmons, L.K., Kumar, R., 2013. Satellite constraints of nitrogen oxide ($NO_x$) emissions from India based on OMI observations and WRF-Chem simulations. Geophysical Research Letters 40, 423–428.

Leip, A., Billen, G., Garnier, J., Grizzetti, B., Lassaletta, L., Reis, S., Simpson, D., Sutton, M.A., de Vries, W., Weiss, F., Westhoek, H., 2015. Impacts of European livestock production: nitrogen, sulphur, phosphorus and greenhouse gas emissions, land-use, watereutrophication and biodiversity. Environmental Research Letters 10, 115004.

Liu, X., Zhang, Y., Han, W., Tang, A., Shen, J., Cui, Z., Vitousek, P., Erisman, J.W., Goulding, K., Christie, P., Fangmeier, A., Zhang, F., 2013. Enhanced nitrogen deposition over China. Nature 494, 459–462.

Ministry of Chemicals and Fertilizers, 2016. Fertilizer Policy. Department of Fertilizers. Government of India. http://fert.nic.in/page/fertilizer-policy.

Norton, R., Davidson, E., Roberts, T., 2015. Position Paper: Nitrogen Use Efficiency and Nutrient Performance Indicators. A publication of the Global Partnership on Nutrient Management. Technical paper 01/2015. http://www.nutrientchallenge.org/sites/default/files/documents/files/NUE%20and%20nutrient%20performance%20indicators_GPNM2015_FINAL.pdf.

Oenema, O., Billen, G., Brentrup, F., Lammel, J., Bascou, P., Dobermann, A., Erisman, J.W., Garnett, T., Hammel, M., Haniotis, T., Hoxha, A., Jensen, L.S., Oleszek, W., Pallière, C., Powlson, D., Quemada, M., Sutton, M.A., Vallejo, A., Van Grinsven, H.J.M., Winiwarter, W., 2015. Nitrogen Use Efficiency (NUE) – an indicator for the utilization of nitrogen in agriculture and food systems. EU Nitrogen Expert Panel. Published by Wageningen University, Alterra, TheNetherlands. http://eunep.businesscatalyst.com/index.html.

Peng, S., Buresh, R.J., Huang, J., Yang, J., Zou, Y., Zhong, X., Wang, G., Zhang, F., 2006. Strategies for overcoming low agronomic nitrogen use efficiency in irrigated rice systems in China. Field Crops Research 96, 37–47.

Raghuram, N., Sangar, S., Abrol, Y.P., 2011. Nitrogen management for sustainable development. Current Science 101 (9), 1126–1127.

Reis, S., Grennfelt, P., Klimont, Z., Amann, M., ApSimon, H., Hettelingh, J.P., Holland, M., LeGall, A.C., Maas, R., Posch, M., Spranger, T., Sutton, M.A., Williams, M., 2012. From acid rain to climate change. Science 338, 1153–1154.

Sutton, M.A., Bleeker, A., 2013. The shape of nitrogen to come. Nature 494, 435–437.

Sutton, M.A., Bleeker, A., Howard, C.M., Bekunda, M., Grizzetti, B., de Vries, W., van Grinsven, H.J.M., Abrol, Y. P., Adhya, T.K., Billen, G., Davidson, E.A., Datta, A., Diaz, R., Erisman, J.W., Liu, X.J., Oenema, O., Palm, C., Raghuram, N., Reis, S., Scholz, R.W., Sims, T., Westhoek, H., Zhang, F.S., with contributions from Ayyappan, S., Bouwman, A.F., Bustamante, M., Fowler, D., Galloway, J.N., Gavito, M.E., Garnier, J., Greenwood, S., Hellums, D.T., Holland, M., Hoysall, C., Jaramillo, V.J., Klimont, Z., Ometto, J.P., Pathak, H. , Plocq Fichelet, V., Powlson, D., Ramakrishna, K., Roy, A., Sanders, K., Sharma, C., Singh, B., Singh, U., Yan, X.Y., Zhang, Y., 2013. Our Nutrient World: The Challenge to Produce More Food and Energy With Less Pollution. Global Overview of Nutrient Management. Centre for Ecology and Hydrology, Edinburgh on Behalf of the Global Partnership on Nutrient Management and the International Nitrogen Initiative, 114 pp.

Sutton, M.A., Drewer, J., Raghuram, N., Kumar, D., Price, A., Smith, J.U., Bhatia, A., Reay, D., Subrahmanyam, D., Ahmad, A., Skiba, U., Dragosits, U., Vieno, M., Bealey, W.J., Carnell, E., Roberts, E., Stott, A., Sohi, S., Moring, A., Hillier, J., Nayak, D.R., Prasana, R., Singh, R., Neeraja, C.N., Voleti, S.R., Kumar, R.M., Surekha, K., Hooda, S., Babu, R., Jain, N., Pandey, P., Ramakrishnan, B., das Saha, N., Pathak, H., 2016. Newton-bhabha virtual centre on nitrogen efficiency of whole-cropping systems for improved performance and resilience in agriculture (NEWS India-UK). In: INI2016 Conference, Nitrogen Science and Policy (2–8 December 2016, Melbourne). http://www.ini2016.com/1432.

Sutton, M.A., Howard, C., Erisman, J.W., Billen, G., Bleeker, A., Grennfelt, P., van Grinsven, H., Grizzetti, B. (Eds.), 2011. The European Nitrogen Assessment: Sources, Effects and Policy Perspectives. Cambridge University Press, 612 pp.

UNEP, 2011. Addressing the Nutrient Challenge: Where We Are, what We Need to Know and what We Need to Do? Intergovernmental Review Meeting on the Implementation of the Global Programme of

Action for the Protection of the Marine Environment from Land-based Activities. Third Session. Manila, 25–27 January 2012. UNEP/GPA/IGR3/INF/7. United Nations Environment Programme.

UNEP, 2012. Manila Declaration on Furthering the Implementation of the global programme of action for the protection of the Marine Environment from land-based activities. In: Intergovernmental Review Meeting on TheImplementation of the GPA. Third Session (Manila, 25–27 January 2012). UNEP/GPA/IGR3/CRP.1/Rev.1.

UNEP, 2013. Drawing down $N_2O$ to protect climate and the ozone layer. In: Alcamo, J., Leonard, S.A., Ravishankara, A.R., Sutton, M.A. (Eds.), A UNEP Synthesis Report. United Nations Environment Programme, Nairobi.

US Department of State, June 17, 2014. Our Ocean Action Plan. Washington, DC. http://www.state.gov/documents/organization/228005.pdf.

US Geological Survey, 2012a. Mineral Commodity Summaries 2012, Phosphate Rock, pp. 118–119. http://minerals.usgs.gov/minerals/pubs/commodity/phosphate_rock/.

US Geological Survey, 2012b. Mineral Commodity Summaries 2012, Potash, pp. 122–123.

Van Grinsven, H.J., Holland, M., Jacobsen, B.H., Klimont, Z., Sutton, M.A., Jaap Willems, W., 2013. Costs and benefits of nitrogen for Europe and implications for mitigation. Environmental Science & Technology 47, 3571–3579.

World Bank, 2016. Data on Gross National Income per capita, Purchasing Power Parity http://data.worldbank.org/indicator/NY.GNP.PCAP.PP.CD.

Westhoek, H., Lesschen, J.P., Rood, T., Leip, A., Wagner, S., De Marco, A., Murphy-Bokern, D., Pallière, C., Howard, C.M., Oenema, O., Sutton, M.A., 2015. Nitrogen on the Table: The Influence of Food Choices on Nitrogen Emissions and the European Environment (European Nitrogen Assessment Special Report on Nitrogen and Food.). Centre for Ecology and Hydrology, UK, 67 pp.

Yadav, Y., Kumar, S., 2006. The Food Habits of a Nation. http://www.thehindu.com/todays-paper/the-food-habits-of-a-nation/article3089973.ece.

Yasuda, M., Tsugita, N., Ito, K., Yamauchi, S., Glomm, W.R., Tsuji, I., Asano, H., 2011. High-efficiency $NO_x$ absorption in water using equipment packed with a glass fiber filter. Environmental Science & Technology 45, 1840–1846.

3

# Concepts for Considerations in the Design of an Indian Integrated Nitrogen Assessment

J. Rudek[1], V.P. Aneja[2], Y.P. Abrol[3]

[1]ENVIRONMENTAL DEFENSE FUND, NEW YORK, NY, UNITED STATES; [2]NORTH CAROLINA STATE UNIVERSITY, RALEIGH, NC, UNITED STATES; [3]ING-SCON, DPS MARG, NEW DELHI, INDIA

**CHAPTER OUTLINE**



## Introduction

Fritz Haber, in his Nobel Prize lecture in 1918, explained that it was the growing demand for food which motivated him to explore the "synthesis of ammonia from the elements" (Erisman et al., 2008.) In collaboration with Carl Bosch, an industrial scale process (known as the Haber-Bosch process) was developed to manufacture large amounts of nitrogen (N) fertilizer. This breakthrough has not only radically changed agriculture but has also more than doubled the release of the reactive nitrogen (Nr) into the environment. Reactive nitrogen, defined as "all biologically, chemically and radiatively active nitrogen compounds in the atmosphere and biosphere" (UNEP and WHRC, 2007), readily transfers between chemical species and environmental media (soil, atmosphere, and water) causing numerous negative consequences to both human health and ecosystem (Aneja et al., 2008, 2009). This active cycling of Nr is known as the Nitrogen Cascade (Galloway et al., 2003).

In addition, about 20% of anthropogenic Nr is released through fossil fuel combustion. Large cities are concentrated sources of Nr due to increased use of fossil fuels for transport and energy as well as incineration of solid waste (Svirejeva-Hopkins et al., 2011).

The Indian Nitrogen Assessment. http://dx.doi.org/10.1016/B978-0-12-811836-8.00003-3

Reactive nitrogen released from these sources has no productive use and so is a “no regrets” target for decreasing environmental Nr pollution loading.

The use of Haber-Bosch Nr has greatly increased agricultural productivity over the last half century with an estimated 40–60% of the global population dependent on it (Erisman et al., 2008). Unfortunately, agricultural uptake of nitrogen is not very efficient and a large fraction of nitrogen fertilizer, often greater than half, is not used by the crops. The excess nitrogen fertilizer is lost to air, groundwater, or surface runoff, resulting in the loading of Nr downstream and downwind (Davidson et al., 2012). Furthermore, Nr removed from farms in food products ends up being processed in whatever human waste facility is being used. Advanced treatment might denitrify 40–50% of the Nr in the human waste stream with the remainder about equally divided between discharge to rivers or land application (Van Breemen et al., 2002; Svirejeva-Hopkins et al., 2011). Less advanced sewage treatment systems likely discharge most of the Nr into river systems. Svirejeva-Hopkins et al. (2011) recommended that in the future sewage processing systems should focus on including treatment that allow for Nr reuse as a fertilizer rather than denitrifying or discharging Nr to rivers. Galloway et al. (2014) reviewed several N-footprint calculators which provide information to consumers, institutions, and governments on their contribution to Nr loss to the environment as a result of dietary, housing, goods and services, and energy use choices.

Unlike other developed countries of the world and China, India lacks a unified study on Nr. Globally, India ranks second in agricultural output, first in cattle population, and second in emissions of ammonia and nitrous oxide from the agricultural sector (Aneja et al., 2012) as a result of a dependence on high levels of fertilizer application. Consumption of fertilizer N increased from 0.6 MT in 1965–1966 to 17.4 MT in 2016 (Tewatia and Chandra, 2016) and the country has emerged as the second largest consumer of fertilizer N in the world. By 2050, input of fertilizer N is expected to increase twofold further enhancing Nr loss into adjacent environments, affecting the health of receiving ecosystems and their inhabitants (Bijai-Singh et al., 2008). In addition, with India developing in leaps and bounds, demand for energy is increasing at a rapid pace leading to a forecast of $NO_x$ emissions increasing to 22,800 Gg in 2020, from 350 Gg in 1990 (Garg et al., 2006). Given the present and future predictions of Nr use and emissions, it is imperative that the flow of Nr in India is critically assessed. Without proper management, Nr losses will cause an ever-increasing cascade of environmental and human health problems in India. With India's burgeoning population and need to improve quality of life for a large percentage of the population, managing Nr in India involves a delicate balance of energy and fertilizer use efficiency so that power and crop yields supply the needs of the populace with minimum Nr loss to air, water, and soil. The Zhang et al. (2015) examination of historical trends in N fertilizer use in India, and its comparison to several other countries, helps highlight the need for reversing current trends of fertilizer N use efficiency in India. The need for sustainable integrated Nr management poses a challenge for the Indian scientific community to provide policy makers with reliable estimates of Nr transfers among ecosystems and to offer feasible strategies to decrease the undesirable loss of Nr to the environment (Aneja et al., 2012).

One of the first steps in developing an Nr management strategy is conducting an assessment of the Nr inputs, and fate and transport within the region under consideration. Such an assessment requires quantification of Nr fluxes at regional and national scales. This chapter provides a suggested outline for an Nr assessment based upon previous Nr assessments and general outlines for scientific assessments.

Following up an Nr assessment with the development of effective management policies is complicated because of the ease with which Nr moves across the air, soil, and water media. A policy to address an issue in one media may inadvertently cause an increased pollutant load in another media (UNEP and WHRC, 2007). UNEP and WHRC (2007) explored case studies of some of the more successful policy efforts in Europe, North America, Latin America, and the Caribbean where the range of concerns necessary for integrated Nr management are considered. Another policy difficulty is setting the targets for agricultural N use given current and future food needs. Zhang et al. (2015) estimated the nitrogen use efficiency that agriculture must achieve to remain sustainable while supplying food and fiber to the growing world population.

## Overview of the Development of an Integrated Science Assessment

The US Environmental Protection Agency (USEPA) published a description of the stepwise process of developing a generic integrated science assessment which can provide a useful framework for designing an Nr assessment for India (US EPA, 2015) (Fig. 3.1). This stepwise process includes literature searches, evaluation of individual study quality, synthesis and integration of the evidence, and development of the scientific conclusions and casual determinations.

The agencies and advisory committees included in the USEPA framework are particular to formulation of air quality standards in the USA, and therefore some of the evaluations would not be needed for an $N_r$ assessment in India. Naturally, the agencies and advisory committees would need to be replaced with the appropriate Indian agencies and advisory committees.

An important initial step in the integrated science assessment process is to identify the key relevant policy and scientific questions to be evaluated by the assessment. The USEPA recommends holding public science and policy issue workshops to seek input on the current state of the science, and to engage stakeholders and experts in discussion of the policy-relevant questions which can develop the key assessment questions. A call for information by the lead agency or organization that invites the public to provide information, especially peer-reviewed articles, on the range of issues associated with Nr in India will also be helpful in the implementation of the first step of the framework; the literature search.

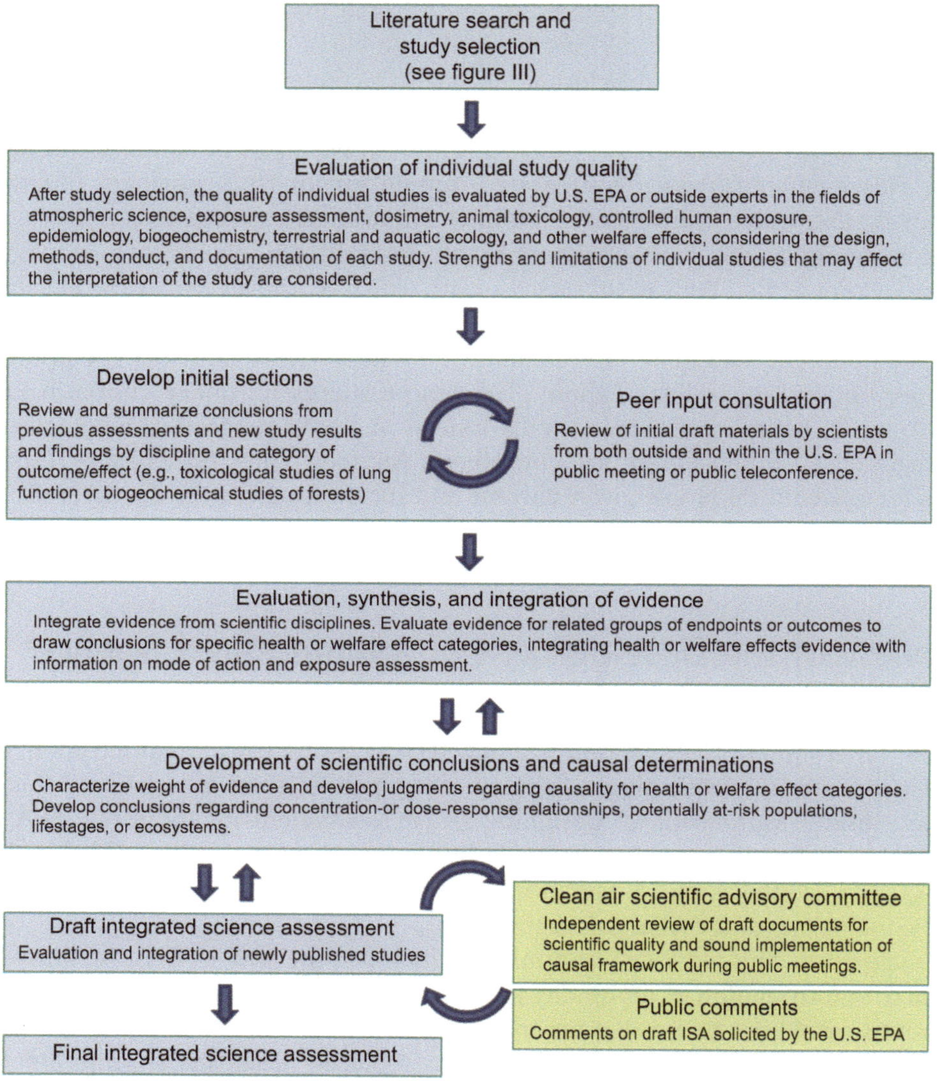


**FIGURE 3.1** Characterization of the general process for Integrated Science Assessment development. (US EPA, 2015).

The USEPA framework presumes that government agency staff will be conducting the reviews and summarization of the literature and studies, and producing the first draft which is then reviewed by subject matter experts. An alternate approach is to assemble subject matter experts into panels to conduct the reviews and summarization, and to have each subject matter panel write the draft report for that subject. The drafts are then made available for public review and comment. The latter approach was used in the USA and European nitrogen assessments (USEPA, 2011; and Sutton et al., 2011b) which are discussed below.

The USEPA recommends that only studies which have undergone peer review and have been published or accepted for publication should be considered for inclusion in the literature review. Selected references should undergo full-text review and be evaluated by considering design, methods, and documentation of each study, but not the study results. The USEPA (2015) includes the following relevant characteristics to evaluate the scientific quality of the literature:

- Were study design, methods, data, and results clearly presented in relation to the study objectives? Were limitations and underlying assumptions of the design and other aspects of the study stated?
- Are the health, ecological, or other welfare effect measurements meaningful, valid, and reliable?
- Were likely covariates or modifying factors adequately controlled or taken into account in the study design and statistical analysis?
- Do the analytic methods provide adequate sensitivity and precision to support the conclusions?
- Were the statistical analyses appropriate, properly performed, and properly interpreted?

After literature review and summarization, the available research needs to be evaluated and synthesized. During the synthesis, there is value in integrating evidence across various approaches or disciplines where coherence can add support to cause–effect associations. Uncertainty should also be considered during evaluation of literature.

Different types of studies are likely to have different types of strengths, limitations, and uncertainties. In laboratory studies, conditions can be better controlled, often leading to less variability in the results, and allowing smaller effects to be detected. However, control conditions may limit the range of responses and large scale processes are difficult to reproduce, so results may not reflect the responses that would occur in the natural environment.

On the other hand, field studies measure biological changes in uncontrolled situation with high natural variability and can provide important information across multiple factors. However, these factors can often be site-specific. Field studies are especially useful for considering pollutant effects at larger geographic scales and higher levels of biological organization. Though because conditions are not controlled, variability is expected to be higher and may mask the effects. The presence of confounding factors can make it difficult to attribute observations to specific causes.

In developing scientific conclusions, the USEPA (2015) discusses aids in judging causality. Effects on ecosystems are often a result of a mix of influences both synergistic and antagonistic. The causality judgment framework provides for a systematic evaluation of the body of evidence which should be informed by peer and public comment and advice. This framework should not be used as a checklist but rather as a means to help weigh the existing evidence inferring causality. In particular, not meeting one or more of the principles does not automatically rule out the causality. Consistency of findings across studies is informed by the repeated observations or associations under different circumstances. However, it should be kept in mind that discordant results could be due to difference in methods, random errors, confounding factors, or study power, and hence should be automatically ruled out.

The USEPA (2015) also offers a framework for evaluating the weight of scientific evidence in determining causal relationships. Ecological effects should be viewed in terms of level of organization, such as cellular, organism, population, community, and ecosystem. While impact to individual organisms may not be of significance, if enough organisms are effected, communities and ecosystems may be disrupted, resulting in detrimental changes in ecosystem function or services.

## Evaluation of Reactive Nitrogen Assessments in the USA and Europe

The Nr assessments conducted both for the USA and Europe in 2011 are compared and contrasted to provide examples of how an Indian assessment might be designed. While we did not include it in our comparison, Gu et al. (2015) represented an additional example of a national Nr assessment for China.

The USEPA assembled a Science Advisory Board (SAB) Integrated Nitrogen Committee to conduct the Nr assessment for the USA (USEPA, 2011). The objectives of the SAB report were to:

- "Identify and analyze from a scientific perspective the problems Nr presents in the environment and links among them;
- Evaluate the contribution an integrated N management strategy could make to environmental protection;
- Identify additional risk management options for USEPA's consideration; and
- Make recommendations to USEPA concerning improvements in nitrogen research to support risk reduction."

The European Nitrogen Assessment (ENA; Sutton et al., 2011b) was organized by the European Science Foundation Research Networking Programme "Nitrogen in Europe" or NinE, to achieve the following objectives:

- To develop the underpinning science that links different forms of N. Hence detailed studies in a single N pollutant are not the first priority. Instead, the priority is for studies examining the interchange and relationships between different N pollutants as N cascades through the environment.
- To develop the science linking N interactions between environmental compartments. Coupled to the above is the need for studies that link N emissions, transformations and impact between soils, plants, air and water and between differ contexts, urban–rural, aquatic–marine, biosphere–troposphere–stratosphere, etc.
- To establish approaches at ranges of scales, from physiological scale, patch scale, landscapes (e.g., 25 $km^2$), regional watersheds (e.g., 10,000 $km^2$) to regions of

Europe and the whole continent. Each of the scales is relevant, but the larger scales that allow explicit assessment of the N cascade are particularly relevant (landscapes to continental).

- To refine methodologies for relating information between different spatial and temporal scales. It is a major scientific challenge to carry information developed at finer scales to larger scale models, while also posing challenges for the different science communities to interact successfully.
- To apply the analyses of NinE in selected case studies across Europe. These may include contrasting case-study landscapes, major watersheds and larger contrasting regions of Europe, including the comparison between source and sink areas for N.
- To establish a meta-database of N research activities and assessment that integrate different N forms, interactions and scales. Such a meta-database would take the widest possible scope, including process analyses and case studies, providing access to both datasets and reports.
- To prepare a major assessment report covering the interlinked problems of N in Europe. The assessment report would be informed by the results of scientific activity of NinE and draw extensively on the NinE meta-database.

In both SAB and ENA processes, chapters were written by a group of experts assembled for the purpose and peer-reviewed externally. The overall structure of the two reports was generally similar with chapters focused on specific aspects of Nr as well as interactions between the different Nr chemical species and media (atmosphere, water, and soil). However, while the SAB report featured a single document with an executive summary contained in the body of the document, the ENA report presented the chapters as separate documents and included technical and policy summary documents. The ENA report also had a more extensive economic analysis compared to the SAB report. The summary and economic chapter documents are likely to be helpful for more easily disseminating information tailored to key audiences during policy-making deliberations.

Both reports started with an overview of the benefits and problems associated with current Nr use (for example, Table 3.1) followed by chapters detailing the sources, transfers, and transformations of Nr in the environment (for example, Fig. 3.2). The N cascade was described and diagrammed (for example, Figs. 3.3 and 3.4), leading to a discussion of the need for an integrated Nr management approach.

The ENA document includes a graphic that creatively summarizes the five key societal threats of excessive Nr; water quality, air quality, greenhouse gas balance, ecosystems and biodiversity, and soil quality. This useful communication tool lists these five threats to make an acronym called the "WAGES" of excess nitrogen. The ENA report also summarizes the cost of economic impacts of excessive Nr in Europe (Tables 3.2 and 3.3) showing that the economic and social impacts are substantial. These estimates are also useful in communicating the importance of Nr management to the public and policy makers.

**Table 3.1** Examples of Impact of Excess Reactive Nitrogen on Human Health and Environment (US EPA, 2011)

| Impact | Cause | Location | Metric | Source |
|---|---|---|---|---|
| **Air** | | | | |
| Visibility decrease | Fine particulate matter | National parks and wilderness areas | Visibility impairment | $NO_y$ and $NH_x$ from fossil fuels and agriculture |
| **Land Ecosystem** | | | | |
| Biodiversity loss | Nitrogen deposition | Grasslands and forests in the USA receiving N deposition in excess of critical load | Decrease in species richness of grasslands and forests | Utilities, traffic, and animal agriculture |
| Forest decline | Ozone and acid deposition | Eastern and western USA | Decreased timber growth; increased susceptibility to disease and pests | Utilities, traffic, and animal agriculture |
| **Land Agriculture** | | | | |
| Crop yield loss | Ozone | Eastern and western USA | $2–5 billion/year | Utilities and traffic |
| **Water** | | | | |
| Acidification of surface waters; loss of biodiversity | Acidification of soils, streams, and lakes is caused by atmospheric deposition of sulfur. $HNO_3$, $NH_3$, and ammonium compounds | Primarily mountainous regions of the USA | Out of 1000 lakes and thousands of miles of streams in the eastern USA surveyed. 75% of the lakes and 50% of the streams were acidified by acid deposition | Fossil fuel combustion and agriculture |
| Hypoxia of coastal waters | Excess nutrient loading, eutrophication, variable freshwater runoff | Gulf of Mexico, other estuarine and coastal waters | Benthic finfish/shell fish habitat loss, fish kills, sulfide toxicity, costs >$50 million annually | N, P from energy and food production |
| Harmful algal blooms | Excessive nutrient loading, climatic variability | Inland and coastal waters | Fish kills, losses of drinking and recreational waters costs >$100 million annually | Excess nutrient (N & P) loading |
| **Human and Environmental Health Damages** | | | | |
| Human mortality | $PM_{2.5}$, $O_3$ and related toxins | US urban and nearby areas | Pollution-related deaths estimated at 28,000–55,000 per year (a range of cardiovascular and respiratory system effects are associated with this pollution) | $NO_y$ and $NH_x$ from fossil fuels and agriculture |

**Table 3.1** Examples of Impact of Excess Reactive Nitrogen on Human Health and Environment (US EPA, 2011)—cont'd

| **Impact** | **Cause** | **Location** | **Metric** | **Source** |
|---|---|---|---|---|
| Total damage to public health and environment | $NO_x$ into air | Chesapeake Bay watershed | \$3.4 billion; 200,000 MT | Mobile sources |
| Total damage to public health and environment | $NH_x$ and nitrate into air and water | Chesapeake Bay watershed | \$1.5 billion; 400,000 MT | Agriculture |

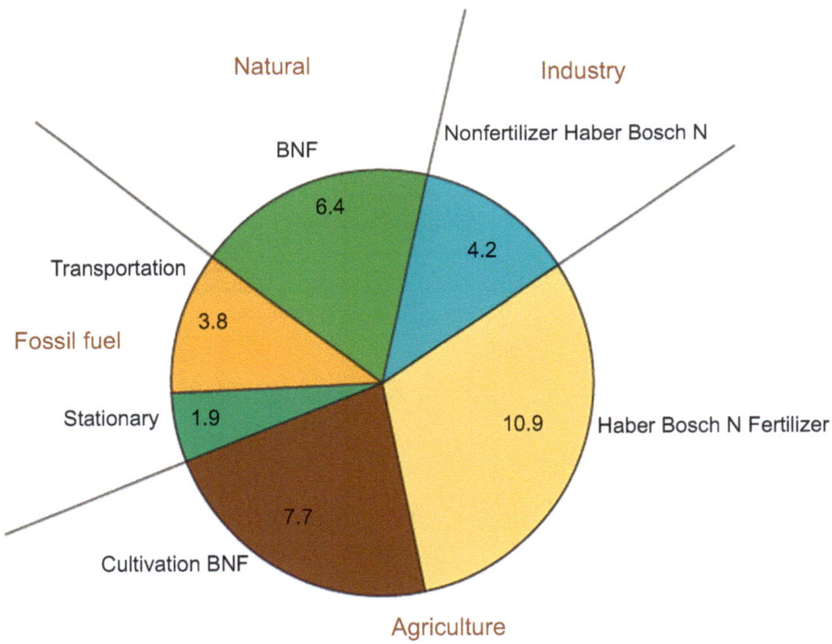


**FIGURE 3.2** . Sources of reactive nitrogen introduced into the United States in 2002 (Tg N year$^{-1}$). BNF, biological nitrogen fixation. (EPA 2011)

In addition to Nr source and transformation discussions, both documents outline existing law and regulations covering various aspects of Nr use. The potential interaction of Nr effects with climate change is also discussed. The ENA report includes a comprehensive summary of the change in radiative forcing due to European Nr emissions to the atmosphere showing the complex interactions of Nr. Interestingly, the overall effects of European Nr emissions are estimated to have a net cooling with uncertainty bounds ranging from substantial cooling to small net warming (Sutton et al., 2011a).

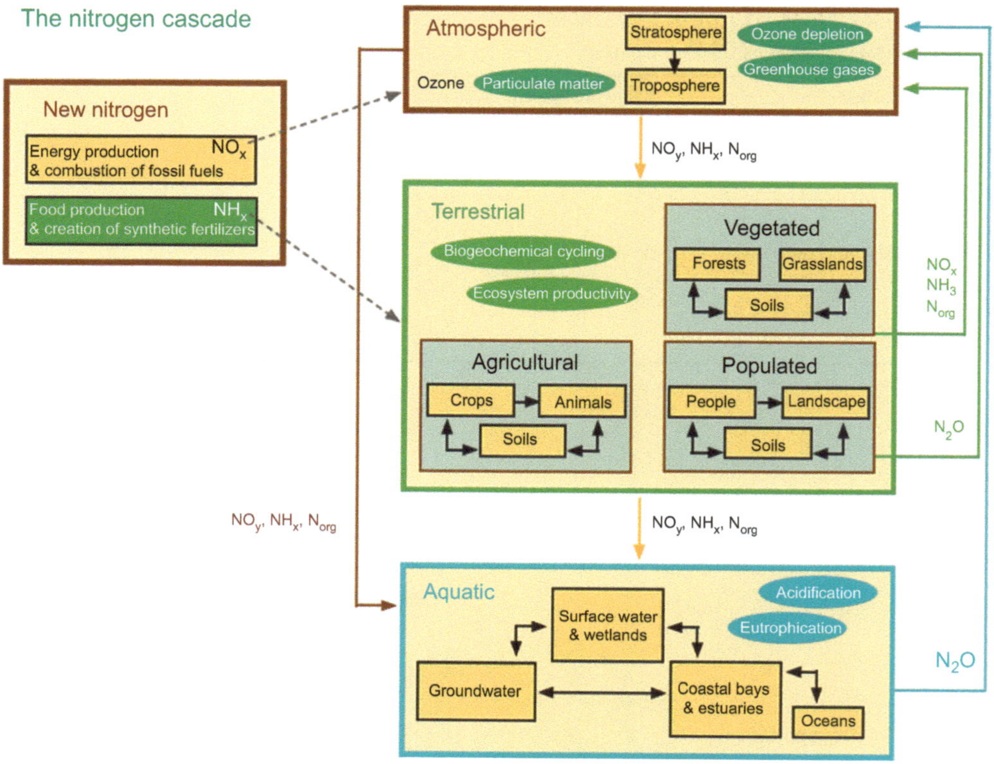


**FIGURE 3.3** The nitrogen cascade (EPA 2011).

Following the subject topic chapters, each document states its major findings and recommended management actions. The SAB report management recommendations are as follows:

1. The Committee estimates that if EPA were to expand its $NO_x$ control efforts for emissions of mobile sources and power plants, a 2.0 Tg N year$^{-1}$ decrease in the generation of Nr could be achieved.
2. The Committee estimates that excess flows of Nr into streams, rivers, and coastal systems can be decreased by approximately 20% (approximately one Tg N year$^{-1}$) through improved landscape management and without undue disruption to agricultural production.
3. The Committee estimates that livestock-derived $NH_3$ emissions can be decreased by 30% (a decrease of 0.5 Tg N year$^{-1}$) by a combination of BMPs (best management practices) and engineered solutions.
4. The Committee recommends that a high priority be assigned to increasing funding for nutrient management. We estimate that adequate financial support for sewage treatment infrastructure upgrades to remove nutrients could decrease Nr emissions by between 0.5 and 0.8 Tg N year$^{-1}$. Additional Nr management from eligible

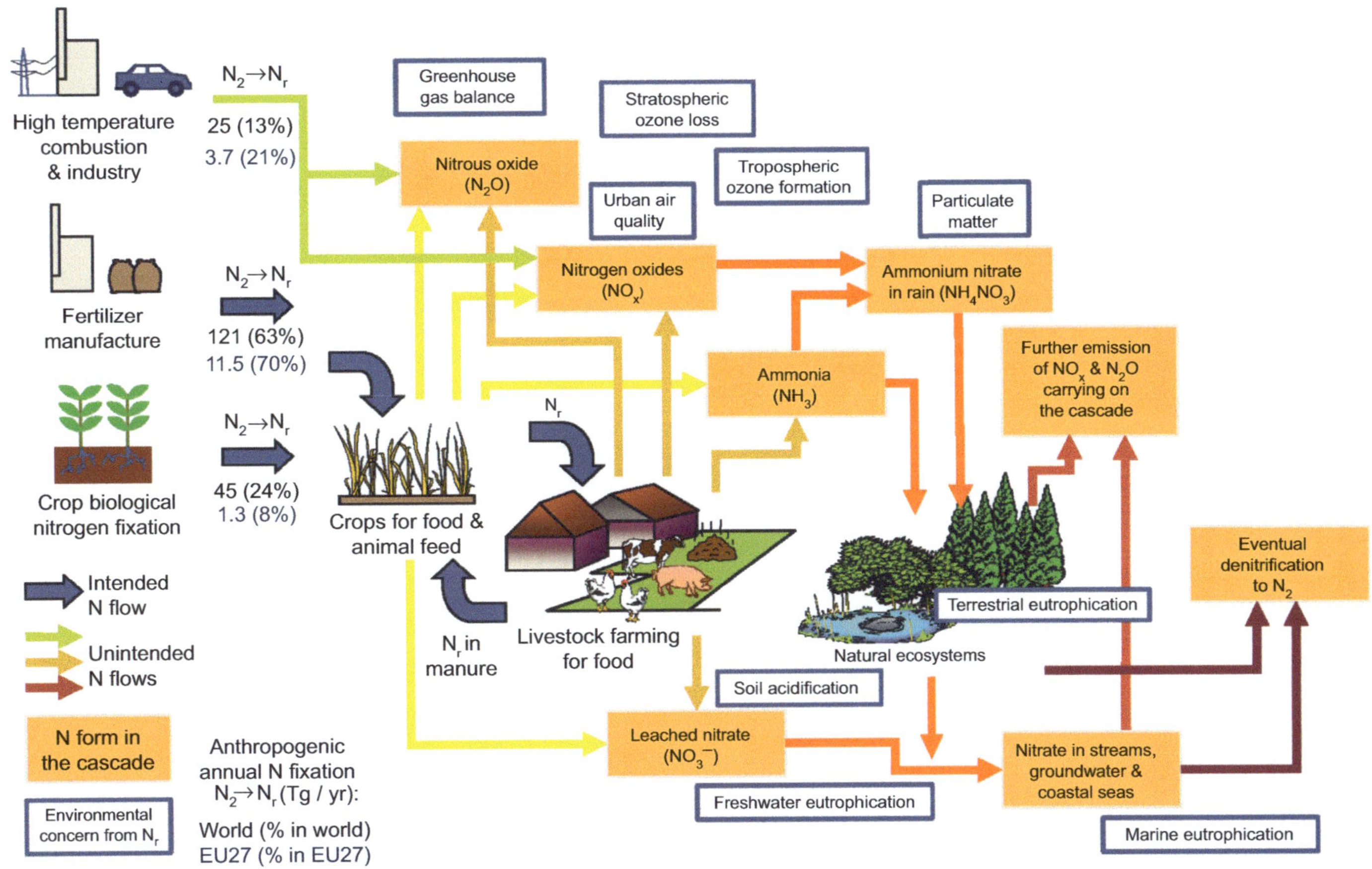


**FIGURE 3.4** Simplified view of the nitrogen (N) cascade, highlighting the major anthropogenic sources of reactive nitrogen (Nr) from atmospheric di-nitrogen ($N_2$), the main pollutant forms of Nr (*filled boxes*) and nine main environmental concerns (*unfilled boxes*). Estimates of anthropogenic N fixation for the world (Tg year for 2005, in straight text) are compared with estimates for Europe (Tg year for 2000, in italic). *Thick arrows* represent intended anthropogenic Nr flows; all other *arrows* are unintended flows. (Sutton et al., 2011a).

**Table 3.2** Estimates of Overall Social Damage Costs in the European Union (EU-27) as a Result of Environmental Nr Emissions (Billions Euros per Year at 2000). Values are Shown Here Rounded to the Nearest 5 billion Euro to Avoid Precision Issues, Explaining Differences With the Sums. The Calculated Value for $N_2O$ Effects on Human Health is 1–2 billion Euros per year (Sutton and van Grinsven, 2011)

| | $NO_x$ Emission to Air | $NH_3$ Emission to Air | $N_r$ Loss to Water | $N_2O$ Emission to Air | Total |
|---|---|---|---|---|---|
| Human health | 35–100 | 5–70 | 0–20[a] | <5 | 40–190 |
| Ecosystems | 5–35 | 5–35 | 15–50[a] | — | 25–115 |
| Climate | — | — | — | 5–10 | 5–10 |
| Total | 40–135 | 10–105 | 15–70 | 5–15 | 70–320 |

[a]The value for health effects is proportionately smaller than the value for ecosystems as not all leaching is associated with health effects (e.g., denitrified during the path from soil to sea).

**Table 3.3** Estimated Cost of Different Nr Threats in Europe per Unit Nr Emitted (Sutton et al., 2011c)

| Effect | Emitted nitrogen form | Emission/ loss to | Estimated cost € per kg $N_r$ emitted |
|---|---|---|---|
| Human health (particulate matter, $NO_2$ and $O_3$) | $NO_x$ | Air | 10–30 |
| Ecosystems (eutrophication, biodiversity) | $N_r$ (inc. Nitrate) | Water | 5–20 |
| Human health (particulate matter) | $NH_3$ | Air | 2–20 |
| Climate (greenhouse gas) | $N_2O$ | Air | 5–15 |
| Ecosystems (eutrophication, biodiversity) | $NH_3$ and $NO_x$ | Air | 2–10 |
| Human health (drinking water) | $N_r$ (inc. Nitrate) | Water | 0–4 |
| Human health (increased ultraviolet radiation from ozone depletion) | $N_2O$ | Air | 1–3 |

stormwater and nonpoint sources could be accomplished through increased support.

The ENA key recommended actions are summarized as a package of seven key actions in four sectors as follows (Sutton et al., 2011c):

*Agriculture*:

1. Improving N use efficiency in crop production.

This includes improving field management practices, genetic potential and yields per Nr input, with the potential to reduce losses per unit of produce, thereby minimizing the risk of pollution swapping.

2. Improving N use efficiency in animal production.

As with crops, this includes management practices and genetic potential, with an emphasis on improving feed conversion efficiency and decreasing maintenance costs, so reducing losses per unit of produce and the extent of pollution swapping.

**3.** Increasing the fertilizer N equivalence value of animal manure.

Increasing fertilizer equivalence values requires conserving the Nr in manure during storage and land application (especially reducing $NH_3$ emissions where much Nr is lost), while optimizing the rate and time of application to crop demand.

*Transport and industry*:

**4.** Low-emission combustion and energy-efficient systems.

These include improved technologies for both stationary combustion sources and vehicles, increasing energy efficiency and use of alternative energy sources with less emission, building on current approaches.

*Waste water treatment*:

**5.** Recycling N (and phosphorus) from waste water systems.

Current efforts at water treatment for Nr in Europe focus on denitrification back to $N_2$. While policies have been relatively successful, this approach represents a waste of the energy used to produce Nr. An ambitious long-term goal should be to recycle Nr from waste waters, utilizing new sewage management technologies.

*Societal consumption patterns*:

**6.** Energy and transport saving.

Against the success of technical measures to reduce $NO_x$ emissions per unit consumption, both vehicle miles and energy use have increased substantially over past decades. Dissuasion of polluting cars and far-distance holidays, and stimulation of energy-saving houses and consumption patterns can greatly contribute to decreasing $NO_x$ emissions.

**7.** Lowering the human consumption of animal protein.

European consumption of animal protein is above the recommended per capita consumption in many parts of Europe. Lowering the fraction of animal products in diets to the recommended level (and shifting consumption to more N-efficient animal products) will decrease Nr emissions with human health co-benefits, where current consumption is over the optimum.

The SAB report followed its management recommendations with a discussion of risk reduction strategies. Four types of management strategies were discussed: command and control, government-based programs for effecting policy, market-based instruments, and voluntary programs.

In the ENA report, the cost effectiveness of various management strategies are discussed in the context of communicating with the public and policy makers. A marginal abatement cost curve case study from the UK is discussed as an example of an important analysis to inform policy makers of the cost and benefit of achieving significant levels of change in Nr losses (Reay et al., 2011). Such an analysis can help identify the most cost-effective management measures. The report also models future scenarios of N using a storyline, similar to that used in the IPCC Special Report in Emission Scenarios, to predict the effect of various management scenarios which provide interesting insights that are a valuable basis for further work (Winiwarter et al., 2011).

## Conclusion

The comparison of the SAB (USA) and ENA (Europe) reports provided a good set of examples from previous Nr assessments to develop the outline for a process to evaluate Nr sources, fate and transport, and management options for India. The ENA report framework, with separately published chapters and chapters devoted to technical and policy summary might prove more beneficial in using the report to influence policy making. The inclusion of an in-depth economic assessment is also a valuable attribute of the ENA report.

The USEPA framework for integrated science assessments provides a good review of the concepts and approaches critical to the implementation of a robust scientific evaluation of the state of a science, a process which is fundamentally different from the proper implementation of an individual experiment, or study with which most researchers are familiar.

## References

Aneja, V.P., Schlesinger, W.H., Erisman, J.W., 2008. Farming pollution. Nature Geoscience 1, 409–411.

Aneja, V.P., Schlesinger, W.H., Erisman, J.W., 2009. Effects of agriculture upon the air quality and climate: research, policy and regulations. Environmental Science and Technology 43, 4234–4240.

Aneja, V.P., Schlesinger, W.H., Erisman, J.W., Behera, S.N., Sharma, M., Battye, W., 2012. Reactive nitrogen emissions from crop and livestock farming in India. Atmospheric Environment 47, 92–103.

Bijai-Singh, Tiwari, M.K., Abrol, Y.P., 2008. Reactive Nitrogen in Agriculture, Industry and Environment in India. Indian National Science Academy, New Delhi, p. 42.

Davidson, E.A., et al., 2012. Excess Nitrogen in the U.S. Environment: Trends, Risks and Solutions. Issues in Ecology. Ecological Society of America, p. 16. Report No. 15.

Erisman, J.W., Sutton, M.A., Galloway, J.N., Klimont, Z., Winiwarter, W., 2008. How a century of ammonia synthesis changed the world. Nature Geoscience 1, 636–639.

Galloway, J.N., Aber, J.D., Erisman, J.W., Seitzinger, S.P., Howarth, R.W., Cowling, E.B., Cosby, B.J., 2003. The nitrogen cascade. BioScience 4, 341–356.

Galloway, J.N., Winiwarter, W., Leip, A., Leach, A.M., Bleeker, A., Erisman, J.W., 2014. Nitrogen Footprints: past, present and future. Environmental Research Letters 9 (115003), 11. http://iopscience.iop.org/article/10.1088/1748-9326/9/11/115003/meta;jsessionid=6308DAB06F928E3C562BFEA5D10032A2.c4.iopscience.cld.iop.org.

Garg, A., Shukla, P.R., Kapashe, M., 2006. The sectoral trends of multigas emissions inventory of India. Atmospheric Environment 40, 4608–4620.

Gu, B., Chang, J., Ge, Y., Vitousek, P.M., 2015. Integrated reactive nitrogen budgets and future trends in China. Proceedings of the National Academy of Sciences 112, 8792–8797. http://www.pnas.org/content/112/28/8792.full.

Reay, D.S., Howard, C.M., Bleeker, A., Higgins, P., Smith, K., Westhoek, H., et al., 2011. Chapter 26: societal choice and communicating the European nitrogen challenge. In: Sutton, M.A., et al. (Eds.), The European Nitrogen Assessment. Cambridge University Press. http://www.nine-esf.org/ENA-Book.

Sutton, M.A., Billen, G., Bleeker, A., Erisman, J.W., Grennfelt, P., van Grinsven, H., et al., 2011a. Technical summary. In: Sutton, M.A., et al. (Eds.), The European Nitrogen Assessment. Cambridge University Press. http://www.nine-esf.org/ENA-Book.

Sutton, M.A., Howard, C.M., Erisman, J.W., Billen, G., Bleeker, A., Grennfelt, P., et al., 2011b. The European Nitrogen Assessment. Cambridge University Press. http://www.nine-esf.org/ENA-Book.

Sutton, M.A., van Grinsven, H., Billen, G., Bleeker, A., Bouwman, A.F., Bull, K., et al., 2011c. Summary for policy makers. In: Sutton, M.A., et al. (Eds.), The European Nitrogen Assessment. Cambridge Univ. Press. http://www.nine-esf.org/ENA-Book.

Svirejeva-Hopkins, A., Reis, S., Magid, J., Nardoto, G.B., Barles, S., Bouwman, A.F., et al., 2011. Chapter 12: nitrogen Flows and fate in urban landscapes. In: Sutton, M.A., et al. (Eds.), The European Nitrogen Assessment. Cambridge University Press. http://www.nine-esf.org/ENA-Book.

Tewatia, R.K., Chandra, T.K., 2016. Trends in fertilizer nitrogen production and consumption in India. In: Abrol, Y.P., Adhya, T.K. (Eds.), The Indian Nitrogen Assessment: Sources of Reactive Nitrogen, Environmental and Climate Effects, and Management Options and Policies. Elsevier, USA.

UNEP, WHRC, 2007. Reactive Nitrogen in the Environment: Too Much or Too Little of a Good Thing. United Nations Environment Programme, Paris. http://www.unep.org/pdf/dtie/Reactive_Nitrogen.pdf.

US EPA, 2011. Reactive Nitrogen in the US: An Analysis of Inputs, Flows, Consequences, and Management Options. https://yosemite.epa.gov/sab/sabproduct.nsf/WebBOARD/INCFullReport/$File/Final%20INC%20Report_8_19_11%28without%20signatures%29.pdf.

US EPA, 2015. Preamble to the Integrated Science Assessments. U.S. Environmental Protection Agency, Washington, DC. EPA/600/R-15/067, 2015. https://cfpub.epa.gov/ncea/isa/recordisplay.cfm?deid=310244.

Van Breemen, N., Boyer, E.W., Goodale, C.L., Jaworski, N.A., Paustian, K., Seitzinger, S.P., et al., 2002. Where did all the nitrogen go? Fate of nitrogen inputs to large watersheds in the northeastern USA. Biochemistry 57/58, 267–293.

Winiwarter, W., Hettelingh, J.-P., Bouwman, A.F., de Vries, W., Erisman, J.W., Galloway, J., et al., 2011. Chapter 24: future scenario of nitrogen in Europe. In: Sutton, M.A., et al. (Eds.), The European Nitrogen Assessment. Cambridge University Press. http://www.nine-esf.org/ENA-Book.

Zhang, X., Davidson, E.A., Mauzerall, D.L., Searchinger, T.D., Dumas, P., Shen, Y., 2015. Managing nitrogen for sustainable development. Nature 528, 51–59. http://www.nature.com/nature/journal/v528/n7580/full/nature15743.html.

4

# Trends in Fertilizer Nitrogen Production and Consumption in India

R.K. Tewatia, T.K. Chanda

*THE FERTILISER ASSOCIATION OF INDIA, NEW DELHI, INDIA*

**CHAPTER OUTLINE**



## Introduction

India has a long history of managing soil fertility by using organic and green manures. Beginning of the production of nitrogenous fertilizer dates back to 1933 with manufacture of ammonium sulfate as a by-product in steel plants. Real impetus to fertilizer

**The Indian Nitrogen Assessment. http://dx.doi.org/10.1016/B978-0-12-811836-8.00004-5**

production and consumption in India came after the beginning of planned era in 1950–51. Concerted efforts were made by government and industry to popularize the use of fertilizers among the farmers. However, there was not much growth in fertilizer production and consumption until 1960. The introduction of fertilizer-responsive high-yielding varieties (HYVs) of rice and wheat in mid-1960s was the turning point in the Indian agriculture. This provided the much needed fillip to Indian fertilizer sector. The country witnessed sharp growth in production and consumption of N fertilizers. The favorable policies formulated during 1970s, which were continued in 1980s and 1990s encouraged rapid growth in domestic production as well as consumption of the fertilizers in general, and N fertilizers in particular. India emerged as the second-largest consumer and producer of the fertilizer-N in the world. The expansion in irrigated area, HYVs, fertilizer distribution network, and fertilizer subsidy support were the main drivers of this remarkable achievement in Indian fertilizer sector.

## Fertilizer Nitrogen Production

The production of ammonium sulfate, as a by-product of steel industry, by Tata Iron and Steel Company Limited, Jamshedpur, in 1933 marked the beginning of the production of nitrogenous fertilizers in India. The first megaplant for manufacturing 355,000 tons per annum of ammonium sulfate was set up at Sindri in 1947 and its production commenced in 1951. Production of urea, using coke oven gas feedstock by Fertilizer Corporation of India (FCI), Sindri, in 1959 revolutionized the fertilizer-N industry. The production of other N-fertilizers, namely ammonium sulfate nitrate by FCI, Sindri, and ammonium chloride by New Central Jute Mills Ltd., Varanasi, was also started in 1959. The production of calcium ammonium nitrate (CAN) fertilizer using electrolysis process was started by FCI (now NFL, National Fertilizers Limited) in 1961 at Nangal.

In addition to straight nitrogenous fertilizers, a variety of NP/NPK complexes came in stream, particularly during 1960s. The production of nitro-phosphate was started by Rashtriya Chemicals and Fertilizers Limited (RCF), Trombay, in 1965. Diammonium phosphate (DAP) production was started in 1967 by Gujarat State Fertilizers and Chemicals Limited (GSFC). It has now become the most popular N- and P-carrier. The production of NPK complex fertilizers by RCF, Trombay, commenced in 1968. At present, 15 grades of complex fertilizers are produced in addition to DAP and mono-ammonium phosphate (MAP). In total, 20 different products (straight nitrogenous and complex fertilizers) are produced as source of N in India. In addition, a few grades of 100% water soluble complex fertilizers are being produced and used for fertigation.

### History and Growth of Nitrogenous Fertilizer Industry

Fertilizer-N industry in India has marched a long journey since its inception in 1933. In initial years, investments were made to set up N production capacity in public sector. A large number of urea plants were set up between 1961 and 1977. Most of these plants were based on naphtha as feed stock. Four gas-based urea plants (at Namrup, Baroda, Kalol, and Trombay) were also commissioned during this period. Between 1977 and 1980, fuel oil/low sulfur heavy stock (LSHS)-based plants were set up by NFL at Bhatinda, Nangal, and Panipat, and by FCI at Sindri. Two coal-based ammonia/urea plants of FCI at Talcher and Ramagundam were commissioned in November, 1980.

After the introduction of Retention Price Scheme (RPS) for nitrogenous fertilizers in November, 1977 that ensured fair return to individual fertilizer producer, many new ammonia/urea plants were set up in private and cooperative sectors. A number of urea plants were commissioned during 1980s and 1990s. Most of these plants were based on natural gas as feedstock. The plants based on feedstock other than gas also converted to gas in recent years. Currently, out of 30 operating plants, 27 plants are based on gas and remaining three plants are based on naphtha. The remaining 3 plants are also ready to use gas but are yet to have gas connectivity. At present, 94% of total urea capacity is based on natural gas. As on November 1, 2015, the share of natural gas in total N production was 78.3%; external $NH_3$ and naphtha had a share of 16.0% and 5.3%, respectively. Out of a total production capacity of 13.55 million tons (Mt) of N, private sector has a share of 47.1%, cooperative sector of 26.8%, and public sector of 26.1%.

Historically, the type of feedstock used for production of nitrogenous fertilizers has been determined by the availability of feedstock and policies of government for its optimum utilization. Naphtha was a preferred feedstock for production of nitrogenous fertilizers till 1975. Thereafter, some urea plants were set up with fuel oil/coal as feedstock during 1977–80. Availability of natural gas in offshore fields of western coast gave fillip to the gas-based fertilizer production. Natural gas as a feedstock has the major share in N production capacity. As on November 1, 2015, share of natural gas in total N production was 78.5%; external $NH_3$ and naphtha had a share of 16.0% and 5.3%, respectively.

## Nitrogen Production Trends

The annual production of N was merely 28,900 tons in 1951–52 and it took almost 10 years for N production to reach the level of 0.11 Mt in 1960–61. After 1960–61, N production grew steadily reaching 0.83 Mt in 1970–71 (Table 4.1). The country witnessed a rapid growth in capacity and production of fertilizer-N between 1970–71 and 2000–01. The decade of 1980s is considered as a golden period which observed record growth in fertilizer production, consumption, as well as in food grain production. The N production increased fivefold (2.16–10.94 Mt) during 1980–81 to 2000–01 and India has emerged as the second-largest producer of N after China.

Since 2000–01, growth rate in N capacity and production has slowed down as no new plant has come up after 1999, except for small additions such as revamp/modification of existing urea plants in recent years. In one decade (2000–01 to 2010–11), N production increased only by 1.24 Mt. However, after a long gap, the country achieved record increase of 1 Mt in N production during 2015–16 due to some changes in policy. Annual

**Table 4.1** Growth in Fertilizer-N Capacity and Production in India (1960–61 to 2015–16)

| Year | N Capacity (Mt) | N Production (Mt) | Capacity Utilization (%) | Share of Urea in N Production (%) |
|---|---|---|---|---|
| 1960–61 | 0.16 | 0.11 | 69.1 | 4.9 |
| 1970–71 | 1.35 | 0.83 | 61.7 | 60.6 |
| 1980–81 | 4.36 | 2.16 | 49.6 | 71.9 |
| 1990–91 | 8.15 | 6.99 | 85.7 | 84.4 |
| 2000–01 | 11.99 | 10.94 | 91.3 | 82.5 |
| 2001–02 | 11.95 | 10.69 | 89.4 | 81.8 |
| 2002–03 | 11.68 | 10.51 | 89.9 | 81.5 |
| 2003–04 | 11.56 | 10.56 | 91.3 | 83.0 |
| 2004–05 | 11.60 | 11.30 | 97.4 | 82.4 |
| 2005–06 | 11.63 | 11.33 | 97.4 | 81.5 |
| 2006–07 | 11.58 | 11.52 | 99.6 | 80.9 |
| 2007–08 | 11.61 | 10.90 | 93.9 | 83.7 |
| 2008–09 | 11.90 | 10.90 | 91.6 | 84.1 |
| 2009–10 | 12.57 | 11.92 | 94.9 | 81.5 |
| 2010–11 | 12.57 | 12.18 | 96.9 | 82.6 |
| 2011–12 | 12.57 | 12.29 | 97.7 | 82.3 |
| 2012–13 | 13.08 | 12.24 | 93.6 | 84.9 |
| 2013–14 | 13.53 | 12.41 | 92.1 | 84.2 |
| 2014–15 | 13.55 | 12.43 | 91.9 | 83.6 |
| 2015–16 | 13.58 | 13.48 | 99.2 | 83.5 |

FAI, 2015. Fertilizer Statistics 2014–15, sixtieth ed. The Fertilizer Association of India, New Delhi.

N production level increased to 13.48 Mt with capacity utilization of 99.2% during 2015–16. Among N fertilizers, urea has emerged as the most dominant product and it has been maintaining a share of 82–84% in total N production since 1980–81 with corresponding share of 83.5% fertilizer-N during 2015–16.

The growth in N capacity and production has not been uniform and there exists a wide gap in production among four major zones (East, West, North, and South). With maximum number of N plants, West zone accounted for 47% of total N production during 2014–15 followed by North zone with 33.5% share. The shares of South zone and East zone in total N production were 13.4% and 6.0%, respectively.

## Improvement in Fertilizer Production Efficiency

Fertilizer production is capital-intensive as well as energy-intensive. It involves deployment of very sophisticated technologies. With ever-increasing cost of energy and raw materials, Indian fertilizer industry has been paying attention to improve efficiency of the plants in terms of improvement in operational efficiency; reduction in energy and water consumption; and maintenance of environmental standards. This has become important not only from the viewpoint of improving economics of production but also for reducing the environmental footprints. The efforts of industry are reflected in significant improvement in performance of all the fertilizer plants.

The energy consumption for ammonia plants has reduced from 12.48 giga calories/ metric ton (Gcal $MT^{-1}$) ammonia in 1987–88 and 8.37 Gcal $MT^{-1}$ ammonia in 2015–16 and that of urea plants from 8.87 Gcal $MT^{-1}$ urea to 5.95 Gcal $MT^{-1}$ during the same period (Chander, 2016). The N in all fertilizer products is derived from ammonia. Hence, efficiency of ammonia consumed per ton of product is a major factor that determines N loss to atmosphere. The average ammonia consumption per ton of urea has reduced from 0.589 in 1990–91 to 0.577 in 2015–16 against the stoichiometry requirement of 0.567. In case of complex fertilizers, the raw material efficiency of nitrogen increased from 93% in 1992–93 to 97.4% in 2014–15. The water consumption of ammonia and urea plants has been reduced by 40% over 25 years. Having achieved a very high level of efficiency comparable to the best in the world, Indian fertilizer-N industry is not resting on its laurels. Industry continues to invest efforts and resources in innovation which bring about improvement, however small these are.

## Fertilizer Nitrogen Consumption

Right from the beginning in early 1930s, N consumption continued to expand in quantum and coverage. However, the growth in consumption remained very slow until 1950s. It took 25 years to reach the N consumption level of 0.21 Mt and per hectare N consumption of 1.4 kg in 1960–61 (Table 4.2). The introduction of fertilizer-responsive HYVs of rice and wheat in mid-1960s boosted the N consumption rapidly. A sevenfold increase in N consumption (0.21–1.48 Mt) was achieved between 1960–61 and 1970–71 with corresponding per hectare consumption of 1.4–8.9 kg.

Implementation of the favorable fertilizer policies during 1970s and 1980s led to a rapid growth in N consumption which increased from 1.48 Mt in 1970–71 to 8.0 Mt in 1990–91. Per hectare fertilizer-N use increased almost fivefold (8.9–43.1 kg) during the corresponding period. Growth in N consumption continued during 1990s but net increase in the consumption was less compared to 1980s. After reaching level of 11.59 Mt in 1999–2000, N consumption remained almost stagnant for half a decade until 2004–05. However, a record increase of 5.6 Mt (11.7–17.3 Mt) in N consumption was witnessed during next 7 years (2004–05 to 2011–12). But the growth in N consumption declined/ remained stagnant in last 5 years of current decade (2011–12 to 2015–16) due to sharp increase in prices of DAP and other NP/NPK complexes and aberrant weather conditions. Growth in N consumption has not been uniform over time and space, and it has varied widely across regions, crops, seasons, farm size, and availability of irrigation facilities.

### Skewed Product Pattern

Although a number of straight nitrogenous and NP/NPK complex fertilizers are available, urea has emerged as the most dominant source of N both in terms of production and

**Table 4.2** Trends in Consumption of Fertilizer-N in India (1960–61 to 2015–16)

| Year | Fertilizer-N Consumption | | NPK Use Ratio | Share of Urea in N Consumption (%) |
|---|---|---|---|---|
| | Total (Mt) | kg $ha^{-1}$ | | |
| 1960–61 | 0.21 | 1.4 | 7.3:1.8:1 | 9.7 |
| 1970–71 | 1.48 | 8.9 | 6.3:2.3:1 | 52.9 |
| 1980–81 | 3.68 | 21.3 | 5.9:1.9:1 | 75.6 |
| 1990–91 | 8.00 | 43.1 | 6.0:2.4:1 | 81.0 |
| 2000–01 | 10.92 | 58.9 | 7.0:2.7:1 | 80.8 |
| 2001–02 | 11.31 | 60.2 | 6.8:2.6:1 | 81.0 |
| 2002–03 | 10.47 | 60.2 | 6.5:2.5:1 | 81.2 |
| 2003–04 | 11.08 | 58.4 | 6.9:2.6:1 | 82.1 |
| 2004–05 | 11.71 | 61.3 | 5.7:2.2:1 | 81.2 |
| 2005–06 | 12.72 | 66.0 | 5.3:2.2:1 | 80.6 |
| 2006–07 | 13.77 | 71.6 | 5.9:2.4:1 | 81.3 |
| 2007–08 | 14.42 | 73.9 | 5.5:2.1:1 | 82.8 |
| 2008–09 | 15.10 | 77.3 | 4.6:2.0:1 | 81.2 |
| 2009–10 | 15.58 | 82.4 | 4.3:2.0:1 | 78.8 |
| 2010–11 | 16.56 | 83.8 | 4.7:2.3:1 | 78.1 |
| 2011–12 | 17.30 | 88.4 | 6.7:3.1:1 | 78.6 |
| 2012–13 | 16.82 | 86.5 | 8.2:3.2:1 | 82.0 |
| 2013–14 | 16.75 | 86.2 | 8.0:2.7:1 | 84.0 |
| 2014–15 | 16.95 | 87.2 | 6.7:2.4:1 | 83.1 |
| 2015–16[a] | 17.44 | 89.7 | 7.3:2.9:1 | 82.1 |

[a]Provisional.

FAI, 2015. Fertilizer Statistics 2014–15, sixtieth ed. The Fertilizer Association of India, New Delhi.

consumption. Share of urea in total N consumption increased from less than 10% in 1960–61 to more than 50% in 1970–71. Urea accounted for 75% of total N consumption in 1980–81 and 81% in 1990–91, and thereafter it has been hovering between 81% and 84%. The prevailing share of urea in total N consumption is much higher compared to other major countries such as Australia (66%), Brazil (57%), Canada (49%), China (69%), United States (22%), as well as the world average (57%). DAP and other NP/NPK complexes accounted for 16% of the total fertilizer-N consumption and share of other straight nitrogenous fertilizers (ammonium sulfate, CAN, ammonium chloride) came down to 1%. N consumption in India is skewed toward urea due to its very low price *vis-à-vis* other P and K fertilizers, and it is not a good sign from nutrient use efficiency point of view.

## Zone-Wise/State-Wise Fertilizer Nitrogen Use

Among four major zones of India, the North zone ranks first both in terms of total as well as per hectare N consumption. North zone accounted for 34.1% of N consumption during 2014–15, followed by West zone (30.7%), South zone (21.4%), and East zone (13.8%). Per hectare N consumption was 133.3 kg in North, 109.1 kg in South, 76.9 kg in East, and 59.5 kg in West zone. With 3.17 Mt of N consumption, Uttar Pradesh is the largest consumer of fertilizer-N in India and it alone accounted for 18.7% of the country's total N consumption in 2014–15. There are eight states, namely Uttar Pradesh, Punjab, Haryana, Andhra Pradesh, Karnataka, Gujarat, Madhya Pradesh, and Maharashtra, having more than 1 Mt of fertilizer-N consumption during 2014–15. In each of 12 districts, namely Guntur, Kurnool, West Godavari, Krishna, East Godavari (Andhra Pradesh), Rangareddy, Karimnagar, Nalgonda, Warangal (Telangana); Bellary (Karnataka), Jalgaon (Maharashtra), and Rajkot (Gujarat), annual N consumption was more than 100,000 tons. The annual consumption of N in Guntur district (172,400 tons) was more than N consumption in states such as Assam (155,900 tons), Jharkhand (91,800 tons), Himachal Pradesh (35,000 tons), Jammu and Kashmir (65,000 tons), Uttarakhand (137,400 tons), and Kerala (104,500 tons).

There is a wide variation in per hectare fertilizer use in various states of the country. The per hectare fertilizer-N use was maximum in Punjab (171.8 kg) followed by Haryana (158.9 kg) and Telangana (145.3) during 2014–15. As against this, per hectare use of N was less than 5 kg in states such as Arunachal Pradesh and Nagaland. Average per hectare N consumption was more than 115 kg in eight states. The growth in N consumption has not been uniform over time. Punjab and Tamil Nadu witnessed rapid growth in N consumption during 1965–90, whereas Andhra Pradesh, Haryana, Karnataka, and Bihar achieved rapid growth during 1990s. Almost all the states recorded remarkable growth in N consumption during the last decade. However, N consumption declined/stagnated in all the states except Orissa in first 4 years of the current decade.

Variation in per hectare N consumption is more pronounced at district level. In the North zone, there are seven districts in Haryana, three districts in Punjab, one district in Uttarakhand wherein N consumption was more than 200 kg $ha^{-1}$ during 2014–15. In the South zone, there are five districts in Andhra Pradesh, two districts in Karnataka, and one district in Tamil Nadu having N consumption more than 200 kg $ha^{-1}$. Similarly in one district in Gujarat (West zone) and six districts in Bihar (East zone), per hectare consumption has crossed 200 kg.

## Nitrogen Use by Crops

Indian agriculture is basically food grains–driven, having 62% of total cropped area under these crops which consumed 69.3% of total N during 2011–12 (Fig. 4.1). Among food grains, rice and wheat are the significant consumers of total N (52.4%; 29.1% in rice and 23.3% in wheat). Oilseed crops accounted for 9.1% of total consumption.

Among cereals, per hectare N use was maximum in wheat (119.2 kg) followed by rice (90.7 kg). Intensity of N use is highest in sugarcane (174.7 kg $ha^{-1}$). Per hectare N use is quite high in some fruit crops such as banana and grapes, and vegetables such as chilies

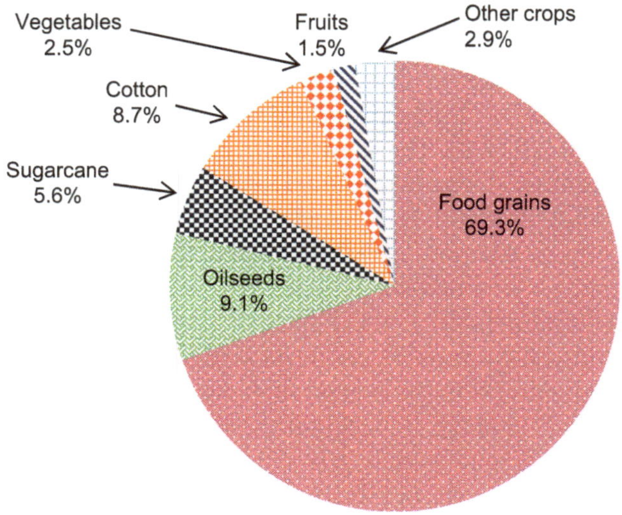


**FIGURE 4.1** Nitrogen use by crops in India. *Based on data presented in Anonymous, 2016. All-India Report on Input Survey, 2011–12. Agricultural Census Division, Ministry of Agriculture and Farmers Welfare, New Delhi.*

and potato. Per hectare N use in oilseeds (53.6 kg) and pulses (52.9 kg) is relatively low as these crops are largely grown under rainfed conditions.

There exists a wide variation in N:P:K use ratios among crops. The NPK use ratio is wide in food grains (6.8:3.0:1.0) due to relatively more use of nitrogen. Among food grain crops, the distortion in NPK use ratio is more in wheat (13.1:5.5:1.0) compared to rice (4.9:2.1:1.0) and maize (5.6:2.6:1.0). Per hectare fertilizer use is relatively higher and more balanced in fruits (2.0:1.3:1.0), vegetables (2.5:1.7:1.0), and plantation crops particularly coffee (1.7:1.1:1.0).

## Farm Size-Wise Nitrogen Use

Contrary to the general impression that small farmers use less fertilizer, total N consumption and intensity of N use is more in case of small and marginal farmers (<2 ha). Per hectare N use was 55.9 kg under large farmers' category (10 ha or above) whereas it was 76.5 kg under small farmers' category and 117.9 kg under marginal farmers during 2011–12 (Fig. 4.2). The NPK use ratio is more balanced in small and marginal farmers' category. NPK use ratio was most favorable in small farmers' category (4.9:2.4:1) and most distorted in large farmers' category (10.5:4.4:1).

## Irrigation Availability

The intensity of fertilizer use is directly related with the availability of irrigation water. The per hectare fertilizer-N use is higher in states such as Punjab and Haryana having higher percentage of irrigated area. Similarly, intensity of N use is more in irrigated crops. The average per hectare N consumption was 115.7 kg in irrigated areas and 48.8 kg

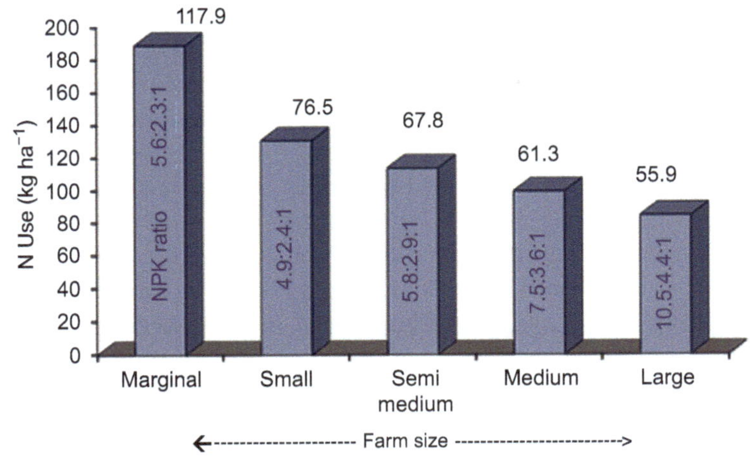


**FIGURE 4.2** Consumption of fertilizer-N according to farm size. *Based on data presented in Anonymous, 2016. All-India Report on Input Survey, 2011–12. Agricultural Census Division, Ministry of Agriculture and Farmers Welfare, New Delhi.*

in unirrigated areas during 2011–12. The extent of variation in per hectare N consumption in unirrigated and irrigated areas was more under large farm holdings (18.3–104.3 kg) than in small (53.1–105.7 kg) and marginal (78.9–155.5 kg) farm holdings.

## Impact of Fertilizer Policies

Implementation of favorable policies during 1970s and 1980s, namely ECA Supply Plan-1977, Fertilizer Movement Control Order-1973, RPS-1977, and Block Delivery Schemes-1980–81 contributed immensely in expanding fertilizer production and consumption base (Tewatia, 2016). Policy environment for fertilizer sector started deteriorating in 1990s due to sudden and ad hoc policy changes. The main objective of these policy changes was to contain the increasing burden of fertilizer subsidy on government exchequer. However, the fertilizer subsidy burden continued to increase as real causes of increase in fertilizer subsidy were not addressed. The increase in fertilizer subsidy was due to stagnant retail price of urea against continuous increase in cost of production/imports. It affected the growth of fertilizer industry.

Slow growth in domestic production of urea has been an area of serious concern as it has increased the country's dependence and consequential vulnerability on imports. India became self-sufficient in domestic production of urea by 2000–01. Import of urea was less than 1 Mt per year up to 2004–05. Thereafter, it started increasing and touched 8.7 Mt during 2014–15 which constituted nearly 30% of the total consumption in 2014–15.

The decontrol of P and K fertilizers in August 1992 resulted in sharp increase in P and K prices which created distortion in N, P, and K prices. Government of India introduced Ad hoc Concession Scheme to restore parity in nutrient prices. This initiative helped in improving NPK use ratio. To take it further, the government moved from product pricing regime toward Nutrient Based Subsidy (NBS) regime, and implemented NBS policy for P and K fertilizers with effect from April 1, 2010. NBS is now applicable on 22 fertilizer products; however, urea accounting for 82% of total nitrogen use was left out of NBS policy and continues to remains so.

After the selective implementation of NBS policy for P and K fertilizers, the prices of these fertilizers increased by 2–3 times whereas urea prices continued to be at very low level due to heavy subsidy. There was serious distortion in urea *vis-à-vis* P and K prices. The price of DAP increased from Rs. 9350 per ton in 2009–10 (pre-NBS) to Rs. 24,000 per ton in 2014–15 (post-NBS). Similarly, the price of muriate of potash (MOP) increased from Rs. 4455 per ton to Rs. 16,800 per ton during the corresponding period. The price ratio of DAP:urea distorted from 1.94:1 in 2009–10 to 4.48:1 in 2014–15 and MOP:urea from 0.92:1 in 2009–10 to 3.13:1 in 2014–15 (Tewatia, 2016). The price ratio of DAP:urea and MOP:urea is likely to narrow down in 2016–17 as there is softening of prices of P and K fertilizers in international market. Based on the present rates of subsidy and retail prices (MRP), in case of urea, farmer is paying just 25% of total cost of production and 75% is paid by Government of India as subsidy but in the case of DAP and MOP the share of subsidy to total cost is 34% and 37%, respectively (Fig. 4.3).

The prevailing very low price of urea has created temptation among farmers to use more urea which has distorted the $N:P_2O_5:K_2O$ consumption ratio; this ratio widened from 4.7:2.3:1 in 2010–11 (which was much closer to ideal 4:2:1 ratio) to 7.3:2.9:1 in 2015–16. The imbalanced nutrient use not only affects crop yields and soil health but also poses threat to environmental pollution. It leads to inefficient use of nutrients which increases risk to atmospheric, surface and ground water pollution. Coating of urea with neem oil may help in improving nitrogen use efficiency marginally, but fertilizer use efficiency can only be optimized by balanced use of all the nutrients. This needs correction in pricing policies for fertilizers.

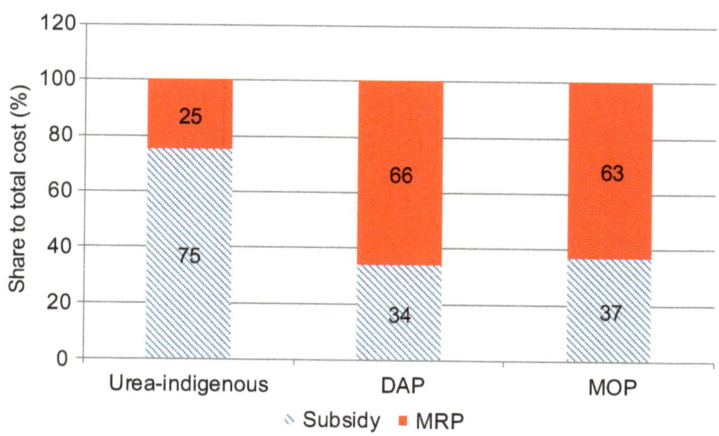


**FIGURE 4.3** Share of subsidy to total cost of urea, diammonium phosphate (DAP) and muriate of potash (MOP)—2015–16.

# Demand Supply Projections of Fertilizer Nutrients by 2030–31

## Demand for Fertilizer Nitrogen

Keeping in view the need for fulfilling the objective of food security, and increase in production of other agricultural products, demand for fertilizers will continue to grow in future. Taking into account the recent policy developments and continuance of current policy, demand projections for fertilizers up to 2030–31 have been prepared. Fertilizer nutrient consumption has been projected based on the gross requirement of food grains by future population and application of appropriate response ratio of fertilizer use to food grains production. Requirement of fertilizer use for crops other than food grains has been derived by taking the share of other crops to total N consumption. Assumption has also been made to move toward more balanced NPK use ratio over the years. NPK use ratio is expected to improve from the current level of 7.3:2.9:1 in 2015–16 to 6.1:2.7:1 in 2020–21 and 3.8:2.2:1 in 2030–31. Based on various assumptions mentioned above and by application of population nutrition method, the projected demand up to 2030–31 has been worked out. The demand for N is projected to increase from 17.44 Mt in 2015–16 to 20.75 Mt in 2021–22 and 23.46 Mt in 2030–31 (Table 4.3).

**Table 4.3** Projected Demand, Supply and Gap of Fertilizer-N (2016–17 to 2030–31)

| Year | Demand[a] (Mt) | Supply[b] (Mt) | Gap (Million Tons) |
|---|---|---|---|
| 2014–15 (actual) | 16.945 | 13.193 | 3.752 |
| 2015–16 (estimated) | 17.437 | 14.235 | 3.202 |
| **Forecasts** | | | |
| 2016–17 | 18.012 | 14.663 | 3.349 |
| 2017–18 | 18.689 | 15.070 | 3.619 |
| 2018–19 | 19.090 | 15.280 | 3.811 |
| 2019–20 | 19.678 | 15.325 | 4.353 |
| 2020–21 | 20.071 | 16.963 | 3.108 |
| 2021–22 | 20.751 | 17.004 | 3.748 |
| 2022–23 | 21.130 | 17.034 | 4.096 |
| 2023–24 | 21.497 | 17.198 | 4.299 |
| 2024–25 | 21.851 | 17.480 | 4.370 |
| 2025–26 | 22.187 | 17.749 | 4.437 |
| 2026–27 | 22.502 | 18.001 | 4.500 |
| 2027–28 | 22.792 | 18.233 | 4.558 |
| 2028–29 | 23.051 | 18.441 | 4.610 |
| 2029–30 | 23.275 | 18.620 | 4.655 |
| 2030–31 | 23.455 | 18.764 | 4.691 |

[a]Based on current pricing policy.
[b]Includes supply from JV OMIFCO.

### Supply of Fertilizer Nitrogen

As mentioned earlier, nitrogen use in Indian agriculture is dominated by a single product—urea. It accounts for 82% use of nitrogen in the country. It is expected that in order to have more balanced use of N, P, and K, use of complex (NP/NPK) fertilizers will be promoted. The share of urea in total N consumption is projected to go down over the years from 82% in 2015–16 to 81% in 2021–22 and 80% in 2030–31.

The future production of fertilizer-N will come from additional production of urea from existing capacities and creation of new capacities. There will also be additional contribution of N from complex fertilizers by maximization of capacity utilization. In medium term, one urea plant is ready for commissioning and three new plants are expected by 2020–21. The degree of self-sufficiency in total N supply has been assumed at 80% for future years in the long run. The aggregate supply of N from domestic production and joint venture OMIFCO is estimated to increase from 14.24 Mt in 2015–16 to 16.96 Mt in 2020–21 and 18.76 Mt in 2030–31.

## Conclusion

India has witnessed phenomenal increase in N production and consumption over the past five decades. Fertilizer-N production has gone up from 0.1 Mt in 1960–61 to 13.5 Mt in 2015–16, and N consumption has increased from 0.6 Mt to 17.4 Mt. during the same period. India ranks second in the world in terms of N production and consumption, next only to China. However, per hectare N consumption is lower compared to many other countries. The growth in N production and consumption has not been uniform over time and space. There exists a wide variation in N consumption among states, crops, and size of farm holdings. Nutrient use pattern is skewed in favor of N (urea). Selective implementation of NBS scheme on P and K fertilizers *w.e.f.* April 1, 2010, leaving urea out of NBS created distortion in urea *vis-à-vis* P and K prices. It distorted NPK consumption ratio. Government of India is paying huge amount of fertilizer subsidy but this instrument of subsidy should be used to promote the balanced use of nutrients. Correction in nutrient (N, P, and K) prices is the need of hour to ensure balanced nutrient use by improving NPK consumption ratio. India has ample scope to increase N consumption further as per hectare N use is still low. Based on current pricing policy and growth trends, it is estimated that the demand for N in India by 2030 would be about 23.46 Mt and supply is expected to be around 18.76 Mt leaving a gap of 4.7 Mt (20%) that has to be met by imports.

## References

Anonymous, 2016. All-India Report on Input Survey, 2011–12. Agricultural Census Division, Ministry of Agriculture and Farmers Welfare, New Delhi.

Chander, S., 2016. Innovations in fertiliser production. Indian Journal of Fertilisers 12 (6), 12–13.

FAI, 2015. Fertiliser Statistics 2014–15, sixtieth ed. The Fertiliser Association of India, New Delhi.

Tewatia, R.K., 2016. Policy developments in nitrogenous fertilisers in India. Indian Journal of Fertilisers 12 (4), 128–136.

SECTION

B

# Nitrogen Processes in the Biosphere

# 5

# Nitrogen Processes in Agroecosystems of India

A. Subba Rao, Pramod Jha, B.P. Meena, A.K. Biswas, Brij Lal Lakaria, Ashok K. Patra

*ICAR-INDIAN INSTITUTE OF SOIL SCIENCE, BHOPAL, INDIA*

**CHAPTER OUTLINE**



## Introduction

Terrestrial ecosystem is the major agroecosystem of India and is predominately controlled by human activities for maximizing the food production with nitrogen (N) as the most widely used input. Among the N sources, the fertilizer N has played a key role in food grain production, although its use efficiency (20–50%) is very low in India. About

**The Indian Nitrogen Assessment. http://dx.doi.org/10.1016/B978-0-12-811836-8.00005-7**

56% of the N fertilizer is used for producing the cereals such as rice, maize, and wheat (IFA, 2002). Reactive nitrogen (Nr) is defined as the fraction of total nitrogen present on the earth that is biologically, photochemically reactive, and radiative in nature (Galloway et al., 2003). Of the two major pools of N, nonreactive N ($N_2$ in the atmosphere) makes up about 99% and the reactive N the rest 1% of the total N. The Nr includes inorganic forms such as $NH_3$, $NH_4^+$, $NO_2^-$, $NO_3^-$, $NO_x$, $N_2O$, and organic forms such as urea, amines, proteins, nucleic acids, and N in soil humus.

For past few decades, human activities are responsible for production of Nr from diatomic N that has exceeded the rate of natural N fixation (Galloway et al., 2004). Globally, more or less 425 Tg of Nr is fixed every year from nonreactive atmospheric $N_2$ in the oceans and on land by natural processes and human activities (Sutton et al., 2013). It is estimated that approximately 67% of fertilizer-N used globally is ultimately converted back to nonreactive atmospheric $N_2$ and nearly 33% remaining reactive N contributes to various transient effects in different ecosystems (Vitousek et al., 1997). Reactive nitrogen is important for growth and reproduction of organisms; however, its buildup from different sources has significantly affected the air and water quality (Aneja et al., 2009), posed threat to human health and stress on terrestrial and aquatic organisms, crop production, and farm profitability. The Nr can cause multiple effects in the atmosphere, terrestrial, and aquatic ecosystems, in addition to affecting human health. Fig. 5.1 illustrates the pathways of conversion of nonreactive nitrogen to Nr in various pools (Galloway et al., 2003).

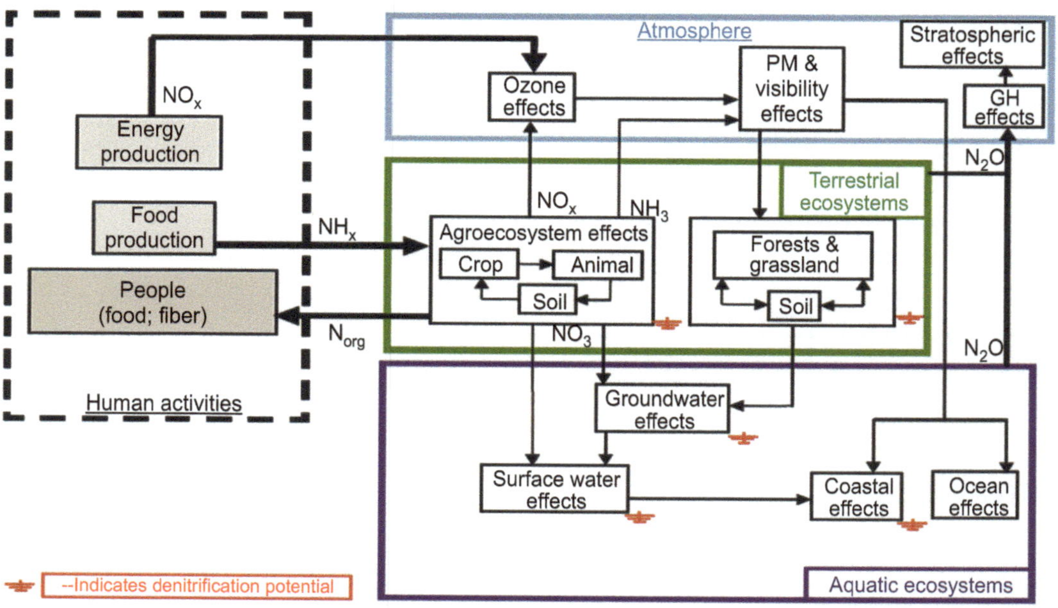


**FIGURE 5.1** The N cycle depicting different N processes. *Adapted from Galloway, J.N., Aber, J.D., Erisman, J.W., Seitzinger, S.P., Howarth, R.W., Cowling, E.B., Cosby, B.J., 2003. The N cascade. Bioscience. 53, 341–356.*

About 97–98% of total N present in the soil is bound to the soil organic matter and while not readily available to plants, it becomes available upon decomposition and mineralization. Inorganic forms of N such as nitrate $NO_3^-$ and ammonium $(NH_4^+)$ constitute 2–3% of the total N readily available to plants. Indian soils have low total N content because of the tropical and subtropical climate prevailing in the country. Total N in Indian soils ranges between 0.02% and 0.1% (Prasad, 2007) barring the forest and hill agroecosystems.

Global consumption of N fertilizers was 107.9 million tons during 2012–13. Out of the total fertilizer-N consumption, 32.7% was consumed in China and 16% in India (FAI, 2013). South and Southeast Asia consumed ~57% of fertilizer N, where rice and wheat are the major crops. In India, nearly 50% of the current food grains production can be attributed to the fertilizer-N application (Fig. 5.2). The consumption of N fertilizer in the country over the years has increased considerably and there exists a direct relationship between the N consumption and gross cropped area as well as food grain production (FAI, 2013).

## Nitrogen Inputs

The major inputs of nitrogen to the nitrogen cycle are from fertilizers, biological fixation, manures, and crop residues. Information on N contribution through irrigation and atmospheric depositions in different agroecosystems of India is very limited.

### Fertilizers

Nitrogen availability is often limited in agroecosystems (Masunga et al., 2016). The main anthropogenic Nr source is fertilizer, which is being produced and applied at ever-increasing rates. After China, India is the largest consumer of fertilizers in the world.

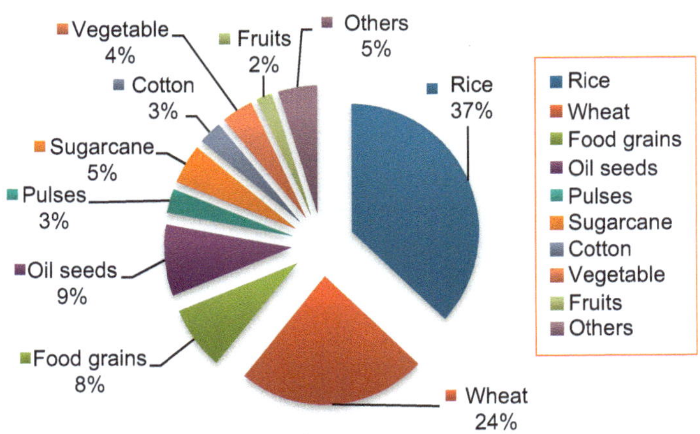


**FIGURE 5.2** Crop-wise consumption of fertilizers in India. *Adapted from Prasad, R., 2009. Efficient fertilizer use—the key to food security and better environment. Journal of Tropical Agriculture 47, 1–7.*

Fertilizer Association of India has projected 28.15, 33.27 and 47.68 million tons production of urea for the years 2021–22, 2031–32, and 2051–52, respectively (FAI, 2015). However, unequal expansion and chronic N deficiency occur due to misuse of fertilizer in most of the Indian soils. In northwest India, fertilizer consumption is as high as 300 kg $ha^{-1}$ $year^{-1}$ in the intensive cropped regions (Galloway et al., 2008). This has led to an imbalance in the fertilizer use in India. Hence, genuine demand for expansion of fertilizer-N use in some areas coexist with concern over the environmental hazards of excessive and inefficient N fertilizer use in other areas, such as pollution of groundwater and surface water bodies. Fertilizer-N use is expected to double by 2050 from the current 150 Tg N $year^{-1}$ used in agricultural system (Galloway et al., 2008). Nr contribution in India is not uniformly distributed within and among regional and national levels or within and between sectors (Galloway et al., 2008). The highest input of nitrogenous fertilizer (Ministry of Chemicals and Fertilizers (2014)) was used in northern zone (141 kg $ha^{-1}$) followed by eastern zone (101 kg $ha^{-1}$) in India. Among the states, the highest consumption occurs in Punjab (188 kg $ha^{-1}$) followed by Haryana (156 kg $ha^{-1}$) and Bihar (154 kg $ha^{-1}$).

Apart from inorganic sources, biological nitrogen fixation (BNF), recycling of N from crop residues, animal manure, atmospheric deposition, and irrigation water contributes 33, 16, 20, and 24 Tg, respectively (Mosier et al., 2004). Of the total N input of 170 Tg, about 50% (85 Tg) is removed by the harvested crops (i.e., the same as fertilizer input), while 37 Tg N either remains in the soil–plant system or is lost through various soil processes such as leaching, runoff, and erosion. A significant proportion is also lost as

**Table 5.1** Soil Surface-N Balance (Tg) for Agricultural Land in India

| Input/Output | Tg N |
|---|---|
| ***Input*** | |
| Synthetic fertilizer | 10.80 |
| N fixation by legumes | 1.14–1.18 |
| Compost and manure | 1.70 |
| Atmospheric deposition | 0.81 |
| Ground water | 0.11 |
| **Total** | **14.56–14.60** |
| ***Output*** | |
| Crop removals and fodder | 9.94 |
| Fossil fuel | 1.90 |
| Erosion loss | 0.06 |
| GHG emission | 0.34–0.81 |
| **Total** | **12.24–12.71** |
| Balance | 1.89–2.32 |

Velmurugan, A., Dadhwal, V.K., Abrol, Y.P., 2008. Regional N cycle: an Indian perspective. Current Science 94 (11), 1455–1468.

ammonia volatilization (AV) from animal wastes, soil, and vegetation (21 Tg); denitrification (14 Tg) and gaseous emission (8 Tg) as nitrous oxide ($N_2O$) or nitric oxide (NO) (Mosier et al., 2004). Total nitrogen balance in soil in agricultural land of India is depicted in Table 5.1 (Velmurugan et al., 2008) which reveals that inorganic N fertilizers are the major input (10.80 Tg) to soil followed by manures (1.53 Tg) and input through BNF (1.14–1.18 Tg). Among the outputs of Nr from the system, harvested crop and fodders are the major contributors (9.94 Tg).

Prasad (2007) reported deficit to surplus N balance in different states of India. The highest positive N balance has been found in Uttar Pradesh (2.5 Tg) and negative N balance was recorded in Orissa (−0.01 Tg) and the northeastern states. In the rice–wheat cropping system, a higher positive N balance was recorded at an application rate of 180 kg N $ha^{-1}$ especially through integrated N approach compared to organics alone (Velmurugan, 2000). Nitrogen availability in India from various N fertilizer sources is low. As per the statistics of the Department of Fertilizers, Ministry of Chemicals and Fertilizers (2013–14), total N availability from different straight and complex fertilizers was 13.21 Tg during 2013–14, of which urea alone contributed 10.44 Tg (79%) and all other sources collectively contributed only 21% share, while their individual share was below 1%.

## Manures and Crop Residues

India has the largest proportion of the world's population of poultry (34.1%) followed by cattle (24.7%), goats (17.1%), and buffalo (13.4%). Organic wastes generated through animals are valuable resources especially for N supply to crops. The estimated annual available nutrient (NPK) contribution through organic sources was about 5 million tons in 2011, and it would increase to 7.75 and 10.25 million tons by 2025 (FAO, 2005) and 2030 (IISS, 2015), respectively. India would also produce around 70 million tons of municipal solid waste annually, which would supply about 210,000–400,000 tons of NPK by 2030. Further, green manuring also contributes to N supply to an area of about 7 million ha of agricultural soils contributing about 80 kg N $ha^{-1}$ (Yadvinder Singh et al., 2010).

## Biological Nitrogen Fixation

BNF is a very valuable alternative to N fertilizer (de Bruijn, 2016). It is a route in which $N_2$ in the atmosphere is converted into ammonia by prokaryotes using an enzyme complex (Nitrogenase), and an N pool is maintained in the ecosystems, which replenishes N losses. BNF occurs naturally in the soil by N fixing bacteria (*Rhizobium* and legumes/*Azotobacter*). It also occurs naturally in the air by means of lightning. N-fixers contribute to N accumulation in long term and bring N supply close to the equilibrium (Vitousek et al., 2002). Prokaryotes can be symbiotic or non-symbiotic microorganisms, but significant biological $N_2$ fixation in agroecosystem is carried out by symbiotic microorganism such as *Rhizobium* which infects the roots of leguminous plants. The total $N_2$ fixation in the world (Terrestrial + oceanic) is estimated to be about 175 Tg of which legumes account for almost half (80 Tg) and industrial N fixation (production of N fertilizer) for the rest half (88 Tg) (Laagreid, 1999). Symbiotic $N_2$ fixation can be around 20–200 kg N $ha^{-1}year^{-1}$. Highly specific and most efficient processes for $N_2$ fixation involve the formation of root nodules on legumes and help the crop by providing N and also substituting the soil N pool (Ladha and Reddy, 2000). The potential contributions by free-living $N_2$-fixing bacteria; by heterotrophic $N_2$-fixation in the rhizosphere of cereals and nonlegumes (often termed as *associative $N_2$-fixation*); and by endophytic $N_2$-fixing bacteria within nonlegumes remain controversial as contributions are difficult to quantify, even under most favorable environments. In early 1980s and 1990s interests were developed in incorporating the N fixation mechanism into non-N-fixing crop plants. Fertilizer N and BNF—the sources of new reactive N—together constitute 44% of the total additions (Prasad et al., 2004).

### Other Sources of Nitrogen

Irrigation water also contributes a sizable amount of N to agricultural soils in India. About 40% of gross cultivated area of the country is irrigated through canal, river, or bore well water, though the information on its N content is scarce. Some report suggests 3 ppm N in irrigation and rain water (Bhandari et al., 2002) but it varies widely depending upon the source of water, agro-ecological region, depth of water table, and season.

Atmospheric deposition is another important factor for transferring N in anthropogenic and terrestrial systems. It occurs in the form of wet deposition through rainwater, through dry deposition in association with suspended particulate matter that is in the air, and Nr gases that interact with the vegetation, soil, and water. Globally, atmospheric deposition accounts for about 12% of the reactive N, entering terrestrial and marine ecosystem. Kulshrestha et al. (2005) reported median concentrations of ammonical-N in the precipitation in rural, suburban, and urban areas of India as 9, 14, and 22 $\mu eq\ L^{-1}$, respectively, while the values for nitrate were 22, 11 and 28 $\mu eq\ L^{-1}$, respectively.

## Nitrogen Cycling

The current N biogeochemical cycle is strongly affected by the agricultural system. The pattern of N cycle changes as an ecosystem develops to its climax state, and it is also dramatically altered by disturbances in management practices (Galloway et al., 2004). Worldwide 120 Tg of N per year is introduced into the atmosphere by human beings for producing food and fodder, and as feedstock for industrial purposes (Galloway et al., 2008). Additionally, 50–70 Tg N is fixed as BNF in various agricultural systems (Herridge et al., 2008). The total N input to the agriculture systems contributes 20–30% of the total applied N to food for human consumption. In a similar manner, a very small fraction of total N input to livestock ends up as meat for human consumption; the remaining portion is recycled by agricultural soils (Billen et al., 2013). In India, the amounts of N lost from the soil–plant–animal system were 2.3 million tons through nitrate leaching, runoff, and erosion of particulate matter; 2.3 million tons through $NH_3$ volatilization; 3.1 million tons through denitrification of nitrate to $N_2$ gas; and 0.2 million tons through emission of $N_2O$ (Pathak et al., 2009). Globally, one-third of the food produced is lost as waste which also contributes to the loss of N in the environment. Agricultural soil contributes around 39–95 Tg of N per year to fresh waters (Billen et al., 2013). A part of it remains in shallow aquifers, whereas another part is lost to the atmosphere by the process of denitrification leading to emission of nitrous oxide ($N_2O$) gas. Some part of it is discharged to coastal waters (40–66 Tg N $year^{-1}$) (Voss et al., 2011), where it enhances eutrophication and hypoxia.

Nitrogen transformations, which occurs after application of N fertilizer to soil the function of prevailing weather conditions, soil organic matter and clay content, and the type of nitrogenous fertilizer applied. The main transformations that occur in soil are as follows:

- Ammonia fixation
- AV
- Nitrification
- Immobilization–mineralization
- Denitrification

The key N transformation processes in N cycle in agroecosystems are elaborated and discussed below.

## Ammonium Fixation

Ammonical fertilizer undergoes larger number of N transformation processes in comparison to fertilizers which contain entire N in nitrate form. When ammonical fertilizer is applied to soil, a part of it is either fixed or adsorbed on to the soil colloids. The $NH_4^+$ ion adsorbed on negatively charged soil colloids is referred to as exchangeable $NH_4^+$ because it can be replaced by other cations in the soil solution. The $NH_4^+$ ion held on the soil matrix prevents it from leaching yet available to the plants. Barring this exchangeable $NH_4^+$, a part of the $NH_4^+$ applied through N fertilizer becomes trapped, or fixed, in the interlayer spaces of 2:1 clay minerals (illite and vermiculite) in soils. This trapped $NH_4^+$ is held very tightly and is not readily available to the plants. The factors which influence fixation of ammonia in soil include the $NH_4^+$ concentration and soil pH. The process of ammonia fixation is more serious when ammonium-containing fertilizers are applied in band. Fixation process is markedly reduced under acidic conditions, possibly because of competition with $H^+$ ion.

## Ammonia Volatilization

AV is a process by which gaseous $NH_3$ is escaped from the soil into the atmosphere, whenever free $NH_3$ is present on or near the soil surface. The AV was accounted for 16–24% in rice and 16–32% in wheat when 120 kg N $ha^{-1}$ was applied to each crop in the rice–wheat cropping system (RWCS) (Pathak et al., 2006). Sharma et al. (2008) reported that AV losses in rice field varied from 1 to 23 kg N $ha^{-1}$ depending upon the source of fertilizer N, soil pH, and N application rate, while in particular RWCS it varied from 60 to 68 kg N $ha^{-1}$. The quantity of ammonia loss as AV depends on ammonia concentration in surface water, weather conditions, soil properties, source of applied N fertilizers (Kang et al., 2014), and also management practices. For example, Pathak et al. (2006) showed that AV loss increased from 5% at 60 kg N $ha^{-1}$ to 41% at 300 kg N $ha^{-1}$ in wheat on Typic Ustochrepts (alluvial soil) of pH 8.1 in Delhi. High AV losses are associated with broadcast application of urea or other ammonium-containing fertilizers. In India and other Asian countries, N is mostly applied as broadcast but in recent years its placement has also received attention. The advantages of placement of N in rice have been well demonstrated with urea or urea super granules (USG) (Mohanty et al., 1999). Humphreys et al. (1992) reported that recovery efficiency of N is lowest in case of broadcast application (37%) and increases with banding (46%) and deep placement (49%) of USG in direct seeded rice in Australia. Therefore, efficient and integrated management practices are effective in reducing AV loss, and increasing grain yield and nutrient use efficiency (NUE) (Cao et al., 2013). More studies based on field experiments are needed to get realistic estimates of AV losses from different agroecosystems.

## Nitrification

Nitrification is the process of enzymatic oxidation of $NH_4^+$ to $NO_3^-$ brought about by specific nitrifying microorganisms, hence the process is quite sensitive to any change in environmental conditions. Nitrification is a dominant pathway of N flow in agriculture systems, where $NO_3^-$ accounts for more than 95% of the total N uptake (Subba Rao et al., 2010). This makes the N cycle prone to leakage of reactive N to the environment, making agricultural systems the worst N polluter (Galloway et al., 2008). The environmental factors greatly influence the nitrification and denitrification and $N_2O$ emission, such as soil moisture, temperature, pH, rainfall, human activities including irrigation, and the type of applied N fertilizers (Baggs et al., 2010). Soil pH greatly influences nitrification and nitrification rate in soils often reaches to its maximum at pH 8.0 to 9.0 (Sahrawat, 2008). Thus, increase in soil pH accelerates the nitrification rate.

## Immobilization–Mineralization

N immobilization is a process in which inorganic soil N is converted to organic forms through biological agent. It occurs when huge loading of carbon-rich materials (high C:N ratio) are applied to soil. Immobilization of N results from uptake of $NO_3$-N and/or $NH_4^+$-N present in the soil by the growing microbes to build proteins (John et al., 2014). The N taken up is assimilated by microbes present in soil to proteins, nucleic acids, and other organic-N constituents of microbial cells and cell wall, and becomes part of the microbial biomass. Nitrogen is released as $NH_4^+$ from microbial biomass after the process of mineralization when the microbes die and decay, the remainder undergoes conversion to more stable organic-N compounds, finally becoming a part of soil organic matter. In general, microbes prefer $NH_4^+$ over $NO_3^-$ ions during immobilization process. Hence, the process of immobilization–mineralization is more severe with ammonical fertilizers than with $NO_3^-$ fertilizers. The process is also affected by weather conditions, soil properties, and nature of residues added to the soil. The amount of N uptake by crops depends on the chemical composition of soil organic matter, i.e., N content, C:N ratio, content of cellulose and hemicelluloses, and lignin and polyphenols (Mohanty et al., 2011). In addition to soil physical and chemical properties, soil microbes also greatly influence the soil immobilization and mineralization.

## Denitrification

As compared to AV, which is purely a chemical process, denitrification is a biological process carried out by facultative microorganisms, which use $NO_3^-$ as an electron acceptor when $O_2$ is the limiting factor such as in rice under submerged and poorly drained environment (De Datta et al., 1991). Bacterial denitrification is the cause for gaseous loss of N under saturated conditions. This process converts the $NO_3^-$ and $NO_2^-$ to the gaseous forms such as $N_2$ and $N_2O$, which ultimately escape from the soil. In general, $NH_4^+$-N is not undergoing denitrification process, but there is an evidence of intermediate $N_2O$ emission during the conversion of $NH_4^+$ to $NO_3^-$ by nitrifying microorganisms. Efforts to predict the loss of N by denitrification had limited success; it is presumed that denitrification occurs in soil aggregates and anaerobic microsites which are heterogeneously distributed in the soil. The high risk of denitrification losses from soils are affected by the factors such as high amount of fertilizers, high labile carbon availability, decreasing oxygen content in the soil, and also nitrate concentration, soil pH, soil organic matter, and temperature of soil, which strongly influence denitrification. Ladha et al. (2005) reported the highest denitrification losses under alternate flooding and drying conditions. Aulakh et al. (2001) estimated that 23–33% of the total N applied through chemical fertilizer is lost via denitrification in rice field. Similarly, Shcherbak et al. (2014) reported that denitrification rate is promoted by N inputs into the agroecosystems, with $N_2O$ emissions positively correlated to fertilizer rates. Although, the magnitude of denitrification losses and $N_2$:$N_2O$ partitioning is largely unknown in agroecosystem, representing a major uncertainty while estimating total N loss and replacement (Fried et al., 2016).

## N Transformations Under Submerged Agroecosystem

In rice agroecosystems nitrogen is lost mainly in the form of AV, nitrous oxide, or $N_2$ emission due to denitrification process. A considerable fraction is also lost through leaching, surface runoff, and soil erosion (Krishna, 2010).

Movement of N in submerged soils is important. In flooded conditions, major fractions of soil N occur as organic-N. Soil organic-N fractions contain amino acids, amino sugars, and humin-N whose proportion depends on soil parent material and agronomic procedures such as addition of FYM, azolla, and green manures. Due to alternate flooding and drainage conditions the total nitrogen content of soil is generally maintained; however a decline in the available pool of N has been experienced in long-term studies. During recent years inorganic N supply has increased enormously, mainly to support high grain yield. N loss from the rice fields can be severe depending on the climatic conditions, soil N status, and cropping pattern. Volatilization of N as $NH_3$, $NO_2$, and $N_2O$ is dependent on atmospheric conditions, rice genotype and its canopy, soil N status, fertilizer-N supply, and farming practices. $^{15}N$ studies have revealed that about 3–8% of fertilizer N applied is lost from rice fields via volatilization. Nitrification inhibitors such as neem-coated urea may curtail loss of N to a certain extent. Detailed studies on nitrogen dynamics in paddy fields revealed that under submerged condition, oxidation of $NH_4$-N derived from $NH_4$ fertilizers in surface layers and subsequent denitrification in subsurface layers together accounted for 30% to 49% N loss from rice cropping zones (CRRI, 2006). Generally, volatilization of $NH_3$-N is the maximum in the first 3 days after urea application and later decreases till 15th day. Floodwater, pH, FYM, and green manure are major factors that affect $NH_3$-N volatilization from paddy fields. A study by Pathak et al. (2005) reported that annually, Indian rice belt contributes 0.038 to 0.048 Tg $N_2O$-N. Among sources of $N_2O$ from rice belt, fertilizer supplied into rice fields contribute 72%, crop residue burning 11%, manure 3%, and indirect emission from soil accounts for remaining 14% (Bhatia et al., 2004). Hence, nitrogen recovery from rice fields is generally low and hardly exceeds 30% to 40% of fertilizer N applied.

## Leaching

Leaching of N is one of the major loss mechanisms that can be as high as 80% depending upon the soil type and crop management practices. It is a serious form of loss from irrigated and well-drained porous soils. Leaching losses of N from various locations in India varied from negligible to as high as 20% in rice and wheat crops. Using crop simulation model Pathak et al. (2006) reported a leaching loss of 10% of applied N in rice–wheat cropping systems of the Indo-Gangetic plains.

Sometimes, N loss via downward leaching can be very high, especially in coarse-textured soils. High intensity rainfall that results in surface runoff, leaching, and percolation may cause 10–12% of N loss from soil (Kundu and Ladha, 2004). Downward movement that results in loss of N from root zone may affect groundwater quality due to increase in $NO_3^-$ contents beyond permissible limits. There have been some reports of nitrate buildup in groundwater in states having high input of nitrogenous fertilizers, such as Punjab, Haryana, Delhi, Maharashtra, and Andhra Pradesh. Sporadic reports of high $NO_3$-N contamination in groundwater from different parts of India, however, may be related more to sewage–sewerage and dumping of organic wastes rather than agricultural N fertilization barring a few sites where the soil is highly coarse textured and hence permeable.

## Plant N Uptake

Plants preferentially absorb N from soil in the form of nitrate ($NO_3^-$) and ammonium ($NH_4^+$) ion. This N is further transported from root to the shoot through xylem. Generally, majority of $NO_3^-$ reduction takes place in the shoots while the roots reduce only a small fraction of the absorbed nitrate to ammonia (Scheurwater et al., 2002). N uptake by crop plants depend on several factors such as (1) contribution of soil and fertilizer toward N supply, (2) plant genetic make up to capture N from soil, (3) root architecture and crop N assimilation efficiency, (4) plant growth and yield potential, (5) ability to retain in the rooting zone, and (6) the duration of crop N uptake. It has been estimated that 50–70% of the N provided to the soil is lost (Hodge et al., 2000). Nitrogen uptake efficiencies of major field crops ranged from 30% to 70%. The rate of N uptake decreases as the nitrate and ammonium concentrations decrease in soil solution (Jackson et al., 1986). N must be supplied at rather high levels to achieve maximum or near maximum yields. This is the reason N-use efficiency invariably decreases at higher levels of N input that are required to achieve maximum yield. Globally, cereal yields and fertilizer-N consumption have increased linearly during the past 40 years and both are highly correlated (Dobermann and Cassman, 2004). It is reported that major cereal crops (rice, wheat, and maize) use 1 kg of N to produce 68 (Witt et al., 1999), 44 (Pathak et al., 2003), and 49 (Janssen et al., 1990) kg of rice, wheat, and maize grain, respectively. In general, more than 50% of the N applied is not utilized by plants (Tilman et al., 2002; Dobermann and Cassman, 2004) and rest of the applied fertilizer is the main culprit for environmental pollution.

Wood et al. (2004) estimated that 50–70% more cereal grain would be required by 2050 to feed 9.3 billion people, requiring 50–70% increased N fertilizer. But, as NUE generally declines with increased fertilizer use, the requirement may be even double (Wood et al., 2004). Therefore, efficient N management through best agronomic practices and use of fertilizer efficient crop genotype are essential for the sustainability of agricultural production for both the economic demand of agriculture and the need to avoid negative environmental consequences of N pollutants. Prasad (2005) reported the values of apparent recovery efficiency of N averaged over several rice and wheat experiments at New Delhi and the global average values for rice, wheat, and maize (Table 5.2). At similar levels of N, values of all N efficiencies in rice were lower in India compared to the global values. The values with respect to all the terms of NUE of N (NUEn) in wheat were higher in India than the global values indicating that in India, N is more efficiently utilized for wheat than rice crop. Therefore, in rice there is scope to increase NUEn (Prasad, 2009).

Fertilizer source is also important to increase N recovery by crops, avoid N loss from volatilization and be matched to the type of placement method to reduce losses and maximize recovery by crops. Urea is the major N source in the RWCS. However, a number of studies have been made to compare different N sources for rice and wheat. For rice ammonium sulfate was better than urea (Yadvinder Singh et al., 1990), while nitrate-containing fertilizer calcium ammonium nitrate (CAN) was inferior to

**Table 5.2** Estimates of Different N-Use Efficiency in Rice and Wheat Crops at New Delhi and at Global Scale

| Dose of N (kg $ha^{-1}$) | PFPn (kg Grain $kg^{-1}$ N Applied) | AEn (kg Grain $kg^{-1}$ N Applied) | REn (%) | PEn (kg Grain $kg^{-1}$ N Taken up by Crop) |
|---|---|---|---|---|
| ***New Delhi*** | | | | |
| Rice | | | | |
| 40–60 | 84.3 | 22.0 | 36.1 | 44.4 |
| 61–120 | 47.7 | 16.2 | 40.2 | 37.7 |
| 121–180 | 32.8 | 13.1 | 31.3 | 40.4 |
| Wheat | | | | |
| 40–60 | 83.7 | 28.8 | 73.8 | 47.8 |
| 61–120 | 50.2 | 20.1 | 57.7 | 42.8 |
| 121–180 | 31.3 | 15.9 | 61.8 | 24.0 |
| ***Global*** | | | | |
| Maize | 72.0 | 24.2 | 65 | 36.7 |
| 123 | | | | |
| Rice | 62.4 | 22.0 | 46 | 52.8 |
| 115 | | | | |
| Wheat | 44.5 | 18.1 | 57 | 28.9 |
| 112 | | | | |
| Average | 20.6 | 54.6 | 55 | 40.6 |

*AEn*, agronomic efficiency of N; *PEn*, physiological efficiency of N; *PFPn*, partial factor productivity of N; *REn*, recovery efficiency of N
Prasad, R., Singh, R.K., Rani, A., Singh, D.K., 2000. Partial factor productivity of nitrogen and its efficiency in rice and wheat. Fertilizer News 45(5), 63–65; Ladha, J.K., Pathak, H., Krupnik, T.J., Six, J., Kessel, C.V., 2005. Efficiency of fertilizer nitrogen in cereal production: retrospect's and prospects. Advances in Agronomy 87, 85–156.

ammonium sulfate, ammonium chloride, and urea (Thind et al., 1984). For wheat most conventional N fertilizer such as ammonium sulfate, CAN, and urea were equally effective (Yadvinder Singh et al., 1990). However some workers reported CAN to be better than urea on calcareous and alkaline soils (Prasad et al., 1992).

## Nitrogen Outputs

Movement of Nr from agricultural field, farm waste, and other domestic and industrial sources eventually into estuaries and coastal waters is a matter of great concern. A major portion of Nr comes from N leakage into environment with negative effects on ecosystem functions and contributes to global warming. Human activity is the primary factor that influenced the global N cycle by introducing unprecedented amounts of Nr into ecosystems, and which is responsible for a large amount of emissions of N in the form of nitrous oxide (40%), nitric oxide (80% or more), and ammonia releases (70%) at global level (Bijay Singh et al., 2008). According to an estimate, out of a total emission of 26.77 million tons of $NO_x$ in Asia, China, and India accounted for 42% and 17% of the emissions respectively in the year 2000 (Bijay Singh et al., 2008). It is also estimated that by 2020, India alone will be contributing $NO_x$ to the tune of 7.1 million tons annually from road and rail transport sectors.

Livestock based agroecosystem impacts air and water quality, ocean health, and greenhouse gas (GHG) emissions on regional to global scales (Eshel et al., 2014). India produces about 78 million tons of cattle and buffalo milk, 40 billion eggs, and 5.9 million tons of meat (including 1.7 million tons of poultry meat) annually (Dikshit and Birthal, 2010). Livestock produce dietary protein from the consumption of proteins produced by crop agroecosystems (Bijay Singh et al., 2008). The conversion of plant protein to animal protein ranges from 5 to 45% depending on the animal species, their ration, and management. The remaining portion is excreted via urine and feces, and can be used as nutrient source for agriculture production. As in crop agroecosystems, most of the reactive N that enters the animal agroecosystem is lost to the environment over the course of a year (Galloway et al., 2003). The $NH_3$ and $N_2O$ from livestock waste for each type of livestock, except for emissions after application as fertilizers, were estimated at different regional levels by multiplying the specific emission factor taking into account the N-excretion value from each livestock and livestock population (Aneja et al., 2010). Emissions of $NH_3$ and $N_2O$ from livestock excretion and cattle contributed the highest $NH_3$ emission at 55.5% of the total emission followed by buffalo (28.1%) from the livestock (Table 5.3).

It is also noteworthy to mention that about same quantity of Nr is added through livestock manures supplying 130 kg N $ha^{-1}$. Among the fertilizers, in terms of gaseous

**Table 5.3** Emission of $NH_3$ and $N_2O$ From Various Fertilizer and Manure Applications in India

| | Emission of $NH_3$ | | Emission of $N_2O$ | |
|---|---|---|---|---|
| **Category** | **Gg year$^{-1}$** | **%** | **Gg year$^{-1}$** | **%** |
| ***Fertilizer*** | | | | |
| Diammonium phosphate | 102.8 | 4.6 | 7.9 | 6.3 |
| Urea | 2044.1 | 92.0 | 114.5 | 90.8 |
| Ammonium sulfate | 0.5 | 0.02 | 0.02 | 0.02 |
| NPK | 73.9 | 3.3 | 3.7 | 2.9 |
| ***Livestock*** | | | | |
| Cattle | 771.9 | 55.5 | 57.4 | 42.3 |
| Buffalo | 331.9 | 23.8 | 38.1 | 28.1 |
| Sheep | 86.1 | 6.2 | 12.9 | 9.5 |
| Goat | 136.6 | 9.8 | 21.1 | 15.5 |
| Camel | 4.4 | 0.3 | 0.7 | 0.5 |
| Pig | 20.2 | 1.5 | 2.4 | 1.8 |
| Horse | 5.2 | 0.4 | 0.7 | 0.5 |
| Mule and ass | 5.8 | 0.4 | 0.7 | 0.5 |
| Poultry | 29.8 | 2.1 | 1.7 | 1.3 |

Adapted from Aneja, V.P., Schlesinger, W.H., Erisman, J.W., Sharma, M., Behera, S.N., Battye, W., 2010. Reactive N emissions from crop and livestock farming in India, In: Pathak, H. (Eds.), ING Bulletins on Regional Assessment of Reactive N, Bulletin No. 3, Scon-Ing, New Delhi, pp I–IV & 1–18.

losses, urea contributed the highest emission of $NH_3$ (92.0%) and $N_2O$ (90.8%). For $N_2O$ emission, cattle accounted for the highest (42.3%), followed by buffalo (28.1%), and goat (15.5%) toward the total pollution from the livestock sector.

## Conclusion

Supply of Nr to crops is indispensable to feed the global population. Higher N demand is due to low NUE. N derived from fertilizers and not taken up by plants may be immobilized in soil organic matter or may be lost to the environment. Thus the economic losses both due to low NUE and Nr input to environment are the serious problems. Agricultural practices, livestock contributions, energy consumption, and other human activities significantly contribute to the rising reactive N concentration in the terrestrial ecosystem as well as in atmosphere. In addition to this, increasing use of N for more food grain production for an ever-growing population would result in increased flow of reactive N, while on the other side N-use efficiency would remain at low level. As a consequence more reactive N is expected to enter into soil, water bodies, and atmosphere. Future research needs to be focused on generation of a large volume of soil fertility data, soil available N pools, and the magnitude of the losses from major soil types and agroecosystems to help prepare cropping system-based integrated N management modules for improving N-use efficiency. Judicious use of N, based on proper understanding of different processes in N cycle, must balance energy use efficiency and environmental safety while optimizing the N resources. NUE is crucial to improve yield and quality of crops, reduce nutrient input cost and improve soil, water, and air quality. An improved NUE in plants can be achieved by careful manipulation of plant, soil, fertilizer, biological and environmental factors, and losses reduced by the matching supply with demand.

## Future Research Needs

- There is meager data available on the amount of reactive N added through atmospheric deposition.
- The data on nitrogen transformation and fluxes on a long-term basis is lacking for critical understanding of N dynamics as they are affected by year to year variation in weather and the heterogeneity of soil conditions.
- To improve the understanding and management of N cycling in both natural and agricultural systems with predictive mathematical models that are lacking on both plot and catchment scale.
- There is a need to focus on ecosystem-based approach to optimize organic and mineral reservoirs with longer mean residence time that can be accessed through microbial and plant-mediated processes.
- There is need to quantify reactive N emission data using isotopic techniques for better understanding of the processes.

# References


Aneja, V.P., Schlesinger, W.H., Erisman, J.W., Sharma, M., Behera, S.N., Battye, W., 2010. Reactive N emissions from crop and livestock farming in India. In: Pathak, H. (Ed.), ING Bulletins on Regional Assessment of Reactive N, Bulletin No. 3. Scon-Ing, New Delhi. I–IV & 1–18.

Aneja, V.P., Schlesinger, W.H., Erisman, J.W., 2009. Effects of agriculture upon the air quality and climate: research, policy, and regulations. Environmental Science & Technology 43, 4234–4240.

Aulakh, M.S., Khera, T.S., Doran, J.W., Bronson, K.F., 2001. Denitrification, $N_2O$ and $CO_2$ fluxes in rice–wheat cropping system as affected by crop residues, fertilizer N and legume green manure. Biology and Fertility of Soils 34, 375–389.

Baggs, E.M., Smales, C.L., Bateman, E.J., 2010. Changing pH shifts the microbial source as well as the magnitude of $N_2O$ emission from soil. Biology and Fertility of Soils 46 (8), 793–805.

Bhandari, A.L., Ladha, J.K., Pathak, H., Padre, A.T., Dawe, D., Gupta, R.K., 2002. Trends of yield and soil nutrient status in a long-term rice-wheat experiment in Indo-Gangetic plains of India. Soil Science Society of American Journal 66, 162–170.

Bhatia, A., Pathak, H., Aggarwal, P.K., 2004. Inventory of methane and nitrous oxide emissions from agricultural soils of India and their global warming potential. Current Science 87 (3), 317–324.

Bijay Singh, Tiwari, M.K., Abrol, Y.P., 2008. Reactive Nitrogen in Agriculture, Industry and Environment in India. Indian National Science Academy, New Delhi.

Billen, G., Garnier, J., Lassaletta, L., 2013. Modelling the N cascade from watershed soils to the sea: from regional to global scales. Philosophical Transactions of the Royal Society of London. Series B 368, 20130123.

Cao, Y., Tiana, Y., Yina, B., Zhu, Z., 2013. Assessment of ammonia volatilization from paddy fields under crop management practices aimed to increase grain yield and N efficiency. Field Crops Research 147, 23–31.

CRRI, 2006. Soil Science and Microbiological Report. CRRI, Cuttack, India.

de Bruijn, F.J., 2016. "Biological N fixation" book summary. Advances in Microbiology 6, 407–411.

De Datta, S.K., Buresh, R.J., Samson, M.I., Obcemea, W.N., Real, J.G., 1991. Direct measurement of ammonia and denitrification fluxes from urea applied to rice. Soil Science Society of American Journal 55, 543–548.

Dikshit, A.K., Birthal, P.S., 2010. India's livestock feed demand: estimates and projections. Agricultural Economics Research Review 23, 15–28.

Dobermann, A., Cassman, K.G., 2004. Environmental dimensions of fertilizer N: what can be done to increase nitrogen use efficiency and ensure global food security? In: Mosier, A.R., et al. (Eds.), Agriculture and the Nitrogen Cycle: Assessing the Impacts of Fertilizer Use on Food Production and the Environment. Scope 65. Island Press, Washington, D.C., USA, pp. 261–278.

Eshel, G., Shepon, A., Makov, T., Milo, R., 2014. Land, irrigation water, greenhouse gas, and reactive N burdens of meat, eggs, and dairy production in the United States. Proceedings of the National Academy Sciences of the United States of America 111 (33), 11996–12001.

FAI, 2013. Fertilizer Statistics 2012–13. The Fertilizer Association of India, New Delhi.

FAI, 2015. Fertilizer Statistics 2014–15. The Fertilizer Association of India, New Delhi.

FAO, 2005. Fertilizer Use by Crop in India. FAO, Rome, p. 45.

Fried, J., Scheer, C., Rowlings, D.W., McIntosh, H.V., Strazzabosco, A., Warner, D.I., Grace, P.R., 2016. Denitrification losses from an intensively managed sub-tropical pasture: impact of soil moisture on the partitioning of $N_2$ and $N_2O$ emissions. Soil Biology and Biochemistry 92, 58–66.

Galloway, J.N., Aber, J.D., Erisman, J.W., Seitzinger, S.P., Howarth, R.W., Cowling, E.B., Cosby, B.J., 2003. The N cascade. Bioscience 53, 341–356.

Galloway, J.N., Townsend, A.R., Erisman, J.W., Bekunda, M., Cai, Z., Freney, J.R., Martinelli, L.A., Seitzinger, S.P., Sutton, M.A., 2008. Transformation of the N cycle: recent trends, questions and potential solutions. Science 320, 889–892.

Galloway, J.N., Dentener, F.J., Capone, D.G., Boyer, E.W., Howarth, R.W., Seitzinger, S.P., Asner, G.P., Cleveland, C.C., Green, P.A., Holland, E.A., Karl, D.M., Michaels, A.F., Porter, J.H., Townsend, A.R., Voosmarty, C.J., 2004. N cycles: past, present, and future. Biogeochemistry 70, 153–226.

Herridge, D.F., Peoples, M.B., Boddey, R.M., 2008. Global inputs of biological N fixation in agricultural systems. Plant and Soil 311, 1–8.

Hodge, A., Robinson, D., Fitter, A., 2000. Are microorganisms more effective than plants at competing for N? Trends in Plant Science 5, 304–308.

Humphreys, E., Chalk, P.M., Muirhead, W.A., White, R.J.G., 1992. N fertilization of dry-seeded rice in South-East Australia. Fertilizer Research 31, 221–234.

IFA, 2002. Fertilizer Use by Crop. International Fertilizer Industry Association (IFA), International Fertilizer Development Centre (IFDC), International Potash Institute (IPI), Potash and Phosphate Institute (PPI), Food and Agriculture Organization (FAO), Rome.

IISS, 2015. Vision 2050. ICAR–Indian Institute of Soil Science, Bhopal, India.

Jackson, W.A., Pan, W.L., Moll, R.H., Kamprath, E.J., 1986. Uptake, translocation, and reduction of nitrate. In: Neyra, C.A. (Ed.), Biochemical Basis of Plant Breeding. CRC Press, Boca Raton, Florida.

Janssen, B.H., Guiking, F.C.T., Van Der Eijk, D., Smaling, E.M.A., Wolf, J., Reuler, H., 1990. A system for quantitative evaluation of the fertility of tropical soils (QUEFTS). Geoderma 46, 299–318.

John, A., Lamb, F., Fernandez, G., Daniel, E.K., 2014. Understanding N in Soils. www.extension.umn.edu/agriculture/nutrient-management/AG-FO-3770-B (Revised 2014).

Kang, Ni, Andreas, P., Henning, K., 2014. Ammonia volatilization after application of urea to winter wheat over 3 years affected by novel urease and nitrification inhibitors. Agriculture, Ecosystems & Environment 197, 184–194.

Krishna, K.R., 2010. Agro-ecosystems of South India: Nutrient Dynamics, Ecology and Productivity. Brown Walker Press, Boca Raton, Florida, USA, 559 pp.

Kulshrestha, U.C., Granat, L., Engardt, M., Roshe, H., 2005. Review of precipitation monitoring studies in India – a search for regional patterns. Atmospheric Environment 39, 7403–7419.

Kundu, D.K., Ladha, J.K., 2004. Sustaining productivity of low land rice soils: issues and options related to N availability. Fertilizer News 49, 65–67.

Ladha, J.K., Pathak, H., Krupnik, T.J., Six, J., Kessel, C.V., 2005. Efficiency of fertilizer nitrogen in cereal production: retrospect's and prospects. Advances in Agronomy 87, 85–156.

Ladha, J.K., Reddy, P.M., 2000. The Quest for N Fixation in Rice. The Philippines. International Rice Research Institute.

Laegreid, M., Bockman, O.C., Kaarstad, O., 1999. Agriculture, Fertilizers and the Environment. Cabi – Norsk,Hydro ASA, Wallingford, UK.

Masunga, R.H., Uzokwe, V.N., Mlay, P.D., Odeh, I., Singh, A., Buchan, D., Stefaan, D.N., 2016. N mineralization dynamics of different valuable organic amendments commonly used in agriculture. Applied Soil Ecology 101, 185–193.

Ministry of Chemicals and Fertilizers, 2014. Indian Fertilizers Scenario 2013. Department of Fertilizers, Ministry of Chemicals and Fertilizers, Government of India, New Delhi, 224 pp.

Mohanty, M., Reddy, S.K., Probert, M.E., Dalal, R.C., Rao, S.A., Menzies, N.W., 2011. Modelling N mineralization from green manure and farmyard manure from a laboratory incubation study. Ecological Modelling 222, 719–726.

Mohanty, S.K., Singh, U., Balasubramanian, V., Jha, K.P., 1999. N deep placement technologies for productivity, profitability and environmental quality of lowland rice systems. Nutrient Cycling in Agroecosystems 53, 43–57.

Mosier, A.R., Wassmann, R., Verchot, L., King, J., Palm, C., 2004. Methane and N oxide fluxes in tropical agricultural soils: sources, sinks and mechanisms. Environment, Development and Sustainability 6, 11–49.

Pathak, H., Aggarwal, P.K., Roetter, R., Kalra, N., Bandyopadhaya, S.K., Prasad, S., Van Keulen, H., 2003. Modeling the quantitative evaluation of soil nutrient supply, nutrient use efficiency, and fertilizer requirements of wheat in India. Nutrient Cycling in Agroecosystems 65, 105–113.

Pathak, H., Li, C., Wassmann, R., 2005. Greenhouse gas emissions from Indian rice fields: calibration and upscaling using the DNDC model. Biogeosciences Discussions, European Geosciences Union 2 (1), 77–102.

Pathak, H., Li, C., Wassmann, R., Ladha, J.K., 2006. Simulation of N balance in the rice–wheat systems of the Indo-Gangetic plains. Soil Science Society of American Journal 70, 1612–1622.

Pathak, H., Mohanty, S., Prasad, R., 2009. Fate of nitrogen in Indian agriculture: environmental impacts, quantification and un-certainities – a review. Proceedings of the National Academy of Sciences, India Section B: Biological Sciences 79, 322–335.

Prasad, R., 2005. Rice-wheat cropping systems. Advances in Agronomy 86, 255–339.

Prasad, R., 2007. N in Indian agriculture. In: Abrol, Y.P., Raghuram, N., Sachde, M.S. (Eds.), Agricultural N Use and its Environmental Implications. I.K. International Publishing, New Delhi, pp. 29–54.

Prasad, R., 2009. Efficient fertilizer use–the key to food security and better environment. Journal of Tropical Agriculture 47, 1–7.

Prasad, R., Singh, R.K., Rani, A., Singh, D.K., 2000. Partial factor productivity of nitrogen and its efficiency in rice and wheat. Fertilizer News 45 (5), 63–65.

Prasad, R., Singh, S., Sharma, S.N., Prasad, M., 1992. Relative efficacy of N source in wheat. Fertilizer News 37 (6), 39–40.

Prasad, V.K., Badarinath, K.V.S., Yonemura, S., Tsuruta, H., 2004. Regional inventory of soil surface nitrogen balances in Indian agriculture (2000–2001). Journal of Environmental Management 73, 209–218.

Sahrawat, A.K., Caroll, R.T., DePauw, M., Taylor, G.I., Good, A.G., 2008. Genetic engineering of improved nitrogen use efficiency in rice by the tissue specific expression of alanine aminotransferase. Plant Biotechnology Journal 6 (7), 722–733.

Scheurwater, I., Koren, M., Lambers, H., Atkin, O.K., 2002. The contribution of roots and shoots to whole plant nitrate reduction in fast- and slow-growing grass species. Journal of Experimental Botany 53 (374), 1635–1642. http://dx.doi.org/10.1093/jxb/erf008. 12096102.

Sharma, C., Tiwari, M.K., Pathak, H., 2008. Estimates of emission and deposition of reactive nitrogenous species for India. Current Science 94 (11), 1436–1446.

Shcherbak, I., Millar, N., Robertson, G.P., 2014. Global meta-analysis of the nonlinear response of soil nitrous oxide ($N_2O$) emissions to fertilizer N. Proceedings of the National Academy Sciences of the United States of America 111, 9199–9204.

SubbaRao, G.V., Nakahara, K., Ishikawa, T., Kishii, M., Kudo, N., Rao, I.M., Ishitani, M., Sahrawat, K.L., Hash, C.T., George, T.S., Berry, W., Lata, J.C., Ito, O., 2010. Nitrification – is it a strategic point of intervention for limiting nitrogen losses from agricultural systems? In: Abrol, Y.P., Chatterjee, S. (Eds.), The Concept of Biological Nitrification Inhibition (BNI). ING Bulletins on Regional Assessment of Reactive Nitrogen.

Sutton, M.A., Bleeker, A., Howard, C.M., Bekunda, M., Grizzetti, B., de Vries, W., van Grinsven, H.J.M., Abrol, Y.P., Adhya, T.K., Billen, G., Davidson, E.A., Datta, A., Diaz, R., Erisman, J.W., Liu, X.J., Oenema, O., Palm, C., Raghuram, N., Reis, S., Scholz, R.W., Sims, T., Westhoek, H., Zhang, F.S.,

Ayyappan, S., Bouwman, A.F., Bustamante, M., Fowler, D., Galloway, J.N., Gavito, M.E., Garnier, J., Greenwood, S., Hellums, D.T., Holland, M., Hoysall, C., Jaramillo, V.J., Klimont, Z., Ometto, J.P., Pathak, H., Plocq Fichelet, V., Powlson, D., Ramakrishna, K., Roy, A., Sanders, K., Sharma, C., Singh, B., Singh, U., Yan, X.Y., Zhang, Y., 2013. Our Nutrient World: The Challenge to Produce More Food and Energy with Less Pollution. Global Overview of Nutrient Management. Centre for Ecology and Hydrology, Edinburgh on behalf of the Global Partnership on Nutrient Management and the International Nitrogen Initiative.

Thind, H.S., Singh, B., Gill, M.S., 1984. Relative efficacy of nitrogenous fertilizer for rice-wheat rotationNitrogen in Soils, Crops and Fertilizers. Bull. Ind. Soc. Soil Sci. 13, 181–184.

Tilman, D., Cassman, K.G., Matsan, P.A., Naylor, R.L., Polaskey, S., 2002. Agricultural sustainability and intensive production practices. Nature 418, 671–677.

Velmurugan, A., 2000. Assessment of Soil Quality Parameters for Sustainable Production in Rice–wheat Cropping System. IARI, New Delhi (Ph.D. thesis).

Velmurugan, A., Dadhwal, V.K., Abrol, Y.P., 2008. Regional N cycle: an Indian perspective. Current Science 94 (11), 1455–1468.

Vitousek, P.M., Cassman, K., Cleveland, C., 2002. Towards an ecological understanding of biological N fixation. Biogeochemistry 57, 1–45.

Vitousek, P.M., Jhon, A.B., Robert, W.H., Gene, E.L., Pamela, A.M., David, W.C., William, H.S., Tilman, D. G., 1997. Human alteration of the global N cycle: sources and consequences. Ecological Applications 7, 737–750.

Voss, M., Baker, A., Bange, H.W., Conley, D., Cornell, S., Deutsch, B., Engel, A., Ganeshram, R., Garnier, J., Heiskanen, A.S., Jickells, T., Lancelot, C., McQuatters-Gollop, A., Middelburg, J., Schiedek, D., Slomp, C.P., 2011. Nitrogen processes in coastal and marine systems. In: Sutton, M.A., Howard, C.M., Erisman, J.W., Billen, G., Bleeker, A., Greenfelt, P., Van Grinsven, H., Grizzetti, B. (Eds.), The European Nitrogen Assessment. Cambridge University Press, Cambridge.

Witt, C., Dobermann, A., Abdulrachman, S., Gines, H.C., Guanghuo, W., Nagarajan, R., Satawathananont, S., Son, T.T., Tan, P.S., Van Tiem, L., Simbahan, G., Olk, D.C., 1999. Internal nutrient efficiencies in irrigated lowland rice of tropical and subtropical Asia. Field Crops Research 63, 113–138.

Wood, S., Henao, J., Rosegrant, M., 2004. The role of N in sustaining food production and estimating future N fertilizer needs to meet food demand. In: Mosier, A.R., Syers, J.K., Freney, J.R. (Eds.), Agriculture and the N Cycle: Assessing the Impacts of Fertilizer Use on Food Production and the Environment, vol. 65. Scope, Paris, France, pp. 245–260.

Yadvinder Singh, Meelu, O.P., Bijay-Singh, 1990. Relative efficiency of calcium ammonium nitrate under different agro-climatic condition – a review. Fertilizer News 35 (4), 41–45.

Yadvinder Singh, Singh, B., Thind, H.S., 2010. Green manure Approaches to Crop Production and Sustainable Agriculture. ING, bulletin on region assessment of reactive nitrogen, SCON, New Delhi, 130 pp.

# 6

# Nitrogen Balances of Intensively Cultivated Rice–Wheat Cropping Systems in Original Green Revolution States of India

D.K. Benbi

*PUNJAB AGRICULTURAL UNIVERSITY, LUDHIANA, INDIA*

**CHAPTER OUTLINE**



## Introduction

Of all the essential nutrients, nitrogen (N) is required in the largest amount and is most often the limiting nutrient. Without consumption of N-fertilizers, it would not have been possible to sustain growing population. More than half of the food consumed by the world's population is produced using N-fertilizer and there is a strong parallel between global population and fertilizer-N use. World population during 1960–2014 increased from 3 to 7 billion, and N consumption and cereal production during this

The Indian Nitrogen Assessment. http://dx.doi.org/10.1016/B978-0-12-811836-8.00006-9

period increased from 11.8 to 113 million tons (mt) and 877 to 2801 mt, respectively. The global population is estimated to cross 9 billion by the year 2050, and the increase in population will occur mostly in underdeveloped countries that are already under stress to meet their food requirements. An additional 1.1 billion tons of cereal and 470 mt of meat per year will be required for the growing population. Nitrogen is the main source of dietary proteins and currently humans ingest about 20 Tg N $year^{-1}$ ($Tg = 10^{12}$ g) in their food (Galloway et al., 2002). An increase in the population will also lead to increase in fertilizer-N demand. With increasing urbanization in developing countries, such as India, demand for quantity and quality of food will increase further.

## Pre-green Revolution Scenario

India occupies an area of 329 mha and ranks 7th in the world. While it has only 2.5% of the world area, it is home to 17% of the world population. Prior to 1950, India had a history of declining per capita food production, food shortages, and even famines. Subsequently, great attention was paid to increasing agricultural production for ensuring national food security. Until the early 1960s, the focus was to bring more area under cultivation to augment food-grain production. In 1950/51, the food-grain production was merely 51 mt, which increased to 72 mt by 1965/66. During this period, the area cultivated with food grains increased from 97 to 115 million hectare (mha), an increase of 18.6%. The productivity during that period was very low (522–710 kg $ha^{-1}$) mainly due to lack of irrigation infrastructure, negligible fertilizer use, and growing of low-yielding non-nutrient-responsive crop varieties. The rate of increase in food-grain production did not match population growth; hence India had to import large quantities of food. The per capita net availability of food grains was 149 kg $year^{-1}$ for a population of about 500 million in 1966. Therefore, pre-green revolution period was characterized by low productivity, lack of inorganic fertilizer use, and dependence on animal manure and biological nitrogen fixation (BNF) as sources of nitrogen.

## Green Revolution in India

Green revolution is a term coined by William Gourd in 1968 to describe a process that leads to improved agricultural productivity. The high-yielding technologies developed in Mexico helped revolutionize cereal production during the 1960s and 1970s, and came to be known as the "Green Revolution." High-yielding varieties of wheat were developed in Mexico by adopting a novel technique of shuttle breeding, which enabled photoperiod sensitivity to be overcome to produce internationally adapted spring wheat germplasm, that eventually spread throughout the world (Borlaug, 2007). The initiatives led by Norman Borlaug, who received the Nobel Prize for Peace in 1970, allowed developing countries, such as India, to overcome chronic food shortage.

The food-grain production in India, which was only 82 mt during 1960/61, more than doubled (170 mt) in 1988/89 and further increased to 264 mt in 2013/14 (Fig. 6.1). The net sown area of 141 mha remained nearly constant over the last 50 years. Because of the increase in cropping intensity, the gross-cropped area increased to 191 mha, of which food grains cover 122 mha. The large increase in food-grain production resulted from an increase in productivity with enhanced use of fertilizers, development of irrigation through bore wells, adoption of new technology with mechanization of farm operations, and greater intensity of cropping. Productivity of food grains increased threefold, from 710 kg $ha^{-1}$ in 1961/62 to 2100 kg $ha^{-1}$ in 2012/13. Area under irrigation coverage increased from 32 to 89 mha. The fertilizer consumption (N, P, and K), which was only 0.34 mt in 1961/62, increased by several orders of magnitude in 1976/77, and to more than 80-fold (28 mt) in 2010/11. Fertilizer-N use also witnessed similar increase—from 0.25 mt in 1961/62 to 2.5 mt in 1976/77 and 16.6 mt in 2010/11. Now India stands second in the world in terms of gross fertilizer consumption with average rate of nutrient application at 141 kg NPK $ha^{-1}$ $year^{-1}$ with N alone accounting for 88 kg $ha^{-1}$. The current fertilizer consumption ratio, however, is highly skewed toward N with N:P:K ratio of 6.7:2.4:1.

Despite achieving self-sufficiency in food-grain production, there is little scope for complacency as India's population is currently growing at 1.2% and the food-grain production has to keep pace with this burgeoning population. It is estimated that by the year 2020 India's cereal requirement will be around 300 mt, and the demand for rice and wheat will be 122 and 103 mt, respectively. This will have to be produced from the shrinking land resource as the per capita availability of gross-sown area has declined

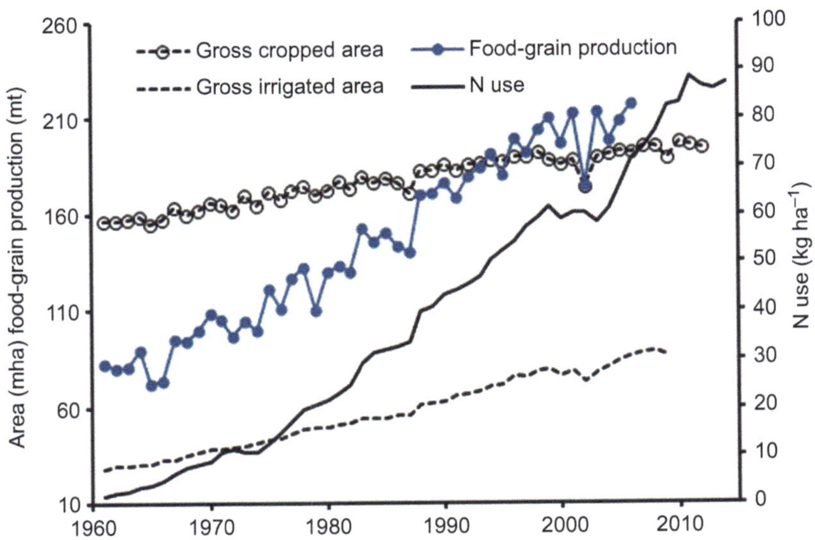


**FIGURE 6.1** Growth in production variables and food-grain production in India during 1960–2012. *Based on Directorate of Economics and Statistics, Govt. of India.*

from 0.36 to 0.16 ha per person. During the green revolution period, the area under rice and wheat had increased at the expense of area under coarse grains and pulses, but overall area under food grains remained almost static. Increase in the area coupled with enhanced productivity of rice and wheat resulted in greater food-grain production. The increase in productivity was greater in wheat than rice. While average wheat yield increased from 1307 to 3117 kg $ha^{-1}$, rice yield increased from 1123 to 2462 kg $ha^{-1}$. Since 1986, rice and wheat together account for 75% or more of the total food-grain production with 36% of the gross-cropped area under the two crops in the country. In addition to improved crop varieties, particularly of rice and wheat, productivity and production of other food grains were also related to fertilizer consumption. Fig. 6.2A shows a strong relationship between food-grain production and fertilizer-N consumption. On an average, each ton of fertilizer N consumed in the country enhanced food-grain production by 10 tons. This highlights the critical role played by N-fertilizers in increasing the productivity. However, fertilizer-N efficiency is declining from ~75 in the late 1960s to ~15 (Fig. 6.2B) at present.

## Growth of Production Factors in Original Green Revolution States

Green revolution technologies in India were originally introduced in the states of Punjab, Haryana, and western Uttar Pradesh (UP). These states are part of the Indo-Gangetic plains (IGP), where Punjab and Haryana fall in Trans IGP and western UP in the middle and upper IGP (Fig. 6.3). These plains are believed to be formed by alluvium brought from the Himalayas by the Indus and Ganges river systems about 7000 years ago (Pal et al., 2009). The soils in the region are generally deep alluvium, sandy loam to loam

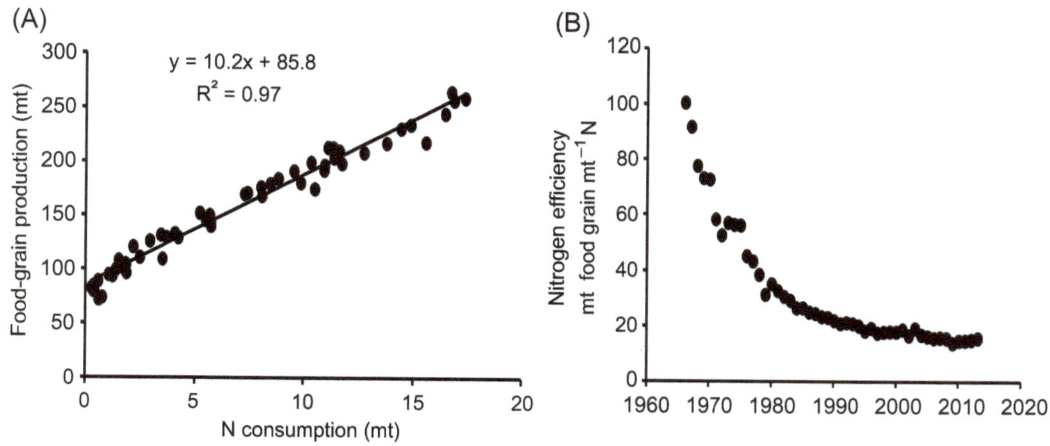


**FIGURE 6.2** (A) Food-grain production and fertilizer-N use in India during 1965–2012. (B) Fertilizer-N efficiency of food-grain production in India during 1965–2012. *(A) and (B) Based on Fertilizer Association of India, 2015. Fertilizer Statistics 2014–15, Fertilizer Association of India, New Delhi, India.*

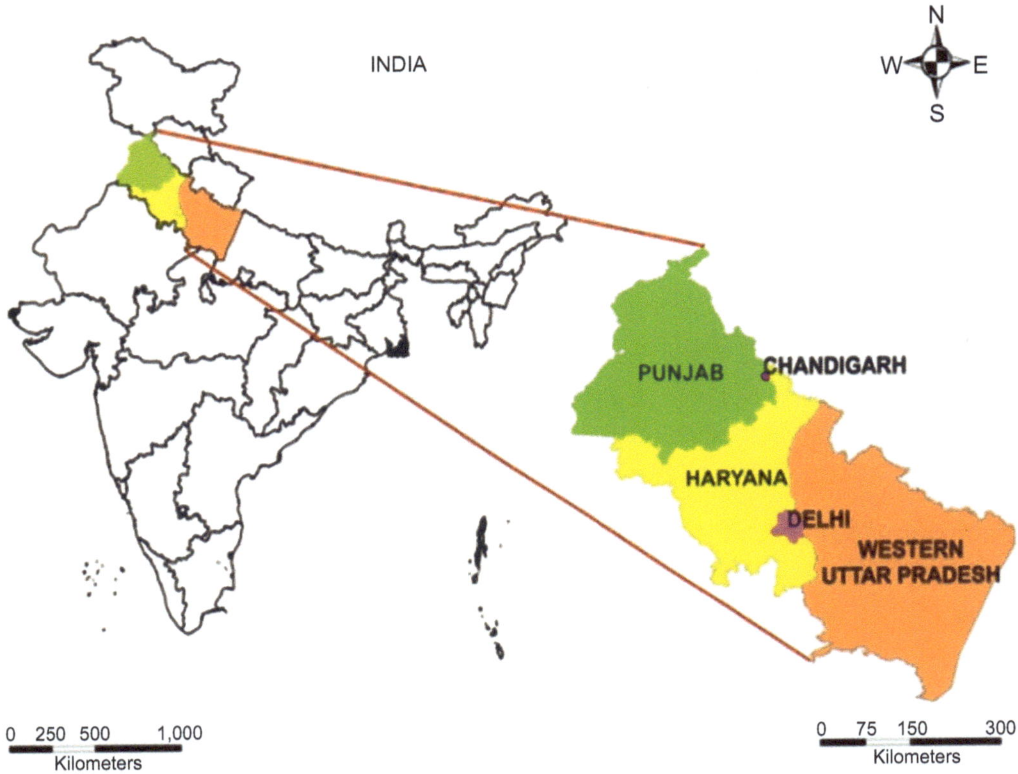


**FIGURE 6.3** The original green revolution states in India.

in texture, alkaline in reaction, and poor in organic matter content. In pre-green revolution period, rice cultivation in this region was limited to a small area that too mostly aromatic basmati variety with productive potential of about 1 t $ha^{-1}$ (Kumar and Nagarajan, 2004). With the ushering of green revolution, rice–wheat emerged as the major cropping system in this region.

Both Punjab and Haryana are intensively cultivated states. During 1966–2012, area under rice increased 10-fold (from 0.28 to 2.85 mha) in Punjab and six-fold (from 0.19 to 1.22 mha) in Haryana. During this period, the area under wheat cultivation in both the states increased 2–3.5 times (from 1.62 to 3.51 mha in Punjab and 0.72–2.5 mha in Haryana). An unprecedented increase in both production and productivity levels of wheat and rice occurred in these states, which helped India meet the food shortages of the 1960s. Being pioneer of India's green revolution, these states sustained both positive and negative effects of green revolution technologies. Initially farmers achieved very high productivity levels but very soon, chemical-based inputs and mechanized farming led to overexploitation of the natural resources, especially groundwater and soil, to the extent that most of the farming enterprises have turned out to be environmentally unsustainable (PSFC, 2013). Water table declined at an alarming rate in the region, especially under intensively cultivated rice–wheat systems (Hira, 2009). During the period of 1993–2003, the water-table fall in central districts of Punjab ranged between 0.3 and 1.0 m annually. By 2006, the water table had sunk to depths as low as 28 m. With the increased production of rice and wheat came the problem of residue management. Most of the rice residue is burnt in situ, which, in addition to causing environmental pollution and health hazards, leads to loss of carbon and other nutrients. Recently, there were reports of declining or stagnating crop yields in these states.

## Fluxes of Reactive Nitrogen

Three major forms of reactive N in soil are organic, ammonium ($NH_4^+$), and nitrate ($NO_3^-$). Under undisturbed conditions, mineral ($NH_4^+ + NO_3^-$) N usually constitutes between 1% and 2% of total soil N. It is, nevertheless, this N that is available for direct uptake by plants. Ammonical N occurs in two forms: ammonium ions ($NH_4^+$) and ammonia gas ($NH_3$). In the soils of this region the equilibrium between $NH_4^+$ and $NH_3$ is in favor of $NH_3$, because of high pH and prevalence of high temperature. $NH_4^+$ can be held on negatively charged sites of clay minerals and organic compounds. This reduces its mobility in soil. The nitrogen forms, such as $NH_3$, $NH_4^+$, $NO_3^-$ and organic N compounds, undergo several transformations in the soil and environment as part of the N cycle. The N cycling processes include N input, N loss, and N transformation processes. Nitrogen inputs include atmospheric wet and dry deposition, BNF by legumes, and industrial fixation of nitrogen by Haber-Bosch process (fertilizer application). The loss processes include erosion and runoff, leaching, ammonia volatilization, and denitrification. Mineralization and immobilization of soil organic N, nitrification, and fixation and release of $NH_4^+$ constitute transformation processes. In addition plant uptake of nitrogen is the most important output process in nitrogen cycle.

## Nitrogen Inputs in Original Green Revolution States

During the green revolution period increase in the area under nutrient-responsive, high-yielding varieties of rice and wheat in Punjab and Haryana resulted in rapid increase in fertilizer-N use. In these two states fertilizer-N use (per hectare) increased from meager 2–8 kg to 160–180 kg. Currently, of the total 16.5 mt of N-fertilizer consumed in India, about 15% is used in the intensively cultivated states of Punjab and Haryana; majority of it being applied to rice and wheat. Productivity of rice and wheat is strongly related to N use in both the states (Fig. 6.4). Interestingly, wheat exhibited similar response to fertilizer N (Eqs. 6.1 and 6.2) in both the states. On an average 1 kg of fertilizer N produced almost 19 kg of wheat grains. However, response of rice to fertilizer N in the two states differed considerably (Eqs. 6.3 and 6.4). While 1 kg of N in Punjab fetched 16 kg rice, it produced only 8 kg in Haryana. This was

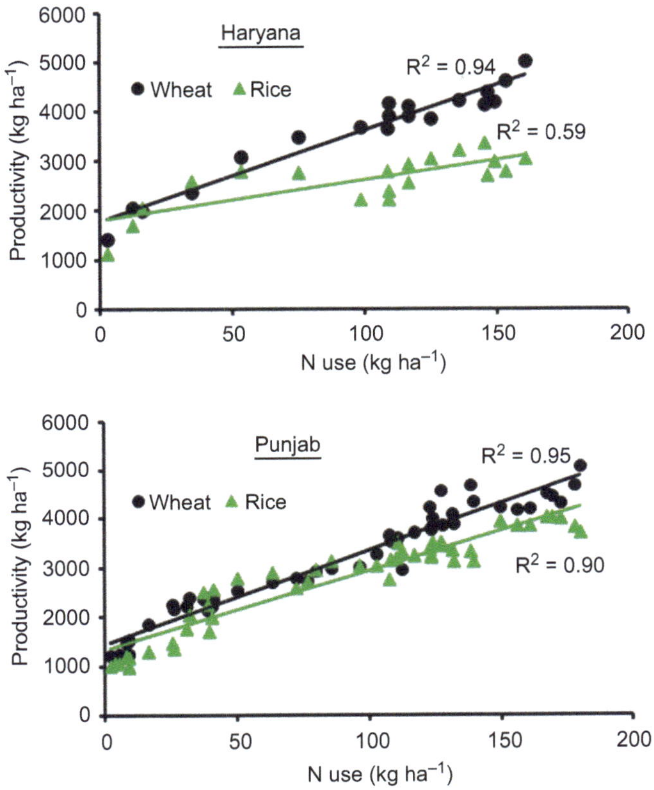


**FIGURE 6.4** Rice and wheat productivity in relation to fertilizer-N use in Punjab and Haryana during 1966–2012. *Based on SAP, 2013. Statistical Abstract Punjab. Economic Advisor to Government of Punjab, Chandigarh, India; SAH, 2014. Statistical Abstract Haryana, 2012–13, Department of Economic and Statistical Analysis Haryana, Chandigarh, India.*

probably because in Haryana fertilizer N is also used in other crops such as oilseeds, sugarcane, and cotton, which occupy substantial area in the state.

$$\text{Haryana: Wheat yield } (\text{kg ha}^{-1}) = 18.5\text{ N} + 1780 \tag{5.1}$$

$$\text{Punjab: Wheat yield } (\text{kg ha}^{-1}) = 19.0\text{ N} + 1458 \tag{5.2}$$

$$\text{Haryana: Rice yield } (\text{kg ha}^{-1}) = 8.2\text{ N} + 1801 \tag{5.3}$$

$$\text{Punjab: Rice yield } (\text{kg ha}^{-1}) = 16.1\text{ N} + 1333 \tag{5.4}$$

The values presented in equations above show long-term averages for crop response to fertilizer-N use. Temporal trends of crop response to fertilizer N show that N use efficiencies (NUEs) for rice and wheat in both the states have been declining (Fig. 6.5). However, during the last decade decline in NUE has decelerated and currently it is

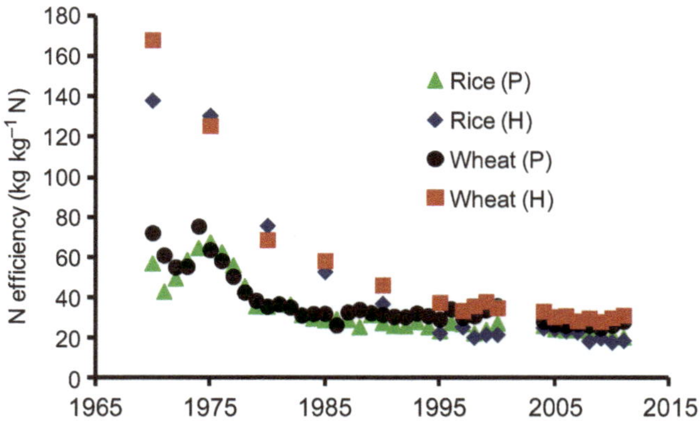


**FIGURE 6.5** Fertilizer-N efficiency of rice and wheat production in Punjab (P) and Haryana during 1966–2012. *Based on SAP, 2013. Statistical Abstract Punjab. Economic Advisor to Government of Punjab, Chandigarh, India; SAH, 2014. Statistical Abstract Haryana, 2012–13, Department of Economic and Statistical Analysis Haryana, Chandigarh, India.*

maintained around 23 and 27 kg grain kg$^{-1}$ of N for rice and wheat in Punjab, and 21 and 30 kg kg$^{-1}$ of N in Haryana, respectively. These NUE values are based only on fertilizer-N input and the contribution from other sources, such as organic manures, atmospheric deposition, and indigenous supply from soil, is not included. Precise information on input of nitrogen through atmospheric deposition and biological fixation is lacking in the study region. The estimated values for the two input processes are presented later in the section on nitrogen balances.

# Losses of Nitrogen

## Gaseous Losses of Nitrogen

Nitrogen is lost from agroecosystems to the atmosphere through gaseous emissions in the form of $NH_3$, NO, $N_2O$, or $N_2$. Nitrification process is central to the flows, transfers, losses, or utilization of N. It influences NUE by crops as well as N losses from soils. Since urea is the main form of fertilizer N used, the rate of nitrification is a key determinant of N losses. Nitrification rates exceeding $NO_3$ utilization by plants can result in increased levels of $NO_3$–N in runoff and groundwater, rendering it unsafe for human consumption. On the other hand, slow rates of nitrification could result in accumulation of $NH_4$–N which may enhance NUE by reducing denitrification and leaching losses. However, reduced nitrification could also increase N losses via $NH_3$ volatilization. Studies in $^{15}$N-labeled urea showed that $NH_3$ volatilization is the main pathway of N loss in wheat on alkaline soils. About 30–42% of applied $^{15}$N was lost through volatilization. However, irrigating the field following urea application reduced losses of applied $^{15}$N from 42% to 16% (Katyal et al., 1985). Modeling studies also showed that $NH_3$ volatilization is the main pathway of N loss in wheat and up to 28% of the added fertilizer N (120 kg N $ha^{-1}$) could be lost through this process, and irrigation management did not influence the losses (Benbi et al., 1991b).

There is increasing evidence that nitrification influences the amount of $N_2O$ and NO released to the atmosphere and could be one of the major contributors. In many rice-growing areas, continuous flooding cannot be maintained and there are periods of alternate drainage and flooding. Ammonium accumulated during flooding can be nitrified rapidly during the aerobic drainage period, and nitrates thus produced are subsequently reduced to $N_2O$ or $N_2$ through denitrification when soils are flooded again. Studies using $^{15}N$ balance approach showed that up to 50% of the applied N could be lost through gaseous emissions when alternating aerobic–anaerobic conditions prevail in rice–fields (Table 6.1).

Measurement of $N_2O$ emission from Indian agriculture started in 1990s. Emission of $N_2O$ is reported to be higher in upland crops, such as wheat grown under aerobic conditions, than in lowland rice grown under flooded anaerobic conditions. Based on the field measurements, $N_2O$–N emission from rice and wheat fields are estimated to be 0.25% and 0.55% of N applied, respectively, with an average value of 0.40% for rice–wheat system. Based on the analysis of other published results, an emission factor of 0.7% has been estimated for $N_2O$–N emitted from various nitrogen additions to the soil (Pathak et al., 2012). These values are much lower than the 1% emission factor used by Intergovernmental Panel on Climate Change based on a simple regression model (Bouwman, 1996). It has been estimated that about 79 Gg $N_2O$–N was emitted from Indian agricultural soils during the year 1994–95 with highest amount emitted from UP (including Uttarakhand) (Table 6.2). Punjab and Haryana with large areas under rice and wheat together contributed about 12 Gg $N_2O$–N. It was observed that inorganic fertilizer is the major source (72%) of nitrous oxide. Other sources such as crop residues and manure contribute 11% and 3%, respectively, to the total emission (Bhatia et al., 2004).

## Nitrate Leaching

The movement of water through soil can result in the transport of N out of the root zone of the plant. This process of N loss, called leaching, usually occurs when nitrogen is in $NO_3^-$ form, and is determined by its concentration in the soil solution and the amount of water percolating through the soil profile. In soil $NO_3^-$ originates from mineralization of

**Table 6.1** Extent of Gaseous N Losses From Flooded Rice Soils in Punjab

| N Application Rate (kg $ha^{-1}$) and Source | N Application Schedule | Gaseous N Loss (% of Applied N) |
|---|---|---|
| 180 (urea) | Basal | 34–50 |
| 180 (urea) | 3 splits | 17–28 |
| 58–116 (urea and $(NH_4)_2SO_4$) | Basal | 46–50 |

Adapted from Rekhi, R.S., Meelu, O.P., Gupta, R.K., 1982. Lysimeter studies on recovery of $^{15}N$-labeled urea in wetland rice. Plant and Soil 66, 57–67; Katyal, J.C., Singh, B., Vlek, P.L.G., Craswell, E.T., 1985. Fate and efficiency of nitrogen fertilizers applied to wetland rice II. Punjab, India. Fertilizer Research 6, 279–290.

**Table 6.2** Emission of Nitrous Oxide (Gg $N_2O$—N Year$^{-1}$) From Agricultural Soils of Different States of India for Base Years 1994 and 2000

| State | 1994 | 2000 | Emissions From Residue Burning (2000) |
|---|---|---|---|
| Chandigarh | 0.03 | – | – |
| Delhi | 0.10 | 0.05 | – |
| Haryana | 4.38 | 6.23 | 0.36 |
| Punjab | 7.69 | 11.09 | 0.82 |
| Uttar Pradesh | 15.53 | 20.98 | 0.89 |
| India | 79.94 | 114.46 | 3.68 |

Adapted from Bhatia, A., Pathak, H., Aggarwal, P.K., 2004. Inventory of methane and nitrous oxide emissions from agricultural soils of India and their global warming potential. Current Science 87, 317–324; Pathak, H., Aggarwal, P.K., Singh, S.D. (Eds.), 2012. Climate Change Impact, Adaptation and Mitigation in Agriculture: Methodology for Assessment and Application. Indian Agricultural Research Institute, New Delhi (302 pp.).

soil organic matter and crop and animal residues, fertilizer N not used by crops, and to a lesser extent from atmospheric deposition. The most important environmental factors that influence $NO_3^-$ leaching include rainfall, evaporation, and temperature. During rainy season as much as 50% of the N applied to maize is leached below a depth of 50 cm and about 50% is leached below 150 cm (Fig. 6.6). Results of a 4-year study using $^{15}N$ in Ludhiana also showed that leaching losses were higher in rice (45–69 kg N ha$^{-1}$) than in wheat (16–22 kg N ha$^{-1}$) (Pathak et al., 2009). High rainfall during the crop season can

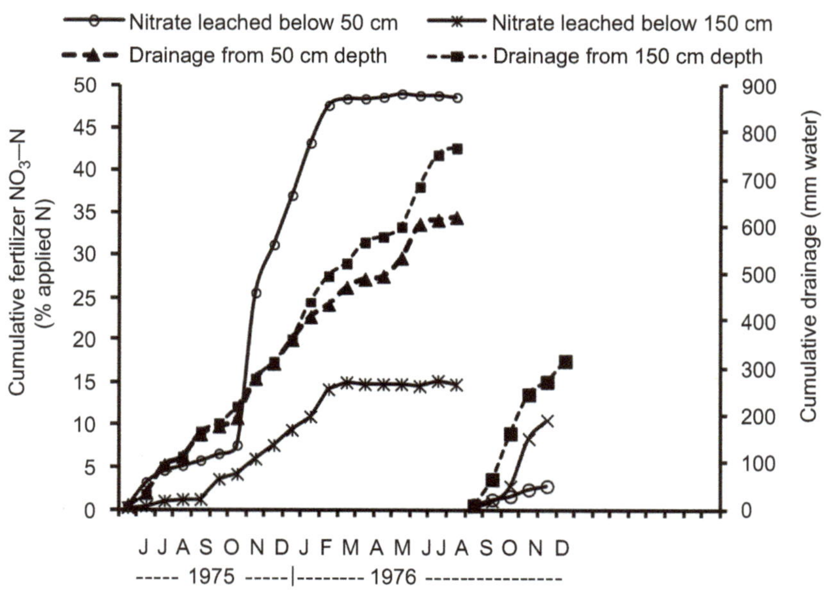


**FIGURE 6.6** Cumulative drainage and fertilizer $NO_3$—N leached below 50- and 150-cm soil depth in maize grown during monsoon season in a coarse-textured soil in New Delhi, India. Alphabets on *x*-axis represent months starting with June. *Adapted from Arora, R.P., Sachdev, M.S., Sud, Y.K., Luthra, V.K., Subbiah, B.V., 1980. Fate of fertilizer nitrogen in a multiple cropping system. In: Soil Nitrogen as Fertilizer or Pollutant. International Atomic Energy Agency, Vienna, pp. 3–22.*

**Table 6.3** Leaching Losses of Nitrogen in Relation to Irrigation Water Management in Wheat at Ludhiana, India

| IW/PAN-E | Leaching (kg N $ha^{-1}$) | Leaching (% of Applied N) | Residual $NO_3$–N (kg $ha^{-1}$) |
|---|---|---|---|
| **Season-I (Crop Sheltered From Rain)** | | | |
| 1.2 | 7.9 | 6.6 | 77.8 |
| 0.8 | 8.9 | 7.4 | 86.2 |
| 0.4 | 11.4 | 9.5 | 93.9 |
| **Season-II (High Rainfall)** | | | |
| 1.2 | 20.3 | 16.9 | 41.2 |
| 0.8 | 18.3 | 15.3 | 44.0 |
| 0.4 | 21.9 | 18.3 | 43.2 |

Adapted from Benbi, D.K., Prihar, S.S., Cheema, H.S., 1991b. A model to predict changes in soil moisture $NO_3$–N content and N uptake by wheat. Fertilizer Research 28, 73–84.

exacerbate leaching losses of $NO_3$–N in wheat (Benbi et al., 1991b). In the absence of rainfall (season-I) the leaching losses comprised less than 10% of the fertilizer N but in a year with high rainfall (season-II) the leaching losses increased up to 18% of the applied N (Table 6.3).

Among the soil factors, texture and structure interact to influence the leaching of $NO_3^-$. Generally, $NO_3^-$ leaches rapidly from sandy than silt and clay soils. Fertilizer-N application greatly influences $NO_3$ leaching in soil with losses increasing proportionally with the rate of fertilizer-N application (Benbi, 1990; Benbi et al., 1991a). In a long-term experiment, Benbi et al. (1991a) found that in a 2.1 m soil profile after 16 cycles of corn–wheat–cowpea cropping, the amount of residual $NO_3$–N was directly related to the amount of fertilizer-N, P, and K application (Fig. 6.7). The $NO_3$–N content in the soil

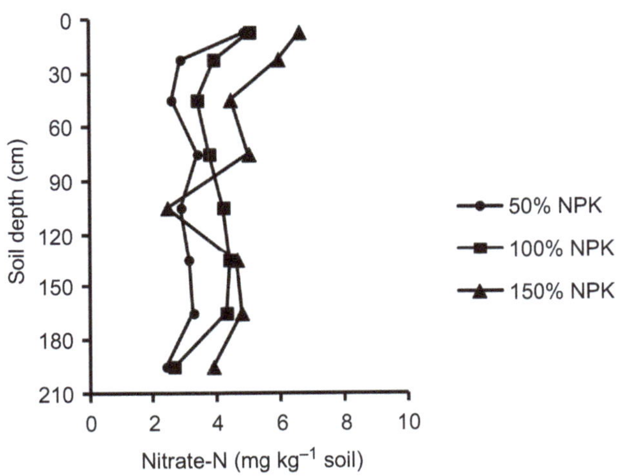


**FIGURE 6.7** Influence of fertilizer application rates on residual $NO_3$–N in soil profile after 16 cycles of corn–wheat–cowpea (fodder) cropping. *Adapted from Benbi, D.K., Biswas, C.R., Kalkat, J.S., 1991a. Nitrate distribution and accumulation in an Ustochrept soil profile in a long-term fertilizer experiment. Fertilizer Research 28, 173–178.*

profile was lowest when half (50% NPK) of the recommended amount of NPK was applied, while the highest amount accumulated under 1.5 times (150% NPK) of the recommended application of fertilizers. Analysis of existing information from these states showed that leaching losses could account for 10–15% of applied N in a rice–wheat system. The $NO_3$ leaching is likely to be higher in the agricultural systems based, to a greater extent, on organic inputs than those in which a larger proportion of the crop's N requirement is met from optimum and well-timed application of inorganic fertilizers.

Measures to reduce $NO_3^-$ leaching include optimal and balanced fertilization, synchronizing N supply with plant demand, manipulation of water applications and rooting depth, appropriate cropping sequence, use of cover crops, and the use of slow-release fertilizers and nitrification inhibitors (Benbi et al., 1991a; Prihar et al., 2000). Benbi et al. (1991a) showed that balanced application of N, P, and K under intensive agriculture can significantly reduce the amount of unutilized nitrates in the soil profile by enhancing N recovery in crop plants. The N recovery averaged 25% for annual application of N alone, 42% for N and P, and 56% for N, P, and K. All N treatments increased the residual soil $NO_3$–N, but the plots receiving N, P, and K where crop yields and N recovery were maximum, had the least residual $NO_3$–N in the 2.1 m soil profile; thus reducing the risk of $NO_3^-$ leaching. Development of suitable irrigation schedules with respect to timing and amount so as to synchronize the $NO_3^-$-rich zone with the moist zone of high-root activity can help efficient N uptake by plant. It has been shown (Pratt, 1984) that where roots have access to the entire soil solution, nitrate is not leached unless excess fertilizer N is added or the soil is over irrigated. Soil-water and nitrogen-dynamics models (e.g., Benbi et al., 1991b) could be used for an efficient on-farm $NO_3^-$ management under both irrigated and rainfed conditions. Slow-release fertilizers, such as urea-formaldehyde, isobutylidene diurea, sulfur-coated urea, control nitrification by slowing down the rate at which $NH_4^+$ is made available for nitrification. However, the use of slow-release fertilizers is limited because of their high cost and possible mismatch between nitrogen availability and crop demand. Nitrification inhibitors, such as dicyandiamide (DCD), N-serve, and calcium carbide, which retard the formation of $NO_3$ by nitrifying bacteria, are known to increase the fertilizer-use efficiency provided that maintaining the added fertilizer N as $NH_4^+$ does not lead to increased ammonia volatilization. The beneficial effect of nitrification inhibitors in reducing N-leaching losses depends on soil type and time and rate of N application.

## GroundWater Pollution

In selected locations in Punjab, Haryana, and western UP, $NO_3$–N concentration in dug-well water or shallow bore well water was several-folds higher than the upper safe limit (Handa, 1987). The average $NO_3$ content in dug-well water was higher in Haryana (99.5 mg $NO_3$ $L^{-1}$) than Punjab (55 mg $NO_3$ $L^{-1}$). However, in both the regions there was great intra-state variability. The $NO_3$–N concentration in some districts of UP is reported to range (mg $L^{-1}$) between 10 and 130 around Lucknow (Sehgal et al.,. 1989), 61 in

Aligarh, 54 in Agra, and 157 in Meerut (Handa, 1987). Bajwa et al. (1993) showed that $NO_3$–N concentration in bore well water in different development blocks of Punjab (n = 239) was well below (average 3.62 ± 1.52) the desired limit in spite of high fertilizer-N use ranging between 151 and 258 kg ha$^{-1}$. However, based on extent of groundwater development, average fertilizer-N consumption (118–163 kg N ha$^{-1}$), and average $NO_3$ in groundwater (55–100 mg L$^{-1}$), Punjab and Haryana have been placed in high-risk category and UP in medium-risk category (Agarwal, 1999). An analysis (Malik, 2000) of the $NO_3$ content of groundwater samples collected by Punjab and Haryana State Ground Water Boards showed that more than 33% of the samples in the two states had $NO_3$ level above the permissible limit (10 mg $NO_3$–N L$^{-1}$) for drinking purposes— about 17% samples had $NO_3$–N level greater than 22 mg L$^{-1}$. Apparently, the very high $NO_3$ content of the groundwater could be because of the sources other than fertilizer.

## Nitrogen Balances

A number of approaches have been used to quantify nutrient fluxes at different scales; the three main approaches include (1) soil-surface balance based on nutrient input from different sources and output or removal in plant parts, (2) farm-gate balance based on accounting for difference between nutrient content of farm inputs and outputs without consideration of nutrients cycling within the farm, and (3) soil-system balance approach where detailed information for all important components of N cycle is required. Soil-system balance, though requires information on various components of N cycle, is the most comprehensive approach and could be used to develop strategies for efficient management of nitrogen. Several studies have quantified different components of N cycle at the field scale but were not extrapolated to develop nitrogen balances at the farm or the regional scales. Results from $^{15}$N studies in the study region showed that only about 21–52% of the applied N is recovered in plants and 16–50% remains unaccounted (Table 6.4). It points toward significant N losses in rice–wheat system.

Preliminary estimates of apparent nitrogen balances based on fertilizer input and crop removal in 1998–99 showed surplus of N in the study region (Table 6.5). This balance sheet neither included nitrogen input from sources other than chemical fertilizers nor considered nitrogen leakages. It has been estimated (Prasad et al., 2004) that in Indian agroecosystems, livestock manure constitutes a major proportion of total inputs

**Table 6.4** Balance Sheet of $^{15}$N in Rice, Wheat and Rice–Wheat Systems

| Crop/Cropping System | Location | Recovery Efficiency (%) | | | References |
|---|---|---|---|---|---|
| | | **Plant** | **Soil** | **Unaccounted** | |
| Rice | Ludhiana | 21–31 | 24 | 46–50 | Katyal et al. (1985) |
| Wheat | Ludhiana | 32–52 | 32–35 | 16–42 | Katyal et al. (1987) |
| Rice–wheat | New Delhi | 36 | 26 | 28 | Goswami et al. (1988) |

**Table 6.5** Apparent Nitrogen Balances ('000 t) for Original Green Revolution States of India in 1998–99

| State | Addition | Removal | Balance |
|---|---|---|---|
| Haryana | 597 | 362 | 235 |
| Punjab | 1081 | 589 | 492 |
| Uttar Pradesh | 2387 | 1497 | 889 |
| India | 10,923 | 9613 | 1310 |

Adapted from Tandon, H.L.S., 2007. Soil nutrient balance sheets in India: importance, status, issues, and concerns. Better Crops—India 1, 15–19.

(44.1%), followed by inorganic fertilizer (32.5%), atmospheric deposition (11.9%), and BNF (11.6%). Considering these sources, Prasad et al. (2004) computed soil-surface nitrogen balances for different states of India for the period of 2000–1. About 35.4 Tg of nitrogen was estimated as input from different sources, with output of nitrogen in harvested crops of about 21.2 Tg. The soil-surface nitrogen balance for agricultural lands of India was found to be surplus by about 14.4 Tg or 54 kg $ha^{-1}$. For the same base year (2000–1), Pathak et al. (2010) estimated a much lower surplus of 1.4 Tg N for the entire country (Table 6.6). In both the studies, high nitrogen surplus was found in the intensively cultivated states of Punjab and Haryana, while Haryana exhibited surplus N balance. Punjab had an average annual surplus of 12.2 kg $ha^{-1}$ of gross-cropped area compared to 10.7 kg $ha^{-1}$ in Haryana. These surface balances are based on a number of assumptions and there are large uncertainties, which must be considered carefully. None of these studies considered indigenous soil N supply or mineralization–immobilization turnover while computing nitrogen balances. The amount of N derived from indigenous resources during a crop cycle ranges between 30 and 100 kg N $ha^{-1}$. Further, as the mineral soils have almost constant C/N ratio, management-induced changes in soil C and N influence the soil N balance. It has been shown that 1 ton of organic C in the plow layer can substitute 4.75 kg fertilizer N (Benbi and Chand, 2007). The fertilizer-N use can be increased by improving recovery in crop plants and reducing the amount of N lost from the system. This will obviously minimize nitrogen surpluses.

**Table 6.6** Soil-Surface Nitrogen Balance for Original Green Revolution States of India in 2000–2001

| | Input (Gg) | | | | Output (Gg) | | |
|---|---|---|---|---|---|---|---|
| State | Fertilizer | Manure | Fixation | Deposition[a] | Crop Removal | Loss[b] | Balance (Gg) |
| Haryana | 714.3 | 38.0 | 59.5 | 89.0 | 403.4 | 432.0 | 65.4 |
| Punjab | 1018.5 | 212.1 | 58.9 | 112.6 | 632.3 | 672.6 | 97.2 |
| Uttar Pradesh | 2282.0 | 69.0 | 259.0 | 400.2 | 12.39.6 | 1416.7 | 354.0 |
| India | 10,823 | 1413 | 2357 | 2311 | 7707 | 7783 | 1414 |

[a]N addition through irrigation and rainfall.

[b]Loss through leaching, volatilization and denitrification assumed 50% of N input.

Adapted from Pathak, H., Mohanty, S., Jain, N., Bhatia, A., 2010. Nitrogen, phosphorus and potassium budgets in Indian agriculture. Nutrient Cycling in Agroecosystems 86, 287–299.

## Future Scenario

The increase in food-grain production with nitrogen consumption during the green revolution period highlights the critical role of fertilizers in ensuring national food security and reiterates what Borlaug said "If high yielding varieties were the catalyst that ignited the green revolution then chemical fertilizers were the fuel that powered its forward thrust". In order to meet the food and nutritional requirement of growing population, India will need to produce 300 mt of food grain—an increase of about 40 mt, by the year 2020. Based on the current figure for partial factor productivity of N an additional 4 mt of nitrogen may be required. Besides environmental effects, it will have significant economic implications. It is important that in addition to chemical fertilizers, alternative sources of nitrogen, such as manure, biological fixation, are tapped.

Overexploitation of groundwater and declining water table in the original green revolution states are also the causes of concern. The trend is likely to continue in future until remedial measures are taken. Water table in central Punjab is projected to fall to 34.2 m in 2016 and to 42.5 m in 2023 from 22.8 m in 2006. Therefore, agricultural production in Punjab, Haryana, and western UP might not be sustainable unless major steps are taken for improved management of groundwater. Water management in these states will be a key factor for ensuring national food security. Rice is a high water and energy consumer crop and puts severe stress on natural resources. Forceful efforts will have to be made to improve water management and develop short-duration, low water-requiring varieties so as to harmonize productivity, environment, and natural-resource conservation. Diversifying to some other low water–requiring crops, such as pulses, oilseeds, in addition to saving water could minimize environmental footprint. Government of India has already chalked out a program on crop diversification in Haryana, Punjab, and western UP (GOI, 2013–14). As per the plan overexploited and critical-development blocks of major rice-growing districts in each state will be identified and at least 5% of area under rice in these blocks would be diverted toward alternate crops. The targeted area for diversion of rice during 2013–14 was 140, 60, and 80,000 ha in Punjab, Haryana, and western UP, respectively.

## References

Agarwal, G.D., 1999. Diffuse agricultural water pollution in India. Water Science and Technology 39, 33–47.

Arora, R.P., Sachdev, M.S., Sud, Y.K., Luthra, V.K., Subbiah, B.V., 1980. Fate of fertilizer nitrogen in a multiple cropping system. In: Soil Nitrogen as Fertilizer or Pollutant. International Atomic Energy Agency, Vienna, pp. 3–22.

Bajwa, M.S., Singh, B., Parminder, S., 1993. Nitrate pollution of ground water under different systems of land management in Punjab. In: Proceedings First Agricultural Science Congress. National Academy of Agricultural Sciences, New Delhi, India, pp. 223–230.

Benbi, D.K., 1990. Efficiency of nitrogen use by dryland wheat in a sub-humid region in relation to optimizing the amount of available water. Journal of Agricultural Science Cambridge 115, 7–10.

Benbi, D.K., Biswas, C.R., Kalkat, J.S., 1991a. Nitrate distribution and accumulation in an Ustochrept soil profile in a long-term fertilizer experiment. Fertilizer Research 28, 173–178.

Benbi, D.K., Chand, M., 2007. Quantifying the effect of soil organic matter on indigenous soil N supply and wheat productivity in semiarid sub-tropical India. Nutrient Cycling in Agroecosystems 79, 103–112.

Benbi, D.K., Prihar, S.S., Cheema, H.S., 1991b. A model to predict changes in soil moisture $NO_3$–N content and N uptake by wheat. Fertilizer Research 28, 73–84.

Bhatia, A., Pathak, H., Aggarwal, P.K., 2004. Inventory of methane and nitrous oxide emissions from agricultural soils of India and their global warming potential. Current Science 87, 317–324.

Borlaug, N.E., 2007. Sixty-two years of fighting hunger: personal recollections. Euphytica. http://dx.doi.org/10.1007/s10681-007-9480-9.

Bouwman, A.F., 1996. Direct emissions of nitrous oxide from agricultural soils. Nutrient Cycling in Agroecosystems 46, 53–70.

Fertilizer Association of India, 2015. Fertilizer Statistics 2014–15. Fertilizer Association of India, New Delhi, India.

Galloway, J.N., Cowling, E.B., Seitzinger, S.P., Socolow, R.H., 2002. Reactive nitrogen: too much of a good thing? Ambio 31, 60–63.

GOI [Government of India], 2013–14. Crop Diversification Program in Haryana, Punjab and Western Uttar Pradesh. Ministry of Agriculture, Department of Agriculture & Cooperation, Crop Division, Krishi Bhavan, New Delhi, India, 8 pp.

Goswami, N.N., Prasad, R., Sarkar, M.C., Singh, S., 1988. Studies on the effect of green manuring on nitrogen economy in a rice-wheat rotation using 15N technique. Journal of Agricultural Science Cambridge 11, 1413–1417.

Handa, B.K., 1987. Nitrate content of groundwater in India. Fertilizer News 32, 11–29.

Hira, G.S., 2009. Water management in northern states and the food security of India. Journal of Crop Improvement 23, 136–157.

Katyal, J.C., Singh, B., Vlek, P.L.G., Craswell, E.T., 1985. Fate and efficiency of nitrogen fertilizers applied to wetland rice II. Punjab, India. Fertilizer Research 6, 279–290.

Katyal, J.C., Singh, B., Vlek, P.L.G., Buresh, R.J., 1987. Efficient nitrogen use as affected by urea application and irrigation sequence. Soil Science Society of America Journal 51, 367–370.

Kumar, R., Nagarajan, S., 2004. Rice-wheat cropping system: food security and sustainability. Current Science 87, 1334–1335.

Malik, R.P.S., 2000. Agriculture and water quality in India towards sustainable management. In: Water Quality Management and Control of Water Pollution. FAO, Rome, Italy, pp. 73–85. Water Reports 21.

Pal, D.K., Bhattacharyya, T., Srivastava, P., Chandra, P., Roy, S.K., 2009. Soils of the Indo-Gangetic Plains: their historical perspective and management. Current Science 96, 1193–1202.

Pathak, H., Ladha, J.K., Singh, Y., Hussain, A., Hussain, F., Munankarmy, R., Gathala, M., Verma, S., Singh, U.K., Nguyen, M., 2009. Resource-conserving technologies in the rice-wheat system of South Asia: field evaluation and simulation analysis. In: Ladha, J.K., et al. (Eds.), Integrated Crop and Resource Management in the Rice–Wheat System of South Asia. International Rice Research Institute, Los Baños, Philippines, pp. 297–318.

Pathak, H., Mohanty, S., Jain, N., Bhatia, A., 2010. Nitrogen, phosphorus and potassium budgets in Indian agriculture. Nutrient Cycling in Agroecosystems 86, 287–299.

Pathak, H., Aggarwal, P.K., Singh, S.D. (Eds.), 2012. Climate Change Impact, Adaptation and Mitigation in Agriculture: Methodology for Assessment and Application. Indian Agricultural Research Institute, New Delhi, 302 pp.

Prasad, V.K., Badrinath, K.V.S., Yonemura, S., Tsuruta, H., 2004. Regional inventory of soil surface nitrogen balances in Indian agriculture, 2000–2001. Journal of Environmental Management 73, 209–218.

Pratt, P.F., 1984. Nitrogen use and nitrate leaching in irrigated agriculture. In: Hauck, R.D. (Ed.), Nitrogen in Crop Production. ASA, CSSA, SSSA, Madison, WI, pp. 319–333.

Prihar, S.S., Gajri, P.R., Benbi, D.K., Arora, V.K., 2000. Intensive Cropping: Efficient Use of Water Nutrients and Tillage. Food Products Press, New York, USA, 311 pp.

PSFC, 2013. Agriculture Policy for Punjab. Punjab State Farmers Commission, Chandigarh, India.

Rekhi, R.S., Meelu, O.P., Gupta, R.K., 1982. Lysimeter studies on recovery of $^{15}$N-labeled urea in wetland rice. Plant and Soil 66, 57–67.

SAH, 2014. Statistical Abstract Haryana, 2012–13. Department of Economic and Statistical Analysis Haryana, Chandigarh, India.

SAP, 2013. Statistical Abstract Punjab. Economic Advisor to Government of Punjab, Chandigarh, India.

Sehgal, U.K., Sehgal, R.K., Kabar, Y.P., 1989. Nitrate pollution of ground water in Lucknow area. In: Proceedings of International Workshop on Appropriate Methodologies for Development and Management of Ground Water Resources in Developing Countries. National Geophysical Research Institute, p. 31. Report No 95, EE 52.

Tandon, H.L.S., 2007. Soil nutrient balance sheets in India: importance, status, issues, and concerns. Better Crops—India 1, 15–19.

# 7

# Efficient Nitrogen Management Under Predominant Cropping Systems of India

B.S. Dwivedi, V.K. Singh, M.C. Meena
*ICAR-INDIAN AGRICULTURAL RESEARCH INSTITUTE, NEW DELHI, INDIA*

**CHAPTER OUTLINE**



The Indian Nitrogen Assessment. http://dx.doi.org/10.1016/B978-0-12-811836-8.00007-0

## Introduction

Majority of Indian soils are inherently poor in nitrogen (N) supply due to their geographic location in subtropical/tropical climates that do not encourage maintenance and accretion of soil organic matter. Sustainable and economically viable crop production in these N-deficient areas, therefore, depends largely on N application in adequate amounts through external sources, in addition to use of other agri-inputs and adoption of appropriate crop-management practices. The dynamic nature of N and its propensity for loss from soil–plant systems further creates a unique and challenging environment for its efficient management. Recovery of applied N in crop plants is extremely low (usually less than 50%), and its enhancement continues to be a major challenge. In fact, fertilizer N recovery efficiency by the first crop is 30–50%. The rest of N either remains in the soil, the recovery of which in the following crops is very limited, or it is lost from the soil–plant system, causing serious disruptions in ecosystem functions (Ladha et al., 2005). Application of N through fertilizers and manures in adequate amounts and its judicious management, therefore, assume great significance for sustaining high productivity on one hand and curbing environment degradation on the other.

Consumption of fertilizer N in India increased from 0.21 million tons (mt) during 1960–61 to 16.9 mt during 2014–15 (FAI, 2015). This increase in fertilizer N consumption is very well correlated with the food grain production during this period. The fertilizer N consumption statistics, however, revealed the following: (1) pace of increase in N consumption is much faster in India compared to that of the world. During 2012–13, 16.8 mt of fertilizer N was used in India which was 15.2% of the world N consumption (FAI, 2015); (2) fertilizer–nutrient consumption ratio ($N:P_2O_5:K_2O$) continued to be too wide (8.0:2.7:1 during 2013–14) as against generally accepted ratio of 4:2:1. This ratio is still wider for high productivity areas such as Punjab, Haryana, and Uttar Pradesh; and (3) diagnostic surveys conducted in intensively cropped areas of Indo-Gangetic plains indicated that farmers' fertilizer practice was skewed toward N with the neglect of K, S, and micronutrients (Dwivedi et al., 2001; Singh et al., 2013).

The N use efficiency (NUE) is directly correlated with water management, N application rate and scheduling, fertilizer N source, and the chemical transformation that take place after N application. In intensively cultivated areas, increasing use of N in order to sustain high yields causes increased N losses to the environment in the form of $NO_3^-$ leaching beyond root zone, $NH_3$ volatilization, surface runoff, nitrification, and denitrification (Dwivedi et al., 2003; Singh et al., 2005). Release of $N_2O$ from soils during denitification contributes to the global warming (Pathak, 2016). Of the total anthropogenic inputs of all forms of N into the global N cycle, about 55% is contributed through fertilizer N losses (Galloway et al., 1995). Therefore, fertilizer N use is considered as the major contributing factor to gaseous N losses and $NH_4^+$ in rain water.

This scenario calls for an efficient use of N and not the increase in application rate, per se, in crops and more specifically in intensive and high-yielding cropping systems. Major cropping systems followed in India are illustrated in Fig. 7.1, which indicate rice–wheat as the most widely practiced cropping system.

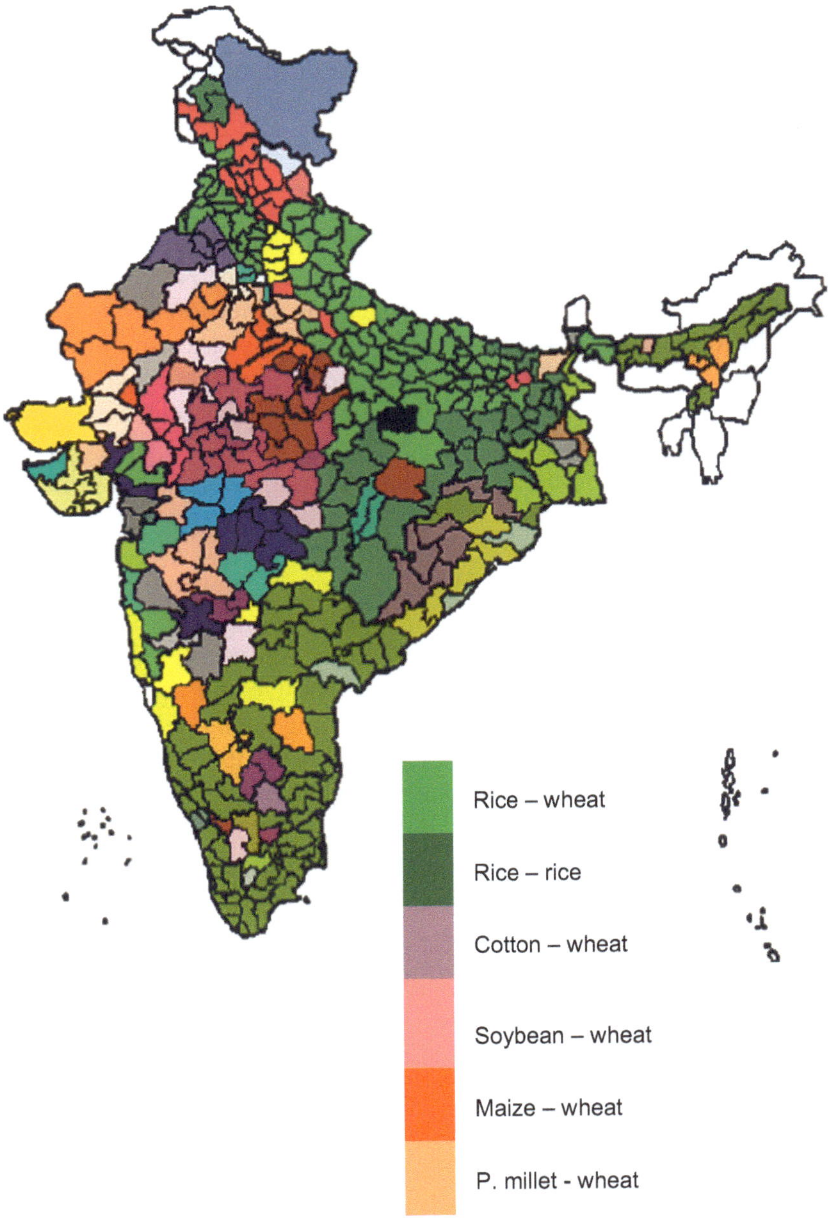


**FIGURE 7.1** Major cropping systems in India.

## Soil N: Transformation and Fertility Status

In soil, N is present as a part of soil organic matter which directly or indirectly regulates the transformation of native and applied N. Nitrogen applied through fertilizers or manures is the major contributor to plant N uptake. The portion of N not consumed by the plant is retained in soil organic matter that gets mineralized subsequently by microbial action and becomes available for plant uptake as also for various pathways of N losses. The N mineralization process starts with formation of ammonia ($NH_3$) out of organic forms of N, which gets converted to ammonium ion ($NH_4^+$). Further conversion of $NH_4^+$ through essentially microbe-mediated process depends largely on soil conditions, especially moisture regime. In an aerobic environment prevailing in upland conditions, $NH_4^+$ is oxidized to nitrite ($NO_2^-$) and finally to nitrate ($NO_3^-$) by nitrification. The $NO_3^-$ ions are taken up by the crops as also lost from root zone by "leaching." Flooding of these soils accentuates the loss of $NO_3^-$ present in the soil through denitrification and leaching. Under anaerobic conditions, $NO_3^-$ gets reduced to $NO_2^-$, then to nitrous oxide ($N_2O$), and finally to $N_2$ gas. Thus, mineralized organic N that undergoes nitrification under aerobic soil conditions gets lost to denitrification before its uptake by the crop. On the other hand, $NH_4^+$ under flooded condition continues to remain available for plant use for longer period.

Ammonium formed from fertilizer or organic matter incorporated into the puddled, anaerobic layer of flooded field is held relatively tightly. Therefore, organic N mineralized after flooding plays important role in determining the rate and time of fertilizer N application. In flooded soils, however, $NH_4^+$ ions present in water are converted to $NH_3$ due to change in water pH resulting from algal activities and lost to the atmosphere ($NH_3$ volatilization). Nevertheless, $NH_4^+$ is attached to the soil cation exchange complex and held by the soil very close to its point of placement. On the contrary, $NO_3^-$ is highly mobile in the soil and thus susceptible to leaching along with the moving water. Therefore, $NH_4^+$ fertilizer is preferred for rice crop to prevent denitrification and leaching.

Information on soil fertility status as generated by soil testing labs is compiled from time to time, which reveals widespread deficiencies of N in Indian soils. In all the compilations, beginning from 1969, more than 90% of Indian soils were categorized as low to medium in available N status (Table 7.1).

## Nitrogen Application and Its Removal Under Different Cropping Systems

Continuous application of suboptimal/unbalanced rates of nutrients to intensive cropping systems, such as rice–wheat, rice–rice, and maize–wheat, leads to lowering crop yields, nutrient use efficiency, and economic returns, in addition to encouraging nutrient mining and other environmental implications. By and large, nutrient input through fertilizers and organics remained far below the removal under high-yielding cropping systems. Results from long-term experiments clearly showed annual removal

**Table 7.1** Nitrogen Fertility Status of Indian Soils as Evaluated Time to Time on the Basis of STL Databases

| | | % Districts in Different N Fertility Classes | | | |
|---|---|---|---|---|---|
| **Year** | **Districts Covered** | **Low** | **Medium** | **High** | **Source** |
| 1969 | 184–226 | 52 | 43 | 5 | Ramamoorthy and Bajaj (1969) |
| 1980 | 365 | 62 | 33 | 5 | Ghosh and Hasan (1980) |
| 2002 | 450 | 63 | 26 | 11 | Motsara (2002) |
| 2011 | 500 | 57 | 36 | 7 | Muralidharudu et al. (2011) |

**Table 7.2** Annual N Addition and Removal Under Important Cropping Systems

| | Addition (kg ha$^{-1}$) | | | | | | |
|---|---|---|---|---|---|---|---|
| **Cropping Systems** | **Fertilizer** | **Other Sources**[a] | **Total** | **Removal (kg ha$^{-1}$)** | **Apparent Balance (kg ha$^{-1}$)** | **Output: Input Ratio** | **Source** |
| Rice–wheat | 300 | 11 | 311 | 244 | 67 | 0.78 | |
| Maize–wheat | 240 | 7 | 247 | 246 | 1 | 0.99 | |
| Pigeonpea–wheat | 135 | 51 | 186 | 260 | −74 | 1.40 | |
| Sesamum–wheat | 155 | 8 | 163 | 149 | 14 | 0.91 | Singh et al. (2015a) |
| Groundnut–wheat | 140 | 33 | 173 | 229 | −56 | 1.33 | |
| Sorghum–wheat | 240 | 9 | 249 | 183 | 66 | 0.73 | |
| Rice–maize | 325 | 19 | 344 | 356 | −12 | 1.04 | |
| Rice–chickpea | 140 | 16 | 156 | 252 | −97 | 1.62 | |
| Rice–lentil | 140 | 16 | 156 | 221 | −65 | 1.42 | |
| Rice–pea | 140 | 16 | 156 | 202 | −46 | 1.30 | |
| Rice–wheat–maize + cowpea fodder | 300 | 21 | 321 | 260 | 62 | 0.81 | Singh et al. (2015b) |
| Rice–chickpea – maize + cowpea fodder | 200 | 20 | 220 | 291 | −71 | 1.32 | |

[a]Includes nutrient addition through rainfall, irrigation water and root + stubbles.

of N beyond 250 kg ha$^{-1}$ in some intensive cropping systems (Table 7.2). Considering present NUE, and N fertility status of soils, fertilizer N application even at recommended rate in most of these systems is not sufficient to meet the crop N demand.

Unbalanced nutrient application by the farmers makes the situation further alarming. Recent surveys conducted in the upper-Gangatic plains indicate that farmers have developed a tendency to use excessive N (because N is comparatively cheaper than P and K), optimal to suboptimal P, and nil or very low K (Fig. 7.2) with sheer neglect of secondary and micronutrients. Only rice farmers apply Zn to rice, that too at suboptimal rate. The indiscriminate use of N not only widens the fertilizer consumption ratio but also accentuates nonsustainability issues by way of promoting nutrient imbalances in soil–plant system, decreasing NUE, enhancing groundwater pollution through $NO_3^-$ leaching, and escalating the cost of production.

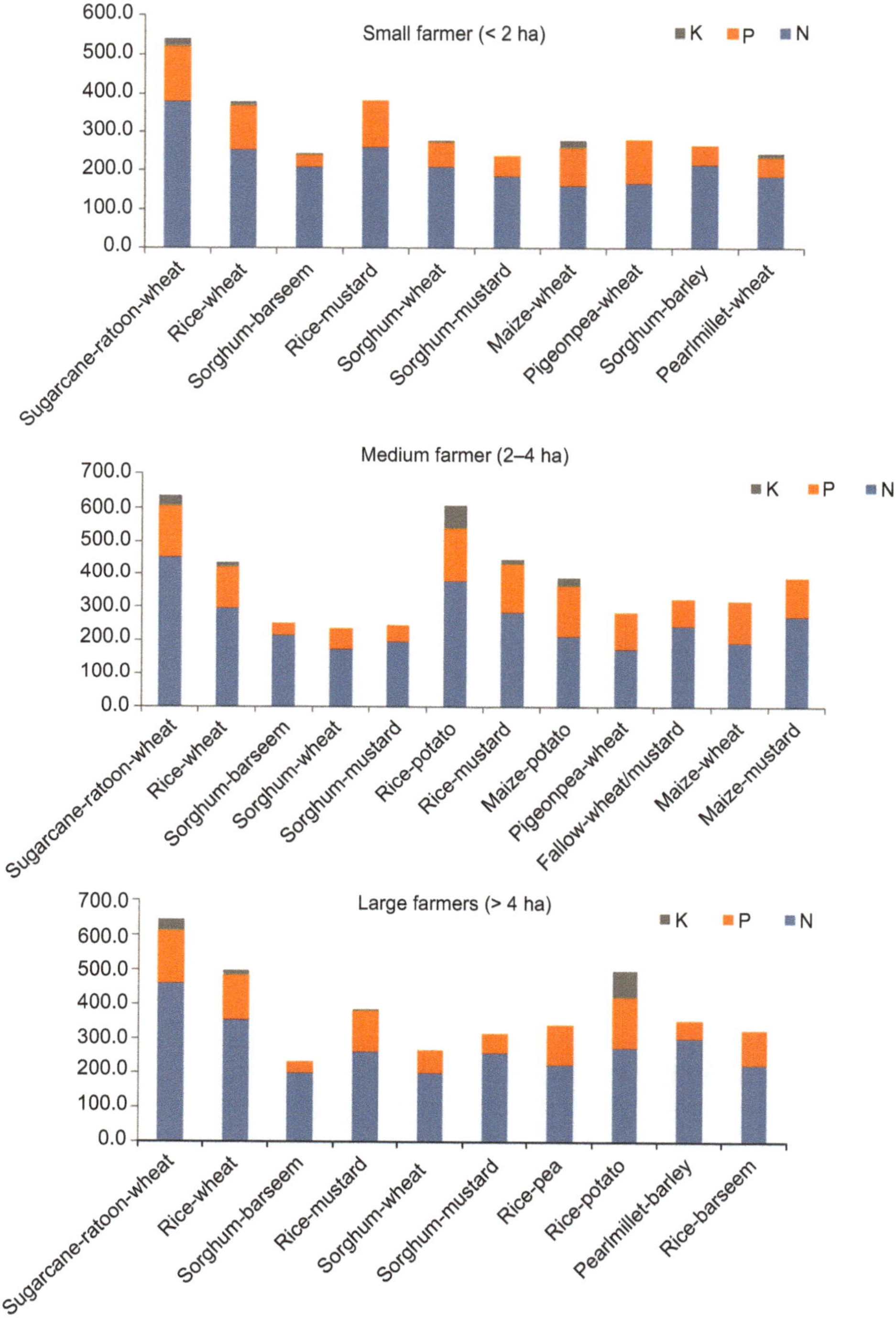


**FIGURE 7.2** Fertilizer use (kg $ha^{-1}$) under different cropping systems in the upper Gangetic plains. *Singh, V.K., 2015. Precision Nutrient Management Using GIS-based Spatial Variability Mapping under Upper and Middle Gangetic Plains of India. ICAR-National Fellow Final Report. Indian Institute of Farming Systems Research, Modipuram. pp. 1–130.*

## Nitrogen Use Efficiency Computations

The use efficiency is measured as output per unit use of an input. The output may be yield, nutrient uptake, economic returns, etc. The computations of NUE often exclude N input from organic manures and consider only fertilizer N input. Some of the most common measures of NUE are computed as under:

- *Agronomic efficiency ($AE_N$; kg kg$^{-1}$ of fertilizer N):* It is the ratio of incremental yield due to fertilizer N input and amount of fertilizer N applied.
- *Recovery efficiency ($RE_N$; %):* It is the ratio of incremental N uptake due to fertilizer N and amount of fertilizer N applied.
- *Partial factor productivity ($PFP_N$, kg kg$^{-1}$ of fertilizer N):* It is the ratio of yield under fertilizer N and amount of fertilizer N applied.

The values of yield, incremental yield or uptake, and fertilizer N are expressed as kg ha$^{-1}$. $PFP_N$ does not take into account the control (no N) yield, and is, therefore, a very useful measure of NUE in the experiments wherein a control plot is not maintained. When $^{15}N$ is used, the recovery efficiency is considered as the true recovery efficiency ($TRE_N$) (Prasad, 2013).

## Management Practices—to Enhance NUE

Research on improving NUE during past decades focused on a better synchronization between the N supply and crop N demand as well as minimizing N losses. Nonetheless, NUE continues to be extremely low, and is presently one of the priority areas for soil fertility and fertilizer management research. The need for intensifying research in this area is presently felt more than ever before due to: (1) projections indicating that 50–70% additional cereal grains will be required by 2050, which would necessitate an increased use of N of similar magnitude if the NUE remains unchanged; (2) bringing urea under nutrient-based subsidy would eventually lead to its cost escalation, resulting in substantial increase in the cost of cultivation due to inefficient use of N; and (3) increased GHG emissions through indiscriminate fertilizer applications.

Extensive on-station and on-farm studies revealed following as the major strategies to ascertain efficient N management in cropping systems:

- Balanced fertilization
- Integrated plant nutrient supply
- Inclusion of legumes
- Precision N management
- N management under conservation agriculture

## Balanced Fertilization

Yield-response studies on different nutrients were conducted across the states under on-farm experiments of AICRP-IFS, wherein all cereal-based systems responded positively to addition of recommended N, P, and K fertilizers (Table 7.3). Across the agro-ecological regions, rice–rice recorded higher rice equivalent yield (REY) of 11,439 kg $ha^{-1}$ with recommended quantity of NPK followed by rice–wheat (8959 kg $ha^{-1}$), rice–greengram (8103 kg $ha^{-1}$), and maize–wheat (8069 kg $ha^{-1}$) systems. Application of recommended nutrients to maize-wheat system recorded 162% yield increase over control (unfertilized) followed by 95% each in rice-wheat and rice–rice systems. The rice–greengram system recorded 72% yield increase with balanced fertilization. Similarly, application of recommended P and K along with N registered an increase in REY to the tune of 55.6, 51.9, 35.7% and 33.5% in maize–wheat, rice–rice, rice–greengram, and rice–wheat systems, respectively, compared to the application of N alone to these systems. On an average, an additional yield (REY) of 2794 kg $ha^{-1}$ can be obtained in cereal-based systems with the application of recommended rates of P and K along with N instead of application of N alone, which continues to be a common practice under larger area of the country. The efficiency parameters measured in terms of productivity, agronomic efficiency, and relative yield response were higher under balanced NPK fertilization. Such improvement

**Table 7.3** Effect of Balanced Fertilization on Agronomic Efficiency of N ($AE_N$) Under Different Rice-Based Cropping Systems

| | | Monsoon Crop | | | | Winter Crop | | | |
|---|---|---|---|---|---|---|---|---|---|
| | | $AE_N$ | | | | $AE_N$ | | | |
| **Cropping System/Agro-Climate Region** | **No of Trials** | **N** | **NP** | **NK** | **NPK** | **N** | **NP** | **NK** | **NPK** |
| **Rice–Rice System** | | | | | | | | | |
| Eastern Himalayan region | 299 | 11 | 16 | 22 | 29 | 13 | 18 | 24 | 31 |
| Southern plateau and hills region | 608 | 11 | 20 | 17 | 28 | 8 | 12 | 16 | 30 |
| East coast plain and hill region | 211 | 9 | 15 | 13 | 23 | 9 | 13 | 14 | 26 |
| Western coast plain and ghat region | 105 | 10 | 21 | 21 | 29 | 7 | 16 | 17 | 23 |
| **Rice–Wheat System** | | | | | | | | | |
| Western Himalayan region | 138 | 10 | 22 | 16 | 28 | 7 | 14 | 11 | 18 |
| Eastern Himalya region | 312 | 9 | 16 | 14 | 22 | 9 | 14 | 13 | 17 |
| Middle Gangetic plain region | 345 | 12 | 23 | 17 | 29 | 8 | 15 | 10 | 18 |
| Upper Gangetic plain region | 234 | 13 | 21 | 18 | 25 | 11 | 16 | 14 | 19 |
| Trans Gangetic plain region | 260 | 4 | 13 | 6 | 17 | 1 | 7 | 3 | 10 |
| Eastern plateau and hills region | 229 | 3 | 9 | 7 | 13 | 3 | 6 | 5 | 10 |
| Central plateau and hills region | 192 | 10 | 22 | 16 | 28 | 7 | 14 | 11 | 18 |
| **Rice-Greengram System** | | | | | | | | | |
| Lower Gangetic plain region | 607 | 4 | 8 | 10 | 27 | 3 | 6 | 9 | 11 |
| East coast plains and hills | 129 | 6 | 9 | 11 | 30 | 3 | 5 | 6 | 9 |

AICRP-IFS, 2012. Annual Report (2011–12) of All India Coordinated Research Project on Integrated Farming System. PDFSR, Modipuram.

in NUE was attributed to increased indigenous nutrient supplying capacity of soil under balanced NPK fertilization (Yadav, 2001).

## Integrated Plant Nutrient Supply

Integrated plant nutrient supply (IPNS) involving conjoint use of different nutrient sources appeared a promising strategy for sustaining high yields, restoration of soil health, and improvement in fertilizer use efficiency as a whole including NUE (Singh et al., 2012; Das et al., 2015; Dwivedi et al., 2016). Roy and Ange (1991) defined IPNS as maintenance or adjustment of soil fertility and plant nutrient supply to an optimum level for sustaining the desired crop productivity through optimization of benefits from all possible resources of plant nutrients in an integrated manner. Goswami (1998) differentiated between integrated nutrient management (INM) and IPNS, and viewed INM as the technical and managerial component for achieving the objective of IPNS under farm situations. Multilocation, long-term experiments (LTEs) established that neither the fertilizers nor the organics in isolation can achieve sustainable production under intensive cropping. Combined use of these nutrient sources had an interactive advantage and proved invariably superior to their sole application (Hegde and Dwivedi, 1993). The IPNS involves monitoring all the pathways of plant nutrient supply in crops and cropping systems, and calls for a judicious combination of fertilizers, organic manures, and biofertilizers (Prasad, 2008). This approach is not new to India, as the pre-green revolution agriculture largely followed IPNS with major share of organics.

Important ingredients of IPNS are fertilizers, organic manures, green manures, crop residues (CR), industrial wastes and by-products, sewage-sludge, and biofertilizers. Despite proven benefits of IPNS, early decades of post-green revolution era did not witness its largescale adoption because of the availability of subsidized fertilizers, time and labor needed for preparation of farmyard manure (FYM) and composts, and higher handling cost of organic manures (Hegde and Dwivedi, 1993). There has been a revival of interest in IPNS since past two decades or so in order to rejuvenate soil health, optimize nutrient input, and enhance fertilizer use efficiency.

### *Bulky Organic Manures in IPNS*

At present, production of organic manures including FYM and composts is estimated at 229.4 mt (FAI, 2015). Less than half of the manurial potential of livestock is utilized, as a large proportion is lost in the form of fuel and droppings in nonagricultural areas. Out of the amount used for agricultural production, substantial quantity of nutrients is lost due to improper handling, storage, and incorporation in soil. Organic manures, in addition to supplying nutrients to the current crop, also leave residual effect on succeeding crops.

The LTEs under the agies of AICRP-IFS indicated that FYM could substitute a part of fertilizer N needs of *kharif* crop in cereal-based cropping systems. At some locations fertilizer needs of the *rabi* crop could be reduced to the extent of 25% by substituting N needs of preceding *kharif* crop through FYM. The NUE, computed as $PFP_N$ in these LTEs increased markedly under IPNS compared with fertilizer NPK application at

**Table 7.4** Recovery Efficiency of Applied N ($RE_N$) in Different Crops as Affected by Long-Term Nutrient Supply With and Without FYM

| | | | $RE_N$ (%) | | | |
|---|---|---|---|---|---|---|
| **Soil Type** | **Location** | **Crop** | **N** | **NP** | **NPK** | **NPK + FYM** |
| Inceptisol | Ludhiana | Maize | 16.7 | 23.5 | 36.4 | 40.2 |
| | | Wheat | 32.0 | 50.6 | 63.1 | 67.8 |
| Alfisol | Palampur | Maize | 6.4 | 34.7 | 52.6 | 63.7 |
| | | Wheat | 1.9 | 35.6 | 50.6 | 72.6 |
| Mollisol | Pantnagar | Rice | 37.5 | 40.7 | 44.4 | 61.7 |
| | | Wheat | 42.4 | 46.1 | 48.4 | 47.9 |

*N*, recommended N; *NP*, recommended N and P; *NPK*, recommended NPK to both crops; *NPK + FYM*, recommended NPK to both crops + FYM (15 t $ha^{-1}$ $yr^{-1}$ up to 2008–09 and 5 t $ha^{-1}$ $yr^{-1}$ thereafter) to *kharif* crop only.

recommended rates. The $PFP_N$ values for rice–rice system varied from 26 to 52 kg grain $kg^{-1}$ N under recommended NPK, which increased to 33–77 kg grain $kg^{-1}$ N with substitution of 25% N through FYM in *kharif* rice and curtailing 25% N in succeeding *rabi* rice. Similarly, $PFP_N$ in other cropping systems was invariably high under FYM, although the magnitude of such increase in $PFP_N$ varied among the cropping systems and the experimental sites (Dwivedi et al., 2016).

Another series of LTEs continuing under AICRP-LTFE since 1971–72 also showed substantial increase in recovery efficiency of applied N ($RE_N$) in maize, rice, and wheat grown on different soils (Table 7.4) (Singh et al., 2012). Besides, FYM and composts, industrial by-products such as spent-wash from distillery, molasses, pressmud etc. from sugar industry, and wastes from other food processing industries could also be included in the IPNS packages. Studies at IIFSR, Modipuram revealed that use of decomposed sulphitation pressmud (SPM) had positive effect on yield stabilization and $PFP_N$ under rice–wheat system. At IARI, integration of SPM with fertilizers proved superior to even FYM in enhancing wheat yield and N availability in the soil (Mandal et al., 2013). In recent years, SPM assumed great importance as an IPNS ingredient in sugarcane–ratoon–wheat and other intensive cropping systems of sugarcane growing areas.

### *Green Manures in IPNS*

Green manuring (GM) with legumes improves soil N via biological N-fixation. Sunnhemp (*Crotolaria juncea*) and *Dhaincha* (*Sesbania aculeata*) are the most common GM crops, although *Sesbania rostrata* has the highest atmospheric $N_2$-fixing potential, and it can completely substitute for urea-N in rice cultivation (Chaudhury and Kennedy, 2004). Crops, such as clusterbean, berseem, Indian clover, are also used occasionally for GM. Long-term studies under AICRP-IFS indicated the possibility to substitute up to 50% of fertilizer N needs of different cropping systems through GM without any yield loss. The $PFP_N$ computed for different cropping systems further revealed an increase by 50% during *kharif* season or 25% each during both seasons due to GM substitution of fertilizer N (Dwivedi et al., 2016). Despite these apparent benefits, area under GM is estimated only as 1.23 million ha (FAI, 2015). Hence, raising short-duration greengram in summer and incorporation of aboveground green biomass after pod picking could be followed as an alternative to conventional GM. Greengram residue incorporation proved as effective as *Sesbania* GM in terms of yield response of rice–wheat system (Hegde and Dwivedi, 1992; Dwivedi et al., 2002).

### *Inclusion of Legumes*

Inclusion of legumes for grain or fodder purpose improved the productivity of rice-wheat system and regenerated soil fertility (Mann and Garrity, 1994). Studies with several leguminous crops revealed that their inclusion in cropping system may meet N demand of succeeding nonlegume crop by 50–60 kg N $ha^{-1}$. Grain yield of cereal crops increased markedly with legume as preceding crop compared with the yield when a cereal crop preceded the main crop (Singh and Dwivedi, 2006). The effect of preceding crops, rice, or cowpea was evaluated in wheat on an Inceptisol, wherein the NUE indices ($AE_N$ and $RE_N$) were significantly greater in wheat following cowpea (fodder or grain crop) compared with those following rice (Yadav et al., 2003; Dwivedi et al., 2003) (Table 7.5). In another study, substitution of rice with pigeon pea enhanced N uptake and NUE in

**Table 7.5** N Use Efficiency in Wheat Grown After Rice or Cowpea on an Inceptisol at Modipuram

| | Agronomic Efficiency ($AE_N$, kg Grain $kg^{-1}$ N) | | | Recovery Efficiency ($RE_N$, %) | | |
|---|---|---|---|---|---|---|
| **Treatment** | **I Year** | **II Year** | **Mean** | **I Year** | **II Year** | **Mean** |
| **After Rice** | | | | | | |
| $N_0$ | — | — | — | — | — | — |
| $N_{60}$ | 20.0 | 19.2 | 19.6 | 47.5 | 38.2 | 42.9 |
| $N_{120}$ | 12.7 | 14.4 | 13.6 | 33.9 | 26.2 | 30.0 |
| $N_{180}$ | 8.7 | 10.2 | 9.5 | 25.9 | 17.9 | 21.9 |
| Mean | 13.8 | 14.6 | — | 35.8 | 27.4 | — |
| **After Cowpea** | | | | | | |
| $N_0$ | — | — | — | — | — | — |
| $N_{60}$ | 33.3 | 34.3 | 33.8 | 50.7 | 61.2 | 56.0 |
| $N_{120}$ | 19.4 | 22.0 | 20.7 | 40.7 | 35.4 | 38.1 |
| $N_{180}$ | 12.8 | 14.4 | 13.6 | 31.2 | 23.4 | 27.3 |
| Mean | 21.8 | 23.6 | — | 40.9 | 40.0 | — |
| **S.E.** | | | | | | |
| Preceding crop (C) | 0.64 | 0.61 | — | 1.72 | 2.00 | — |
| N rates (N) | 0.79 | 0.75 | — | 2.11 | 2.45 | — |
| C × N | 1.12 | 1.06 | — | 2.98 | 3.47 | — |

Yadav, R.L., Singh, V.K., Dwivedi, B.S., Shukla, A.K., 2003. Wheat productivity and N use-efficiency as influenced by inclusion of cowpea as a grain legume in a rice-wheat system. Journal of Agricultural Sciences 141, 213–220.

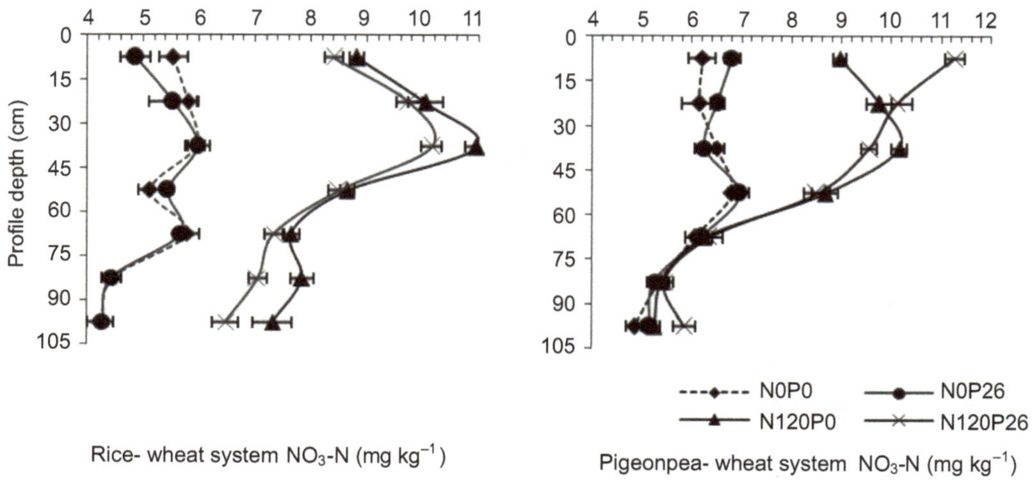


**FIGURE 7.3** Effect of NP fertilization and cropping system on the distribution of nitrate-N in soil profile after wheat harvest. Bars indicate standard error of mean, $n = 4$.

wheat (Singh et al., 2005), and thus helped curtailing fertilizer N rates. The advantage of inclusion of pigeon pea could be further enhanced by induced defoliation (ID) through foliar spray of urea. Studies at IARI revealed that the green foliage retained at maturity in extra short–duration varieties of pigeon pea could be recycled through ID using foliar spray of 10% urea solution (w/v in water) at physiological maturity, resulting in addition of 1–1.2 t $ha^{-1}$ additional leaf litter, thus adding more than 40 kg N $ha^{-1}$ (Meena et al., 2012; Mandal et al., 2013). Further, inclusion of a legume crop in rice–wheat system could minimize the adverse effects of continuous puddling on the soil compaction, thus ensuring better establishment of wheat and thereby enhancing nutrient absorption and NUE, and minimizing $NO_3^-$ leaching (Fig. 7.3) (Singh et al., 2005).

### *Crop Residues in IPNS*

About 141 million tons of cereal residues are available for recycling in to the soil, after meeting competitive demands. Farmers in north-west India often prefer in situ burning of rice residues which has serious environmental implications besides loss of nutrients. There are large amounts of residues of other crops such as potato, sugarcane, vegetables which are practically wasted in most cases. CR, in addition to supplementing nutrients, play significant role in improving soil properties and thereby increasing productivity and fertilizer use efficiency. Incorporation of CR as a component of IPNS to substitute 50% of fertilizer N improved the $PFP_N$ compared with sole fertilizer application in multilocation LTEs (Fig. 7.4). In order to minimize cereal CR-induced immobilization of N and other nutrients, CR incorporation may be supplemented with small amount of fertilizer N or alternatively CR could be composted using efficient microbial strains. Good quality compost can be obtained from the CR such as rice straw within a short time of 8–10 weeks.

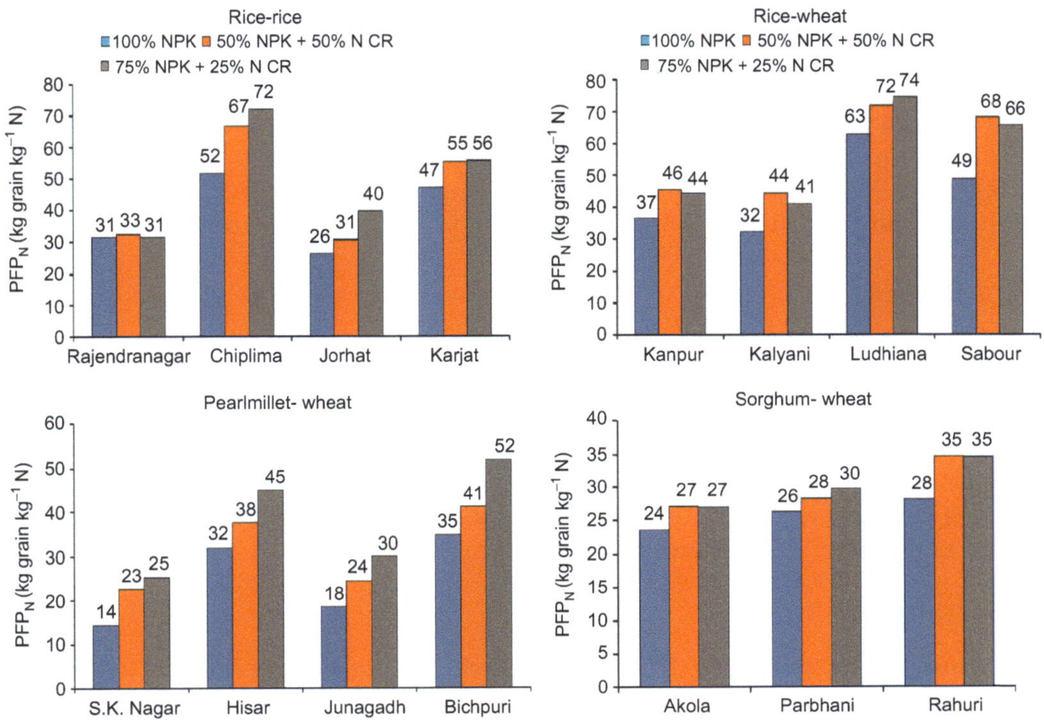


**FIGURE 7.4** Effect of INM involving crop residues (CR) on partial factor productivity of N ($PFP_N$) under different cropping systems.

### *Biofertilizers in IPNS*

Biofertilizers can help curtailing fertilizer N requirement of crops. While legume-*Rhizobium* symbiosis can meet more than 80% of the N demand of legumes, *Azotobacter* and *Azospirillum* inoculants usually contribute 20–25 kg N $ha^{-1}$, respectively, under field conditions. On the other hand, average N-fixation through BGA and *Azolla* is estimated at 25–30 kg N $ha^{-1}$ and 30–40 kg N $ha^{-1}$, respectively (Hegde and Dwivedi, 1993). The N input through *Azolla* can be enhanced by its dual cropping. Inconsistent response to biofertilizers is often associated with their substandard quality and negligence during production, storage, transport, and usage. Over the years, production of biofertilizers has been constantly increasing in India from mere 2000 tons in 1992–93 to 65.5 ×1000 tons in 2013–14 (FAI, 2015).

## Precision N Management for Enhancing NUE

Precision N management (PNM) refers to N management strategies that encourage better utilization of applied N, enhance NUE and minimize N losses. Fertilizer N recovery efficiencies ($RE_N$) in researcher-managed experiments for major grain crops range from 46% to 65%, compared to on-farm $RE_N$ of 30–40%. The NUE is often highest at the lower parts of the yield-response curve where fertilizer inputs are lowest, but effectiveness of fertilizers in increasing crop yields and optimizing farmer profitability should not be sacrificed for the sake of NUE alone. There must be a balance between NUE and crop productivity. The NUE could be optimized by bringing in precision management following '4 R' principle, which implies right fertilizer source, right rate, right time, and right method of placement. The '4 Rs' are discussed in brief hereunder:

## Right Source

The source of fertilizer N plays significant role in regulating N losses and influencing NUE. Nitrogen fertilizers predominantly contain N in ammonium, nitrate, or amide form. Ammonium and amide form are more susceptible to volatilization loss than fertilizers containing $NO_3$-N (Ladha et al., 2005). Controlled-release fertilizers have the potential to reduce losses of N from the system. Efficient fertilizer N management involving nitrification inhibitors and modified urea materials has been documented (Prasad and Power, 1995; Prasad, 2013).

## Right Rate and Time

One approach to increasing NUE is to match the rate and timing of N fertilizer application with the N demand of the crop. Unlike fixed N-scheduling as usually prescribed, this approach requires in situ monitoring of crop N status so as to take a decision on N application. At least, three decision gadgets namely leaf color chart (LCC), chlorophyll meter (SPAD) and GreenSeeker are available for in situ monitoring of leaf N status. A chlorophyll meter can provide a quick estimate of the leaf N status, but it is relatively expensive. The LCC, on the other hand, is inexpensive, simple, and easy to use tool to monitor the relative greenness of leaf as an indicator of crop N status (Shukla et al., 2004; Yang et al., 2003). Use of LCC, however, requires determination of critical LCC values that simultaneously optimize the grain yield and NUE ($AE_N$ and $RE_N$) for a group of varieties exhibiting similar plant-type and growth duration (e.g., traditional long duration, semidwarf, short-duration, hybrid rice). Considering $AE_N$ of 20 kg $kg^{-1}$ N and $RE_N$ of 50% as optimum NUE for rice, critical LCC values were judged as: LCC $\leq$3 for Basmati, LCC $\leq$4 for inbred (Saket 4), and LCC $\leq$5 for Hybrid (PHB 71) rice (Shukla et al., 2004). Fertilizer N-scheduling based on LCC proved superior to conventional practice, i.e., application of 120 kg N $ha^{-1}$ in three-splits (Bijay-Singh et al., 2003; Shukla et al., 2004). In certain cases, even a saving of fertilizer N (up to 30 kg N $ha^{-1}$) was recorded with the use of LCC.

Besides LCC and SPAD, optical sensors could also be used for rationalizing N application rate and time. The N stress in plants could be measured by near infrared (NIR) response of the canopy (Ma et al., 1996). Spectral indices such as normalized-difference vegetation index may help predicting photosynthetic efficiency and productivity potential of a crop (Raun et al., 2001). Hand-held GreenSeeker optical sensors have been used to evaluate N status of rice and wheat to develop real-time N prescriptions (Bijay-Singh et al., 2011, 2015). Depending on growing conditions, site-specific real-time N management in rice could be adopted following (1) fixed threshold SPAD and LCC score, (2) dynamic threshold SPAD and LCC score, (3) optical sensor, and (4) fixed-time variable rate options (Bijay-Singh, 2016).

### Right Method

Fertilizer N application methods have direct bearing on its utilization by the crop and suitability of fertilizer placement strategy in a given situation depending on a number of factors, namely plant genetics, tillage practices, spacing, crop rotation, and weather variability (Majumdar et al., 2014). Ladha et al. (2005) suggested that the common practice of surface broadcasting N fertilizers could entail large N losses in rice, particularly through $NH_3$ volatilization from the system resulting in a reduction in the NUE. In contrast, deep placement of urea or USG improves NUE (Majumdar et al., 2016). Humphreys et al. (1992) noted that recovery efficiency of N was 37% for broadcasting, 46% for band placement, and 49% for deep placement of USG in direct-seeded rice in Australia. Under CA-based maize–wheat system, band placement of 80% of the recommended N through slow release urea materials as basal dressing proved superior to conventional practice, i.e., basal application of one-third of recommended N in terms of NUE and yield improvement (Meena and associates, personal communication). In fact, deep placement of prilled or large-sized urea (i.e., urea briquettes, USG) is often constrained due to nonavailability of appropriate farm machinery. Sporadic studies under vegetables and cereal-based cropping systems revealed that the NUE ($RE_N$) could be enhanced to as high as 80% when N is applied through fertigation (Singh et al., 2014).

## Decision Support Tools for Enhancing NUE

In India, more than 84% of the farm holdings belong to marginal and small farmers, which exhibit substantial spatial variability owing to variable crop-management practices (cropping systems, cultivars, input use, irrigation, etc.) adopted by the farmers. Site-specific N prescription for these small holders is remotely possible unless a decision support system is developed to address spatial variability in indigenous N supply (INS) and link the same with crop N demand to arrive at rational N prescription. In recent years, International Plant Nutrition Institute (IPNI), in collaboration with CIMMYT and National Agricultural Research System (NARS), developed a tool named "Nutrient Expert (NE)" for wheat, maize, and rice. This tool maintains the background databases on geospatial variability in INS and relationship between crop N status and productivity. Evaluation of NE in predominant cropping systems revealed marked increase in grain yield of target crops, in addition to curtailing fertilizer N rates wherever farmers used excessive N (Table 7.6). In response to few simple questions related to crop

**Table 7.6** Fertilizer N Use and Yield Gain Under Nutrient Expert-Guided N Application *vis-à-vis* Farmers' Fertilizer Practice (FFP)

| Cropping System | | N Use (kg $ha^{-1}$) | | | % Increase in Yield Over FFP |
|---|---|---|---|---|---|
| | | FFP | NE | Difference (FFP-NE) | |
| Rice–wheat system (5)[a] | Rice | 168 | 119 | 49 | 13–27 |
| | Wheat | 153 | 113 | 40 | 9–35 |
| Rice–rice system (3) | Rice (*kharif*) | 141 | 119 | 22 | 5–32 |
| | Rice (*rabi*) | 154 | 115 | 39 | 9–23 |
| Maize–wheat system (3) | Maize | 69 | 113 | −44 | 24–46 |
| | Wheat | 101 | 115 | −14 | 16–38 |

[a]Number of experimental sites.

management, NE provides N rate, time of application, and also apportioning of N into different sources (organic/inorganic) depending on their availability.

# Efficient N Management Under Conservation Agriculture

The three pillars of conservation agriculture (CA), i.e., minimum soil disturbance, permanent organic soil cover, and diversified crop rotations including legumes, do influence the soil nutrient dynamics. When tillage is reduced, greater CR accumulate on the soil surface minimizing wind and water erosion and improving the quality of the soil. CR on the soil surface increase water infiltration and reduce evaporation losses, reduce nutrient losses through erosion, and also lower the surface temperature. Retention of CR on surface helps to check the evaporation, improve soil water content, and biological and chemical processes of nutrient transformation through moderation of temperature and moisture in the soil. Application of mulch @ 10 t $ha^{-1}$ with 60 kg N $ha^{-1}$ registered significantly higher tuber yield and WUE than 120 kg N $ha^{-1}$ without mulching, indicating saving of 60 kg N $ha^{-1}$ in former case (Acharya and Kapur, 2001). The $AE_N$ and $RE_N$ were significantly improved under precision land-leveling with raised-bed planting compared to conventional practices (Jat et al., 2011). Studies on harnessing the potential benefits of resource conserving technologies, revealed that in the pigeon pea–wheat system pigeon pea on permanent raised–bed (PRB) had greater N recycling (11–23% N) through its residue comprising root, stubble, and leaf litter. The economic optimum doses of fertilizer N for wheat in the pigeon pea–wheat system were smaller (128 kg N $ha^{-1}$) under PRB as compared to flat bed (FB) (152 kg N $ha^{-1}$) owing to an increased NUE in the former case (Singh et al., 2010). Also, in the treatments that received both N and P, postharvest $NO_3$-N content below 20 cm soil depth was lower compared to those receiving N or P alone.

Unlike CR incorporation in conventional practice, retention of CR on soil surface as advocated under CA is an advantageous option, as the relatively less soil-residue contact reduces the pace of CR decomposition and also that of immobilization of soil available N. Surface-retained residues are slowly mineralized compared to soil incorporated CR, and thus help maintaining plant available N for larger periods (Yadvinder-Singh et al., 2010). Recent studies at IARI on development of N management protocols under CA indicated significant increase in $AE_N$ and $RE_N$ under CA compared with conventional practice in maize–wheat cropping system (IARI, 2015). Nonetheless, broadcasting of urea under CA-based systems may enhance the possibility of volatilization losses of N (Jat et al., 2016). Hence, protocols for deep placement of urea under CA need to be developed and perfected. CA involving CR retention and inclusion of legumes could prove the best management practice for restoration of soil organic matter and curtailing fertilizer N demand of cropping systems.

## Other Important Agronomic Practices for Enhancing NUE

For higher N recovery and crop productivity, optimum amount of irrigation water and its scheduling is extremely important. Studies in sandy loam soils of north-western IGP indicated that intermittent submergence, i.e., irrigation for 02 days after disappearance of pond water resulted in greater NUE and rice productivity compared with continuous submergence. Also, $NO_3$-N losses beyond effective root zone were lower under intermittent submergence (Singh et al., 2003). As percolating water moves faster in coarse-texture soils, the N loss as $NO_3$-N through leaching is more severe in sandy loam/ loam soils which are predominantly managed under rice–wheat system in Punjab, Haryana, and western Uttar Pradesh.

Choice of N efficient genotypes could be an important strategy for crop production. An ideal N efficient genotype would be one that absorbs relatively large amount of N from soil and other indigenous sources and fertilizers, produces high grain yield per unit of absorbed N, and stores relatively small amount of N in straw. Rice cultivars having high N harvest index may be considered for achieving higher yields. Yang (1987) reported that total N uptake in hybrid rice was greater than that of inbred rice especially from transplanting to tillering and from panicle emergence to grain filling stage. Hybrid rice takes about 15–20% of its total N uptake after heading and consistently respond to N fertilizer application at flowering compared to inbred cultivars. Therefore, hybrid rice had relatively greater $AE_N$. The primary factors contributing to a higher NUE in hybrids are higher N recovery efficiency, greater root N absorption potential, greater shoot N use capacity, and greater N remobilization efficiency. In addition, adoption of other fertilizer best management practices, such as timely sowing/planting, adoption of recommended crop protection measures, and timely weed control, are extremely essential for enhancing NUE in crops and cropping systems.

## Conclusion

The perpetual problem of low NUE in crops and cropping systems not only leads to poor exploitation of the yield potential of high–yielding varieties and hybrids but also raises the cost of production and serious environmental concerns. Studies conducted so far established the advantage of balanced fertilization, IPNS, inclusion of legumes, precision N management, and CA in improving NUE. Nonetheless, focused research in following areas would be of great help in developing efficient N management practices:

- Fertilizer product research needs to be intensified so as to come out with sustained N release materials to meet crop N demands and minimize N losses.
- Development of efficient farm machinery for deep placement of N fertilizers should be taken up on priority. The tools and machines available, if any, should be validated under diverse conditions for need-based improvement.
- Water soluble fertilizers (WSF) being imported or locally manufactured are seldom included in the on-going research program of ICAR and SAUs, whereas these are the ideal fertilizers for fertigation. Performance of WSFs in improving NUE has to be evaluated vis-à-vis conventional fertilizers.
- PNM is a relatively less-researched area. Tools and techniques for PNM need to be validated and popularized for rationalizing N use.
- Fertilizer N management protocols for CA-based cropping systems should be developed considering the dynamics of native and applied N. At present, N rates prescribed for conventional-till systems are followed under CA.

The lessons learned so for in the area of improving NUE in cropping systems amply suggest that this issue cannot be addressed through conventional compartmentalized research. Scientists belonging to at least five disciplines, namely soil science, agronomy, crop improvement, microbiology, and crop physiology should join hands to deliver a tangible outcome.

## References

Acharya, C.L., Kapur, O.C., 2001. Using organic wastes as compost and mulch for potato (*Solarium tuberosum*) in low water-retaining hill soils of north-west India. The Indian Journal of Agricultural Sciences 71, 306–309.

AICRP-IFS, 2012. Annual Report (2011–12) of All India Coordinated Research Project on Integrated Farming System. PDFSR, Modipuram.

Bijay-Singh, Singh, Y., Bains, J.S., 2003. Real-time nitrogen management using chlorophyll meter and leaf colour chart (LCC) in rice and wheat. In: Singh, Y., Singh, B., Singh, J. (Eds.), Nutrient Management for Sustainable Rice-Wheat Cropping System. NATP, Indian Council of Agricultural Research, New Delhi and Punjab Agricultural University, Ludhiana, India, pp. 115–124.

Bijay-Singh, Sharma, R.K., Kaur, J., Jat, M.L., Martin, K.L., Singh, Y., Singh, V., Chandna, P., Choudhary, O.P., Gupta, R.K., Thind, H.S., Singh, J., Uppal, H.S., Khurana, H.S., Kumar, A., Uppal, R.K., Vashistha, M., Raun, W.R., Gupta, R., 2011. Assessment of the nitrogen management strategy using an optical sensor for irrigated wheat. Agronomy for Sustainable Development 31, 589–603.

Bijay-Singh, Singh, V., Purba, J., Sharma, R.K., Jat, M.L., Singh, Y., Thind, H.S., Gupta, R.K., Choudhary, O. P., Chandna, P., Khurana, H.S., Kumar, A., Singh, J., Uppal, H.S., Uppal, R.K., Vashistha, M., Gupta, R. K., 2015. Site-specific nitrogen management in irrigated transplanted rice (*Oryza sativa*) using an optical sensor. Precision Agriculture 16, 455–475.

Bijay-Singh, 2016. Site-specific nitrogen management in cereals in India. Indian Journal Fertilisers 12 (4), 46–57.

Chaudhury, A.T.M.A., Kennedy, I.R., 2004. Prospects and potentials for systems of biological nitrogen fixation in sustainable rice production. Biology and Fertility of Soils 39, 219–227.

Das, D., Dwivedi, B.S., Meena, M.C., Singh, V.K., Tiwari, K.N., 2015. Integrated nutrient management for improving soil health and crop productivity. Indian Journal of Fertilisers 11 (4), 64–83.

Dwivedi, B.S., Shukla, A.K., Singh, V.K., Yadav, R.L., 2001. Sulphur Fertilisation for Sustaining Productivity of Rice-wheat System in Western Uttar Pradesh, Tech. Bull. 2001–1. PDCSR, Modipuram, pp. 1–35.

Dwivedi, B.S., Shukla, A.K., Singh, V.K., 2002. Integrated Nutrient Management in Transplanted Rice-wheat System. PDCSR Annual Report (2001–02). PDCSR, Modipuram.

Dwivedi, B.S., Shukla, A.K., Singh, V.K., Yadav, R.L., 2003. Improving nitrogen and phosphorus use efficiencies through inclusion of forage cowpea in the rice-wheat system in the Indo-Gangetic Plains of India. Field Crops Research 84, 399–418.

Dwivedi, B.S., Singh, V.K., Meena, M.C., Dey, A., Datta, S.P., 2016. Integrated nutrient management for enhancing nitrogen use efficiency. Indian Journal of Fertilisers 12 (4), 62–71.

FAI, 2015. Fertiliser Statistics, fourth ed. The Fertiliser Association of India, New Delhi.

Galloway, J.N., Schlesinger, W.H., Levy II, H., Michaels, A., Schnoor, J.L., 1995. Nitrogen fixation: anthropogenic enhancement environmental response. Global Biogeochemical Cycles 9, 235–252.

Ghosh, A.B., Hasan, R., 1980. Nitrogen fertility status of soils of India. Fertilizer News 25 (11), 19–24.

Goswami, N.N., 1998. Some thoughts on concept, relevance and feasibility of IPNS under Indian condition. In: Integrated Plant Nutrient Supply System for Sustainable Productivity, Bulletin No. 2. IISS, Bhopal, pp. 3–9.

Hegde, D.M., Dwivedi, B.S., 1992. Nutrient management in rice-wheat cropping system in India. Fertiliser News 37 (2), 27–41.

Hegde, D.M., Dwivedi, B.S., 1993. Integrated nutrient supply and management as a strategy to meet nutrient demand. Fertiliser News 38 (12), 49–59.

Humphreys, E., Chalk, P.M., Muirhead, W.A., White, R.J.G., 1992. Nitrogen fertilization of dry-seeded rice in southeast Australia. Fertilizer Ressearch 31, 221–234.

IARI, 2015. Annual Report 2015-16, Indian Agricultural Research Institute, New Delhi - 110 012, India.

Jat, M.L., Saharawat, Y.S., Gupta, R.K., Khosla, R., 2011. Layering precision land leveling and furrow irrigated raised bed planting: productivity and input use efficiency of irrigated bread wheat in Indo-Gangetic plains. American Journal of Plant Sciences 2, 578–588.

Jat, H.S., Jat, R.K., Singh, Y., Parihar, C.M., Jat, S.L., Tetarwal, J.P., Sidhu, H.S., Jat, M.L., 2016. Nitrogen management under conservation agriculture in cereal-based systems. Indian Journal of Fertilisers 12 (4), 76–91.

Ladha, J.K., Pathak, H.P., Krupnik, T.J., Six, J., van Kessel, C., 2005. Efficiency of fertiliser nitrogen in cereal production: retrospect and prospect. Advances in Agronomy 87, 85–156.

Ma, B.L., Morrison, M.J., Dwyer, L.M., 1996. Canopy light reflectance and field greenness to assess nitrogen fertilization and yield of maize. Agronomy Journal 88, 915–920.

Majumdar, K., Dey, P., Tewatia, R.K., 2014. Current nutrient management approaches-issues and strategies. Indian Journal of Fertilisers 10 (5), 14–27.

Majumdar, K., Sanyal, S.K., Dutta, S.K., Satyanarayana, T., Singh, V.K., 2016. Nutrient mining: addressing the challenges to soil resources and food security. In: Singh, U., Praharaj, C.S., Singh, S.S., Singh, N.P. (Eds.), Biofortification of Food Crops. Springer, India, pp. 177–198.

Mandal, N., Dwivedi, B.S., Meena, M.C., Singh, D., Datta, S.P., Tomar, R.K., Sharma, B.M., 2013. Effect of induced defoliation in pigeonpea, farmyard manure and sulphitation pressmud on soil organic carbon fractions, mineral nitrogen and crop yields in a pigeonpea – wheat cropping system. Field Crops Research 154, 178–187.

Mann, R.A., Garrity, D.P., 1994. Green manures in rice-wheat cropping systems in Asia. In: Ladha, J.K., Garrity, D.P. (Eds.), Green Manure Production Systems for Asian Rice Lands. International Rice Research Institute, Los Banos, Philippines, pp. 27–42.

Meena, M.C., Dwivedi, B.S., Singh, D., Sharma, B.M., Kumar, K., Singh, R.V., Kumar, R., Rana, D.S., 2012. Effect of integrated nutrient management on productivity and soil health in pigeonpea – wheat cropping system. Indian Journal of Agronomy 57, 333–337.

Motsara, M.R., 2002. Available NPK status of Indian soils as depicted by soil fertility maps. Fertilizer News 47 (8), 5–21.

Muralidharudu, Y., Sammi Reddy, K., Mandal, B.N., Subba Rao, A., Singh, K.N., Sonekar, S., 2011. Gis-based Soil Fertility Maps of Different States of India. AICRP-STCR, IISS, Bhopal, pp. 1–224.

Pathak, H., 2016. Is nitrogen fertilizer a polluter or producer? The answer lies in its use efficiency. Indian Journal of Fertilisers 12 (4), 106–117.

Prasad, R., Power, J., 1995. Nitrification inhibitors for agriculture, health and strategies for balanced fertilization. Fertiliser News 49 (12), 73–80.

Prasad, R., 2008. Integrated plant nutrient supply (IPNS) for sustainable agriculture. Indian Journal of Fertilisers 4 (12), 71–90.

Prasad, R., 2013. Fertiliser nitrogen, food security, health and the environment. Proceedings of Indian National Science Academy 79 (4), 997–1010. Spl. Issue, Part B.

Ramamoorthy, B., Bajaj, J.C., 1969. Available nitrogen, phosphorus and potassium status of Indian soils. Fertilizer News 14 (8), 25–36.

Raun, W.R., Johnson, G.V., Stone, M.L., Solie, J.B., Lukina, E.V., Thomason, W.E., Schepers, J.S., 2001. In season prediction of potential grain yield in winter wheat using canopy reflectance. Agronomy Journal 93, 131–138.

Roy, R.N., Ange, A.L., 1991. In: Integrated Plant Nutrition System (IPNS) and Sustainable Agriculture. FAI, New Delhi. SV/1-1–SV/1-12.

Shukla, A.K., Ladha, J.K., Singh, V.K., Dwivedi, B.S., Balasubramanian, V., Gupta, R.K., Sharma, S.K., Singh, Y., Pathak, H., Pandey, P.S., Padre, A.T., Yadav, R.L., 2004. Calibrating the leaf color chart for nitrogen management in different genotypes of rice and wheat in a system's perspective. Agronomy Journal 96, 1606–1621.

Singh, V.K., Dwivedi, B.S., 2006. Yield and nitrogen use-efficiency in wheat, and soil fertility status as influenced by substitution of rice with pigeon pea in a rice–wheat cropping system. Australian Journal of Experimental Agriculture 46, 1185–1194.

Singh, V.K., Dwivedi, B.S., Shukla, A.K., Yadav, R.L., 2003. Reduce Water and Nutrient Losses in Rice Grown in Different Soils. NATP Final Report (PSR 7). Indian Council of Agricultural Research, New Delhi.

Singh, V.K., Dwivedi, B.S., Shukla, A.K., Chauhan, Y.S., Yadav, R.L., 2005. Diversification of rice with pigeonpea in a rice–wheat cropping system on a Typic Ustochrept: effect on soil fertility, yield and nutrient use efficiency. Field Crops Research 92, 85–105.

Singh, V.K., Dwivedi, B.S., Shukla, A.K., Mishra, R.P., 2010. Permanent raised bed planting of the pigeonpea–wheat system on a Typic Ustochrept: effects on soil fertility, yield, and water and nutrient use efficiencies. Field Crops Research 116, 127–139.

Singh, M., Dwivedi, B.S., Datta, S.P., 2012. Integrated nutrient management for enhancing productivity, nutrient use efficiency and environmental quality. In: Soil Science in the Service of Nation. ISSS, New Delhi, pp. 55–67.

Singh, V.K., Dwivedi, B.S., Buresh, R.J., Jat, M.L., Majumdar, K., Gangwar, B., Govil, V., Singh, S.K., 2013. Potassium fertilization in rice-wheat system across northern India: crop performance and soil nutrients. Agronomy Journal 105, 471–481.

Singh, V.K., Mishra, R.P., Rani, M., Singh, S.K., 2014. Water and nutrient interaction. I. Irrigated agriculture. In: Prasad, R., Kumar, D., Rana, D.S., Shivay, Y.S., Tewatia, R.K. (Eds.), Textbook of Plant Nutrient Management. Indian Society of Agronomy, pp. 249–267.

Singh, V.K., Shukla, A.K., Singh, M.P., Majumdar, K., Mishra, R.P., Rani, M., Singh, S.K., 2015a. Effect of site-specific nutrient management on yield, profit and apparent nutrient balance under predominant cropping systems of Upper Gangetic Plains. Indian Journal of Agricultural Sciences 4, 22–76.

Singh, V.K., Shukla, A.K., Dwivedi, B.S., Singh, M.P., Majumdar, K., Kumar, V., Mishra, R.P., Rani, M., Singh, S.K., 2015b. Site-specific nutrient management under rice-based cropping systems in Indo-Gangetic plains: yield, profit and apparent nutrient balance. Agricultural Research. http://dx.doi.org/10.1007/s40003-015-0179-1.

Yadav, R.L., Singh, V.K., Dwivedi, B.S., Shukla, A.K., 2003. Wheat productivity and N use-efficiency as influenced by inclusion of cowpea as a grain legume in a rice-wheat system. Journal of Agricultural Sciences 141, 213–220.

Yadav, R.L., 2001. On-farm experiments on integrated nutrient management in rice-wheat cropping systems. Experimental Agriculture 37, 99–113.

Yadvinder-Singh, Gupta, R.K., Singh, J., Singh, G., Singh, G., Ladha, J.K., 2010. Placement effects on rice residue decomposition and nutrient dynamics on two soil types during wheat cropping in rice-wheat system in north-western India. Nutrient Cycling in Agroecosystems 88, 471–480.

Yang, W.H., Peng, S., Huang, J., Sanico, A.L., Buresh, R.J., Witt, C., 2003. Using leaf color charts to estimate leaf nitrogen status of rice. Agronomy Journal 95, 212–217.

Yang, X., 1987. Physiological Mechanisms of Nitrogen Efficiency in Hybrid Rice. Ph.D. Dissertation). Zhejiang Agricultural University, Hangzhou, China.

# 8

# Nitrogen Inputs From Biological Nitrogen Fixation in Indian Agriculture

D.L.N. Rao[1], D. Balachandar[2]

[1]INDIAN INSTITUTE OF SOIL SCIENCE, BHOPAL, MADHYA PRADESH, INDIA; [2]TAMIL NADU AGRICULTURAL UNIVERSITY, COIMBATORE, TAMILNADU, INDIA

CHAPTER OUTLINE



## Introduction

The process by which the highly inert atmospheric nitrogen ($N_2$) is converted into ammonia ($NH_3$) is known as nitrogen fixation (Postgate, 1998). Nitrogen is broken down only at high temperature and pressure in the presence of a catalyst to form ammonia in fertilizer factories. However, in nature, the reaction is catalyzed by the enzyme nitrogenase present in the cells of certain prokaryotes at ambient temperature and pressure. Natural or synthetic nitrogen fixation is essential to sustain all forms of life as nitrogen is required to biosynthesize the building blocks of all life forms. Microorganisms specifically bacteria that fix nitrogen are called diazotrophs. Some

The Indian Nitrogen Assessment. http://dx.doi.org/10.1016/B978-0-12-811836-8.00008-2

higher plants and termite-like organisms form symbiotic association with diazotrophs for their nitrogen requirement. *Clostridium pasteurianum* was the first diazotrophic bacterium discovered by S. Winogradsky in 1893, followed by *Azotobacter chrococcum* by the Dutch Microbiologist M.W. Beijerinck in 1901. Symbiotic nitrogen fixation was discovered by the German scientists H. Hellriegel and H. Wilfarth during 1886–1888.

Biological nitrogen fixation (BNF) occurs with the help of nitrogenase in the diazotrophs. The BNF is coupled to the hydrolysis of 16 equivalents of ATP (adenosine triphosphate) and is accompanied by the coformation of one molecule of hydrogen. The reaction of BNF is as follows:

$$N_2 + 8H^+ + 8e^- \rightarrow 2NH_3 + H_2$$

The nitrogenase generated ammonium is assimilated into glutamate through glutamine synthetase/glutamate synthase pathway. Nitrogenase is highly sensitive to oxygen and many bacteria cease production of the enzyme in the presence of oxygen. Some diazotrophs exist only in anaerobic condition while others with high respiration rates draw down oxygen levels or detoxify the oxygen by using special binding proteins, e.g., leghemoglobin (De Bruijn, 2015).

Prior to the advent of Haber-Bosch process in 1909, the industrial fixation of $N_2$ into $NH_3$, almost all the reactive nitrogen in the biosphere was generated and recycled by microorganisms (Galloway, 1998). Though the chemical nitrogen fertilizers synthesized through Haber-Bosch process have quadrupled the productivity of agricultural crops, they led to unprecedented environmental degradation (Galloway et al., 2004). The world nitrogen fertilizer demand increased from 105 million tons in 2011 to 112.9 million tons in 2015 at an annual growth rate of 1.7%. According to FAO estimates (FAO, 2011), the Asian continent alone consumes 68% (73.4 million tons) of global nitrogen fertilizer of which India shared 25% (18.4 million tons) followed by China (24%). Almost 50% of this fertilizer is consumed by maize, rice, and wheat (16%, 16% and 18% respectively). Though nitrogen fertilizers revolutionized the agriculture in last century, they also left severe environmental footprints. Extensive use of chemical nitrogen in intensive agriculture, particularly in an unbalanced way coupled with a reduction in recycling of organic residues and manures, impaired the soil physical properties, reduced the soil biodiversity, and organic carbon levels leading to the fertilizer-driven crop responses (Robertson et al., 2000; Galloway et al., 2008; Schlesinger, 2009). Thus, use of chemical fertilizers alone (without addition of organics) in tropical agriculture stalled the yield of major food crops, induced climate forcing, and deteriorated soil biodiversity and fertility. It is estimated that about 67% of fertilizer N used globally is eventually converted back to nonreactive atmospheric $N_2$ and that nearly 33% as reactive N contributes to its transient effects in different ecosystems (Vitousek et al., 1997). Hence, the ecofriendly and natural fixation process of BNF is now re-gaining importance since any effort of increasing its contribution in food production will reduce the consumption of chemical fertilizers and that will directly increase the revenue of the farmers with indirect benefits such as improved soil health and reduced adverse environmental effects. The accurate determination of global inputs of biologically fixed nitrogen (BNF-N) is thus essential for effective management of nitrogen in the agricultural fields.

## Global N and BNF—Historical Trends

The total amount of nitrogen in the atmosphere, soil, and water of earth is approximately $4 \times 10^{9}$ Tg (Mackenzie and Mackenzie, 1998). However, more than 99% of this nitrogen is not available to more than 99% of the living organisms. The global N can be divided in to nonreactive N ($N_2$) and reactive N (Nr) including biologically, photochemically, and radioactively active N compounds. Nr compounds include inorganic reduced forms such as $NH_3$, $NH_4^+$, inorganic oxidized forms, e.g., $NO_x$, $HNO_3$, $N_2O$, $NO_3$, and organic compounds, e.g., urea, amines, proteins, and nucleic acids (Galloway et al., 2003). In the prehuman world, two processes that accumulated Nr from $N_2$ were lightning and BNF. At that time, the denitrification and BNF were approximately equal; hence, the reactive nitrogen did not accumulate in the environmental reservoirs (Ayres et al., 1994). However, the anthropogenic activities doubled the reactive N entering the biosphere. The production of Nr by humans is now much greater than the production from all natural terrestrial systems. The Haber-Bosch process which converts the $N_2$ to $NH_3$, followed by combustion of fossil fuels which converts $N_2$ to $NO_x$ and BNF that converts $N_2$ to $NH_3$ through the cultivation of legumes, cereals, and many other crops, and subsequently the organic N are the three major human interventions that increase the Nr in the biosphere. Among the three activities (Galloway et al., 2003), the crop cultivation contributed 33 Tg $year^{-1}$ of Nr, while the fossil fuel combustion had a share of 25 Tg $year^{-1}$, and the Haber-Bosch process contributed 100 Tg $year^{-1}$ in 2002. Between 1860 and 2000, the anthropogenic Nr creation rate increased from 15 Tg N per year to 165 Tg per year, with about five times more Nr from food production than from energy production (Galloway et al., 2002). After 2010, the anthropogenic Nr got raised to 210 Tg $year^{-1}$ of which 120 Tg $year^{-1}$ came through Haber-Bosch process for fertilizers and 60 Tg Nr $year^{-1}$ through BNF from cultivation of crops. The rate of Haber-Bosch mediated Nr increase is much faster than agrosystem-based BNF-Nr, which has some environmental concerns (Fowler et al., 2013). BNF through crop plants, particularly legumes, is thus gaining renewed attention as the other two processes lead to severe environmental effects.

## Agrosystem and Nr

About 75% of Nr, which is created globally by humans, is added to agroecosystem to sustain the food production. The balance Nr is met by cultivation-induced BNF. Smil (2002) estimated that nearly 40% of the human population depends on the food produced using fertilizers made by the Haber-Bosch process, which is believed to have increased further in recent past. In agroecosystem, indigenous source of N in the soil, BNF, recycling of crop residues, animal manures, and the applied synthetic fertilizers are the major contributors for Nr. However, the annual crop yield is determined primarily by the amount of Nr added, especially the synthetic fertilizers. On a global scenario, about 120 Tg N from newly synthesized reactive N (Fowler et al., 2013) and about 50 Tg of

N from previously created (e.g., crop residues, animal manures, and atmospheric deposition) were added to the crop agrosystem (Smil, 2002). Of the 170 Tg N per year, 33 Tg and 16 Tg of N were utilized by animals and humans respectively through plant produce and the remaining is deposited as part of materials in crop system; lost to the atmosphere and water. Smil (1999) reported that only 4 Tg of N out of 170 Tg N per year was accumulated in the crop agro-system and the remaining was lost through air and water systems. Any increase in either natural ecosystem–based or cultivation-dependent BNF-created Nr will certainly reduce the usage of human-created Nr and thus can avoid the environmental and health concerns.

## BNF-Nr

Quantification of the magnitude of terrestrial Nr-creating capacity of BNF is challenging because of various uncertainties and variability in estimates. The methodological differences, uncertainty, or lack of information about the spatial coverage of important nitrogen-fixing species and locational variations have been identified as critical gaps in our understanding of natural BNF at large scale (Cleveland et al., 1999). In addition to these, in the potential area where BNF-Nr is important, especially in the tropical regions of Asia, Africa, and South America, there are virtually very little data on BNF. Galloway et al. (2004) stated that the terrestrial natural BNF will decline from 107 Tg N year$^{-1}$ to 98 Tg year$^{-1}$ in 2050 due to decline of natural vegetation and urbanization. In contrast, the crop cultivation–dependent BNF may increase from 33 Tg year$^{-1}$ to 50 Tg in 2050 due to intensive agriculture. However, these values vary in several reports. For example, Galloway et al. (1995) estimated the global nitrogen fixation through agricultural system was about 43 Tg (ranged between 32 and 53 Tg); Smil (1999) estimated 33 Tg (ranged between 25 and 41 Tg); and Fowler et al. (2013) estimated it to be around 60 Tg year$^{-1}$. It is obvious that as the global agricultural intensification is in increasing trend, the BNF-Nr would also be increasing annually. However, the Haber-Bosch N fixation is also to increase from 120 Tg year$^{-1}$ to 165 Tg in 2050. By means of soil biological activities, crop rotation with legumes, increasing the organic matter content of soil, minimal tillage, and soil conservation practices, it is possible to increase the BNF contribution. If we can increase the level of cultivation-dependent BNF-Nr annually, it is possible to maintain the Haber-Bosch N production at the same level without further increase. This is the only feasible way to minimize the disturbance of N-cycling cascades.

## World Agriculture and BNF-Nr

Computing the annual cultivation–dependent BNF from world agriculture is a difficult exercise due to poor availability of complete data. Food and Agriculture Organization compiles annual data on area, production, and productivity of important crops. As the crop-BNF differs by locality, soil type, and cultivation methods, potentially high errors are possible. However, the agents of BNF in agriculture are essentially the same though the species may be different. The main problems in calculating agriculture-BNF are the uncertainty in nitrogen-fixation intensity per unit area, estimation bias, and difficulties in scaling up to the global level (Herridge et al., 2008). However, total N balancing method, differentiating plant N accumulation between $N_2$ fixing crop plant with non-$N_2$ fixing crop plant, and calculating percentage N derived from atmosphere (%Ndfa) using $^{15}N$ isotope are the most common methods used to calculate the BNF-derived N annually by the crops. Herridge et al. (2008) calculated using %Ndfa method that food legumes of global area can fix about 2.95 Tg of N $year^{-1}$. Among the crops, chick pea, 0.60 Tg; common beans, 0.58; pea, 0.57 Tg; other pulses, 0.47 Tg are the predominant contributors. Compare to oilseed legumes, the food legumes have less contribution. Soybean and groundnut together contribute 6.3 times higher BNF annually (18.5 Tg $year^{-1}$). Soybean, the predominant BNF-contributing legume of world agriculture shared 76.6% of total legume crops' BNF (16.44 Tg $year^{-1}$). Groundnut contributes 2.06 Tg of BNF-N annually which represents 10% of the total contribution. Altogether, the legume food crops fix 21.45 Tg of nitrogen per year, globally. Salvagiotti et al. (2008) in a meta-analysis of 637 data sets (site–year–treatment combinations) worldwide on soybean N uptake and BNF pointed out that data from experiments in which soybean was fertilized with <10 kg N $ha^{-1}$ showed that the relative contribution of $N_2$-fixation was 58%. BNF ranged from 0 to 337 kg N $ha^{-1}$, averaging 111 kg N $ha^{-1}$.

There has been a major re-assessment of the contribution of nonsymbiotic nitrogen fixation recently by Ladha et al. (2016) who constructed a top-down global N budget for maize, rice, and wheat for a 50-year period (1961–2010) and showed that cereals harvested a total of 1551 Tg of N, of which 48% was supplied through fertilizer N and 4% came from net soil depletion. The rest of 48% (737 Tg) of crop N, equal to 29, 38, and 25 kg $ha^{-1}$ $year^{-1}$ for maize, rice, and wheat, respectively, is contributed by other sources. Nonsymbiotic $N_2$ fixation was the major source of this N, which is 370 Tg or 24% of total N in the crop, corresponding to 13, 22, and 13 kg $ha^{-1}$ $year^{-1}$ for maize, rice, and wheat, respectively. Manure (217 Tg or 14%) and atmospheric deposition (96 Tg or 6%) are the other sources of N. Crop residues and seed contribute marginally. Ladha et al. (2016) argued that this scaling-down approach to estimate the contribution of nonsymbiotic $N_2$ fixation is robust because it focuses on global quantities of N in sources and sinks that are easier to estimate, in contrast to estimating N losses per se, because losses are highly soil-, climate-, and crop-specific. Their estimate suggests that maize, rice, and wheat together added 383 Tg of N through BNF from 1961 to 2010 with an annual contribution of 7.66 Tg $year^{-1}$. Among the three cereals, rice added 4.0 Tg, followed by wheat (3.51 Tg $year^{-1}$), and maize (2.04 Tg $year^{-1}$).

Herridge et al. (2008) related the annual global biomass production of forage and fodder legumes using N balancing method and derived about 12–15 Tg of BNF/year. Using %Ndfa method, the sugarcane endophytes BNF contribution comes out to be about 0.5 Tg of N $year^{-1}$. The recent rough estimate of BNF-Nr of all the agro-systems of world is about 50–70 Tg $year^{-1}$, which is higher than the values reported earlier (31.5 Tg $year^{-1}$) (Galloway et al., 1995).

**Table 8.1** Estimate of Total BNF Contribution by Food Legumes of India by Ndfa Method (area and yield as per Agricultural Statistics, 2015, India)

| Crops | Area (mha) | Yield (mt) | Productivity (kg ha$^{-1}$) | Shoot dry matter[a] (Tg) | Shoot N[b] (Tg) | Crop N[c] (Tg) | %Ndfa[d] | Crop N fixed (Tg) |
|---|---|---|---|---|---|---|---|---|
| Pigeon pea | 3.88 | 3.29 | 849 | 13.16 | 0.29 | 0.38 | 58 | 0.22 |
| Chickpea | 10.22 | 9.88 | 967 | 28.23 | 0.56 | 0.74 | 63 | 0.47 |
| Lentil | 1.42 | 1.13 | 797 | 3.23 | 0.06 | 0.08 | 63 | 0.05 |
| Others | 9.81 | 4.97 | 507 | 14.20 | 0.28 | 0.37 | 53 | 0.20 |
| *Total Pulses* | *25.33* | *19.27* | *780* | *58.82* | *1.20* | *1.58* | *59* | *0.94* |
| Groundnut | 5.53 | 9.67 | 1750 | 24.18 | 0.56 | 0.74 | 68 | 0.50 |
| Soybean | 12.20 | 11.99 | 983 | 29.98 | 0.90 | 1.18 | 68 | 0.80 |
| *Total legumes* | *43.06* | *40.93* | *1171* | *112.97* | *2.65* | *3.49* | *65* | *2.24* |

*mha*, Million hectare; *mt*, million tons; *Tg*,Teragram.

[a]Harvest index of respective crops used to convert grain yield to shoot dry matter.

[b]Shoot N based on N content (%) of crop.

[c]Ground biomass N was derived assuming 24% shoot N allocation below ground (Rochester et al., 1998).

[d]Global nitrogen-fixing potential of each legume crop in terms of Ndfa (nitrogen derived from air).

# Indian Agriculture and BNF-Nr

## Food Crops

The food grain legumes including groundnut and soybean are cultivated on an area of about 43.06 million ha (Agricultural Statistics, 2015) with a total grain yield of 40.93 million tons (Table 8.1). By using average N percent, harvest index and % Ndfa of the respective legume crops, the BNF contribution by legumes being cultivated in India was calculated. To avoid the underestimation of BNF, it was assumed that 24% of the total plant N is allotted to belowground parts which include root, nodule, exudate, and rhizodeposition (Rochester et al., 1998). By adopting this fractionation method, soybean BNF was successfully quantified by Salvagiotti et al. (2008). Among the grain legumes, chickpea (gram) contributed 0.47 Tg of N year$^{-1}$, followed by pigeon pea (0.22 Tg year$^{-1}$). Other legumes including peas, cowpea contributed to 0.20 Tg year$^{-1}$ (Table 8.1). The total BNF contributed by pulses worked out to 0.94 Tg. Soybean, the predominant oilseed legume crop contributed 0.80 Tg of N per year, while groundnut contributed 0.50 Tg year$^{-1}$. The total contribution of legume crops amounted to 2.24 Tg of N per year. However, if we calculate the BNF contribution through average annual nitrogen-fixing capacity of each legume adopted from the data of various sources, the annual nitrogen fixation by Indian food legumes came to 2.81 Tg year$^{-1}$ (ranged between 2.06 and 3.56 Tg year$^{-1}$) (Table 8.2). In this method, pulses contributed 1.39 Tg. The range between 2.24 and 2.81 Tg year$^{-1}$ may be the total annual BNF-Nr added by Indian food legumes.

Measurements of legume BNF in long-term field experiments are very few in India. Rawat et al. (2013) in a 8 year soybean–wheat rotation experiment in Vertisol showed

**Table 8.2** Alternate Estimate of Total BNF Contribution by Food Legumes of India Based on Nitrogen-Fixing Potential of Crop

| | | BNF (kg ha year$^{-1}$) | | | BNF (Tg year$^{-1}$) | | |
|---|---|---|---|---|---|---|---|
| **Crops** | **Area (mha)** | **Minimum** | **Maximum** | **Mean** | **Minimum** | **Maximum** | **Mean** |
| Pigeon pea | 3.88 | 40 | 80 | 60 | 0.155 | 0.310 | 0.233 |
| Chickpea | 10.22 | 40 | 80 | 60 | 0.409 | 0.818 | 0.613 |
| Lentil | 1.42 | 30 | 50 | 40 | 0.043 | 0.071 | 0.057 |
| Others | 9.81 | 40 | 60 | 50 | 0.392 | 0.589 | 0.490 |
| *Total Pulses* | *25.33* | | | | *0.999* | *1.788* | *1.393* |
| Groundnut | 5.53 | 60 | 100 | 80 | 0.332 | 0.553 | 0.442 |
| Soybean | 12.20 | 60 | 100 | 80 | 0.732 | 1.220 | 0.976 |
| *Total legumes* | *43.06* | | | | *2.063* | *3.561* | *2.811* |

Data adapted from Smil, V., 1999. Nitrogen in crop production: an account of global flows. Global Biogeochemical Cycles 13, 647–662 except for other legumes.

that fertilized and uninoculated soybeans (nodulated by native rhizobia) removed 158 kg N ha$^{-1}$ (1.8 Mg ha$^{-1}$ seed yield) whereas fertilized and inoculated soybeans removed 173 kg N ha$^{-1}$ for a 2.0 Mg ha$^{-1}$ seed yield and BNF amounted to ~92–100 kg N ha$^{-1}$ year$^{-1}$. The soil N increment in 0–15 cm soil was 27.8 kg ha$^{-1}$ year$^{-1}$ after soybean harvest and 33.2 kg N ha$^{-1}$ year$^{-1}$ after wheat harvest in the fertilized uninoculated plots averaged over 8 years. In the fertilized and inoculated plots (*Bradyrhizobium* to soybean and *Azotobacter* to wheat) the soil N increments were 40.8 and 34.0 kg N ha$^{-1}$ year$^{-1}$ after soybean and wheat harvest, respectively. These experimentally measured values of soybean BNF in uninoculated soybean (92 kg N ha$^{-1}$ year$^{-1}$) and those used in this analysis (Table 8.2, mean 80 kg N ha$^{-1}$ year$^{-1}$) are in fair agreement and on the lower side.

By adopting N-fraction method (Ladha et al., 2016), the annual nonsymbiotic nitrogen fixation by maize, rice, and wheat grown in Indian subcontinent was calculated. All the three crops contributed 1.51 Tg year$^{-1}$, which is almost two-third of the legume contribution. Among the three crops, rice contributed 0.76 Tg, wheat 0.65 Tg, and maize 0.10 Tg year$^{-1}$ (Table 8.3). However, if the average N-fixing potential data were used for annual BNF quantification, the rice and wheat contributed about 0.69 Tg and 0.28 Tg, respectively, when the N fixation data of both crops derived from long-term experiments in north India (Rao and Gill, 2000) was used. When global values (Ladha et al., 2016) were used, these estimates increased to 0.97 and 0.41, respectively, while maize contributed 0.12 Tg (Table 8.4). The N-fraction method gave a total of 1.51 Tg year$^{-1}$, while the % Ndfa method gave a similar estimate of 1.50 Tg of N year$^{-1}$ for cereal food crops, thus supporting the BNF rates used in calculations.

The current scenario of BNF in agricultural systems, particularly in the Indian context, was recently reviewed (Rao, 2014). Measurements of cereal BNF in long-term experiments are very few in India. Rao and Gill (2000) conducted a 10-year experiment in which rice–wheat were grown for 6 years following the cultivation of a

**Table 8.3** Estimate of Total BNF Contribution by Cereal Crops of India (Area and Yield as per Agricultural Statistics, 2015, India) Using N-Fraction Method

| Estimates | Maize | Rice | Wheat |
|---|---|---|---|
| Production (mt) | 24.35 | 106.54 | 95.91 |
| Area (Mha) | 9.43 | 43.95 | 31.19 |
| Yield (kg $ha^{-1}$) | 2583 | 2424 | 3075 |
| Harvest index | 0.494 | 0.481 | 0.423 |
| Crop dry matter (Tg) | 49.29 | 221.50 | 226.74 |
| Grain N (%) | 1.34 | 1.13 | 1.84 |
| Straw N (%) | 0.80 | 0.69 | 0.56 |
| Grain N (Tg)[a] | 0.33 | 1.20 | 1.76 |
| Straw N (Tg)[a] | 0.20 | 0.79 | 0.73 |
| Total crop N (Tg) | 0.53 | 2.00 | 2.50 |
| Fertilizer N recovery (%)[b] | 0.56 | 0.36 | 0.48 |
| Ndf (Tg)[c] (Crop N × Fert N recovery) | 0.30 | 0.72 | 1.21 |
| Others[d] (Tg) (total crop N—Ndf) | 0.23 | 1.28 | 1.29 |
| BNF (% in non-fertilizer N) | 44 | 59 | 50 |
| Total BNF (Tg) | 0.10 | 0.76 | 0.65 |

[a]% N x area.

[b]Adopted from global N fertilizer use efficiency data.

[c]Ndf refers to nitrogen derived from fertilizer.

[d]N contributed by the soil, manure, crop residues, and BNF.

**Table 8.4** Alternative Estimate of Total BNF Contribution by Cereal and Other Nonlegume Crops of India (Including Sugarcane) Based on Nitrogen-Fixing Potential

| | | BNF (kg ha $year^{-1}$) | | BNF (Tg $year^{-1}$) | |
|---|---|---|---|---|---|
| Crops | Area (mha) | Estimate A[a] | Estimate B[b] | Estimate A | Estimate B |
| Maize | 9.43 | 13.0 | 13.0 | 0.12 | 0.12 |
| Rice | 43.95 | 22.0 | 15.6 | 0.97 | 0.69 |
| Wheat | 31.19 | 13.0 | 9.1 | 0.41 | 0.28 |
| Coarse cereals | 25.67 | 5.0 | 5.0 | 0.13 | 0.13 |
| Sugarcane | 5.01 | 40.0 | 40.0 | 0.20 | 0.20 |
| Cotton | 11.69 | 5.0 | 5.0 | 0.06 | 0.06 |
| Oilseeds[c] | 10.8 | 5.0 | 5.0 | 0.05 | 0.05 |
| Total | | | | 1.94 | 1.53 |

[a]Non-symbiotic BNF of rice, wheat and maize, and other cereals adopted from Ladha et al. (2016); sugarcane BNF from Boddey et al. (2003).

[b]Non-symbiotic BNF of rice and wheat from Rao and Gill (2000) and for others same as in estimate A.

[c]Oilseeds other than soybean and groundnut (mustard, rapeseed, sunflower, safflower, sesamum, castor, and linseed).

fast-growing perennial N-fixing *Sesbania sesban* for 4 years in north-west India in an alluvial reclaimed alkali soil. They observed that the residual benefits of symbiotic N fixation by *Sesbania* in the form of additional N uptake by rice and wheat (30.6 and 12.2 kg N $ha^{-1}$ $year^{-1}$ each year) had stabilized after 2 years (four crops). Consistently for next 4 years, additional N uptake in rice and wheat was observed amounting to 15.6 kg in rice $ha^{-1}$ $year^{-1}$ and 9.1 kg $ha^{-1}$ $year^{-1}$ in wheat which is due to nonsymbiotic N fixation. These values used in our calculations are fairly close to the figures of 22.0 and 13.0 $ha^{-1}$ $year^{-1}$ reported by Ladha et al. (2016) for rice and wheat, respectively.

## Sugarcane

Sugarcane a C4 crop is known for harboring associative nitrogen-fixing organisms, especially *Gluconacetobacter* spp. These Proteobacteria are natural habitants in soil, enter through roots and profusely colonize the plants, form endophytic association, and fix atmospheric nitrogen. Smil (1999) reported that the world's 20 Mha of sugarcane area can contribute 2.0 Tg of BNF at rate of 100 kg BNF-N/ha. The evidence for such high quantity of N-input can be attributed to the isolation of large and diversified $N_2$-fixing bacteria from inside and outside of sugarcane (Boddey et al., 2003). However, several other studies concluded that such high rates are not evident by $^{151}N$ tracer studies (Herridge et al., 2008). From these experiments, it was revealed that nearly 20% of the total N of sugarcane (200 kg N ha $year^{-1}$) could be from BNF, which accounts for 40 kg N ha $year^{-1}$. In such scenario, the world sugarcane production contributes about 0.8 Tg of N fixed per year. By analogy, Indian sugarcane fields (about 5 Mha) could contribute nearly 0.20 Tg of N fixation annually.

## Fodder Crops

To meet the protein and carbohydrate requirement of the livestock, diversified herbaceous and tree legumes are cultivated for forage to livestock under integrated farming system (Peoples and Baldock, 2001). Nutrition-rich legumes and grasses can either be grazed or harvested and pan feed. Even though large number of legume genera and species are used for forage and green manuring in tropical and subtropical regions, there are few estimates of BNF by these crops. The methodological difficulty in measuring the BNF of larger biomass of these crops is the bottleneck for accurate contribution to the global scenario (Danso et al., 1993). Likewise, accurate estimation of global nitrogen-fixation contribution from forage and fodder is also challenging as the statistics on area and productivity of these crops are difficult to obtain. Smil (1999) reported that nearly 100–120 Mha of land is cultivated for forage and green manuring globally. By assuming average BNF rates of 200 kg N $ha^{-1}$ for alfalfa; 150 kg N $ha^{-1}$ for clovers; 100 kg N $ha^{-1}$ for other legumes; and 10 kg N $ha^{-1}$ for grasses, the annual BNF-N contribution from grasses, legume fodders, forage crops, and green manures amount to about 12 Tg annually. However, Peoples and Baldock (2001) and Carlsson and Huss-Danell (2003) used aboveground N content of these crops and N-fixing capability of these crops, and estimated that the annual BNF contribution from these crops could be around 25 Tg which is twice the estimate of Smil (1999). Herridge et al. (2008) concluded that the global contribution of forage and fodder may fall between 12 and 25 Tg N annually, as there is no accurate data available for its precise quantification.

In India, fodder legumes and grasses are cultivated on a total land area of about 8 Mha, predominantly with forage sorghum, 2.6 Mha; clover, 1.9 Mha; alfalfa, 1 Mha; other legume fodders, 1.9 Mha; pearl millet, 0.9 Mha; and fodder grasses, 0.89 Mha. Using the mean N uptake (kg N $ha^{-1}$) data of forages and grasses (IGFRI, Jhansi, India) berseem: 240–264; lucerne: 216–264; fodder cowpea: 161–181; fodder sorghum: 128–160; Napier grass: 288–360; Guinea grass: 288–360; para grass: 240–300; oats: 120–144, and %Ndfa values of 0.7 for legumes and 0.1 for cereals, the BNF values used are depicted in Table 8.5. With the per ha values optimized from the data of Herridge et al. (2008), the annual BNF contribution of Indian forage and fodder crops comes around 0.61 Tg per year. As criticized by Herridge et al. (2008), the value calculated would be the aboveground N content of the biomass harvested, omitting the belowground contribution and from mulching the crop residues. However, the Indian forage and fodder could contribute about 5% of the world BNF of forage and fodder. The legume trees grown naturally in nonagricultural lands also contribute significantly to the global N fixation. Even though, biological nitrogen-fixing capability of these trees is available (Danso et al., 1992), the data on the area under these tree legumes in different agroecosystem are not available to allow for reasonable accounting.

## Agroforestry

Agroforestry, the practice of growing trees in crop farming provides sustainable farm inputs, improves soil conservation, sequesters carbon, and increases the biodiversity and soil health. The trees involved in agroforestry are multipurpose utility such as for shade, fodder, fuel wood, fruit, vegetables, and medicines. Leguminous trees fix considerable amount of nitrogen and thereby increase the Nr with no cost to the environment. A wide range of nitrogen-fixing trees are used in agroforestry with a potential objective of raising soil nitrogen levels and consequently improving the growth and production of nonnitrogen-fixing trees and agricultural crops (Rao, 2002). The total forest cover in India is about 675,538 sq km, which is about 20.55% of total geographical area. The agroforestry accounts for 111,554 sq km of India which is 3.39% of the geographical area of India. Hence, in India, nearly 16% of the total forest cover is supported by agroforestry. *Acacia nilotica, Areca catechu, Mangifera indica, Azadiracta indica, Borassus flabelliformis, Eucalyptus*, and *Cocus nucifera* are the predominant tree species which occupy nearly 35% of the agroforestry cover of India. As per the forest survey of India (2013), nearly 24% of the agroforests of India are covered by legume trees and rest would be nonleguminous trees. The nitrogen-fixing tree legumes through their indeterminate type of nodules fix nearly 22–45 kg of N $ha^{-1}$ (Danso et al., 1992). Fast-growing nitrogen-fixing fodder trees such as *Leucaena* and *Sesbania* have very high biomass production and make a substantial contribution to

**Table 8.5** Contribution of BNF From Fodder Crops and Agroforestry Systems of India

| Fodder Crops | Area ($\times 10^3$ ha$^{-1}$)[a] | BNF[b] (kg N ha$^{-1}$) | N Fixed ($\times 10^3$ kg year$^{-1}$) |
|---|---|---|---|
| **Fodder Legumes** | | | |
| Egyptian clover | 1900 | 168 | 319,200 |
| Alfalfa | 1000 | 151 | 151,000 |
| Sweet clover, Persian clover | 10 | 150 | 1500 |
| Fenugreek, fodder cowpea, cluster bean, rice bean, lablab bean, horsegram, mothbean, Subabul (Leucaena) | 655 | 113 | 74,015 |
| **Fodder Cereals** | | | |
| Fodder sorghum | 2600 | 13 | 33,800 |
| Pearl millet | 900 | 12 | 10,800 |
| Oat | 100 | 12 | 1200 |
| Barley, makchari, chara | 30 | 12 | 360 |
| **Fodder Grasses** | | | |
| Napier grass | 100 | 29 | 2900 |
| Guinea grass | 100 | 29 | 2900 |
| Para grass | 100 | 24 | 2400 |
| Setaria grass, deenanath grass, ruzi grass, signal grass, dharaf grass, baffel grass, birdwood grass, sewan grass, sudan grass | 530 | 24 | 12,720 |
| Sub-total | | | 612795 or 0.61 Tg |
| **Agroforestry** | | | |
| Nitrogen-fixing tress | 2789 | 100 | 278,900 |
| Non-nitrogen-fixing trees | 8367 | 15 | 125,505 |
| Sub-total | | | 404405 or 0.40 Tg |
| Total | | | **1017200 or 1.02 Tg** |

[a]The data on annual forage crop area of India are from Hand Book of Agriculture (2014).
[b]The nitrogen-fixing potential of these forage crops are based on 70%Ndfa of fodder legumes and 10%Ndfa from grasses using N uptake data (IGFRI, India).

nitrogen budgets of soils. Rao et al. (1990) reported destructive harvesting of 4-year old perennial *Sesbania sesban* plantations' estimated soil N balance and showed that it could fix 260–332 kg N ha$^{-1}$ year$^{-1}$. The slow and medium-growing trees have much lower N-fixing rates and so we adopted the BNF capacity of legume trees of 100 kg N ha$^{-1}$ and nonnitrogen-fixing trees of 15 kg N ha$^{-1}$(Danso et al., 1992). Using these values, we estimated that the Indian agroforestry could contribute nearly 0.4 Tg of nitrogen to the Nr.

## Indian Scenario for Global BNF

When considering the human inputs to the global nitrogen cycle, BNF provides an important reference, as this is the primary nonanthropogenic input of Nr

(Vitousek et al., 2002; Fowler et al., 2013). Even though the modern agricultural practice is considered as the primary reason for enhanced anthropogenic Nr creation (through Haber-Bosch process for synthetic fertilizer N, 120 Tg year$^{-1}$), the agricultural crops in turn fix atmospheric nitrogen into ammonia to a tune of about 60 Tg of N year$^{-1}$ without significant harm to the environment (Vitousek et al., 2013). This value is about half of the chemical Nr synthesis and efforts should be made to increase the BNF-Nr and to decrease the HB-Nr. It is well known that under high levels of available N in soil, BNF activity is adversely affected as the plant is able to meet its N need from soil N and BNF process is inhibited. When considering BNF inputs in global agriculture, cereal food crops (rice, wheat, maize, sorghum, barley), legume grain crops (grams, peas, beans), leguminous oilseed crops (soybean and groundnut), sugarcane, legume and grass fodder crops, and agroforestry all contribute varying amounts to net BNF inputs. Among them, soybean and groundnut (oil legumes) contribute the maximum share of 18.5 Tg year$^{-1}$ (34%) followed by cereals (12.5 Tg year$^{-1}$). Forage legumes shared 22% of global annual BNF (7.7 Tg year$^{-1}$). Altogether, the global agricultural system accounts for 54.45 Tg of N year$^{-1}$, which is synthesized ecofriendly through diazotrophs associated with the crops (Fig. 8.1A). Of this value, Indian agriculture contributes 5.2 Tg of BNF-Nr year$^{-1}$ (Table 8.6), which accounts for 9.5% of the annual BNF-Nr. Among different agro-systems, cereal food crops including rice, wheat, maize, and other coarse cereals shared 32% of Indian agricultural BNF-Tn (1.64 Tg year$^{-1}$). As the cereals are the major cultivable crops of India, it is obvious that the contribution from cereals to BNF is quite appreciable. The food legumes including chickpea, pigeon pea, blackgram, greengram, lentils, peas contributed 0.94 Tg (18%); the oilseed legumes (soybean and groundnut) contributed 1.30 Tg year$^{-1}$ (25%) and shared 43% of BNF of India, with addition of another 11% from forage and fodder crops, predominantly by legumes (Fig. 8.1B). The commercial crops—sugarcane, cotton, and nonlegume oilseeds—had a little share 6% (0.31 Tg year$^{-1}$) while agroforestry system of India also had a share of 8% (0.41 Tg). Comparing the proportion of various crop systems to global agriculture-BNF, Indian agriculture differed in the following aspects: (1) higher contribution from low efficient nitrogen-fixing (nonsymbiotic and associative symbiotic) crops such as cereals; (2) higher contribution from food legumes and lower contribution from oilseed legumes; and (3) lower contribution from fodder crops and agroforestry system, though they have high potential.

## Conclusion and Way Forward

To feed the ever-increasing human population, the productivity of major food crops has to be increased without compromising the environmental safety. The use efficiency of chemical fertilizer N is low. However integration with organic manures or biologically fixed nitrogen (through grain, fodder, and tree legumes) has a potential to reduce the use of chemical fertilizer N and improve the efficiency of applied N thereby reducing the environmental footprint of Nr. There is an enormous scope to increase the contribution of BNF from fodder legumes and agroforestry systems. Exploring and improving the

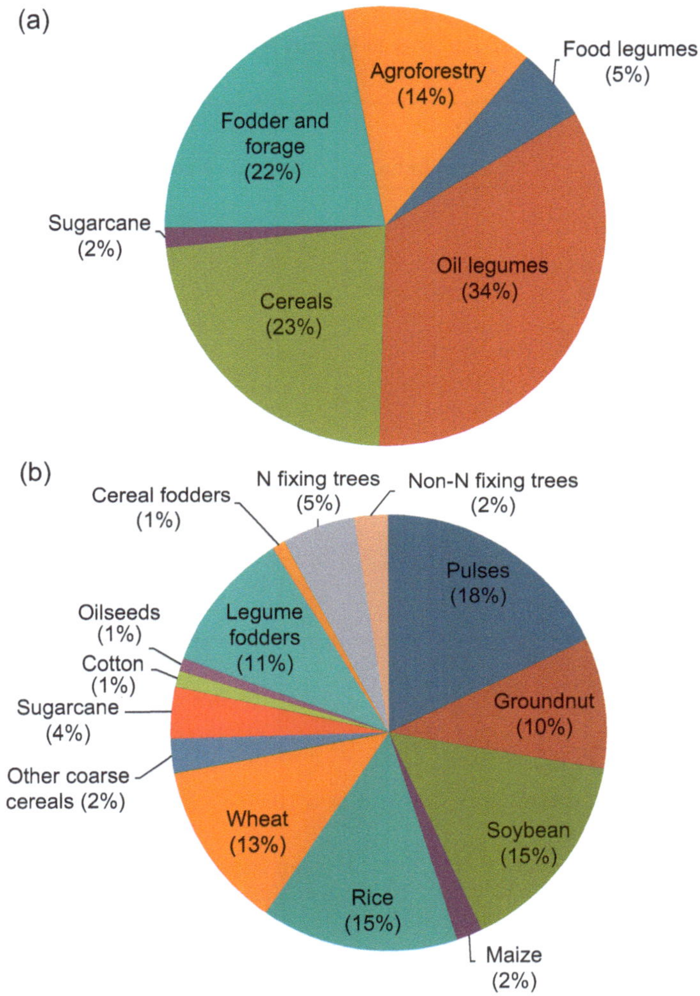


**FIGURE 8.1** Crop-wise share of annual biological nitrogen fixation in world (A) and Indian (B) agriculture.

biological nitrogen-fixing capability of the diazotrophs associated with agricultural crops to increase the BNF-Nr should be an important goal for future Nr management. Even though we have calculated that about 10% of global agricultural BNF is contributed by Indian agriculture, more precise data are needed by more extensive surveys, precise quantification in various agrosystems and regional scaling. Nitrogen-fixation potential of different agro-systems need to be quantified using $^{15}N$ isotope abundance and new modeling techniques, extensive microbial inoculation programs particularly of rhizobia for legumes are needed to improve the BNF contribution and to reduce fertilizer-N consumption. More intensive and internationally coordinated research programs are crucial for maximizing the BNF in Indian agriculture so as to sustain the environmental health without compromising food production.

**Table 8.6** Summary of BNF Contribution (Tg) in Indian Agriculture Using Ndfa and N-Fixing Potential Methods (2014 Area)

| | | Biological Nitrogen Fixation (Tg) | |
|---|---|---|---|
| **Crop** | **Area (mha)** | **Ndfa Method** | **BNF Potential** |
| Pulses | 25.33 | 0.94 | 1.39 |
| Groundnut | 5.53 | 0.50 | 0.44 |
| Soybean | 12.20 | 0.80 | 0.98 |
| Maize | 9.43 | 0.10 | 0.12 |
| Rice | 43.95 | 0.76 | 0.69–0.97 |
| Wheat | 31.19 | 0.65 | 0.28–0.41 |
| Other coarse cereals | 25.67 | (0.13)[a] | 0.13 |
| Sugarcane | 5.01 | (0.20)[a] | 0.20 |
| Cotton | 11.69 | (0.06)[a] | 0.06 |
| Oilseeds[b] | 10.80 | (0.05)[a] | 0.05 |
| Fodder legumes | 3.57 | (0.55)[a] | 0.55 |
| Fodder cereals | 4.46 | (0.06)[a] | 0.06 |
| N-fixing trees | 2.79 | (0.28)[a] | 0.28 |
| Non-N-fixing trees | 8.37 | (0.13)[a] | 0.13 |
| Total | | 5.20 | 5.35–5.76 |

[a]Area method using BNF potential.
[b]Other than soybean and groundnut.

# References


Agricultural Statistics, 2015. Agricultural Statistics at a Glance—2014. Oxford University Press, New Delhi, India.

Ayres, R.U., Schlesinger, W.H., Socolow, R.H., 1994. Human impacts on the carbon and nitrogen cycles. In: Socolow, R., Andrews, C., Berkhout, R., Thomas, V. (Eds.), Industrial Ecology and Global Change. Cambridge University Press, Cambridge, MA, pp. 121–155.

Boddey, R.M., Urquiaga, S., Alves, B.J., Reis, V., 2003. Endophytic nitrogen fixation in sugarcane: present knowledge and future applications. Plant Soil 252, 139–149.

Carlsson, G., Huss-Danell, K., 2003. Nitrogen fixation in perennial forage legumes in the field. Plant Soil 253, 353–372.

Cleveland, C.C., Townsend, A.R., Schimel, D.S., Fisher, H., Howarth, R.W., Hedin, L.O., Perakis, S.S., Latty, E.F., Von Fischer, J.C., Elseroad, A., 1999. Global patterns of terrestrial biological nitrogen ($N_2$) fixation in natural ecosystems. Global Biogeochemical Cycles 13, 623–645.

Danso, S., Bowen, G., Sanginga, N., 1992. Biological nitrogen fixation in trees in agro-ecosystems. Plant Soil 141, 177–196.

Danso, S., Hardarson, G., Zapata, F., 1993. Misconceptions and practical problems in the use of 15N soil enrichment techniques for estimating $N_2$ fixation. Plant Soil 152, 25–52.

De Bruijn, F.J., 2015. Biological Nitrogen Fixation. John Wiley, New Jersey, UK.

FAO, 2011. Current World Fertilizer Trends and Outlook to 2015. Food and Agricultural Organizations of United Nations, Rome, Italy, p. 41.

Forest Survey of India, 2013. India State of Forest Report. Ministry of Environment & Forests, Government of India, Dehradun, India, p. 252.

Fowler, D., Coyle, M., Skiba, U., Sutton, M.A., Cape, J.N., Reis, S., Sheppard, L.J., Jenkins, A., Grizzetti, B., Galloway, J.N., 2013. The global nitrogen cycle in the twenty-first century. Philosophical Transactions of the Royal Society of London B Biological Sciences 368, 20130164.

Galloway, J.N., Schlesinger, W.H., Levy, H., Michaels, A., Schnoor, J.L., 1995. Nitrogen fixation: anthropogenic enhancement-environmental response. Global Biogeochemical Cycles 9, 235–252.

Galloway, J.N., Cowling, E.B., Seitzinger, S.P., Socolow, R.H., 2002. Reactive nitrogen: too much of a good thing? Ambio 31, 60–63.

Galloway, J.N., Aber, J.D., Erisman, J.W., Seitzinger, S.P., Howarth, R.W., Cowling, E.B., Cosby, B.J., 2003. The nitrogen cascade. Bioscience 53, 341–356.

Galloway, J.N., Dentener, F.J., Capone, D.G., Boyer, E.W., Howarth, R.W., Seitzinger, S.P., Asner, G.P., Cleveland, C., Green, P., Holland, E., 2004. Nitrogen cycles: past, present, and future. Biogeochemistry 70, 153–226.

Galloway, J.N., Townsend, A.R., Erisman, J.W., Bekunda, M., Cai, Z., Freney, J.R., Martinelli, L.A., Seitzinger, S.P., Sutton, M.A., 2008. Transformation of the nitrogen cycle: recent trends, questions, and potential solutions. Science 320, 889–892.

Galloway, J.N., 1998. The global nitrogen cycle: changes and consequences. Environmental Pollution 102, 15–24.

Hand Book of Agriculture. 2014. Indian Council of Agricultural Research, New Delhi, p. 1617.

Herridge, D.F., Peoples, M.B., Boddey, R.M., 2008. Global inputs of biological nitrogen fixation in agricultural systems. Plant Soil 311, 1–18.

Ladha, J.K., Tirol-Padre, A., Reddy, C., Cassman, K., Verma, S., Powlson, D., vanKessel, C., Richter, DdB., Chakraborty, D., Pathak, H., 2016. Global nitrogen budgets in cereals: a 50-year assessment for maize, rice, and wheat production systems. Scientific Reports 6, 19355.

Mackenzie, F.T., Mackenzie, J.A., 1998. Our Changing Planet: An Introduction to Earth System Science and Global Environmental Change. Prentice Hall, New Jersey, p. 600.

Peoples, M., Baldock, J.A., 2001. Nitrogen dynamics of pastures: nitrogen fixation inputs, the impact of legumes on soil nitrogen fertility, and the contributions of fixed nitrogen to Australian farming systems. Animal Production Science 41, 327–346.

Postgate, J., 1998. Nitrogen Fixation, third ed. Cambridge University Press, Cambridge UK.

Rao, D.L.N., Gill, H.S., 2000. Residual effects of *Sesbania sesban* forestry on yield and nitrogen uptake by rice and wheat in a reclaimed alkali soil in Haryana. In: Abrol, I.P., Bronson, K.F., Duxbury, J.M., Gupta, R.K. (Eds.), Long Term Soil Fertility Experiments in Rice-Wheat Cropping Systems. Rice-Wheat Consortium Paper series 6, Rice-Wheat Consortium for the Indo-Gangetic Plains, New Delhi, India, pp. 139–148.

Rao, D.L.N., Gill, H.S., Abrol, I.P., 1990. Regional experience with perrenial *Sesbania* in India. In: Macklin, B., Evans, D.O. (Eds.), Perennial *Sesbania* Species in Agroforestry Systems, Nitrogen Fixing Tree Association, Special publication (90–91), Hawaii, USA, pp. 189–198.

Rao, D.L.N., 2002. Nitrogen fixation by tree legumes. In: Kannaiyan, S. (Ed.), Biotechnology of Biofertilizers. Kluwer Academic Publishers, Dordrecht, Netherlands, pp. 165–178.

Rao, D.L.N., 2014. Recent advances in biological nitrogen fixation in agricultural systems. Proceedings of the Indian National Science Academy 80, 359–378.

Rawat, A.K., Rao, D.L.N., Sahu, R.K., 2013. Effect of soybean inoculation with *Bradyrhizobium* and wheat inoculation with *Azotobacter* on their productivity and N turnover in a Vertisol. Archives of Agronomy and Soil Science 59, 1559–1571.

Robertson, G.P., Paul, E.A., Harwood, R.R., 2000. Greenhouse gases in intensive agriculture: contributions of individual gases to the radiative forcing of the atmosphere. Science 289, 1922–1925.

Rochester, I.J., Peoples, M.B., Constable, G.A., Gault, R.R., 1998. Faba beans and other legumes add nitrogen to irrigated cotton cropping systems. Australian Journal of Experimental Research 38, 253–260.

Salvagiotti, F., Cassman, K.G., Specht, J.E., Walters, D.T., Weiss, A., Dobermann, A., 2008. Nitrogen uptake, fixation and response to fertilizer N in soybeans: a review. Field Crops Research 108, 1–13.

Schlesinger, W.H., 2009. On the fate of anthropogenic nitrogen. PNAS 106, 203–208.

Smil, V., 1999. Nitrogen in crop production: an account of global flows. Global Biogeochemical Cycles 13, 647–662.

Smil, V., 2002. Nitrogen and food production: proteins for human diets. Ambio 31, 126–131.

Vitousek, P.M., Aber, J.D., Howarth, R.W., Likens, G.E., Matson, P.A., Schindler, D.W., Schlesinger, W.H., Tilman, D.G., 1997. Human alteration of the global nitrogen cycle: sources and consequences. Ecological Applications 7, 737–750.

Vitousek, P.M., Cassman, K., Cleveland, C., Crews, T., Field, C.B., Grimm, N.B., Howarth, R.W., Marino, R., Martinelli, L., Rastetter, E.B., 2002. Towards an ecological understanding of biological nitrogen fixation. Biogeochemistry 57, 1–45.

Vitousek, P.M., Menge, D.N., Reed, S.C., Cleveland, C.C., 2013. Biological nitrogen fixation: rates, patterns and ecological controls in terrestrial ecosystems. Philosophical Transactions of the Royal Society of London B Biological Sciences 368, 20130119.

9

# Nitrogen Management Paradigm in Horticulture Systems in India

A.N. Ganeshamurthy, T.R. Rupa, D. Kalaivanan, T.K. Radha
*ICAR-INDIAN INSTITUTE OF HORTICULTURAL RESEARCH, BENGALURU, INDIA*

**CHAPTER OUTLINE**



## Introduction

Nitrogen (N), a highly mobile plant nutrient when applied to a cropping system but not removed in harvested products is at risk of eventually leaving the field in gaseous or liquid form. Hence there is a need to understand all aspects of N flow in crop production systems, be it agriculture or horticulture. It is more so in horticulture because of a higher level of fertilizer use and a greater imbalanced use of nutrients. The cost of fertilizer is small compared to the cost of lost yield. Due to this monetary imbalance, the farmers' motivation to avoid overfertilization is small. In overfertilized crops, large amounts of N remain in soil after harvest. This includes residual soil mineral N or N present in crop residues. Both sources of N may have a harmful effect on the environment. This unnecessary surplus of N contradicts one of the key aims of sustainability. Regulators are interested in N "mass balance." Therefore, it is necessary to understand the N mass balance to help navigate the changing threat to environment from excess N and to convey message to farmers on the changing regulatory landscape (Ganeshamurthy, 2016).

**The Indian Nitrogen Assessment. http://dx.doi.org/10.1016/B978-0-12-811836-8.00009-4**

During the green revolution period, horticulture was not in the forefront. But later it picked up and the production of horticulture today (280.4 Mt) has exceeded the agriculture production (265 Mt). Initially, fertilizers were viewed as supplemental nutrient sources. During the green revolution period, emphasis was given to the importance of temporal plant diversity, maintenance of soil organic matter (SOM), long-term experimentation, and the need to understand and manage soil biological processes (Swaroop and Ganeshamurthy, 1998). It took a different turn when it was thought that fertilizers could support high yields without additions of organic amendments and horticultural intensification began with high emphasis on fertilizer application ignoring organic amendments (Scotti et al., 2015). This transition represented a fundamental paradigm shift. The focus of soil fertility management became the relatively small, ephemeral, soluble inorganic N and P pools rather than nutrient reservoirs with longer mean-residence times. With the aim to reduce the cost of production, the researchers made efforts to develop such technologies that minimize fertilizer costs while maximizing yields and profitability through the development of soil tests for plant-available nutrients combined with fertilizer trials conducted for a single growing season (Kemmler and Melicornet, 1989). Predicting plant-available soil N has proven to be more challenging, and efforts to develop a soil test for N continue. Though the pre-sidedress nitrate test is the most widely recommended soil N test currently used to assess plant-available N just before the exponential growth phase of the crop, it is not been practiced in India. The impacts of N losses from horticultural lands on aquatic ecosystems became apparent, prompting horticulturists to think about how best to achieve yields without harming the environment (Raghupathi and Ganeshamurthy, 2013). The application of ecological principles to develop horticultural production systems based on internal, biologically driven processes in an effort to reduce external inputs while achieving adequate yields was not directly promoted, but it took a different turn in promoting organic and other forms of horticulture production. In contrast, an integrated nutrient management strategy based on ecological concepts is intensively experimented across the crops and cropping systems in the country with the main aim of saving the fertilizer rather than addressing the environment safety issues. The problem of improved fertilizer use efficiency has been viewed mainly as a consequence of temporal asynchrony and spatial separation between applied nutrients and the crop. As a result, nutrient management research continues to emphasize the improved delivery of fertilizers such as split fertilizer applications, drip fertigation, placement, and modified N fertilizers such as neem-coated urea. Despite many concentrated efforts, mass balances indicate annual N and P inputs consistently exceed harvested exports by 40 to >100% resulting in substantial leakage of these nutrients to the environment.

A study using $^{15}N$ on the percent N derived from the fertilizer in different plant parts of Robusta banana (*Musa acuminata*) applied at various stage of the crop showed that much of the N was partitioned to vegetative parts except at shooting when fruits showed considerable percent N derived from fertilizer (Fig. 9.1). Even application at early vegetative stage contributed significantly to the N requirement of fruits, though

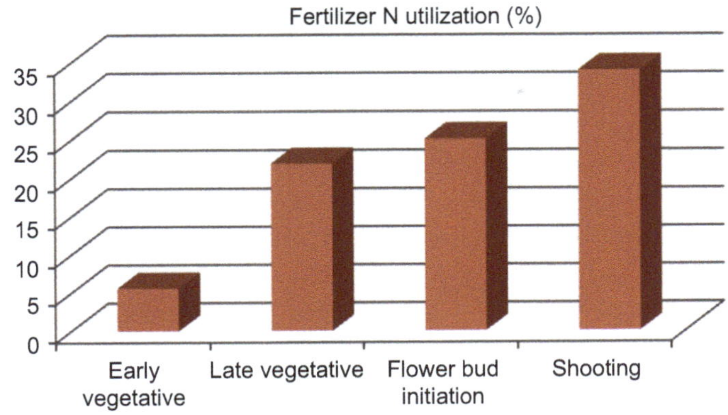


**FIGURE 9.1** Partitioning, utilization, and uptake of fertilizer N in "Robusta" banana (Keshava Murthy and Kotur, 1998).

application at shooting contributed to the maximum. However, total N recovery from all the split applications (22.3%) was rather low (Keshava Murthy and Kotur, 1998). The N derived from fertilizer revealed that broadcasted fertilizer N contributed only 6.8%, while fertilizer placed appropriately 30 cm. below the surface contributed 60% (Keshava Murthy and Kotur, 1998).

Concerted efforts were made in India to inhibit nitrification to prevent N loss through use of inhibitors. To overcome N losses, coating/blending urea with neem oil and other were tried. It was found that melicans, or bitters, present in neem (*Azadirachta indica* L.) products, when blended with urea, inhibit nitrification and volatilization culminating in reduced leaching losses in soil (Devakumar and Goswami, 2002; Suri et al., 2004). This lowered rates of nitrification result in accumulation of ammoniacal and other mineralized N, and facilitate its subsequent availability to the crops (Singh et al., 1989). Coating urea with neem oil leads to delayed dissolution and hydrolysis of urea to ammonia resulting into continuous and steady supply of N (Singh and Singh, 1989; Vyas et al., 1991; Upadhyay and Patel, 1992). Nitrification of the ammonia evolved was also inhibited by neem oil leading to longer persistence of applied urea and better supply of N, and its utilization by the crop at later stages (Biddappa and Sarkunanan, 1981). According to Prasad et al. (1999), neem products act as dual-purpose inhibitors of both ammonia volatilization and concurrent nitrification. Studies carried out by Kotur et al. (2007) showed 21% and 9% higher yields of tomato and French bean when N was applied as prilled urea coated with neem oil rather than plain prilled urea.

## Biogeochemical Processes in Horti-Ecosystems

Currently horticultural crops in India are heavily dependent on fertilizers for nutritional needs. This has led to a series of management changes such as use of drip irrigation, fertigation, plastic mulch, herbicides, and restructured horticulture ecosystems and uncoupled N cycle. Excess irrigation is one of the root causes of N loss in horticulture system. Hence irrigation scheduling and techniques needed to be integrated into strategies to improve N fertilizer efficiency to avoid nitrate leaching as well as denitrification losses. Drip irrigation or broadly known as micro-irrigation can save 30–70% of the irrigation water and increase yields by 25–80%. It was shown that in chili 100% application of recommended dose of fertilizers as water soluble fertilizers + liquid biofertilizers + panchagavya (a mixture of five products from cow such as milk, curd, cow dung, cow's urine and ghee in a proper ratio and allow it to ferment, the resulting product is known as Panchagavya) + humic acid (biostimulants) increased the N use efficiency both in monsoon (33.75 kg $kg^{-1}$ NPK $ha^{-1}$) and in summer seasons. The corresponding N use efficiency through soil application recommended dose fertilizer (RDF) was 18.83 kg $kg^{-1}$ NPK $ha^{-1}$ (Table 9.1) (Ciba and Manikandan, 2012).

Capsicum, tomato, and brinjal receiving 9, 18, or 30 g urea/plant through fertigation outyielded plants receiving sidedressed urea (15 g/plant at planting and 15 g/plant at first harvest). Mulching with black polyethylene sheets, especially in combination with overhead irrigation, considerably increased the total recovery of N applied to tomato (Goyal et al., 1985). Monoculture is the dominant system today and even simple rotations are disappearing. Noncash crops such as cover crops and intercrops have little or no cash value per se. Nevertheless these crops help to maintain internal cycling of nutrients through a variety of mechanisms which have been ignored. Specifically, with exploitation of underground water, horticulture cropping intensity has increased, and farmers are growing vegetables throughout the year. For example, in some parts of southern India tomato after tomato, chili after chili, cucurbits after cucurbits is a very common. Hence it is imperative to identify fertilizer-efficient cropping sequences. Of the six cropping sequences popular in this region, it was found that two cropping sequences, namely French bean–tomato–onion and Okra–cabbage–brinjal, were superior in utilizing applied N. The former cropping sequence was more efficient in using applied N to

**Table 9.1** Effect of Fertigation on Nitrogen Use Efficiency (kg $kg^{-1}$ NPK $ha^{-1}$) of Chilies (Ciba and Manikandan, 2012)

| Treatments | Nitrogen (kg $kg^{-1}$ NPK $ha^{-1}$) | |
|---|---|---|
| | Kharif | Summer |
| T1—RDF through soil application (120:60:30 kg NPK/ha) | 18.83 | 17.67 |
| T2—75% RDF as urea + DAP(as basal) + KCl | 20.08 | 19.42 |
| T3—75% RDF as water soluble fertilizers | 25.42 | 24.00 |
| T4—100% RDF as urea + DAP(as basal) + KCl | 23.25 | 22.00 |
| T5—100% RDF as water soluble fertilizers | 29.67 | 28.75 |
| T6—T2 + Liquid biofertilizers + Panchagavya + Humic acid | 27.92 | 27.67 |
| T7—T3 + Liquid biofertilizers + Panchagavya + Humic acid | 32.00 | 31.42 |
| T8—T4 + Liquid biofertilizers + Panchagavya + Humic acid | 30.08 | 28.92 |
| T9—T5 + Liquid biofertilizers + Panchagavya + Humic acid | 33.75 | 32.58 |

**Table 9.2** Nitrogen Utilization (%) in Two Vegetable Cropping Sequences (Kotur and Keshava Murthy, 1999)

| First Crop | Second Crop | Third Crop | Total |
|---|---|---|---|
| **French Bean–Tomato–Onion** | | | |
| 22.20 | 4.48 | 1.57 | 28.25 |
| 30.31 | 11.14 | 0.71 | 42.16 |
| 14.19 | 6.72 | 0.54 | 21.45 |
| **Okra–Cabbage–Brinjal** | | | |
| 9.46 | 2.03 | 1.74 | 13.30 |
| 24.47 | 2.48 | 0.56 | 27.51 |
| 20.20 | 0.62 | 1.91 | 22.73 |

the first crop, than the latter. Tomato–onion–French bean sequence was the best (42.16% N utilization) due to better use of N applied to the first crop (tomato being the best user, at 30.31%) and residual N by onion (11.14%) (Table 9.2) (Kotur and Keshava Murthy, 1999). Thus highly efficient cropping sequences need to be adopted in different situations to attain high utilization of applied N fertilizer.

A combination of crops instead of a single crop helps in economizing the N use and enhances its use efficiency as well. Koturet al. (2010) demonstrated that on an Alfisol N use efficiency of crop combinations, namely capsicum (onion)–watermelon (radish)–okra (French bean), was drastically reduced to 6.44–19.21% from 10.85 to 37.16%, compared to any of the solo (main) crops (Table 9.3).

Increased cropping intensity and over exploitation of the soil without sufficient organic inputs led to deterioration of soil health, leaching of N, ground water getting polluted with nitrates, and a spurt in pest and disease attack. Similarly in perennial crops, a shift toward high density orchards with little dependence on organic inputs and heavy dependence on inorganic fertilizers has exhausted the soil. Many examples of such a heavy exhaustion of soil are available. Pomegranate farmers are taking three to four year crops under high density with very heavy inorganic inputs and high levels of other agrochemicals and abandon the crop as by fourth year the crop becomes susceptible to wilts and blights and become uneconomical to continue. Similarly ultra-high density papaya is grown only for 12–18 months and uprooted as it becomes uneconomical to maintain further because of poor yield and quality both due to heavy exhaustion of the soil and attack by diseases such as papaya ringspot virus. Such examples are very common in high density mango, apple, and guava (Raghupathi and Ganeshamurthy, 2013; Ganeshamurthy, 2014). Ultimately, the soil nutrient cycle, especially the N and C cycle, is worst hit due to loss of carbon with heavy tillage, increased soil erosion, and depletion of SOM stocks (Ganeshamurthy et al., 2016), and increases the susceptibility of these ecosystems to nutrient overloading and nutrient losses (Singh and Ryan, 2015).

**Table 9.3** Crop Yield and N Fertilizer Utilization Under Solo Crops and Their Combinations (Koturet al., 2010)

| Treatment | Capsicum (Onion) | Watermelon (Radish) | Okra (French Bean) |
|---|---|---|---|
| **Yield[a] (kg/1.8 m$^2$)** | | | |
| Solo crop-1 | 6.253 | 15.539 | 5.644 |
| Solo crop-2 | 2.683 | 8.919 | 2.425 |
| Crop combination | 4.371 | 15.583 | 5.454 |
| SEm (±) | 0.1019 | 0.2537 | 0.1380 |
| CD ($p = .05$) | 0.3526 | 0.8778 | 0.4774 |
| **Fertilizer N Utilization (%)** | | | |
| Solo crop-1 | 10.85 | 23.70 | 22.76 |
| Solo crop-2 | 25.25 | 37.16 | 23.10 |
| Crop combination | 6.60 | 19.12 | 6.44 |
| SEm (±) | 0.590 | 0.843 | 0.821 |
| CD ($p = .05$) | 2.042 | 2.916 | 2.839 |

[a]Yield expressed as capsicum-equivalent in capsicum (onion), as watermelon-equivalent in watermelon (radish), and as okra-equivalent in okra (French bean) crop combinations based on prevalent wholesale market prices.

In these situations, microbial diversity is affected in the soil and the overall biologically mediated retention of added N becomes irrelevant.

Both under high density orchards and high intensive vegetable and flower cultivation, the arrangement of roots, soil aggregates, and pores create tremendous microscale spatial heterogeneity. As a result, environmental conditions favoring microbial processes and N-limiting or C-limiting conditions frequently co-occur within the soil matrix. C-limitation rarely occurs in the rhizosphere. But heavy use of herbicides and other pesticides coupled with reduced additions of organic amendments drastically reduces the microbial population itself (Wasim Aktar et al., 2009). Thus the decomposers in soil are usually C-limited (Koch et al., 2001) and less numerous (Rouatt et al., 1960). Under normal spaced orchards and plantations and less intensive annual horticultural systems, microbial assimilation of N occurs primarily in the rhizosphere, exactly the same location as plant uptake, while processes contributing to nutrient losses predominate in bulk soil (Smith and Tiedje, 1979). Increased denitrification can also occur in the rhizosphere under conditions of "Fertilizer N treadmill" which is common in intensive horticulture such as high density orchards, intensive vegetable and flower farms, and protected cultivation (Smith and Tiedje, 1979).

In modern horticulture systems (except organic production) N availability exceeds the capacity of the ecosystem to cycle or store the nutrients in internal reservoirs that can be accessed by plants or microorganisms. The degree of N saturation is greatest and is lost through leaching and denitrification and in some situations through $NH_3$ volatilization. Hence these production systems will not approach a balanced steady state.

The N content of horticulture produce is generally lower than those of agricultural crops. As a consequence the proportion of fertilizer N exported out of the systems through fruits and vegetables is low. Hence under high intensive horticulture systems, amount of free N available for loss through leaching, denitrification, and erosion is large.

## Toward an Ecosystem-Based Approach to Improving N Use Efficiency

In India, horticulture is a major component in watershed management. There is however, a need to develop a landscape-level strategy involving horticulture and agriculture landscapes, water bodies, wetlands to effectively protect sensitive natural ecosystems as these form the key strategy in ecosystem-scale N management (Wani et al., 2011; Pathak et al., 2013). A complimentary approach needs to be adapted to minimize nitrogen leakage to the environment by adjusting plant communities, enhancing nutrient use efficiencies through various management approaches, and develop such cropping systems that attain steady state where yields and soil reservoirs are maintained with nutrient inputs that are approximately equal harvested exports from the system.

A typical example of such a conceptual framework for an ecosystem-based approach is outlined in Fig. 9.2. The concept revolves around enhancing retention time of nitrogen in the system for a longer-time access of microorganisms and plants so that N is not let out into water bodies. This should address issues and processes such as soil microbial dynamics, soil aggregate development, microbially controlled processes that control N transformations, biological weathering with the goal of enhancing the internal cycling capacity, and reduce the need for chronic additions of surplus N. One of the main approaches is to diversify the sources of N input such as legume crops, use of N fixers, and use of recycled organic residues. Mineral nitrogen sources must be managed to enhance assimilation of N in biologically regulated sinks through both plant and microbially driven processes. The focus is more toward enhancing and expanding the sinks for the excess applied N, and will have long-term and cascading impacts on the internal cycling capacity in horticultural systems. The system should be developed in such a way that a greater proportion of total N inputs is accounted for, either as harvested exports or as N stored in the soil, in diversified rotations using low C:N residues as N sources compared to simplified rotations managed with inorganic N fertilizers (e.g., a sweet potato–based cropping system or a cover crop of sweet potato in orchards). It has been shown in orchards that green manuring increased the humic and fulvic acid fractions of SOM (Srivastava et al., 1984; Azam et al., 1985; Singh et al., 1992). Green manuring with *Sesbania* has been found to contribute more toward synthesis of fulvic acid as compared to humic acid fraction (Srivastava et al., 1984). This further favorably affects the soil microbial biomass. Azam et al. (1985) noted that soil biomass N increased by about 90% and 10% of its N was found in microbial biomass. When nutrient sources are manipulated, a smaller proportion of the nutrients added as organic residues is taken up by the

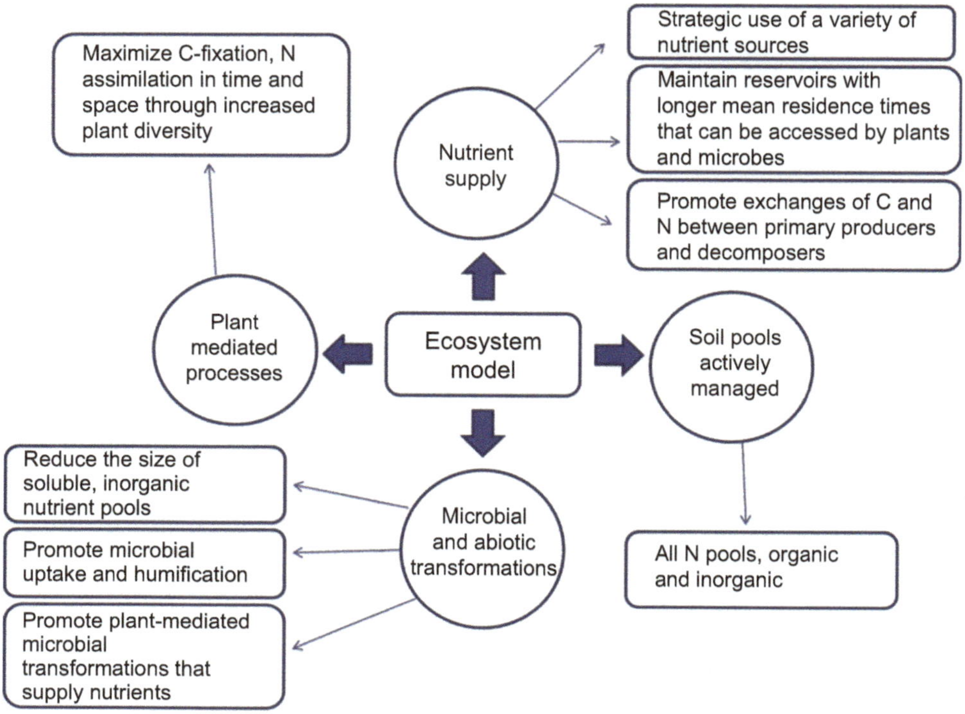


**FIGURE 9.2** Conceptual framework for an ecological approach to N management.

crop. However, a greater proportion is retained in various soil pools and is available to the crop in subsequent seasons. These concepts suggests that nutrient mass balance and soil storage reservoirs must be considered in conjunction with the current approach of estimating the proportion of fertilizer N harvested as crop yield to accurately judge the efficacy of added nutrients. Research data on these types are limited. There are numerous examples of research that target a single process either in natural or managed ecosystems that could contribute to an integrated strategy for nutrient management. Microbial interventions can enhance N use efficiency through both N fixation and enhancing the N retention time. *Azotobacter* culture is shown to enhance the N use efficiency (Wavhal et al., 1985). Similarly, *Arbuscular Mycorrhizal* Fungi (AMF) is of great significance for fruit crops. It forms a symbiotic association with the roots of plants and helps them in transfer of the nutrients from the soil to the root system. It was shown that AMF has the ability to absorb N with their extraradical hyphae and to supply the N to their host plants through intraradical hyphae (Govindarajulu et al., 2005). This helps to scavenge N from larger volumes of soil and prevents leakage of N to the water bodies and escape through gaseous losses.

## Using Plant Diversity to Restore Ecosystem Functions

The modern intensive horticulture is primarily a monoculture system. This is one of the root causes for the disturbance and deterioration of the ecosystem functions of the horticulture systems. To restore ecosystem functions, there is a need to rethink on diversifying the modern horticultural systems. Reinstating traditional rotations alone will not serve the purpose. There is also a need to consider the interrelationships between plants and their associated microbes and understand the mechanisms which regulate myriad processes that ultimately control ecosystem fluxes of C and N (Hooper and Vitousek, 1997; Fierer et al., 2001). There is scope to introduce plant diversity within the existing horticulture systems. While doing so there is need to select such plant diversity based on the capacity of a species to contribute to the ecosystem processes. Conservation horticulture in the existing fruit orchards and other perennial plantation systems through introduction of cover crops, mixed crops, and intercrops is a way toward achieving these goals (Ganeshamurthy et al., 2016). Selection of such crops should be based on their suitability to seasons, duration such as annual or perennial, plant characteristics such as N fixer or non-N fixer, litter chemistry, root exudates, allelopathic effects, rhizosphere environment enhanced disease suppression, reduced weed competition and herbicide requirements, and pesticide reductions due to beneficial arthropod communities. Such interventions can increase yields in systems without creating N saturated systems such as the intensive monoculture systems. Cover crops and intercrops can reduce leaching losses and surface runoff of N and yield additional profits to farmers without incurring any significant sacrifice in the main crop yield. Recent report on long term legume cultivation and cultivation of legumes in perennial orchard crops has shown potential for scavenging nitrate that had leached below the root zone (Ganeshamurthy, 2009; Ganeshamurthy et al., 2016). Significant addition of litter and other plant residues and their in situ decomposition enhances internal N and other nutrient supply through additions to labile SOM pools, such as particulate organic matter decomposed during subsequent growing seasons (Ganeshamurthy et al., 2016).

## Restoration of Ecosystem Function Through Plant–Microbial Interactions

Rhizosphere is the primary site for the plant–microbial interactions. The evolutionary function of root exudates is to develop codependency between the decomposers and primary producers—the plants in this case. This codependency has developed through evolution in return for nutrients supplied by microorganisms to get their supply of C by primary producers. This evolutionary symbiosis has not been exploited in horticultural systems (Naeem et al., 2000).

Ectomycorrhizal symbiosis is assumed to benefit plant growth. Once the mycorrhiza infects the plant root, it works lifelong for the main plant continuously and in return gets the soluble C from the plants. However, experiments show inconsistent benefits for plants and volatility of individual partnerships, which calls for a reevaluation of the presumed role of this symbiosis. Franklin et al. (2014) reported that plants abruptly switch from a mixed strategy with both mycorrhizal and nonmycorrhizal roots to a purely mycorrhizal strategy as soil N availability declines. In line with observations in field-scale isotope labeling experiments, their model explains why ectomycorrhizal symbiosis does not alleviate plant N limitation. Instead, market mechanisms may generate self-stabilization of the mycorrhizal strategy via N depletion feedback, even if plant growth is ultimately reduced. Franklin et al. (2014) suggested that this feedback mechanism maintains the strong N limitation and this may also have the capacity to eliminate or even reverse the expected positive effect of rising $CO_2$ on tree growth in strongly N-limited environment. The situation in modern horticulture systems is different from other strongly N-limited situations such as forest, semiarid savannas, and dry land horticulture systems. The production practices appear to have inadvertently reduced diversity, function, and efficiency in these symbioses, shifting a mutualistic, beneficial association to almost a parasitic relationship. While such partner discrimination has been confirmed for ectomycorrhizal symbiosis, Wyatt et al. (2016) reported that mutualistic relationships can be selected to actively restrict their partners' ability to directly obtain resources. This prediction arises because restriction increases the partner's comparative advantage when engaging in the mutualism, and therefore improves the restricting individuals' terms of trade. The combined effect of restriction and the maintenance of cooperation can lead to either an increase or a decrease in the restricted partner's fitness. These results emphasize how mutualists will continue to be favored to maximize their own fitness, even at the cost of their partner (Hamilton and Frank, 2001; Wyatt et al., 2016).

Tomato and pepper use only a small proportion of the N available from inorganic sources. Brinjal, however, is very effective in making use of plant nutrients already available in the soil. All species prefer $NO_3$-N. High concentrations of $NH_4$-N in the growing medium greatly reduce growth and productivity, especially in tomato and capsicum (Hegde, 1997). Although tomato requires large quantities of available nutrients to achieve high yields, it can utilize only a small percentage of the inorganic N available in a large volume of soil. Poor root development could explain low fertilizer N use by tomato in an N-rich topsoil (Jackson and Bloom, 1990). Unlike tomato and pepper, brinjal is very effective in making use of plant N already available in the soil (Shukla and Naik, 1993).

Crop varieties developed in recent time require very high nutrient inputs as the rate of biomass production is very high, and so is the nutrient demand. The exploitable rhizosphere area being restricted many a time, the rate of nutrient required by the plant may not be sufficient to produce such high yields. Under such situations the scope for mycorrhizal role in meeting the increased nutrient demands of these modern varieties is high. Utilizing this concept, an enhanced mycorrhizal population under an intensive horticulture system with very high yield potentials and a nutrient-saturated condition may benefit both partners and in turn reduces the loss of N from the system. Indian

Horticultural Research Institute, Bengaluru has taken a step forward by developing a technology wherein the spore production by mycorrhiza has been enhanced by 100 folds. This method of mycorrhiza production has paved the way to supply sufficient inoculum for the plants to get increased level of mycorrhizal association and thereby enhancing the symbiotic association (Pannerselvam, Ganeshamurthy and Selvakumar, 2014. IIHR patent 3817CHE/2/14).

## Microbially Mediated Processes

It is estimated that the amount of N present in the plants equals to that held by the microbial biomass in the soil. The nature, quality, and quantity of this microbially held N depends upon the crop, cropping system, tillage, quality and quantity of organics addition, fertilizer N addition, water management, herbicides and other agrochemicals used, etc. Under intensive horticulture in India, this component is under a rude shock and the quantity of N held in microbial biomass may not be significantly high as compared to 630 kg $ha^{-1}$ reported in the literature (Whitman et al., 1998). This is because the microbial biodiversity in intensive horticulture in India is affected due to continuous use of agrochemicals, addition of relatively excess N, and reduced application of organic amendments. Application of very heavy doses of fertilizer N and pesticides to chili crop adversely affected the populations of fungi, actinomycetes, and ammonifiers along with a drastic change in the eubacterial community profile and diversity over normal rates (Table 9.4) (Malhotra et al., 2015). The situation may be more severe in very intensive systems such as grapes, high density pomegranate, banana and other fruit crops, and crops grown under protected cultivation structures.

Conservation horticulture with reduced tillage, biofertilizer application, and addition of organic residues in the present day intensive horticulture systems can rekindle the

**Table 9.4** Microbial counts (cfu $g^{-1}$ soil), glomalin content (SE $\pm$ mean), and most probable numbers (MPN, $g^{-1}$ soil) of various functional groups of bacteria in soils of chili field with various rates of fertilizer and pesticide application (Jaya Malhotra et al., 2015)

| Parameter | Normal Input | High Input |
|---|---|---|
| Bacteria $\times 10^5$ | $39.5 \pm 1.0$ | $31.0 \pm 0.4$ |
| Fungi $\times 10^4$ | $15.8 \pm 1.2$ | $6.8 \pm 0.09$ |
| Actinomycetes $\times 10^5$ | $66.8 \pm 2.8$ | $38.8 \pm 2.5$ |
| Glomalin (mg $g^{-1}$ soil) | $10.6 \pm 0.2$ | $10.7 \pm 0.1$ |
| Associative $N_2$ fixers $\times 10^4$ | 1.7 (0.6–3.9) | 1.7 (0.6–3.9) |
| Ammonifiers $\times 10^4$ | 7.9 (2.3–22.0) | 2.1 (0.7–4.1) |
| Ammonium oxidizers $\times 10^5$ | 2.2 (0.7–4.4) | 1.1 (0.2–2.4) |
| Nitrite oxidizers $\times 10^2$ | 0.20 (0.02–0.99) | 0.18 (0.01–0.65) |

*cfu*, colony-forming units; *CL*, confidence limits in parenthesis.

system functions, and enhance the microbial processes and the quantity of microbially held N and carbon. A shift toward conservation horticulture would shift the microbial community dominance toward fungal dominance (Ganeshamurthy et al. 2016), and such increased abundance of fungal decomposers can lead to increased C and N retention (Holland and Coleman, 1987). This would reduce the exit of excess N from the system even under a saturated N situation. With the introduction of new biofertilizer consortiums by the Indian Institute of Horticultural Research (e.g., Arka Microbial Consortium, Arka Actino Plus, and Arka Fermented cocopeat), a change has occurred in the farmers attitude toward N and other chemical fertilizer application.

Application of these biofertilizers along with organic amendments would convert a greater proportion of metabolized C and soluble N to biomass (Chandra, 2005; Aoyama et al., 2000; Fliessbach and Mader, 2000). In addition to other channels of N exit from the horticultural systems, denitrification causes significant losses of N. However, horticulture systems are generally dominant in fungal population than other microbial groups. In such a situation, dissimilatory nitrate reduction to ammonium exceeds denitrification, and the resulting reduction in $NO_3$ availability to denitrifiers and leaching may contribute to N conservation in these ecosystems. Horticulture systems generally have a greater $NH_4$:$NO_3$ ratio. Though occurrence of organisms capable of performing dissimilatory nitrate reduction to ammonium has yet to be determined, this channel may be an important N-conserving process in horticulture ecosystems.

## Conclusion

Accomplishing enhanced nutrient use efficiency, high crop productivity, high quality produce, and environmental protection has become a major challenge in Indian intensive horticulture. But high yield sustainability and reduced environmental impacts are in conflict with each other. Nitrogen is the mineral nutrient that horticultural crops require in the greatest amount. In spite of the increased cost and conceivable adverse effects of $NO_3$ toxicity in the groundwater, inorganic sources of N fertilizers are being widely used. A major portion of this reactive-N, applied as fertilizer to horticulture crops/cropping systems, escapes into the environment with cascading negative effects on ecosystem functions and contributes to global warming. In this respect, green manure crops, organic manures have immense potential to partially substitute for inorganic N. Other than N applied as fertilizer, plant-available N is derived from SOM through mineralization. The available N status of soil varies with cropping systems and management practices. The quantity of N fertilizer to be applied to horticultural crops depends on the yield potential of the cultivar/variety, cropping system, soil nutrient status, and the growing conditions. As vegetative and reproductive stages overlap in these crops, the plants require nutrients even up to fruit ripening. Various approaches to efficient use of N would be to adopt conservation horticulture, enhance soil organic carbon, breed for cultivars having high N use efficiency, use of crop genotypes efficient in N acquisition, rhizosphere management, appropriate fertilizer source, split application of fertilizers, slow and controlled-release N fertilizers that extend the time of N availability for plant uptake, subsurface placement of N fertilizers, integrated use of inorganic, organic nutrient sources and microbial inoculants, multiculture systems, addition of litter, green manuring, fertigation, mulching, promote N use efficiency, and crop productivity. These practices also reduce N losses, thus making them environment friendly.

## Future Projections

- Efforts must be made to develop various strategies to synchronize N fertilizer application with crop N requirements to enable quick uptake and reduce N residence time in soil, thereby limiting losses due to denitrification and leaching. Geneticists and plant breeders must look for traits that enhance both external and internal N use efficiency.
- Future work should focus toward strategic use of diverse organic sources of N and enhancing and expanding the sinks for the excess applied N so that a greater proportion of total N inputs are accounted for, either as harvested exports or as N stored in the soil.

## References

Aoyama, M., Angers, D.A., N'Dayegamiye, A., Bissonnette, N., 2000. Metabolism of C-labeled 13 glucose in aggregates from soils with manure application. Soil Biology and Biochemistry 32, 295–300.

Azam, F., Malik, K.A., Sajjad, M.I., 1985. Transformation in soil and availability to plants of $^{15}N$ applied as inorganic fertiliser and legume residues. Plant and Soil 86, 3–13.

Biddappa, C.C., Sarkunanan, V., 1981. Effect of urea blended with neem cake on the nitrogen transformation in three typical rice soils of Orissa. Mysore Journal of Agricultural Sciences 15, 388–433.

Chandra, K., 2005. Organic Manures. Regional Centre of Organic Farming, Hebbal, Banglaore.

Ciba, C., Manikandan, N., 2012. Effect of drip fertigation on water, nutrient and agronomic use efficiency of chilli (*Capsicum annuum L.*)cv. KKM-1. Haryana Journal of Horticultural Sciences 41 (1 and 2), 64–67.

Devakumar, C., Goswami, B.K., 2002. Nematicidal principles from neem-isolation bioassay of some melicans. Pesticide Research 4, 79–84.

Fierer, N., Schimel, J.P., Cates, R.G., Zou, J., 2001. Influence of balsam poplar tannin fractions on carbon and nitrogen dynamics in Alaskan taiga floodplain soils. Soil Biology and Biochemistry 33, 1827–1839.

Fliessbach, A., Mader, P., 2000. Microbial biomass and size-density fractions differ between soils of organic and conventional agricultural systems. Soil Biology and Biochemistry 32, 757–768.

Franklin, O., Nasholm, T., Hogberg, P., Hogberg, M.N., 2014. Forests trapped in nitrogen limitation – an ecological market perspective on ectomycorrhizal symbiosis. The New Phytologist 2032, 657–666.

Ganeshamurthy, A.N., Ravindra, V., Panneerselvam, P., Sathyarahini, K., Bhat, R.M., 2016. Conservation horticulture in mango orchards: comparative effects of conventional and conservation management practices on soil properties of an Alfisol under seasonally dry tropical savanna climate. Journal of Agricultural Science 8, 173–188.

Ganeshamurthy, A.N., 2009. Soil changes following long-term cultivation of pulses. The Journal of Agricultural Science 147, 699–706.

Ganeshamurthy, A.N., 2014. Ground water decline putting perennial horticultural crops in peril: urgent policy intervention needed. NAAS News 14, 13–14.

Ganeshamurthy, A.N., March 9–10, 2016. Fertiliser Systems of Key Horticultural Crops in India. NPK Fertilisers India 2016 (Shangri-La's Eros Hotel, New Delhi, India).

Govindarajulu, M., Pfeffer, P.E., Jin, H., Abubaker, J., Douds, D.D., Allen, J.W., Bucking, H., Lammers, P.J., Shachar-Hill, Y., 2005. Nitrogen transfer in the arbuscular mycorrhizal symbiosis. Nature 435, 819–823.

Goyal, M.R., Rivera, L.E., Santiago, C.L., 1985. Nitrogen fertigation in drip irrigated peppers, tomatoes and eggplant. In: Drip/Trickle Irrigation in Action, Congress by American Society of Agricultural Engineers, Fresno, CA, Michigan, USA, 1, pp. 388–392.

Hamilton, E.W., Frank, D.A., 2001. Can plants stimulate soil microbes and their own nutrient supply? Evidence from a grazing tolerant grass. Ecology 82, 2397–2402.

Hegde, D.M., 1997. Nutrient Requirements of Solanaceous Vegetable Crops. Extension Bulletin/Food and Fertiliser Technology Centre, pp. 1–9, 0379–7587; 441.

Holland, E.A., Coleman, D.C., 1987. Litter placement effects on microbial and organic matter dynamics in an agroecosystem. Ecology 62, 425–433.

Hooper, D.U., Vitousek, P.M., 1997. The effects of plant composition and diversity on ecosystem processes. Science 277, 1302–1305.

Jackson, L.E., Bloom, A.J., 1990. Root distribution in relation to soil nitrogen availability in field grown tomatoes. Plant and Soil 128, 115–121.

Kemmler, G., Malicornet, H., 1989. Fertiliser Experiments the Need for a Long-term Approach, second ed. revised edition. International Potash Institute, Worblaufen – Bem/Switzerland, pp. 1–50.

Keshava Murthy, S.V., Kotur, S.C., 1998. Effect of ammonium sulphate and superphosphate on the comparative efficiency of N and P absorption and N utilization in "Robusta" banana *(Musa paradisiaca)* using labelled fertilisers. Indian Journal of Agricultural Sciences 68, 765–768.

Koch, B., Worm, J., Jensen, L.E., Hojberg, O., Nybroe, O., 2001. Carbon limitation induces sigma dependent gene expression in *Pseudomonas fluorescence* in soil. Applied and Environmental Microbiology 67, 3363–3370.

Kotur, S.C., Keshava Murthy, S.V., 1999. Fertiliser management in vegetable crops. Indian Horticulture 43, 23–25.

Kotur, S.C., Shilpashree, Y.P., Sheshshayee, M.S., Ramesh, P.R., 2007. Nitrogen use efficiency in tomato (*Lycopersicon esculentum L.*) and French bean (*Phaseolus vulgaris* L.) as influenced by coating of urea with neem oil and graded levels of nitrogen. Journal of Horticultural Sciences 2, 119–122.

Kotur, S.C., Ramesh, P.R., Anjaneyulu, K., Ramachandran, V., 2010. Assessment of fertiliser use efficiency under enhanced crop intensity of vegetables due to inter-cropping using $^{15}N$ and $^{32}P$ labelled fertilisers. Indian Journal of Agricultural Sciences 80, 325–326.

Malhotra, J., Aparna, K., Dua, A., Sangwan, N., Trimurtulu, N., Rao, D.L.N., Lal, R., 2015. Microbial and genetic ecology of tropical Vertisols under intensive chemical farming. Environmental Monitoring and Assessment 187, 1–17.

Naeem, S., Hahn, D.R., Schuurman, G., 2000. Producer-decomposer co-dependency influences biodiversity effects. Nature 403, 762–764.

Pathak, P., Chourasia, A.K., Wani, S.P., Sudi, R., 2013. Multiple impact of integrated watershed management in low rainfall Semi-Arid region: a case study from Eastern Rajasthan. India Journal of Water Resource and Protection 5, 27–36.

Prasad, R., Singh, D.K., Singh, R.K., 1999. Ammonia volatilization loss in rice-wheat cropping system and ways to minimize it. Fertiliser News 44, 53–56.

Raghupathi, H.B., Ganeshamurthy, A.N., 2013. Deterioration of irrigation water quality. Current Science 110, 764–766.

Rouatt, J.W., Katznelson, H., Payne, T.M.B., 1960. Statistical evaluation of rhizosphere effect. Soil Science Society of America Proceedings 24, 271–273.

Scotti, R., Bonanomi, G., Scelza, R., Zoina, A., Rao, M.A., 2015. Organic amendments as sustainable tool to recovery fertility in intensive agricultural systems. Journal of Soil Science and Plant Nutrition 15, 333–352.

Shukla, V., Naik, L.B., 1993. Agro-techniques for solanaceous vegetables. In: Advances in Horticulture, 5. Malhotra Publishing House, New Delhi, India, pp. 364–399.

Singh, B., Ryan, J., 2015. Managing Fertilisers to Enhance Soil Health, first ed. International Fertiliser Industry Association, Paris, France, pp. 1–24 www.fertiliser.org/Library.

Singh, M., Singh, T.A., 1989. Comparison of neem oil coated urea with some other coated urea fertilisers on an alkaline calcareous soil. Journal of Indian Society of Soil Science 37, 314–318.

Singh, J.S., Raghubanshi, A.S., Singh, R.S., Srivastava, S.C., 1989. Microbial biomass acts as a source of plant nutrients in tropical forest and savanna. Nature 338, 499–500.

Singh, Y., Singh, B., Khind, C.S., 1992. Nutrient transformations in soils amended with green manures. In: Stewart, B.A. (Ed.), Advances in Soil Science, 20. Edwards Brothers, Inc., Ann Arbor, MI, USA, p. 282.

Smith, M.S., Tiedje, J.M., 1979. The effect of roots on soil denitrification. Soil Science Society of America Journal 43, 951–955.

Srivastava, L.L., Ahmed, N., Mishra, B., 1984. Studies on nitrogen and carbon status of soils as affected by crop residues, farmyard manure and green manures. In: Nitrogen in Soils, Crops and Fertilisers. Indian Society of Soil Science, New Delhi. Bull no. 13:277–284.

Suri, I.K., Prasadand, R., Devakumar, C., 2004. Neemcoating of urea – present status and future trends. Fertiliser News 49, 21–24.

Swaroop, A., Ganeshamurthy, A.N., 1998. Emerging nutrient deficiencies in intensive cropping systems and remedial measures for sustainable high productivity. Fertiliser News 43, 37–50.

Upadhyay, D.N., Patel, G.R., 1992. Nitrogen management in rice. Journal of Indian Society of Soil Science 40, 383–385.

Vyas, B.N., Godrej, N.B., Mistry, K.B., 1991. Development and evaluation of neem extract as a coating for urea. Fertiliser News 36, 19–27.

Wani, S.P., Anantha, K.H., Sreedevi, T.K., Sudi, R., Singh, S.N., D'Souza, M., 2011. Assessing the environmental benefits of watershed development: evidence from the Indian Semi-Arid tropics. Journal of Sustainable Watershed Science & Management 1, 10–20.

Wasim Aktar, M.D., Sengupta, D., Chowdhury, A., 2009. Impact of pesticides use in agriculture: their benefits and hazards. Interdisciplinary Toxicology 2, 1–12.

Wavhal, K.N., Rane, D.A., Kamble, K.B., Choudhary, S.M., 1985. Nutritional studies in pomegranate. In: Third National Workshop on Arid Zone Fruit Research. MPAU, Rahuri.

Whitman, W.B., Coleman, D.C., Wiebe, W.J., 1998. Prokaryotes: the unseen majority. Proceedings of the National Academy of Sciences of the United States of America 95, 6578–6583.

Wyatt, G.A., Kiers, E.T., Gardner, A., West, S.A., 2016. Restricting mutualistic partners to enforce trade reliance. Nature Communications. http://dx.doi.org/10.1038/ncomms10322.

# 10

# Management and Use Efficiency of Fertilizer Nitrogen in Production of Cereals in India—Issues and Strategies

Bijay-Singh

*PUNJAB AGRICULTURAL UNIVERSITY, LUDHIANA, PUNJAB, INDIA*

**CHAPTER OUTLINE**



## Introduction

In the years to come, the Indian agriculture is going to witness a very delicate balance between inputs and outputs of reactive nitrogen (N). The demand of burgeoning population for enough and high-protein food will require substantial N inputs into agro-ecosystems to ensure high production levels. Increasing N inputs often leads to increased crop yields (Liu et al., 2010), but field experiments consistently demonstrate that the proportion of added N taken up by the crop plants tends to decline with increasing N inputs (Cassman, 1999). Thus cropping systems receiving high N inputs may lead to diminishing food production returns as fertilizer N inputs increase (Cassman et al., 2003; Tilman et al., 2002). Whether the proportion of N inputs captured in harvest grows, shrinks, or remains unchanged will have important impacts on both

The Indian Nitrogen Assessment. http://dx.doi.org/10.1016/B978-0-12-811836-8.00010-0

food production and N pollution. During last five decades, the share of mineral fertilizers in supplying N inputs in cropping systems in most of the countries including India has increased several folds, but the response of agricultural systems to the increased fertilizer N inputs has evolved differently in different countries. While some countries have improved their agro-environmental performances, in others the increased fertilization has produced low agronomical benefits and high losses to the environment. One of the objectives of this chapter, therefore, is to analyze the time trends in N inputs, N outputs in the crop yield, N balance, and N use efficiency in cropping systems in India.

To ensure minimal leakage of N from agriculture to other ecosystems, fertilizer N will have to be used efficiently to produce more food with less N inputs. Therefore, fertilizer N management strategies in cereal-based cropping systems in India will have to be formulated keeping in view the dual challenge of substantially increasing agricultural productivity and sustainability. Issues and management strategies pertaining to fertilizer N use in cereals can be best evaluated based on criteria for productivity, profitability, environmental impact, and sustainability. These criteria coincide with those used in the approach of ecological intensification for intensive crop production systems, which aims to satisfy the increasing demand for food, feed, fiber, and fuel while meeting acceptable standards of environmental quality (Cassman, 1999). Therefore, another objective of this paper is to analyze prevailing and emerging trends in fertilizer N management in cereal systems in India for enhancing fertilizer N use efficiency as well as crop productivity.

## Time Trends in Input, Output, and Balance of Nitrogen in Cropping Systems and Nitrogen Use Efficiency in India

Using FAO data bases, Lassaletta et al. (2014a) calculated N balances for annual total N inputs to cropland (sum of N in manure, synthetic fertilizer, symbiotic fixation and atmospheric deposition, in kg N $ha^{-1}$ $year^{-1}$) and the N removed in crop yield (expressed in harvested vegetal protein in kg N $ha^{-1}$ $year^{-1}$) for 124 countries, including India, during 1961–2009. Total annual crop production was calculated from the harvested yield of primary crops and their N content as reported by Lassaletta et al. (2014b). The area under cropping systems was estimated by summing up the area of all individual crops. Annual fertilizer N application data were obtained from the resources module of the FAOSTAT database and the International Fertilizer Association (www.fertilizer.org/). The biological N fixation was estimated following yield-based approach assuming that yield of N fixing crops included in the FAOSTAT database best aggregates variables associated with crop, soil, and climatic conditions including available N, soil moisture, vigor of stand, and other management factors influencing $N_2$ fixation: N fixed = %Ndfa × Y × BGN/NHI, where % Ndfa is the percentage of N uptake derived from $N_2$ fixation, Y is the yield (expressed in kg N $ha^{-1}$ $year^{-1}$), NHI is the N harvest index, defined as the ratio of the harvested material to the total above-ground N production, and BGN is a multiplicative factor expressing the total $N_2$ fixation including below-ground contributions associated with roots, nodules, and rhizo-deposition via exudates and decaying root cells and hyphae. To estimate the animal excretion factors, different ratios for dairy and for other cattle stocks were worked out using the dairy stocks provided in the livestock primary module of the FAOSTAT database. The proportion of N excreted used as manure applied onto cropland was taken from the estimates of Sheldrick et al. (2003) at the regional level for each type of animal. As proposed by Liu et al. (2010), 30% of the available manure was considered to be lost during management and storage before reaching the crop. Deposition of oxidized and reduced N compounds onto croplands was calculated from the Global News database (Seitzinger et al., 2010) by extrapolating linearly between available years. National annual average of input of N per ha was worked out and taken as input per ha into the surface of cropland. Using the data pertaining to N removed in crop yield and input of N to cropping systems (Lassaletta et al., 2014a, supplementary material 2 results, http://www.stacks.iop.org/ERL/9/105011/mmedia), N balance was calculated as the difference between input and output, and N use efficiency was expressed as percentage of total N input going into yield.

The data pertaining to total N input, N output in crop yield, N balance, and N use efficiency in India are plotted in Fig. 10.1 to show 50-year time trends. During 1961 to 2009, while total N input into cropping systems through organic manure, mineral fertilizers, symbiotic fixation, and atmospheric deposition increased from 30.3 to 135.2 kg N $ha^{-1}$ $year^{-1}$ (346%), N in harvestable output from the cropping systems increased from 16.6 to 42.7 kg N $ha^{-1}$ $year^{-1}$ (157%). That means the N balance, the difference between total N input and output in the crop yield, increased from 3.7 to

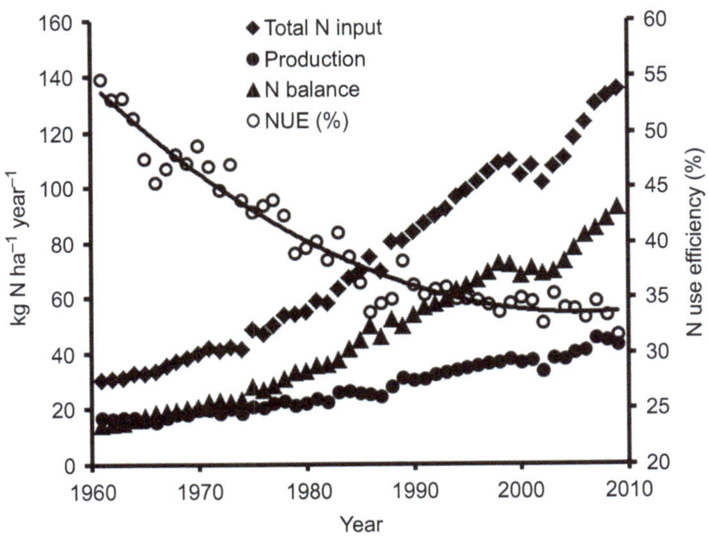


**FIGURE 10.1** Total N input, N output in crop yield, N balance, and N use efficiency in cropping systems in India during 1961–2009. *Data taken from Lassaletta, L., Billen, G., Grizzetti, B., Anglade, J., Garnier, J., 2014a. 50 year trends in nitrogen use efficiency of world cropping systems: the relationship between yield and nitrogen input to cropland. Environmental Research Letters 9, 105011.*

92.5 kg N $ha^{-1}$ $year^{-1}$ (575%). It suggests that unless substantial portion of the N balance ends up in becoming a part of the soil organic N, it can potentially contribute to environmental pollution by reaching surface and groundwater bodies or atmosphere. Not considered in straw N, which, as most straw is removed from the field in India or burned, is not returned to the soil. The continuously decreasing trend in N use efficiency since 1961 (Fig. 10.1) confirms that increasing amounts of N applied in cropping systems in India are not being efficiently translated into yield. One reason for this could be the low soil organic matter content of Indian soils. Agricultural soils contain a large pool of N that exists in organic combinations and it is distinguished from applied N as soil N. Generally, fertilizer N applications in agro-ecosystems rely on the premise that fertilization, rather than soil N supply, is the major source for crop uptake. A $^{15}$N-recovery experiment conducted in 13 diverse tropical agro-ecosystems and with six different crops convincingly showed that soil N is the primary source of N for crops (Dourado-Neto et al., 2010). Contribution of fertilizer to the mean total crop N uptake of $147 \pm 6$ kg N $ha^{-1}$ was only 21% $\pm$ 1%; soil N contributed 79% $\pm$ 1%. Calculating N uptake by the difference method, however, yields greater N fertilizer recovery than the $^{15}$N method (Bijay-Singh et al., 2001).

Integrative values of annual N transferred to crop yield in cropping systems (Y) and total N applied (F) as worked out by Lassaletta et al. (2014a) for the period of 1961–2009 can be considered as overall indicators of the agronomical and environmental performances of N in cropping systems in India. The Y/F ratio, or partial factor productivity of N ($PFP_N$) (Dobermann, 2007) is a measure of its N use efficiency, while the $F - Y$ difference is the regional N surplus (or N balance) representing the potential for hydrological or gaseous losses of N to the environment. Plots of Y, Y/F, and $F - Y$ as functions of F are shown in Fig. 10.2. The plot between Y and F was fitted to a function

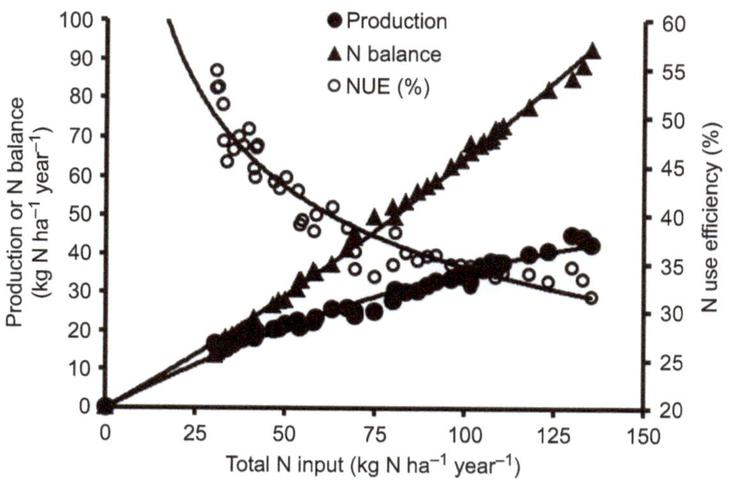


**FIGURE 10.2** Nitrogen output in crop yield, N balance, and N use efficiency as functions of total N input in cropping systems in India during 1961–2009. *Data taken from Lassaletta, L., Billen, G., Grizzetti, B., Anglade, J., Garnier, J., 2014a. 50 year trends in nitrogen use efficiency of world cropping systems: the relationship between yield and nitrogen input to cropland. Environmental Research Letters 9, 105011.*

satisfying the three properties: (1) the function intercept should be zero; (2) the slope of the function should be 1 at low fertilization; and (3) the function should reach a plateau at high fertilization. The first two properties reflect the fact that, in the long run, harvest cannot exceed N restitutions to the soil, and that the effect of low fertilization in strongly N-limited systems is characterized by an N use efficiency close to 1. The third property expresses the classical law of diminishing return and the fact that, in constant technical agronomical context, some other limiting factor will always impose a ceiling to production at saturating N availability. A hyperbolic function of the form $Y = Y_{max} \times F/(F + Y_{max})$ provided the best fit to the data. The term $Y_{max}$ represented the yield value reached at saturating N fertilization, as well as the value of fertilization at which a definite fraction of this maximum yield is reached. The $Y_{max}$ for India as plotted in Fig. 10.2 worked out to be 51 kg N $ha^{-1}$ $year^{-1}$ based on the data for the period 1961–2009 and 59 kg N $ha^{-1}$ $year^{-1}$ when based on the data for the period 1995–2009 (Lassaletta et al., 2014a, supplementary material 2, http://www.stacks.iop.org/ERL/9/105011/mmedia).

The two $Y_{max}$ values, defined as the protein yield that could be obtained from cropland at a maximum N fertilization rate with the corresponding cropping practices, characterize the past and current agricultural potential, respectively. Nearly the same $Y_{max}$ values for India for the two periods suggest that although fertilizer N rates are 2–4 times higher in India, N use efficiency has not changed substantially since the Green Revolution era and the N management needs to be greatly improved. Better irrigation management would also contribute to improved N use efficiency. In sharp contrast to the scenario in India, at a global scale, a shift during 1980s from one Y/F relationship characterized by a $Y_{max}$ of about 70 kg N $ha^{-1}$ $year^{-1}$occurred to an improved one with $Y_{max}$ of 110 kg N $ha^{-1}$ $year^{-1}$.

The relationship between Y and F gets translated into a regularly decreasing trend of N use efficiency for cropping systems in India (Fig. 10.2). Following the concept of yield gap (difference between the actual farmers' yield of a particular crop cultivar and the potential yield which could be obtained in the same place in the absence of limitation by nutrient and water, and with efficient control of pests and diseases) introduced by van Ittersum and Rabbinge (1997) and van Ittersum et al. (2013) to characterize the performance of agriculture of a given region, Lassaletta et al. (2014a) defined N-based yield gap as the relative difference between the maximum crop yield at saturating N fertilization (as deduced from the apparent relationship between N yield and N input) and the actual observed farm's yield. Calculated as $(Y_{max} - Y)/Y_{max}$, a dimensionless indicator of the degree of N limitation of current agricultural yields, the value for India turned out to be less than 0.3 (compared to more than 0.75 for North America) indicating low N use efficiency and no benefit in terms of yield to be expected from simple increase of N fertilization in the absence of radical agronomical improvement of the cropping system. In several developing countries, crop yields do not respond to increased inputs due to poor agronomical practices (George, 2014) and it may partly be true for India as well. If long-term variations of the soil organic N pool are not taken into consideration, the soil N balance or surplus is defined as the excess N brought to cropland over the N export by crop harvest (Figs. 10.1 and 10.2). It is an indicator of environmental losses of N. The N surplus (expressed in kg N $ha^{-1}$ $year^{-1}$) is strongly related to the N use efficiency.

## Nitrogen Use Efficiency and Fertilizer Nitrogen Management in Cereals

During 2011–12, food-grain driven cropping systems consumed 69.3% of the total N consumed in India (Anonymous, 2016). Given the variability encountered in agricultural systems, it is not an easy task to efficiently manage nutrients in cropping systems. For example, to achieve high $PFP_N$ (kg harvest product per kg N applied), managing the N supply from soil and other indigenous sources is as important as maximizing the fertilizer N use efficiency. According to Dobermann (2007), management in modern cereal production systems should aim to achieve agronomic efficiency of 20–35 kg grain $kg^{-1}$ N applied and it can be achieved only at recovery efficiencies of 50–70% which in turn depends on crop demand for N, supply of N from indigenous sources, fertilizer rate, timing, product, and mode of application.

$PFP_N$, defined as an aggregate N use efficiency index that includes contributions to crop yield derived from uptake of indigenous soil N, fertilizer N uptake efficiency, and the efficiency with which N acquired by the plant is converted to grain yield, has been decreasing exponentially in India since 1960s (Fig. 10.2). A decrease in $PFP_N$ is an expected consequence of the adoption of N fertilizers because yields move higher along a fixed-N response function, unless other factors shift the response function up. $PFP_N$, or N use efficiency expressed as percentage of total N applied (F) transferred to crop yield (Y), seems to have leveled off around 34% in recent years (Fig. 10.2) in India, possibly due to better extension services as compared to most of the other developing nations. On the other hand, the overall observed global trend is a distinct decrease of N use efficiency from 68% to 45% during 1961–80 but stabilizing around 47% during the last 30 years (Lassaletta et al., 2014a). To meet cereal demand in 2025 at a modest pace of increased N consumption, the global $PFP_N$ in cereals needs to increase at a rate of 0.1%–0.4% $year^{-1}$ (Dobermann and Cassman, 2005). Following improved fertilizer N management strategies in large areas, average cereal $PFP_N$ increased from 36 kg $kg^{-1}$ in 1981/85 to 44 kg $kg^{-1}$ by 2001–02 (+23%, 1.1% $year^{-1}$) in the UK. Similarly in USA, $PFP_N$ in maize increased from 42 kg $kg^{-1}$ in 1980 to 57 kg $kg^{-1}$ in 2000 (+36%, 1.6% $year^{-1}$) (Dobermann and Cassman, 2002). Knowing and managing the N supply from soil and other indigenous sources and maximizing the agronomic efficiency are equally important components for achieving high $PFP_N$ (Dobermann, 2007). In India, efforts to improve awareness of farmers about fertilizer best management practices, decision of Government of India to coat all subsidized urea (domestically produced and imported) supplied to farmers in the country with neem oil, an indigenous nitrification inhibitor, and linking fertilizer recommendations to soil health cards issued to large number of farmers should help increase fertilizer

N use efficiency in future. However, these efforts will revolve around intensive research input during next few years. For example, benefits in terms of increased N use efficiency and reduced urea application to crops will be possible only if new recommendations for applying neem-coated urea are worked out or it is used following site-specific N management strategies.

## Blanket Fertilizer Nitrogen Management Recommendations

In majority of the cereal growing regions in India, blanket recommendations for N management are practiced. Optimum crop yields can be obtained with the fixed-rate method, provided the nutrient application level is high enough to compensate for a possible low supply of the nutrient from the soil in any field in the region. Blanket fertilizer recommendations cannot ensure economy and efficiency of fertilizer N use in a given region because variation in soil fertility is not taken into account and there will be wastage of fertilizers in some fields. Ladha et al. (2003) analyzed 30 long-term experiments on rice–wheat cropping system from South Asia to investigate the extent of yield stagnation or decline when recommended rates of N, P, and K were applied for 7–23 years at different locations. Long-term experiments for yields of rice and wheat showed stagnation in 72% and 85% cases and a significant decline in 22% and 6% cases, respectively.

Several efficient N management strategies have been worked out to improve fertilizer N use efficiency in cereals. These include optimum fertilizer N rates, improved methods and timings of application and placement, use of specially formulated forms of fertilizer including those with urease and nitrification inhibitors, integrated use of fertilizer, manures, and/or crop residues to achieve better synchronization between the supply and requirement of N by cereals. Nevertheless, best management practices for fertilizer N in cereals do not simply consist of a universal set of recommendations. These are designed to achieve four main objectives of cropping systems management: productivity, profitability, sustainability, and environmental health. The phrase "right source at the right rate, right time, and right place" implies that each fertilizer N management practice is right in terms of the goals of sustainable production. A balance among the four rights is essential as it helps avoid too much emphasis on one at the expense of overlooking the others. For example, rate may sometimes be overemphasized, owing to its direct relation to cost. Source, time, and place are more frequently overlooked and hold opportunity for improving performance. The four aspects of fertilizer N management—source, rate, time, and place are completely interconnected and are also linked to the full set of management practices for the cropping system. None of the four can be right if any one of them is wrong. For a given situation, there can be more than one right combination of source, rate, timing, or placement, but when one of the four changes, the others may as well.

Farmers in India have developed a tendency to rely primarily on N fertilizers to maximize rice or wheat yields, rather than targeting achievable yields determined by local agronomic, economic, and environmental conditions. High levels of N fertilizers without appropriate balance with fertilizer P and K result in negative effects on cereal yields, the soil and the environment, and increased incidence of crop lodging, weed competition, and pest attacks. As intensive cereal systems remove large quantities of N, P, and K, an imbalanced application of nutrients leads to mining from the soil of the nutrient not supplied in adequate amounts. Yield levels are then determined by the amount of nutrient available in the least amount. In a long-term experiment on rice–wheat cropping system at Faizabad in India, Kumar and Yadav (2001) found that 20 years after continuously applying different combinations of N, P, and K rates to both the crops, the highest rate of yield decline in rice and wheat was found when 120 kg $ha^{-1}$ N was applied with no P and K. The lowest rate of decline was observed when N, P, and K were applied at 40, 35, and 33 kg $ha^{-1}$, respectively.

## Evolution in the Understanding of Researchers Regarding Better Management of Fertilizer Nitrogen in Cereals

With the advent of Green Revolution era in mid-1960s, high yielding varieties of cereals and N fertilizers were introduced in South Asia. Up till 1980s, focus was on achieving increased N fertilizer use efficiency by reducing the losses of applied N from the soil–plant system because the high recovery efficiency was considered as the key parameter of success. A significant research outcome of this period was the recommendation of applying fertilizer N to cereals in two or three split doses before flowering of the crop begins. Rates of fertilizer N were preset and uniform for regions with similar climate and landform, and became a basis of the blanket recommendations.

After 1990s, the trend in fertilizer N management research started shifting toward matching the application of fertilizer N to field-specific needs of cereals. The focal message which emerged from about two decades' research efforts was to feed the crop needs (Buresh, 2007). This evolution in the understanding of researchers was based on the recognition of the importance of managing N to accommodate the variability in field-, season-, and variety-specific needs of cereal crops for supplemental N. This change in the fertilizer N management scenario has resulted in a shift in the key parameter of success from increased recovery efficiency of fertilizer N to increased agronomic efficiency. With increasing awareness about the role of field-to-field and temporal variability in soil N supply in reducing fertilizer N use efficiency in cereals, guidelines were evolved for adjusting rates and timings to site-specific and crop growth stage–specific needs. The split application of fertilizer N remains an essential component of site-specific fertilizer N recommendations for cereals. As early application of fertilizer N often exceeds the demand of young crop plants for N, guidelines are now becoming available to ensure application of only moderate amounts of early N to match crop needs. A key component of recent advances in N management in cereals consists of using gadgets such as leaf color chart, chlorophyll meter, and optical sensors to apply varying rates and distribution of fertilizer N within the growing season to match crop demand for supplemental N.

## Site-Specific Fertilizer Nitrogen Management in Cereals

Due to large field-to-field variability of soil N supply, broad-based blanket recommendations for fertilizer N for cereals cannot help increase N use efficiency beyond a limit (Adhikari et al., 1999). Among many factors that influence N use efficiency, one potentially important factor is the uncertainty faced by the farmer in deciding the amount of fertilizer N to be applied (Lobell, 2007). This uncertainty can be reduced by knowing the N supplying capacity of the soil. Applying quantity of fertilizer N calculated by considering soil N supply resulted in the increase in N use efficiency by 30%–40%, and 7% increase in grain yield in 179 irrigated rice fields in Asia (Dobermann et al., 2002). Thus, improvement in the synchrony between crop N demand and N supply from soil and/or the applied N fertilizer is likely to be the most promising strategy to increase N use efficiency in cereal crops in India (Varinderpal-Singh et al., 2010). An important element of site-specific N management is the development and use of diagnostic tools and models that can assess "real" N need of crop plants. Recently, some non-invasive optical methods based on leaf greenness, absorbance, and/or reflectance of light by the intact leaf have been developed. These include use of chlorophyll meters (SPAD meter), leaf color charts (LCC), ground-based remote sensors, and digital, aerial, and satellite imageries. Over the last decade, chlorophyll meter, LCC, and hand-held GreenSeeker optical have been introduced in India to improve N use efficiency in cereal-based cropping systems. The LCC has the added advantage of being inexpensive.

Bijay-Singh et al. (2002), Maiti et al. (2004), Nagarajan et al. (2004), and Khurana et al. (2007) demonstrated that SPAD meter–based N management leads to increased N use efficiency. Experiments carried out by Balasubramanian et al. (1999) and Bijay-Singh et al. (2002) indicated that soils producing a grain yield of 3 t $ha^{-1}$ or more in no-N control plots do not need basal N application if SPAD-based N management starts two weeks after transplanting of rice. As rice seedlings need about 7 days to recover from transplanting shock, N uptake within 2 weeks of transplanting is very small. Hussain et al. (2000) evaluated fertilizer management in rice following sufficiency index approach. Sufficiency index defined as the SPAD value of the plot in question divided by that of a well-fertilized reference plot or strip was monitored at 7–10 day interval, and whenever it was less than the critical value of 0.90, 30 kg N $ha^{-1}$ was broadcasted up to the time of 50% flowering. Rice yields obtained for different cultivars were similar to those obtained in the fixed-time N application treatment but with 30 kg less N $ha^{-1}$. A small overfertilized plot needs to be established to follow sufficiency index approach but it has the advantage of being self-calibrating for different soils, seasons, and cultivars.

Real-time N management using LCC requires the identification of an optimal leaf color that must be maintained throughout the season to obtain high yields. As shade 4 on the LCC represents greenness equivalent to SPAD value somewhere between 35 and 37, it was found to be the threshold value for inbred rice varieties prevalent in northern India (Bijay-Singh et al., 2002; Varinderpal-Singh et al., 2007; Yadvinder-Singh et al., 2007; Thind et al., 2010; Maiti et al., 2004). For direct wet-seeded rice grown under northwest Indian conditions, LCC shade 3 was found to be the threshold value (Bijay-Singh et al., 2006). Shukla et al. (2004) established LCC 3, 4, and 5 as the critical values for *basmati*, inbred, and hybrid rice cultivars.

Farmers prefer less frequent monitoring of leaf color as they are strongly accustomed to apply fertilizer N at critical growth stages as per blanket recommendations. The fixed-time option involves monitoring of leaf color using LCC at the growth stages critical for adequate supply of N, and upward or downward applications of fertilizer N can then be adjusted based on the leaf color. Bijay-Singh et al. (2012) worked out a combination of fixed and adjustable rates of fertilizer N at critical stages of transplanted rice. A dose of 30 kg N $ha^{-1}$ at transplanting as prescriptive N management proved to be adequate for achieving high yields of rice. Corrective N management consisting of adjustable N doses was worked out as application of 45, 30, or 0 kg N $ha^{-1}$ depending upon leaf color to be <LCC shade 4, between LCC shade 4 and 5, or ≥LCC shade 5 both at maximum tillering and panicle initiation stages, and 30 kg N $ha^{-1}$ only if leaf color is less green than LCC shade 4 at initiation of flowering.

Unlike in rice, where N can be top-dressed any time during the growth period, N application in wheat has to be linked with irrigation event. Varinderpal-Singh et al. (2012) worked out a fertilizer N management strategy consisting of application of a basal dose of 25 kg N $ha^{-1}$ at planting, 45 kg N $ha^{-1}$ at first irrigation, and 30 or 45 kg N $ha^{-1}$ at second irrigation stage depending on the color of the leaf to be ≥LCC shade 4 or < LCC shade 4, which resulted in high yield levels as well as improved agronomic and recovery efficiencies of fertilizer N in wheat. In north-western India, fertilizer N is applied to irrigated wheat in two split doses at planting and at crown root initiation stages of the crop, but farmers have a tendency to apply another dose of N at maximum tillering stage to avoid the risk of N deficiency. Bijay-Singh et al. (2002, 2013) found that greenness of wheat leaves at maximum tillering stage was a function of N applied at planting and crown root initiation stages, and that wheat grain yield at maturity was determined by the level of greenness of leaves at maximum tillering stage as measured by SPAD meter. Grain yield response to N application at maximum tillering regressed against SPAD readings at maximum tillering revealed that an application of 30 kg N $ha^{-1}$ will increase wheat yield by 1.0 or 0.5 t $ha^{-1}$ when color of the wheat leaves was equivalent to or less than SPAD value of 32.5 or 42.5, respectively. It should serve as an easy and useful criterion for applying a supplementary dose of fertilizer N at maximum tillering stage of wheat.

Using the GreenSeeker optical sensor, the amount of additional N fertilizer required by rice or wheat was determined by taking the difference in estimated N uptake between estimates of yield potential with no added fertilizer N and with fertilizer N application and an efficiency factor. Bijay-Singh et al. (2011) showed that the optical sensor–guided fertilizer N applications resulted in high yield levels and high N use efficiency in wheat. Bijay-Singh et al. (2015) found that high yields along with high N use efficiency in rice can be achieved applying a moderate amount of fertilizer N at transplanting, and enough fertilizer N to meet the high N demand during the period between active tillering and panicle initiation before applying an optical sensor–guided fertilizer N dose at panicle initiation stage of rice.

Varinderpal-Singh et al. (2011) established and evaluated threshold leaf color to guide in-season need-based fertilizer N topdressings in maize in north-western India. Evaluation of the established threshold leaf greenness revealed that fertilizer N management using LCC shade 5 starting from six-leaf (V6) stage to before R1 stage resulted in improved agronomic and N recovery efficiency in different maize genotypes. There was no response to fertilizer N application at R1 stage. Matching fertilizer N supply with crop demand using threshold LCC shade 5 saved 25–50% fertilizer N as compared to blanket recommendation of 120 kg N $ha^{-1}$.

## Conclusion

The N removal by the crop as governed by the attained harvestable yield levels is the most important factor influencing fertilizer N use efficiency. The N output in the yield component of the cropping systems at saturating N fertilization levels works out to be low (51–59 kg N $ha^{-1}$ $year^{-1}$), based on last 50 or 15 years data, suggesting that although N fertilizer inputs have gone up dramatically, N use efficiency from the management practices in India has not changed substantially since the Green Revolution era and needs to be greatly improved. There exists a large variation in N use efficiencies in cereals because of following blanket recommendations for N in fields differing greatly in N supplying capacity of the soil. There is an evolutionary trend in the efforts to develop fertilizer N recommendations for enhancing N use efficiency in cereals. Improved synchronization between the supply and the uptake of N by the crop has been made in the form of a shift from blanket fertilizer N recommendations to site-specific need-based fertilizer N management scenarios. It is now being increasingly realized that rather than enhancing only the recovery efficiency of fertilizer N, fertilizer N management strategies should lead to high agronomic efficiency coupled with high yield levels. Improvement in N use efficiency will also become possible by the balanced use of N, P, and K fertilizers in cereals, and with more judicious water management.

## References

Adhikari, C., Bronson, K.F., Panaullah, G.M., Regmi, A.P., Saha, P.K., Dobermann, A., Olk, D.C., Hobbs, P. R., Pasuquin, E., 1999. On-farm N supply and N nutrition in the rice–wheat system of Nepal and Bangladesh. Field Crops Research 64, 273–286.

Anonymous, 2016. All-India Report on Input Survey: 2011-12. Agricultural Census Division, Ministry of Agriculture and Farmers Welfare, New Delhi.

Balasubramanian, V., Morales, A.C., Cruz, R.T., Abdulrachman, S., 1999. On-farm adaptation of knowledge-intensive nitrogen management technologies for rice systems. Nutrient Cycling in Agroecosystems 53, 59–69.

Bijay-Singh, Bronson, K.F., Yadvinder-Singh, Khera, T.S., Pasuquin, E., 2001. Nitrogen-15 balance as affected by rice straw management in a rice-wheat rotation in Northwest India. Nutrient Cycling in Agroecosystems 59, 227–237.

Bijay-Singh, Yadvinder-Singh, Ladha, J.K., Bronson, K.F., Balasubramanian, V., Jagdeep-Singh, Khind, C. S., 2002. Chlorophyll meter- and leaf color chart-based nitrogen management for rice and wheat in northwestern India. Agronomy Journal 94, 821–829.

Bijay-Singh, Gupta, R.K., Yadvinder-Singh, Gupta, S.K., Jagmohan-Singh, Bains, J.S., Vashistha, M., 2006. Need-based nitrogen management using leaf color chart in wet direct-seeded rice in northwestern India. Journal of New Seeds 8, 35–48.

Bijay-Singh, Sharma, R.K., Jaspreet-Kaur, Jat, M.L., Martin, K.L., Yadvinder-Singh, Varinderpal-Singh, Chandna, P., Choudhary, O.P., Gupta, R.K., Thind, H.S., Jagmohan-Singh, Uppal, H.S., Khurana, H.S., Ajay-Kumar, Uppal, R.K., Vashistha, M., Raun, W.R., Gupta, R., 2011. Assessment of the nitrogen management strategy using an optical sensor for irrigated wheat. Agronomy for Sustainable Development 31, 589–603.

Bijay-Singh, Varinderpal-Singh, Yadvinder-Singh, Thind, H.S., Ajay-Kumar, Gupta, R.K., Kaul, A., Vashistha, M., 2012. Fixed-time adjustable dose site-specific fertilizer nitrogen management in transplanted irrigated rice (*Oryza sativa* L.) in South Asia. Field Crops Research 126, 63–69.

Bijay-Singh, Varinderpal-Singh, Yadvinder-Singh, Thind, H.S., Ajay-Kumar, Satinderpal-Singh, Choudhary, O.P., Gupta, R.K., Vashistha, M., 2013. Supplementing fertilizer nitrogen application to irrigated wheat at maximum tillering stage using chlorophyll meter and optical sensor. Agricultural Research 2, 81–89.

Bijay-Singh, Varinderpal-Singh, Purba, J., Sharma, R.K., Jat, M.L., Yadvinder-Singh, Thind, H.S., Gupta, R.K., Choudhary, O.P., Chandna, P., Khurana, H.S., Kumar, A., Jagmohan-Singh, Uppal, H.S., Uppal, R.K., Vashistha, M., Gupta, R.K., 2015. Site-specific nitrogen management in irrigated transplanted rice (*Oryza sativa*) using an optical sensor. Precision Agriculture 16, 455–475.

Buresh, R.J., 2007. Fertile progress. Rice Today 6 (3), 32–33.

Cassman, K.G., 1999. Ecological intensification of cereal production systems:Yield potential, soil quality, and precision agriculture. Proceedings of the National Academy of Sciences of the United States of America 96, 5952–5959.

Cassman, K.G., Dobermann, A., Walters, D.T., Yang, H., 2003. Meeting cereal demand while protecting natural resources and improving environmental quality. Annual Review of Environment and Resources 28, 315–358.

Dobermann, A., Cassman, K.G., 2002. Plant nutrient management for enhanced productivity in intensive grain production systems of the United States and Asia. In: Progress in Plant Nutrition: Plenary Lectures of the XIV International Plant Nutrition Colloquium. Springer, Netherlands, pp. 153–175.

Dobermann, A., Cassman, K.G., 2005. Cereal area and nitrogen use efficiency are drivers of future nitrogen fertilizer consumption. Science China Life Sciences 48, 745–758.

Dobermann, A., Witt, C., Dawe, D., Gines, H.C., Nagarajan, R., Satawathananont, S., Son, T.T., Tan, P.S., Wang, G.H., Chien, N.V., Thoa, V.T.K., Phung, C.V., Stalin, P., Muthukrishnan, P., Ravi, V., Babu, M., Chatuporn, S., Kongchum, M., Sun, Q., Fu, R., Simbahan, G.C., Adviento, M.A.A., 2002. Site-specific nutrient management for intensive rice cropping systems in Asia. Field Crops Research 74, 37–66.

Dobermann, A., 2007. Nutrient use efficiency - measurement and management. In: Fertiliser Best Management Practices - General Principles, Strategy for Their Adoption and Voluntary Initiatives vs Regulations. International Fertiliser Industry Association (IFA), Paris, France, pp. 1–28.

Dourado-Neto, D., Powlson, D., Bakar, R.A., Bacchi, O.O.S., Basanta, M.D.V., Keerthisinghe, G., Ismaili, M., Rahman, S.M., Reichardt, K., Safwat, M.S.A., Sangakkara, R., 2010. Multiseason recoveries of organic and inorganic nitrogen-15 in tropical cropping systems. Soil Science Society of America Journal 74, 139–152.

George, T., 2014. Why crop yields in developing countries have not kept pace with advances in agronomy. Global Food Security 3, 49–58.

Hussain, F., Bronson, K.F., Yadvinder-Singh, Bijay-Singh, Peng, S., 2000. Use of chlorophyll meter sufficiency indices for nitrogen management of irrigated rice in Asia. Agronomy Journal 92, 875–879.

Khurana, H.S., Phillips, S.B., Bijay-Singh, Dobermann, A., Sidhu, A.S., Yadvinder-Singh, Peng, S., 2007. Performance of site-specific nutrient management for irrigated, transplanted rice in Northwest India. Agronomy Journal 99, 1436–1447.

Kumar, A., Yadav, D.S., 2001. Long-term effects of fertilizers on the soil fertility and productivity of a rice-wheat system. Journal of Agronomy and Crop Science 186, 47–54.

Ladha, J.K., Pathak, H., Padre, A.T., Dawe, D., Gupta, R.K., 2003. Productivity trends in intensive rice-wheat cropping systems in Asia. In: Ladha, J.K., Hill, J.E., Duxbury, J.M., Gupta, R.J., Buresh, R.J. (Eds.), Improving the Productivity and Sustainability of Rice-Wheat Systems: Issues and Impacts, American Society of Agronomy Special Publication, 65. ASA, CSSA, SSSA, Madison, WI, USA, pp. 45–76.

Lassaletta, L., Billen, G., Grizzetti, B., Anglade, J., Garnier, J., 2014a. 50 year trends in nitrogen use efficiency of world cropping systems: the relationship between yield and nitrogen input to cropland. Environmental Research Letters 9, 105011.

Lassaletta, L., Billen, G., Grizzetti, B., Garnier, J., Leach, A.M., Galloway, J.N., 2014b. Food and feed trade as a driver in the global nitrogen cycle: 50-year trends. Biogeochemistry 118, 225–241.

Liu, J., You, L., Amini, M., Obersteiner, M., Herrero, M., Zehnder, A.J., Yang, H., 2010. A high-resolution assessment on global nitrogen flows in cropland. Proceedings of the National Academy of Sciences of the United States of America 107, 8035–8040.

Lobell, D.B., 2007. The cost of uncertainty for nitrogen fertilizer management: a sensitivity analysis. Field Crops Research 100, 210–217.

Maiti, D., Das, D.K., Karak, T., Banerjee, M., 2004. Management of nitrogen through the use of leaf color chart (LCC) and soil plant analysis development (SPAD) or chlorophyll meter in rice under irrigated ecosystem. The Scientific World Journal 4, 838–846.

Nagarajan, R., Ramanathan, S., Muthukrishnan, P., Stalin, P., Ravi, V., Babu, M., Selvam, S., Sivanatham, M., Dobermann, A., Witt, C., 2004. Site-specific nutrient management in irrigated rice systems of Tamil Nadu, India. In: Doberman, A., Witt, C., Dawe, D. (Eds.), Increasing Productivity of Intensive Rice Systems through Site-Specific Nutrient Management. Science Publishers, Inc., Enfield, NH and International Rice Research Institute, Los Baños, Philippines, pp. 101–123.

Seitzinger, S.P., Mayorga, E., Bouwman, A.F., Kroeze, C., Beusen, A.H.W., Billen, G., Harrison, J.A., 2010. Global river nutrient export: a scenario analysis of past and future trends. Global Biogeochemical Cycles 24, GB0A08.

Sheldrick, W., Syers, J.K., Lingard, J., 2003. Contribution of livestock excreta to nutrient balances. Nutrient Cycling in Agroecosystems 66, 119–131.

Shukla, A.K., Ladha, J.K., Singh, V.K., Dwivedi, B.S., Balasubramanian, V., Gupta, R.K., Sharma, S.K., Singh, Y., Pathak, H., Pandey, P.S., Padre, A.T., Yadav, R.L., 2004. Calibrating the leaf color chart for nitrogen management in different genotypes of rice and wheat in a systems perspective. Agronomy Journal 96, 1606–1621.

Thind, H.S., Bijay-Singh, Pannu, R.P.S., Yadvinder-Singh, Varinderpal-Singh, Gupta, R.K., Vashistha, M., Jagmohan-Singh, Kumar, A., 2010. Relative performance of neem (*Azadirachta indica*) coated urea *vis-a-vis* ordinary urea applied to rice on the basis of soil test or following need based nitrogen management using leaf colour chart. Nutrient Cycling in Agroecosystems 87, 1–8.

Tilman, D., Cassman, K.G., Matson, P.A., Naylor, R., Polasky, S., 2002. Agricultural sustainability and intensive production practices. Nature 418 (6898), 671–677.

van Ittersum, M.K., Rabbinge, R., 1997. Concepts in production ecology for analysis and quantification of agricultural input-output combinations. Field Crops Research 52, 197–208.

van Ittersum, M.K., Cassman, K.G., Grassini, P., Wolf, J., Tittonell, P., Hochman, Z., 2013. Yield gap analysis with local to global relevance - a review. Field Crops Research 143, 4–17.

Varinderpal-Singh, Yadvinder-Singh, Bijay-Singh, Baldev-Singh, Gupta, R.K., Jagmohan-Singh, Ladha, J. K., Balasubramanian, V., 2007. Performance of site specific nitrogen management for irrigated transplanted rice in northwestern India. Archives of Agronomy and Soil Science 53, 567–579.

Varinderpal-Singh, Bijay-Singh, Yadvinder-Singh, Thind, H.S., Gupta, R.K., 2010. Need based nitrogen management using the chlorophyll meter and leaf colour chart in rice and wheat in South Asia: a review. Nutrient Cycling in Agroecosystems 88, 361–380.

Varinderpal-Singh, Yadvinder-Singh, Bijay-Singh, Thind, H.S., Ajay-Kumar, Vashistha, M., 2011. Calibrating the leaf colour chart for need based fertilizer nitrogen management in different maize (*Zea mays* L.) genotypes. Field Crops Research 120, 276–282.

Varinderpal-Singh, Bijay-Singh, Yadvinder-Singh, Thind, H.S., Gobinder-Singh, Satwinderjit-Kaur, Kumar, A., Vashistha, M., 2012. Establishment of threshold leaf colour greenness for need-based fertilizer nitrogen management in irrigated wheat (*Triticum aestivum* L.) using leaf colour chart. Field Crops Research 130, 109–119.

Yadvinder-Singh, Bijay-Singh, Ladha, J.K., Bains, J.S., Gupta, R.K., Jagmohan-Singh, Balasubramanian, V., 2007. On-farm evaluation of leaf color chart for need-based nitrogen management in irrigated transplanted rice in northwestern India. Nutrient Cycling in Agroecosystems 78, 167–176.

# 11

# Plant Nitrogen Use Efficiency

V. Jain[1], Y.P. Abrol[2]

[1]INDIAN COUNCIL OF AGRICULTURAL RESEARCH, NEW DELHI, INDIA;
[2]ING-SCON, DPS MARG, NEW DELHI, INDIA

**CHAPTER OUTLINE**



## Introduction

Nitrogen is the most widely used fertilizer in India and its consumption has grown dramatically. In 1950–51, nitrogen (N) fertilizer consumption was only 0.55 Mt (55,000 t), which increased to 17.4 Mt in 2015–2016 and India has emerged to be the second largest producer/consumer of N fertilizer in the world (FAI, 2016).

Nitrogen use efficiency (NUE) for cereals, which are the major consumers of N fertilizers, is around 33%. This results in losses of N in various forms to soil, air, and water, and is of a concern for both economic and environmental reasons. Projected 50–70% enhancement in cereal production by 2050 will necessitate doubling the N fertilizer consumption (FAI, 2016). It is therefore imperative that every effort is made to improve the nitrogen use efficiency of crop plants.

In the above context, as in-depth analysis was conducted over a number of years with seedlings as well as field-grown wheat plants, main shoot, and uniculm wheat. Series of observations to determine the demands by different plant parts, over the growth and development period, and for different genotypes is summarized below.

**The Indian Nitrogen Assessment. http://dx.doi.org/10.1016/B978-0-12-811836-8.00011-2**

## Uptake and Assimilation of Nitrate

Nitrate-N along with some exchangeable ammonia is a prime source of nitrogen for the crop plants. Nitrate ions are taken up by the plant roots through a low affinity transporter system (LATS) which probably includes a signaling role and helps induce the expression of transporter/assimilatory genes of the nitrate assimilation pathway. The major amount of nitrate, especially when the soil concentration is below 1 mM, is taken up by the roots via inducible high affinity transporter system (iHATS) and constitutive high affinity transporter (cHATS) system (Crawford and Glass, 1998). Once nitrate ($NO_3^-$) enters into the plant roots its reduction is catalyzed by enzyme of the nitrate assimilation pathway, i.e., nitrate reductase (NR) (Crawford, 1995). In most of the higher plants the reduction of nitrate to nitrite is done by NADH:NR (NADH as an electron donor). This enzyme has low Km for the substrate nitrate and is located in cytoplasm. Subsequently, the enzyme nitrite reductase (NiR), located in chloroplasts, which utilizes reduced ferrodoxin ($Fd_{red}$) as an electron donor ,wherein six electrons are transferred from reduced ferrodoxin to nitrite (Sechley et al., 1992), converts nitrite ($NO_2^-$) to ammonium ($NH_4^+$).

Ammonium ions generated from different sources, i.e., nitrate reduction, direct absorption by roots, photorespiration, or metabolism of other compounds are incorporated into organic form through GS/GOGAT cycle (Freeman et al., 1990). The first enzyme of this cycle glutamine synthetase (GS) converts amino acid glutamate to glutamine in the presence of ATP, $NH_4^+$, and divalent cations ($Mg_2^+/Mn_2^+/Co_2^+$) as cofactors. GS is present in cytoplasm ($GS_1$) as well as in chloroplasts ($GS_2$). Finally, enzyme glutamate synthase (GOGAT) catalyzes the reductive transfer of glutamine to 2-oxoglutarate resulting in two molecules of glutamate, of which one molecule enters back into the GS-GOGAT pathway and the other goes into the biosynthetic pathway. Depending on the type of reductant used, there are two forms of GOGAT enzyme: Fd-GOGAT and NADH-GOGAT which occur in the chloroplasts in leaves and plastids in roots, respectively (Kamachi et al., 1992).

## Contribution of Different Plant Parts

In cereals, the role of roots is not solely limited to water and nutrient uptake. Cereal roots can also very effectively reduce nitrate. However, major contribution toward reduction of nitrate is from leaves, especially in light. The leaves consist of blade and cylindrical sheath, wrapped around the subtending leaf. The internodes are present which increase in length from basal to most distal. In addition, awns, ears, and grains also contribute to nitrate reduction when green. The contribution of the individual plant parts was assessed and it was shown that nitrate is primarily assimilated by the leaf blades. The leaf sheaths, internodes, and reproductive parts collectively reduced about 23% while roots contribute about 9–10% of the total nitrate reduced (Abrol et al., 1976; Nair and Abrol, 1979). The maximum NR activity in leaf blades was followed by sheaths, spike components, and internodes. The ontogenic pattern in the various stages correlated with soil N concentrations. Percent contributions of various plant parts in nitrate reduction in wheat as well as barley are given in Table 11.1.

## Contribution of Upper Three Leaf Blades

Among the leaf blades, the upper three leaves (8th, 9th, and 10th (flag) contributed more than 70% to the grain N and more than 50% of the total reduced N per plant. This was calculated on the basis of NR integration method as well as leaf blade removal method (see Table 11.2). Maximum reduction of up to 23% N per plant was observed when flag leaf blades were removed, and when all the three upper leaf blades were removed the reduced N per plant decreased by 51% (Abdin et al., 1994) along with substantial reduction in grain N. Assessment by graphical integration method reduced N decline by 46% (Table 11.2). When only the flag leaf blade was left on the main shoot and lower two leaf blades were removed the flag leaf blades could compensate for the other two leaves that were active in the post-anthesis stage (Abdin et al., 1994).

The flag leaf blade (at position 9), when intact on the shoot, reduced 8% more N in the absence of 7th and 8th leaf blades. No such compensation was observed toward reduced N when either 7th or 8th leaf blade was intact, and when others were removed. The compensation observed in case of flag leaf blade was due to much higher nitrate reductase activity in these blades as compared to other lower leaves.

### Pattern of Nitrate Assimilation

Nitrate reductase activity in the sequentially formed leaf blades was high in the first formed ones and declined in the subsequently formed ones. Ontogenetical pattern of NR activity correlated with the soil nitrate levels (Pokhriyal and Abrol, 1980), which were also very low during the later stages of growth (Figs. 11.1A and B).

This was, therefore, one of the important reasons for the very low activity of NR in the later-formed leaf blades. These upper leaves had higher weight, were larger, and remained green for a much longer time as compared to the lower leaves indicating that there might be large unrealized potential for nitrate assimilation in these upper leaves

**Table 11.1** Nitrate Reduced by Various Plant Parts in Wheat and Barley

| | Percent Contribution | |
|---|---|---|
| **Plant part** | **Wheat[a]** | **Barley[b]** |
| Leaf blades | 72.6 | 79.9 |
| Sheaths | 9.8 | 10.1 |
| Internodes | 11.2 | 4.0 |
| Developing ears | 6.6 | 6.0 |

[a]Abrol et al. (1976).
[b]Chatterjee et al. (1979).

**Table 11.2** Contribution of Upper Three Leaf Blades Toward Total Reduced Nitrogen as Obtained in the Defoliation Experiment and When Calculated on the Basis of Integration of NR Activity of the Individual Leaf Blade

| | Reduced N (μmoles Plant$^{-1}$)[b] | |
|---|---|---|
| **Leaf Blade Position** | **Defoliation Experiment** | **Calculated by Integrating In Vivo NR Activity**[a] |
| Control | | |
| 6th leaf | 5612 (100) | 4474 ± 26.0 (100) |
| 7th leaf | 672 (12) | 534 ± 14.0 (12) |
| 8th leaf | 814 (15) | 624 ± 14 (14) |
| 9th leaf | 1296 (23) | 897 ± 35.3 (20) |
| 7 + 8 + 9 leaves | 2782 (50)[c] | 2055 (46) |

Values in parentheses represent the percentage contribution by the individual leaf blades toward total reduced nitrogen plant$^{-1}$.
[a]Values are corrected by dividing by a factor of 1.6 (Abrol et al., 1976).
[b]Total reduced nitrogen plant$^{-1}$.
[c]The value increased to 31% when both 7th and 8th leaf blades were removed, suggesting a compensatory increase of 8% in total nitrogen harvest by flag leaves.
Abdin, M.Z., Kaim, M.S., Bansal, K.C., Abrol, Y.P., 1994. Nitrate assimilation and nitrogen harvest in uniculm"Gigas" wheat (*Triticum aestivum* L.). Journal of Agronomy and Crop Science 173, 225–229.

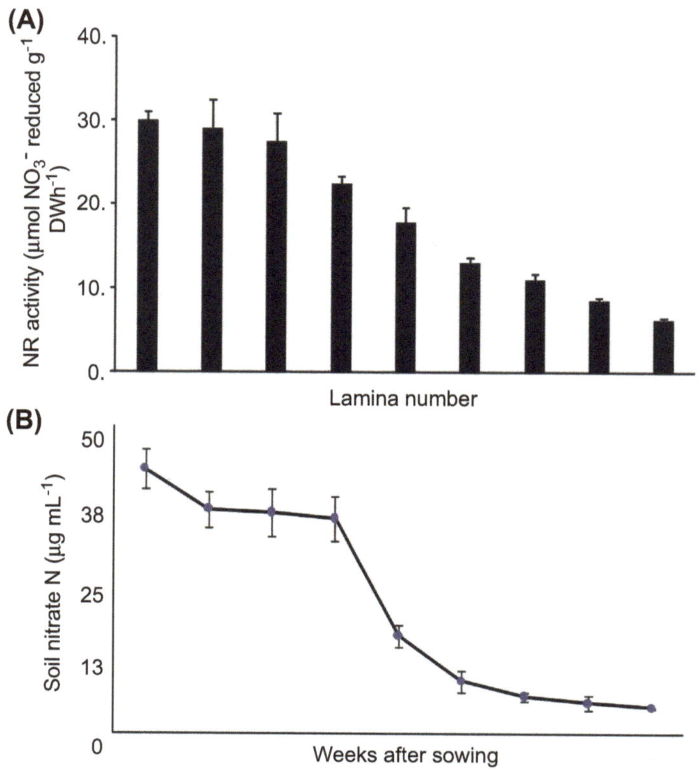


**FIGURE 11.1** (A): *In vivo* nitrate reductase activity in the leaves on the main shoot of field-grown wheat. The activity was determined in the fully expanded leaf blades (Mean ± SE) (B): Soil nitrate-N (μg mL$^{-1}$) at 50 mean cm depth. Each value is a mean of three replicates. Vertical bars indicate SE of means. *Abrol, Y.P., 1990. Pattern of nitrate assimilation and grain nitrogen yield in field grown wheat (*Triticum aestivum*). In: van Beusichem, M.L. (Eds.), Plant Nutrition – Physiology and Applications. Kluwer Academic Publishers, pp. 773–778.*

which mainly contributes to the total plant as well as grain N. Total shoot analysis for NR, nitrate ($NO_3^-$), and total nitrogen revealed similar picture except that peak N accumulation took place during shoot elongation/flag leaves expansion phase (Kaim et al., 1983).

## Potential Nitrate Reductase Activity

Taking into consideration the dry weight and duration for which the leaf blades remain green, it was assumed that the upper leaf blades may have large unutilized potential of N reduction which could be realized if nitrate was available during the later stages of plant growth, i.e., during flowering and grain formation. Hence, each of the fully expanded leaf blades from the main shoot was incubated in Hoagland's solution containing 15 mM nitrate. The NR activities in the upper leaves enhanced manifold and only very slight increase was observed in the NR activities of lower leaf blades. Hence, the activity of the enzyme nitrate reductase could be enhanced and maintained in the upper leaves, which is crucial during reproductive phase if nitrate concentration is maintained in the soil (Abrol, 1990).

## Variation in Nitrate Reductase Activity Among Genotypes

Genotypic differences in NR activity have been reported in various cereals (Ahmad et al., 2007). In wheat genotypes more than twofold variability exists in NR activity (Abdin and Abrol, 1997). These genetic differences in NR activity are also visible in reduced N per plant and in some genotypes higher grain N was also observed. The high nitrate reductase (HNR) genotypes had greater nitrate content and higher activity of nitrate reductase as compared to LNR genotypes. The comparison of 15 days old seedlings of two genotypes clearly showed that HNR genotypes had better growth and this might be due to better nitrogen harvest (Table 11.3).

The HNR genotypes maintained higher activity of the NR enzyme even under lower soil N levels as compared to LNR genotypes (Abdin et al., 1996). On a dry weight basis,

**Table 11.3** Comparison of Dry Weight, Nitrate Content, and Nitrogen in 15-day-old Seedlings of LNR and HNR Wheat Genotypes. The Values Represent Mean of Three Replicates and Values in Parentheses Represent Least Significant Difference at 1% Probability

| Parameter | LNR | HNR | Difference |
|---|---|---|---|
| Nitrate (μmol plant$^{-1}$) | 9.716 (0.0943) | 23.187 (2.0628) | 2.39 fold |
| Dry weight (g plant$^{-1}$) | 0.017 (0.0025) | 0.036 (0.0057) | 2.12 fold |
| Nitrogen (mg plant$^{-1}$) | 1.010 (0.1217) | 2.223 (0.2808) | 2.20 fold |

Anjana, U.S., Abrol, Y.P., Iqbal, M., 2011. Modulation of nitrogen utilization efficiency in wheat genotypes differing in nitrate reductase activity. Journal of Plant Nutrition 34, 920–933.

large differences of up to 2- to 2.5-fold in the activity of NR have been observed (Ahmad et al., 2007). Even the leafy vegetables like spinach had large variation in NR activity as well as accumulation of nitrate in leaves (Anjana et al., 2007). However, the inability to observe such large differences in N utilization efficiency under field conditions among contrasting genotypes were due to the low soil N levels toward the later stages of crop growth that limited the nitrate assimilating capacity of even HNR genotypes. When the potential of upper leaf blades of two contrasting genotypes of wheat was compared, by incubating them in the Hoagland solution containing 15 mM nitrate, the enhancement in NR activity was more in HNR genotypes as compared to LNR genotypes. The HNR genotypes maintained higher NR activity even at very low concentrations of nitrate in the feeding medium, thereby suggesting the scavenging property of HNR genotype as far as N acquisition is concerned.

## Biochemical Basis of Differential NR Activity in Cultivars

Series of biochemical investigations were conducted to understand the reasons for the differences in the NR activity among the genotypes in crop plants. The pattern of N uptake, accumulation, assimilation, and its partitioning between vegetative parts and developing grains was investigated to explore the ways to maximize grain N and enhance the nitrogen use efficiency. The complete N assimilation and mobilization was investigated to understand the physiological and biochemical basis of split application of N fertilizers (Anjana et al., 2011; Jain et al., 2011).

The detailed analysis during kinetic studies in wheat seedlings revealed that the uptake of nitrate from the medium (Das et al., 2006; Anjana et al., 2011) was high in HNR genotypes as compared to LNR genotypes. These genotypes differed in Km and $V_{max}$ of the nitrate uptake from the medium (Table 11.4).

It was noted that the affinity of NR enzyme was higher for nitrate in "HNR" as compared to the LNR genotype. With the increasing concentration of the nitrate in the medium, the nitrate reductase activity in HNR genotype increased more than in LNR

**Table 11.4** The Values of Km and $V_{max}$ of the Various Enzymes of the N Assimilation Pathway in Contrasting Wheat Genotypes

| | LNR | | HNR | |
|---|---|---|---|---|
| **Enzyme** | **Km** | **$V_{max}$** | **Km** | **$V_{max}$** |
| NR | 0.2699 | 0.8615 | 0.1921 | 1.1586 |
| NiR | 0.393 | 3.54 | 0.182 | 7.052 |
| GS | 3.636 | 0.315 | 2.1015 | 0.447 |
| GOGAT | 0.892 | 0.0343 | 0.631 | 0.459 |
| GDH | 0.466 | 0.0271 | 0.278 | 0.0302 |

Anjana, U.S., Abrol, Y.P., Iqbal, M., 2011. Modulation of nitrogen utilization efficiency in wheat genotypes differing in nitrate reductase activity. Journal of Plant Nutrition 34, 920–933.

genotype. Similarly, the Km for nitrate reductase (NiR), GS, and glutamate synthetase (GOGAT) was lower and $V_{max}$ was higher in HNR genotypes as compared with LNR genotypes. The downstream enzymes of N assimilation pathway had higher affinities for their substrates in HNR genotypes.

The studies in contrasting genotypes also showed that these genotypes differ not only in the activity of enzyme nitrate reductase, but in the activity of all the downstream enzymes of N assimilation pathway. The studies were conducted in 15-day-old seedlings of contrasting genotypes (Anjana et al., 2011) and field-grown plants of wheat (Jain et al., 2011). High NR activity was also associated with higher nitrate content of the genotypes (Tables 11.3 and 11.5). The activities of the enzymes of ammonia assimilation pathway were also greater in HNR genotypes studied as compared to LNR genotypes.

The studies conclusively showed that higher nitrate content in HNR genotypes is due to its better nitrate uptake capacity. The biomass of HNR genotypes was more compared to LNR genotypes. Hence it is possible that high NR activity is reflected in the plant biomass. The studies conducted in 15 days old seedlings (Anjana et al., 2011) as well as in field-grown wheat plants (Jain et al., 2011) showed that the N assimilating enzymes between HNR and LNR genotypes differed by similar magnitude. Hence it was shown that the downstream enzymes of the pathway function in accordance with the enzyme nitrate reductase in a coordinated manner. It was also observed that the transcript levels of NR were higher in HNR genotypes (Jain and Abrol, 2005).

# Physiological Basis of Split Application/Slow-Release N Fertilizers

Nitrogenous fertilizers as per the management practices are invariably applied at early growth stages of the crop and the N concentration in soil at this stage reaches to 100 mM and subsequently declines to as low as 10 mM during flag leaf blade emergence (Ahmad et al., 1999). One of the approaches to enhance NUE was to identify the genotypes which use N naturally in a more efficient manner, i.e., HNR genotypes. The genetic studies

**Table 11.5** The Comparison of Activities of Nitrate and Ammonia Assimilating Enzymes in HNR and LNR Genotypes

| Parameter | LNR | HNR | Difference |
|---|---|---|---|
| NRA (μmol $NO_2^{-1}$ $h^{-1}$ $mg^{-1}$ protein) | 0.159 (0.0025) | 0.448(0.0057) | 2.82-fold |
| NiR (μmol $NO_2^{-}h^{-1}$ $mg^{-1}$ protein) | 1.628 (0.0245) | 3.016 (0.0401) | 1.85-fold |
| GS (μmol γ-GHA $min^{-1}$ $mg^{-1}$protein) | 0.115 (0.0696) | 0.172 (0.01601) | 1.50-fold |
| GOGAT (μmol NADH $min^{-1}mg^{-1}$protein) | 0.0052 ($49 \times 10^{-5}$) | 0.0078 ($114 \times 10^{-5}$) | 1.49-fold |
| GDH (μmol NADH $min^{-1}mg^{-1}$protein) | 0.0132 (0.0002) | 0.0262 (0.0006) | 1.97-fold |

The NR and NiR activities are in vivo, GS, GOGAT activities are in vitro LSD (0.05) values are shown in parentheses

Anjana, U.S., Abrol, Y.P., Iqbal, M., 2011. Modulation of nitrogen utilization efficiency in wheat genotypes differing in nitrate reductase activity. Journal of Plant Nutrition 34, 920–933.

revealed that the level of enzyme is heritable. The second approach was to employ such a management strategy that the nitrate was available for crop growth even at the later stages of growth. It was observed that split application of nitrogenous fertilizers improved the yield and biomass of the crop plants. To reap the advantage of split application of N it was important to ascertain the contribution of individual leaves since major amount of nitrate is assimilated by the leaf blades. The total nitrate reduced by the leaf blades is shown in Table 11.1. As the successive leaves are formed, the total nitrate assimilated by these leaves is reduced indicating that their potential is not being utilized fully. Split application of N enhanced the in vivo activity of the flag leaves due to increased nitrate availability that was known to limit the NR activity in upper leaves during the later stages of wheat growth (Kumar et al., 1981; Abrol, 1990).

According to the prevalent practice, at that time N was applied as basal dose or maximum in two splits, i.e., half of the total dose at sowing and other half at the time of tiller formation (~30 days after sowing) to ensure good tillering. Hence, series of investigations were conducted in an effort to improve the management practices so as to increase the availability of nitrate during later stages of growth (Abrol et al., 1999). It was shown that by good management practices it was possible to enhance the N utilization under field conditions, and the potential of upper leaves for N assimilation could also be increased. Same amount of N at 120 kg $ha^{-1}$ applied in four splits of 30 kg N $ha^{-1}$ each, last split during reproductive phase, instead of two splits of 60 kg N $ha^{-1}$ each, significantly enhanced the nitrate assimilation capacity of the upper leaves. When this was assessed in the contrasting genotypes, i.e., HNR and LNR, magnitude of the increase of NR activity was high in "HNR" genotypes as compared to LNR genotypes as shown in figure below (Fig. 11.2).

The split application of N fertilizer also increased N mobilization toward grains and total N content of the vegetative parts declined prior to anthesis, and grain N and grain protein percentage was enhanced in wheat (Ahmad et al., 2007) (Table 11.6).

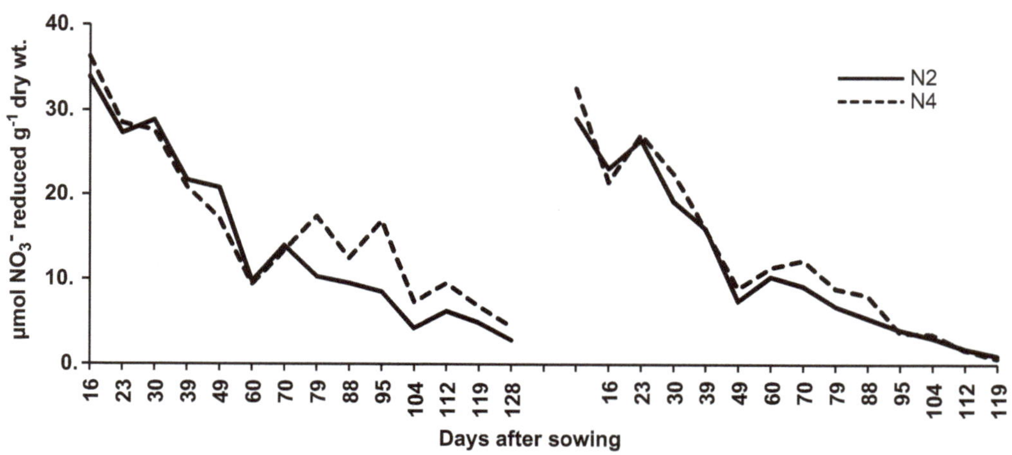


**FIGURE 11.2** In vivo nitrate reductase activity (µmol $NO_3^-g^{-1}$ $DWh^{-1}$) in the fully expanded laminae in wheat ("HNR"&"LNR") genotypes. The nitrogen was applied in two (solid line) and four (broken line) splits.

**Table 11.6** Effect of splitting N fertilizer Dose (120 kg N $ha^{-1}$) on Nitrogen and Protein Accumulation in Wheat Cultivars Differing in NR Activity. $S_2$-N Fertilizer in Two Splits and $S_4$- N Fertilizer Applied in Four Splits

| | | N Accumulated (mg $Plant^{-1}$) | | |
|---|---|---|---|---|
| **Genotype** | **Frequency of Splitting** | **Whole Plant** | **Grains** | **Grain Protein (%)** |
| LNR | $S_2$ | 90.0 | 80.2 | 10.2 |
| | $S_4$ | 107.8 | 81.0 | 10.8 |
| HNR | $S_2$ | 122.5 | 82.5 | 9.6 |
| | $S_4$ | 179.9 | 84.5 | 11.3 |

Abrol (1990).

Hence, the full genetic potential of HNR and LNR genotypes was assessed under non-limiting supply of nitrogen. The HNR genotype possessed greater nitrate content and high NR activity. The higher nitrate content in HNR was due to better uptake of the available nitrate by these genotypes as compared to LNR genotypes. The activities (Jain et al., 2011) and Km $V_{max}$ (Anjana et al., 2011) of all the other downstream enzymes of the nitrate assimilation pathway were higher in HNR genotype. Better growth of HNR genotypes was, therefore the result of better N harvest than LNR genotype. Flag leaf blades of the HNR genotype had higher activities of the enzymes of nitrate assimilation pathway, i.e., nitrite reductase, GS, and glutamate synthase as compared to the activities of these enzymes in LNR genotypes when nitrate-N was non-limiting. It was also observed that N mobilization was higher in HNR genotypes (Jain et al., 2011).

## Conclusion

HNR and LNR genotypes did not differ much in their N utilization efficiency under field conditions because of the decline in nitrate assimilation due to the sharp decrease in soil nitrate levels at that stage when N fertilizers were applied as basal dose or in two splits (Abrol, 1990). The plants therefore could express their full genetic potential. Application of N in four splits maintained the nitrate levels in the soil during reproductive stage of plant growth. Application of N fertilizers enhanced NR activity at later stages of growth with yield improvement. Enhancement was more in HNR genotypes (Abrol et al., 1999).

Many studies have suggested that the NR activity alone could be considered as an index of reduced N availability in the plant and plants with high NR activity possess greater potential for accumulation of reduced N if soil nitrate is available throughout the growing period. However, studies in contrasting wheat genotypes with large variation in NR activity showed that there is a difference in amount of N assimilated, and HNR genotypes had much higher potential of harvesting N and mobilize the same to grains due to well-coordinated system of N uptake and utilization.

# References

Abdin, M.Z., Abrol, Y.P., 1997. Effect of split nitrogen application on grain nitrogen and nitrogen harvest of wheat genotypes. In: Ando, T., Matsumoto, H., Moni, S., Seskya, S. (Eds.), Plant Nutrition for Sustainable Food Production and Environment. Kluwer Academic Publishers, pp. 635–638.

Abdin, M.Z., Bansal, K.C., Kumar, P.A., Abrol, Y.P., 1996. Effect of split nitrogen application on growth and yield of wheat (*Trticum aestivum* L.) genotypes with different N assimilation potential. Journal of Agronomy and Crop Science 176, 88–92.

Abdin, M.Z., Kaim, M.S., Bansal, K.C., Abrol, Y.P., 1994. Nitrate assimilation and nitrogen harvest in uniculm"Gigas" wheat (*Triticum aestivum* L.). Journal of Agronomy and Crop Science 173, 225–229.

Abrol, Y.P., Chatterjee, S.R., Anandakumar, P.A., Jain, V., 1999. Improvement in nitrogen use efficiency – physiological and molecular approaches. Current Science 76, 1357–1364.

Abrol, Y.P., 1990. Pattern of nitrate assimilation and grain nitrogen yield in field grown wheat (*Triticum aestivum*). In: van Beusichem, M.L. (Ed.), Plant Nutrition – Physiology and Applications. Kluwer Academic Publishers, pp. 773–778.

Abrol, Y.P., Kaim, M.S., Nair, T.V.R., 1976. Nitrogen assimilation, its mobilization and accumulation in wheat (*Triticum aestivum* L.) grains. Cereal Research Communications 4, 431–440.

Ahmad, A., Abrol, Y.P., Abdin, M.Z., 1999. Effect of split application of sulphur and nitrogen on growth and yield attributes of *Brassica* genotypes differing in their time of flowering. Canadian Journal of Plant Science 79, 175–180.

Ahmad, A., Jain, V., Abrol, Y.P., 2007. Physiological and biochemical aspects of nitrogen use efficiency in crop plants. In: Abrol, Y.P., Raghuram, N., Sachdev, M.S. (Eds.), Agricultural Nitrogen Use and its Environmental Implications. I.K. International Publishing House Pvt. Ltd., pp. 305–326

Anjana, U.S., Abrol, Y.P., Iqbal, M., 2011. Modulation of nitrogen utilization efficiency in wheat genotypes differing in nitrate reductase activity. Journal of Plant Nutrition 34, 920–933.

Anjana, U.S., Iqbal, M., Abrol, Y.P., 2007. Are nitrate concentrations in leafy vegetables within safe limits? Current Science 92, 355–360.

Chatterjee, S.R., Pokhriyal, T.C., Abrol, Y.P., 1979. Nitrate assimilation and grain protein accumulation in barley (*Hordeum vulgare* L.). Proceedings of the American Society of Plant Physiology S–264.

Crawford, N.M., 1995. Nitrate: nutrient and signal for plant growth. Plant Cell 7, 859–868.

Crawford, N.M., Glass, A.D.M., 1998. Molecular and physiological aspects of nitrate uptake in plants. Trends in Plant Science 3, 389–395.

Das, R., Jain, V., Smitha, A., Mrinal, B., Srivastava, G.C., 2006. Kinetics of nitrate uptake system in wheat genotypes. Indian Journal of Plant Physiology 11, 160–165.

FAI, 2016. Trends in fertilizer production and consumption in India. Indian Nitrogen Assessment (in press).

Freeman, J., Marquez, A.J., Wallsgrove, R.M., Sarrelainen, R., Forde, B.G., 1990. Molecular analysis of barley mutants efficient in chloroplast glutamine synthetase. Plant Molecular Biology 14, 297–311.

Jain, V., Abrol, Y.P., 2005. Regulation of nitrate reductase activity by phytohormones in wheat seedlings. Journal of Plant Biology 32, 43–48.

Jain, V., Khetarpal, S., Das, R., Abrol, Y.P., 2011. Nitrate assimilation in contrasting wheat genotypes. Physiology and Molecular Biology of Plants 17, 137–144.

Kaim, M.S., Nair, T.V.R., Abrol, Y.P., 1983. Nitrogen economy of the main shoot of wheat (*Triticum aestivum* L.) grown at two soil nitrogen levels. II. In vivo nitrate reductase activity. Indian Journal of Plant Nutrition 2, 38–50.

Kamachi, K., Yamaya, T., Hayakawa, T., Mass, T., Ojima, K., 1992. Changes in cytosolic glutamine synthase polypeptide and its mRNA in a leaf blade of rice plants during natural senescence. *Plant Physiol.* 98, 1323–1329.

Kumar, P.A., Grover, H.L., Abrol, Y.P., 1981. Potential for $NO_3^-$ reduction in wheat (*Triticum aestivum* L.). Journal of Plant Nutrition 3, 843–885.

Nair, T.V.R., Abrol, Y.P., 1979. Physiological studies on nitrate assimilation and grain protein accumulation in wheat (*Triticum aestivum* L.). The Journal of Agricultural Science 93, 473–484.

Pokhriyal, T.C., Abrol, Y.P., 1980. Nitrate assimilation in relation to total reduced N at various stages of growth in *Cicer arietinum* L. Experimental Agriculture 16, 127–135.

Sechley, K.A., Yamaya, T., Oaks, A., 1992. Compartmentation of nitrogen assimilation in higher plants. International Review Cytology 134, 85–163.

# 12

# Nitrogen Nutrition in Crops and Its Importance in Crop Quality

M. Maheswari[1], A.N.G. Murthy[2], A.K. Shanker[1]

[1]ICAR-CENTRAL RESEARCH INSTITUTE FOR DRYLAND AGRICULTURE, HYDERABAD, TELANGANA, INDIA; [2]ICAR-INDIAN INSTITUTE OF HORTICULTURAL RESEARCH, BENGALURU, KARNATAKA, INDIA

**CHAPTER OUTLINE**



## Introduction

Nitrogen (N) availability is the major limiting factor for increasing productivity of agricultural systems. Judicious application of N to crops and cropping systems, through N fertilizers, is thus one of the important elements for producing adequate food to meet the demand of growing human population. Adequate N availability throughout the crop growth and development can ensure high yield and productivity. Over the last 40 years, the global use of mineral N fertilizers has increased sevenfold (Francois et al., 2014) corresponding with the doubling of agricultural food production and gives an indication of the role of N fertilization in global food production. Nitrogen fertilization is one of the most practical ways to maintain and improve productivity and nutritional quality of crops for human consumption. However, in this context, it needs to be mentioned that balanced nutrition of the major and minor nutrients is the key to maintain crop yields as well as nutritive quality.

The Indian Nitrogen Assessment. http://dx.doi.org/10.1016/B978-0-12-811836-8.00012-4

Deficiency of N limits crop growth; on the other hand, excess N is mostly lost to the environment resulting in pollution. There is often a huge disparity between N supply and demand in farming systems. Efficient use of N is thus crucial for enhancing productivity and profitability of the cropping system as well as to decrease the risk of N loss to the environment. Apart from re-mobilization and transport, uptake and assimilation of N at reproductive stage also contributes toward accumulation of N in the grain. Thus, application of the same amount of N in more than two splits under field conditions increases the N availability at later stages of growth (Abrol et al., 1999). Generally, the N use efficiency (NUE) of crops is low and it depends on several factors including climatic conditions, soil type, crop genotype, rootstock, root development, fertilizer N source, placement, rate and timing of fertilizer application, irrigation system, and type of associated crops. Nitrogen in excess quantities is also toxic and can have adverse effects on the quality of produce. It is therefore necessary to bring out the state of the art of N nutrition in crops and its importance in crop quality highlighting the current strategies and future perspectives.

## Metabolic Aspects of N Uptake, Assimilation and Crop Quality

Nitrogen utilization by crop plants involves two major processes, i.e., uptake and assimilation. Nitrate is the major N form in which it is taken up by plants in general, although it is urea which is applied predominantly to the soil in India. Subsequent to the conversion of urea to nitrate in the soil, nitrate is taken up by the plants and further assimilated through a well-documented metabolic pathway involving the enzymes nitrate reductase and nitrite reductase leading to ammonium (Abrol et al., 2007). Ammonium is then incorporated further into amino acids through the GS/GOGAT pathways (Naik et al., 1982; Abrol et al., 1988; Ahmad et al., 2007).

The major constraint in nitrate assimilation is that the conversion of nitrate to nitrite is mediated by the enzyme nitrate reductase which has been observed to be correlated with grain yield as well as grain protein yield (Nair and Abrol, 1979; Ahmad et al., 2007). Further, remarkable variability in the enzyme level has been reported across the plant organ, genotypes, and nitrogen levels. Careful selection of genotypes coupled with the management of nutritional regime may thus be important in enhancing crop yield and quality.

Nitrogen use efficiency (NUE) is a complex multigenic trait with numerous intricate physiological and biochemical processes involved and in turn affected by various environmental, abiotic and biotic, factors. It is often expressed in terms of agronomic efficiency (kg grain $kg^{-1}$N applied), physiological efficiency (kg grain $kg^{-1}$ N taken by crop), chemical efficiency or apparent recovery (percentage of applied N taken by the crop), efficiency ratio (kg grain $kg^{-1}$ N uptake, without considering unfertilized control), and also sometimes as partial factor productivity (kg grain $kg^{-1}$ N applied without considering unfertilized control). The NUE is thus a function of soil to supply adequate amount of N, ability of plant to acquire and transport in the root and shoot, and to remobilize to others parts of the plant.

Understanding N uptake and assimilation is necessary to improve the NUE of crops through various management strategies. Utilization of phenotypic and genotypic variability for QTL mapping for precise analysis and genetic enhancement of the complex trait of NUE have been the focus in major food crops such as rice (Vijayalakshmi et al., 2013), wheat (Mahjourimajd et al., 2016), maize (Mueller and Vyn, 2016; Yu et al., 2015), and sorghum (Malleswari, 2013). Role of the regulatory elements and signal transduction need to be focused upon for improving the plant NUE. Development of N efficient genotypes could help in enhancing crop productivity and quality as well as decreasing reactive N losses.

The emission of $NH_3$ from plants is also an important phenomenon contributing to N losses albeit at low quantities. Stomatal $NH_3$ emission occurs when the partial pressure of $NH_3$ in the apoplastic air is higher than that in the ambient air (Farquhar et al., 1980). Ammonia is a product of several metabolic pathways. In addition to nitrate reduction and photo respiratory N cycle (Keys et al., 1978), increased proteolytic activity is also an important source of $NH_3$, especially during plant senescence. The elevated tissue $NH_4^+$ concentrations during the later stages of crop growth provide an increased potential for $NH_3$ losses and such losses have been reported from various field crops (Maheswari et al., 1992; Mattsson and Schjoerring, 2003; Hayashi et al., 2008; Flechard et al., 2013). Predictably, N losses are also higher at high N application level.

Crop quality is defined by a range of physical and compositional properties of the economically important produce, the thresholds of which are fixed according to the end-user requirements. Nitrogen is a component of amino acids, protein, and nucleic acids and is also an integral component of many other compounds and enzymes. Crops are the basic sources of several essential nutrients including carbohydrates, lipids, amino acids, proteins, vitamins, and some mineral elements which are now known to be of great significance in maintaining human health (Wang et al., 2008; Graham et al., 2001).

## Nitrogen and Crop Quality

The application of fertilizer N results in increased growth, higher biomass, and yield, and facilitates metabolism to produce high amounts of protein in plant tissue. Nitrogen directly influences the amino acid composition of protein and thereby the nutritional quality of the economic produce. Quality of food grains is a complex phenomenon and is influenced by several factors among which the effect of N fertilization seems to be the most significant. Along with the nutritional qualities involving amino acid composition etc., the structural and functional properties of the produce in several crops are also regulated by N fertilization and protein content.

## Cereals

In cereals, increase in N affects the proportion of amino acids lysine and threonine. Kernel quality and strength increases with increased N which in turn facilitates milling properties. It is well known that N increased the protein, fiber, and ash contents in wheat (Abrol and Uprety, 1969; Nair and Abrol, 1979; Sardana, 2000). It was observed that application of the same amount of N in wheat in more than two splits under field conditions increased the N availability at later stages of growth and eventually resulted in a significant improvement in the nitrate assimilatory activity of the upper laminae, and enhancement in the total plant N harvest and grain protein content (Abrol, 1981; Abrol et al., 1999). Increased levels of protein due to N fertilization resulted in shorter baking mixing times of flour in winter wheat varieties (Blumenthal et al., 2000). Xue et al. (2016) reported that late N fertilization enhanced grain protein content and the relative abundance of certain high molecular weight glutenin subunits. The N fertilization effect was more marked on the wheat cultivar whose baking quality was more dependent on protein concentration, and gliadin and glutenin contents were better correlated with loaf volume than the overall protein concentration.

Nitrogen nutrition has a major influence on grain quality in rice, though its influence on milling characteristics, translucence, and color (Blumenthal et al., 2008) as well as protein contents are variable (Prakash et al., 2002; Champagne et al., 2007). Application of N significantly increases protein content in rice generally as indicated by a plethora of reports. Organic N nutrition is also shown to increase grain protein content in rice (Yadav et al., 2013). Rice is the primary source of dietary energy and protein for a large population especially in Asia. Early work on improved rice varieties demonstrated that N fertilization significantly increased grain protein content (De Datta et al., 1972). Subsequent research efforts indicated the importance of rate and timing of N-fertilizer application to achieve high yields and high grain quality. Nitrogen fertilization is reported to have a favorable effect on milling quality by modifying protein distribution in rice grains (Leesawatwong et al., 2005). Greater amounts of storage proteins glutelin and prolamin primarily reflect the increase in rice grain protein (Cagampang et al., 1966). Improvement in protein quality by decrease in phytic acid content and increase in protein fractions as well as total protein is also reported with N fertilization (Ning et al., 2009). Nitrogen application significantly affected protein profile, textural and pasting properties of rice cultivars (Kaur et al., 2016). They also reported increase in glutelins and decrease in the apparent content which is related to better cooking qualities. Other quality traits, such as translucence is also often associated with protein content.

Maize, an important cereal crop, contains mainly starch, protein, and oil as well as vitamins and minerals in its kernels (Eckhoff and Paulsen, 1996; Wang et al., 2008). Quality protein maize (QPM) holds superior nutritional and biological value (Prasanna et al., 2001). Many studies showed improvement in protein concentration with increased N rates in maize (Mason and Mason, 2002; Saha et al., 2008; Arun Kumar et al., 2007; Bhatt, 2012; Jeet et al., 2012; Blumenthal et al., 2008).

Tsai et al., 1983 demonstrated that protein concentration of maize grain increased with N supply and there was a preferential deposition of zein, which contains lower amounts of most limiting essential amino acids lysine and tryptophan, over other endosperm proteins. However, since N fertilization increases the size of the germ which has a better amino acid balance than the endosperm (Bhatia and Rabson, 1987), this could be compensated. Increased N fertilization showed to improve the content of essential amino acids lysine and tryptophan in QPM maize (Jeet et al., 2012).

In sorghum, Kaufman et al., 2013 reported that the size and hardness of the kernels were influenced by N rates. The protein content increased as the N rate increased, while the protein digestibility values and starch granule size distributions were not affected by N rate. In sweet sorghum, a smart biofuel crop, a linear significant increase in juice yield and potential ethanol yield was recorded up to 90 kg N $ha^{-1}$ (Sawargaonkar and Wani, 2016).

## Oil Seeds

Increased N supply in oil seeds generally improved seed yield and protein concentration, but decreased the oil content (Wang et al., 2008; Blumenthal et al., 2008). In general, it is well documented that there is a negative relationship between protein and oil content. During seed development, availability of N may affect the fatty acid synthesis, thus resulting in not only the final oil content but also composition. Hence, N fertilization in terms of rate and timing is crucial in influencing the oil yield per unit area as well as value of the particular oil seed crop. In addition to N-fertilizer effects on oil concentration, variation in the composition of sunflower oil has also been reported by Bahl et al. (1997) who found a decrease in additional N but an increase in the ratio of oleic acid and linoleic acid. Interaction of N with other major nutrients such as P may lead to more variability in fatty acid composition. This kind of information may be helpful in managing the fatty acid composition in oil seeds. Other oil seeds such as soybean and peanuts both of which are nodulating legumes are reported to be more responsive to N levels and oil content in these seems to be less negatively impacted by N levels (Bishnoi and Dutt, 1980; Pawar et al., 1982; Nagre et al., 1991). However, decreasing oil concentration at higher N rates has been reported in these crops also (Jadhav et al., 1994). Further, a strong interaction between N and S for seed and oil production has been reported in oilseed crops (Abdin et al., 2003; Jamal et al., 2006; Sattar et al., 2011; Kumar et al., 2016).

## Fodder Crops

In fodder crops such as teosinte, sorghum, sudan grass, and pearl millet improvement of crude protein, ether extract, and digestible dry matter yields were reported upon N application (Mohan et al., 2015; Abichandani et al., 1970; Walli and Relwani, 1970; Lee and Seo, 1988; Singh et al., 1988). Ayub et al. (2002) reported that quality parameter of forage sorghum, such as crude protein, ether extract, and ash content, increased with an increase in the level of N fertilization. However, neutral and acid detergent contents decreased with increase in N levels. Beneficial effect of N on protein content of fodder oat is also well reported (Pradhan and Mishra, 1994; Midha et al., 1999; Ratan et al., 2016).

## Horticultural Crops

Nitrogen fertilization has significant impact on horticultural crops with remarkable effects on harvested fruit quality, quality retention, and shelf life. Recent research on plant nutritional effects on vegetable quality is also intensified. Low nitrogen levels usually result in reduced protein content and inferior quality vegetables. In tomato, balanced nutrition of NPK recorded in the highest total soluble solids content. Higher N supply produced the highest number of ripe fruits per plant (Kumar et al., 2013). Increased application of N often decreases the concentration of vitamin C, soluble sugar, soluble solids, while increasing the titratable acidity in tomato as reviewed by Wang et al. (2008). The quality of vegetables, such as tomato, zucchini, cabbage, lettuce, and potato, was documented to be improved by N fertilization. In these vegetable crops, it was generally seen that the level of N producing the highest yield in most cases also produced the best quality edible plant part. For tomato, high N levels produced firm fruit with the high total soluble solids and dry matter content. For zucchini, medium N levels produced fruit with the highest dry matter content. For cabbage and lettuce, the crispness of heads was reduced at low and high N levels (Huett, 1988).

Higher applied N fertilizer along with microbial inoculants, especially *azospirillum*, improved quality of the head cabbage (Kumari et al., 2015; Choudhary and Choudhary, 2005) and cauliflower (Sable and Bhamare, 2007). In ginger (*Zingiber officinale* Rose.), increased N fertility improved the leaf and rhizome nutrient content as well as essential oil, carbohydrate, and protein contents up to 120 kg N $ha^{-1}$ (Singh et al., 2016).

In leafy vegetables, accumulation of nitrate due to excessive application of N fertilizers attracted a lot of research attention. It is essential to manage balanced N fertilization and careful selection of genotypes based on the nitrate assimilation ability so as to avoid any associated health hazards (Anjana et al., 2007).

In guava (Sharma et al., 2013) N-fertilizer application resulted in high TSS, total sugars, and minimum physiological loss in weight. Total soluble solids, fruit color, texture and firmness were negatively affected by high N-fertilizer levels in apple (Nava et al., 2008; Sharma, 2016). It was also found in apple that the total dissolved solids, total acidity, fruit weight, length, juice pH, and diameter were not affected by increased ammonium fertilization while, firmness decreased as the proportion of $NH_4^+$ increased (Mohammad Sokri et al., 2015). In grapes, N application showed effects on the quality of the fruit and also composition of anthocyanin. High N application delayed fruit maturation and berries had increased arginine and decreased anthocyanin content with relatively more abundant acylated anthocyanins compared to berries of vines supplied with low N (Hilbert et al., 2015). Decrease in anthocyanin content during maturation was more in the skin of berries from high N vines indicating that in grapes high N application affected the anthocyanin biosynthetic pathway both quantitatively as well as qualitatively and enhanced its degradation during maturity of berries.

## Strategies to Improve Crop Quality With Efficient Use of N

One of the most important limiting factors for crop production is N nutrition in crops. Proper use of N by judicious and timely application in the crops results in an increase in biomass yields and also increases in the yield of compounds that have N as their major constituent, and important one among them is the protein content. In addition to this the concentration of N in the plant tissue is high. The amount of N supplied affects the composition of protein in terms of amino acid composition which in turn affects the quality of protein. In general, in cereal crops high N dosage reduced the ratio of lysine and threonine, consequently decreasing the biological value of the protein. Increased N application in general has a positive effect on the kernel quality and strength, resulting in improved milling properties of the grain. In oil seed crops, unlike cereals, there is a different relationship with N and quality, wherein protein levels increase under high N fertilization while oil content decreases. There is a degree of variation seen in the relationship between N content and quality in sugar crops. Specifically, in sugar beet crop, the outcome of high dosage of N usually decreases the sucrose concentration on fresh weight basis and undesirable metabolites such as alpha-amino-N, invert sugars, and salts are known to increase, affecting the sucrose extraction efficiency. Nitrogen dosage variations in potatoes mainly affect tuber size, dry matter, and sugar contents. Nitrogen dosage in potato can be varied according to the end product that is required, e.g., staple food purpose, french fries or chips, as each of these end product will need different quality parameters. The following points are to be considered for evolving a strategy to increase NUE and quality in crops under N nutrition: (1) a seamless integration of genomic and phenomic approaches involving meta-analysis of all the relevant studies related to agronomic and crop physiological QTLs; (2) a molecular and mechanistic approach involving mapping of the network of pathways involved in N metabolism including uptake, assimilation, and recycling and their relationship with growth development and yield under varying N fertilization levels; (3) the third approach should be a modeling framework approach wherein crop simulation models should be developed that can predict responses under different N levels that help in judicious management of N fertilization.

## Conclusion and Future Outlook

Nitrogen is often the most important factor limiting crop productivity. Nitrogen directly influences the amino acid composition of protein and thereby its nutritional quality. Metabolic partitioning, root-shoot allocation, and re-mobilization have to be the focal points for the research on improving NUE and crop quality. Strategies to increase NUE and simultaneously enhance the quality have to account for the complexities of metabolic coordination and the multigenic responses involved. Novel strategies involving omics approach may be required to tackle the issues of balanced N nutrition and quality. Also understanding of processes related to the assimilation and re-mobilization of N and source–sink relationships, and key regulators of the nutritional quality at metabolic and molecular levels is essential. Excessive amount of N fertilizers applied to vegetable crops leads to accumulation of nitrate, especially in the leafy vegetables, thereby causing health concerns. Research efforts have to be intensified with an integrative approach involving transcriptomics, functional genomics, metabolic and whole-plant assessment, quantitative genetics, and soil nutrient dynamics to enhance crop quality with efficient use of applied nitrogen.

## References

Abdin, M.Z., Ahmad, A., Khan, N., Khan, I., Jamal, A., Iqbal, M., 2003. Sulphur interaction with other nutrients. In: Abrol, Y.P., Ahmad, A. (Eds.), Sulphur in Plants. Kluwer Academic Publishers, Dordrecht. The Netherlands, pp. 359–374.

Abichandani, C.T., Manikar, N.D., Mishra, M.N., Karnail, G.T., Gill, A.S., Maurya, R.K., Gopichandra, 1970. Effect of Top Dressing of N on Fodder Yield of Teosinte. Annual Report. I.G.F.R.I. Jhansi, p. 50.

Abrol, Y.P., Uprety, D.C., 1969. Effect of soil fertility levels on the protein quality of wheat grains. Indian Journal of Plant Physiology 12, 211–215.

Abrol, Y.P., Dhar, D., Kumar, P.A., 1988. Nitrogen metabolism. In: Sen, S.P. (Ed.), Plant Physiological Research in India. Society for Plant Physiology and Biochemistry, pp. 75–83.

Abrol, Y.P., Chattergee, S.R., Kumar, P.A., Jain, V., 1999. Improvement in nitrogen use efficiency: physiological and molecular approaches. Current Science 76, 1357–1364.

Abrol, Y.P., Raghuram, N., Sachdev, M.S., 2007. Agricultural Nitrogen Use and its Environmental Implications. I.K. International publishing House Pvt. Ltd., New Delhi, p. 552.

Abrol, Y.P., 1981. Physiological analysis of nitrogenous fertilizer utilization. In: Proc. Annual Seminar: Fertilizr Use Efficiency. Fertilizer Assoc. of India, New Delhi, pp. 11–20.

Ahmad, S., Fazil, I.S., Jamal, A., Iqbal, M., Abdin, M.Z., 2007. Interactive effect of sulphur and nitrogen on nitrate reductase and ATP-sulfurylase activities in relation to seed yield from *Psoralea corylifolia* L. Journal of Plant Biology 50, 351–357.

Anjana, Omer, S., Iqbal, M., Abrol, Y.P., 2007. Are nitrate concentrations in leafy vegetables within safe limits ? Current Science 92, 355–360.

Arun Kumar, M.A., Gali, S.K., Patil, R.V., 2007. Effect of levels of NPK on quality of sweet corn grown on vertisols. Karnataka Journal of Agricultural Sciences 20, 44–46.

Ayub, M., Nadeem, M.A., Tanveer, A., Husnain, A., 2002. Effect of different levels of nitrogen and harvesting times on growth, yield and quality of sorghum fodder. Asian Journal of Plant Sciences 4, 304–307.

Bahl, G.S., Pasricha, N.S., Ahuja, K.L., 1997. Effect of fertilizer nitrogen and phosphorus on the grain yield, nutrient uptake and oil quality of sunflower. Journal of the Indian Society of Soil Sciences 45, 292–296.

Bhatia, C.R., Rabson, R., 1987. Relationship of grain yield and nutritional quality. In: Olason, R.A., Frey, K. L. (Eds.), Nutritional Quality of Cereal Grains. Genetics and Agronomic Management. American Society of Agronomy, Madison, WI, pp. 11–44.

Bhatt, P.S., 2012. Response of sweet corn hybrid to varying plant densities and nitrogen levels. African Journal Research 7, 6158–6166.

Bishnoi, K.C., Dutt, R., 1980. Effect of rhizobim isolates inoculation methods and nitrogen levels on nodulation and quality of soybean. Indian Journal of Agronomy 25, 544–545.

Blumenthal, J.M., Baltensperger, D.D., Shelton, D.R., Nelson, L.A., Binford, G.D., Geleta, S., 2000. Nitrogen and variety effects in winter wheat. In: Abstracts of the Third International Crop Science Congress, Hamburg, Germany.

Blumenthal, J.M., Baltensperger, D.D., Cassman, K.G., Mason, S.C., Pavalista, A.D., 2008. Importance and effect of Nitrogen on crop quality and health. In: Nitrogen in the Environment:Sources, Problems and Management, pp. 51–69.

Cagampang, G.B., Cruz, L.J., Espiritu, S.G., Sanitago, R.G., Juliano, B.O., 1966. Studies on the extraction and composition of rice proteins. Cereal Chemistry 43, 145–155.

Champagne, E.T., Bett-Garber, K.L., Grimm, C.C., McClung, A.M., 2007. Effects of organic fertility management on physiochemical properties and sensory quality of diverse rice cultivars. Cereal Chemistry 84, 32–327.

Choudhary, R.K., Choudhary, D.N., 2005. Effect of different levels of nitrogen and phosphorus on growth yield and quality of hybrid cabbage. Haryana Journal of Horticultural Science 34, 145–146.

De Datta, S.K., Obcemea, W.N., Jana, R.K., 1972. Protein content of rice grain as affected by nitrogen fertilizer and some triazines and substituted ureas. Agronomy Journal 64, 785–788.

Eckhoff, S.A., Paulsen, M.R., 1996. Maize. In: Henry, R.J., Kettelewell, P.S. (Eds.), Cereal Grain Quality. Champman & Hall, London, pp. 77–112.

Farquhar, G.D., Firth, P.M., Wetselar, R., Weir, E., 1980. On the gaseous exchanges of ammonia between leaves and environment : determination of ammonia compensation point. Plant Physiology 66, 710–714.

Flechard, C.R., Massad, R.S., Loubet, B., Peronne, E., Simpson, D., Bash, J.O., Cooter, E.J., Nemitz, E., Sutton, M.A., 2013. Advances in understanding, models parameterizations of biosphere – atmosphere ammonia exchange. Biogeosci. 10, 5183–5225.

Francois, G., Durand, J.-L., Louarn, G., 2014. Quantifying crop responses to nitrogen and avenues to improve nitrogen-use efficiency. Crop physiology–Applications for genetic improvement and agronomy. Academic Press 2, 161–206.

Graham, R.D., Welch, R.M., Bouis, H.E., 2001. Addressing micronutrient malnutrition through enhancing the nutritional quality of staple foods: principles, perspectives and knowledge gaps. Advances in Agronomy 70, 77–142.

Hayashi, K., Nishimura, S., Yagi, K., 2008. Ammonia volatilization from a paddy field following applications of urea: rice plants are both an absorber and an emitter for atmospheric ammonia. Sciences of the Total Environment 390, 485–494.

Hilbert, G., Soyer, J.P., Molot, C., Giraudon, J., Milin, M., Gaudillere, J.P., 2015. Effects of nitrogen supply on must quality and anthocyanin accumulation in berries of cv. Merlot. VITIS-Journal of Grapevine Research 42, 69.

Huett, D.O., 1988. Effect of nitrogen on the yield and quality of vegetables. In: Research and Development Conference on Vegetables, the Market and the Producer, 247, pp. 205–210.

Jadhav, P.J., Jadhav, A.S., Bachchhav, S.M., 1994. Effects of N, row spacing and plant densities on the yield and quality of soybean. Journal of Maharashtra Agricultural Universities 19, 75–77.

Jamal, A., Fazili, I.S., Ahmad, S., Abdin, M.Z., 2006. Interactive effect of nitrogen and sulphur on yield and quality of groundnut (*Arachis hypogea* L.). Korean Journal of Crop Science 51, 519–522.

Jeet, S., Singh, J.P., Kumar, R., Prasad, R.K., Kumar, P., Kumari, A., Prakash, P., 2012. Effect of nitrogen and sulphur levels on yield, economics and quality of QPM hybrids under dryland condition of eastern Uttar Pradesh, India. Journal of Agricultural Science 4, 9.

Kaufman, R.C., Wilson, J.D., Bean, S.R., Presley, D.R., Blanco-Canqui, H., Mikha, M., 2013. Effect of nitrogen fertilization and cover cropping systems on sorghum grain characteristics. Journal of Agricultural and Food Chemistry 61, 5715–5719.

Kaur, A., Ghumman, A., Singh, N., Kaur, S., 2016. Effect of different doses of nitrogen on protein profiling, pasting and quality attributes of rice from different cultivars. Journal of Food Science Technology 53, 2452–2462.

Keys, A.J., Bird, I.F., Cornelius, M.J., Lea, P.J., Wallsgrove, R.M., Miflin, B.J., 1978. Photorespiratory nitrogen cycle. Nature 275, 741–743.

Kumar, M., Meena, M.L., Kumar, S., Maji, S., Kumar, D., 2013. Effect of nitrogen, phosphorus and potassium fertilizers on the growth, yield and quality of tomato var. Azad T-6. The Asian Journal of Horticulture 8, 616–619.

Kumar, V., Nath, P., Kumar, R., Kumar, V., Verma, J.K., Naresh, R.K., 2016. Interactive effect of sulphur and nitrogen on growth, yield and quality of Indian mustard (*Brassica Juncea* L.). International Journal of Science Nature 7, 57–61.

Kumari, C., Mankar, A., Karuna, K., Solankey, S.S., Singh, V.K., 2015. Effect of different levels of nitrogen and microbial inoculants on yield and quality of cabbage (*Brassica oleracea* var *capitata*) cv Pride of India. Indian Journal of Agricultural Science 85, 515–518.

Lee, J.K., Seo, S., 1988. Effect of N fertilization levels on the regrowth, carbohydrate reserves and dry matter yield of sorghum-sudangrass hybrid and other forage crops. Korean Journal of Animal Science 30, 441–445.

Leesawatwong, M., Jamjod, S., Kuo, J., Dell, B., Rerkasem, B., 2005. Nitrogen fertilizer increases seed protein and milling quality of rice. Cereal Chemistry 82, 588–593.

Maheswari, M., Nair, T.V.R., Abrol, Y.P., 1992. Ammonia metabolism in the leaves and ears of wheat (*Triticum aestivum* L.) during growth and development. Journal of Agronomy and Crop Science 168, 310–317.

Mahjourimajd, S., Taylor, J., Sznajder, B., Timmins, A., Shahinnia, F., Zed, R., Khabaz-Saberi, H., Kuchel, H., Mamoru, O., Langridge, P., 2016. Genetic basis for variation in wheat grain yield in response to varying nitrogen application. Public Library of Science One 11 (7), e0159374.

Malleswari, G., 2013. Identifying the Genes Underlying low N Tolerance & nitrogen Use efficiency (NUE) of sorghum. In: Plant and Animal Genome XXI Conference. Plant and Animal Genome.

Mason, S.C., D'Coz-Mason, N.E., 2002. Agronomic practices influence maize grain quality. Journal of Crop Production 5, 75–91.

Mattsson, M., Schjoerring, J.K., 2003. Senescence—induced changes in apoplastic and bulk tissue ammonia concentrations of ryegrass leaves. New Phytologist 160, 489–499.

Midha, L.K., Panwar, K.S., Sharma, S.K., 1999. National Commission on Agriculture (1976). Forage Research 25, 99–102.

Mohammad Sokri, S., Babalar, M., Barker, A.V., Lesani, H., Asgari, M.A., 2015. Fruit quality and nitrogen, potassium, and calcium content of apple as influenced by nitrate: ammonium ratios in tree nutrition. Journal of Plant Nutrition 38, 1619–1627.

Mohan, S., Singh, M., Kumar, R., 2015. Effect of nitrogen, phosphorus and zinc fertilization on yield and quality of *kharif* fodder—a review. Agricultural Reviews 36, 218–226.

Mueller, S.M., Vyn, T.J., 2016. Maize plant Resilience to N stress and post-silking N capacity changes over time: a review. Frontier in Plant Science. http://dx.doi.org/10.3389/fpls.2016.00053.

Nagre, K.T., Deshmukh, S.B., Bhalerao, P.D., Thorve, P.V., 1991. Effect of sowing dates and fertilizers on growth, yield and quality of soybean varieties. PKV Research Journal 15, 82–84.

Naik, M.S., Abrol, Y.P., Rama Rao, C.S., Nair, T.V.R., 1982. Nitrate assimilation and grain protein accumulation in wheat (*Triticum aestivum* L.). Journal of Agricultural Science (Cambridge) 93, 473–484.

Nair, T.V.R., Abrol, Y.P., 1979. Physiological studies on nitrate assimilation and grain protein accumulation in wheat (*Triticum aestivum* L.). Journal of Agricultural Science (Cambridge) 83, 473–484.

Nava, G., Dechen, A.R., Nachtiga, G.R., 2008. Nitrogen and potassium fertilization affect apple fruit quality in Southern Brazil. Communication in Soil Science and Plant Analysis 39, 96–107.

Ning, H., Liu, Z., Wing, Q., Lin, Z., Chen, S., Li, G., Wan, S., Ding, Y., 2009. Effect of nitrogen fertilizer application on grain phytic acid and protein concentrations in japonica rice and its variations with genotypes. Journal of Cereal Science 50, 49–55.

Pawar, V.P., Sarnaik, N.T., Sondge, V.D., 1982. Effect of nitrogen and phosphate on yield and quality of soybean (*Glycine max* (L.) Murrill) varieties. Journal of Maharashtra Agricultural Universities 7, 90.

Pradhan, L., Mishra, S.N., 1994. Beneficial effect of nitrogen on crude protein content and yield. Indian Journal of Agronomy 39, 233–236.

Prakash, Y.S., Bhadoria, P.B.S., Amitava, R., Rakshit, A., 2002. Relative efficacy of organic manure in improving milling and cooking quality of rice. International Rice Research 27, 43–44.

Prasanna, B.M., Vasal, S.K., Kassahun, B., Singh, N.N., 2001. Quality protein maize. Current Science 81, 10–25.

Ratan, N., Singh, U.N., Pandy, H.C., 2016. Yield and quality of oat (*Avena sativa* L.) as influenced by nitrogen and varieties in Bundelkhand region (U.P.) India. Agricultural Science Research Journal 6, 27–30.

Sable, P.B., Bhamare, V.K., 2007. Effect of biofertilizers (*Azotobacter* and *Azospirillum*) alone and in combination with reduced levels of nitrogen on quality of cauliflower cv snowball 16. Asian Journal of Horticulture 2, 215–217.

Saha, S., Gopinath, K.A., Mina, B.L., Gupta, H.S., 2008. Influence of continuous application of inorganic nutrients to a Maize-Wheat rotation on soil enzyme activity and grain quality in a rainfed Indian soil. European Journal of Soil Biology 44, 521–531.

Sardana, V., 2000. Grain quality of wheat as influenced by application of nitrogen, climate and cultural practices: a review. Agriculture Review 21, 116–120.

Sattar, A., Cheema, M.A., Wahid, M.A., Saleem, M.F., Hassan, M., 2011. Interactive effect of sulphur and nitrogen on growth, yield and quality of canola. Crop Environment 2, 32–37.

Sawargaonkar, G.L., Wani, S.P., 2016. Nitrogen response of sweet sorghum genotypes during rainy season. Current Science 110, 1699–1703.

Sharma, A., Wali, V.K., Bakshi, P., Jasrotia, A., 2013. Effect of integrated nutrient management strategies on nutrient status, yield and quality of guava. Indian Journal of Horticulture 70, 333–339.

Sharma, L.K., 2016. Effect of nutrients sprays on growth, yield and fruit quality of apple under cold desert condition of Himachal Pradesh. Journal of Applied & Natural Science 8, 207–300.

Singh, V., Singh, A.K., Verma, S.S., Joshi, Y.P., 1988. Effect of nitrogen fertilization on yield and quality of multicut tropical forages. Tropical Agriculture 65, 129–131.

Singh, M., Masroor, M., Khan, A., Naeem, M., 2016. Effect of nitrogen on growth, nutrient assimilation, essential oil content, yield and quality attributes in *Zingiber officinale* Rose. Journal of Saudi Society of Agricultural Sciences 15, 171–178.

Tsai, C.Y., Warren, H.L., Huber, D.M., Bressan, R.A., 1983. Interactions between the kernel N sink, grain yield and protein nutritional quality of maize. Journal of the Science of Food and Agriculture 34, 255–263.

Vijayalakshmi, P.T., Vishnu Kiran, Y., Venkateswara Rao, B., Srikanth, I., Subhakara Rao, B., Sailaja, K., Surekha, et al., 2013. Physiological approaches for increasing nitrogen use efficiency in rice. Indian Journal of Plant Physiology 18, 208–222.

Walli, T.K., Relwani, L.L., 1970. Effect of seed rates and nitrogen levels on fodder yields of teosinte. Indian Journal of Agronomy 15, 400.

Wang, Z.H., Li, S.-X., Malhi, S., 2008. Effects of fertilization and other agronomic measures on nutritional quality of crops. Journal of the Science of Food and Agriculture 88, 7–23.

Xue, C., Rücker, S., Koehler, P., Obenauf, U., Mühling, K.H., 2016. Late nitrogen application increased protein concentration but not baking quality of wheat. Journal of Plant Nutrition and Soil Science 179, 591–601. http://dx.doi.org/10.1002/jpln.201500569.

Yadav, S.K., Singh, Y., Kumar, R.P., Yadav, M.K., Singh, K., 2013. Effect of organic nitrogen sources on yield quality and nutrient uptake of rice (*Oriza sativa*) under different cropping system. Vegetos 26, 58–66.

Yu, P., Li, X., White, P.,J., Li, C., 2015. A large and deep root system underlies high nitrogen-use efficiency in maize production. PLoS One 10 (5), e0126293.

# 13

# Nitrogen Dynamics in Grasslands

P.K. Ghosh, S.K. Mahanta, S.N. Ram

*ICAR-INDIAN GRASSLAND AND FODDER RESEARCH INSTITUTE, JHANSI, INDIA*

**CHAPTER OUTLINE**



## Introduction

Grasslands/pastures are the largest land users globally, occupying 3.2 billion ha of the 4.9 billion ha of available agricultural land (Steinfeld et al., 2006). Grass produces large amount of plant biomass, since the leaf-area index (the ratio of the total area of leaf to the area of ground covered) is higher when compared to other crops and also has the ability to regrow after cut-and-carry harvesting or direct grazing by animals. Majority of the species, except those of the semiarid areas, are perennial in nature and grow when the temperature is above 5°C with adequate moisture. The number of grass-growing days depends mainly on climate, partly because it is harvested as vegetative growth and its response to fertilizer N is greater when compared to other crops. The huge response in terms of herbage yield, and hence the amount of milk or meat produced per

The Indian Nitrogen Assessment. http://dx.doi.org/10.1016/B978-0-12-811836-8.00013-6

unit area of land, has encouraged using large amounts of fertilizer N. However, mineralization of soil organic matter (SOM) is the major N source in extensive grassland systems. For intensive pastures/grasslands, fertilizer N is an important source, of which only 30% is recovered by plant protein and entered into the animal system while the remaining 70% is lost to the environment in reactive N forms (Galloway et al., 2009). Nitrogen-use efficiency in grassland systems (meat or milk protein produced/kg plant protein N intake) ranges from 5% to 10%, depending on whether milk or meat is the output. Grazing animals typically retain about 5% of the N intake (from the grass consumed) in their bodies and excrete the rest through urine and dung, which becomes an N source for the pasture (Subbarao et al., 2013a). Much of this N, however, is lost as $NO_3^-$ leaching and gaseous N emissions ($N_2O$, NO, and $N_2$), causing ecological damage and economic loss (Herrero et al., 2011; Subbarao et al., 2013b).

The rates of fertilizer N recommended for intensively farmed grasslands in temperate regions are generally in the range of 200–400 kg N $ha^{-1}$ depending on soil type, weather, and pasture management practices. Although the intensive production of milk and meat is based mainly on fertilizer N usage, it is possible, though often more difficult, to achieve similar rates of production from grass–legume pastures without applying any fertilizer N. Such pastures/grasslands are dependent on symbiotic $N_2$ fixation. But the problem with grass–legume pasture is that the growing season is shorter for legumes compared to grasses. Moreover, in many areas, it is difficult to maintain an adequate proportion of legumes in grass–legume pasture over a number of years. This is the reason why in most countries, grass-based pastures receiving fertilizer N are much more widespread than grass–legume pastures. The only exception is New Zealand where well-managed grass–legume pastures often produce herbage yields of 10–15 t dry matter (DM) $ha^{-1}year^{-1}$ (Daly, 1990), similar to the yields produced by N fertilized all-grass pastures.

## Status of Indian Grasslands

In general, Indian grasslands/grazing lands are considered as fragile ecosystems and are ranked as class IV and V in their land capability classification. But these grassland resources play a major role in the agricultural economy of the country, since they are used as pasture/forage resources for livestock. They are also associated with the livelihood of thousands of people as grasses are used as fuel, shelter, and various traditional activities. However, the estimates of grasslands and grazing resources in India vary from 3.7% to as much as 12% of the total area (Krishnan et al., 2012). Although grasslands in India are the least understood and the most underestimated natural habitats, they are also most mistreated and ignored ecosystems in the country. On account of increasing grazing pressures, the present condition of grassland/grazing lands is far from pleasing. There is even gradual shrinkage in area due to pressures on land for other activities. Today they are considered as mere assemblage of vegetation that provides grazing and browsing to animals. The carrying capacity of grasslands/grazing lands is 0.20–1.47

adult cattle units (ACU)/ha depending on the factors such as climate, soil, rain, and geographical location, but the present stocking rates are much higher, often leading to deterioration of such resources. Moreover, other factors such as free-grazing practices, lack of management, extremes of temperature, variable precipitation, and scarcity of moisture in arid and semiarid situations are also important. Indeed, grass covers in India are largely governed by climatic factors and especially by latitudes resulting in the occurrence of *Sehima-Dichanthium* cover in tropical regions; *Dichanthium-Cenchrus-Lasiurus, Phragmites-Saccharum-Imperata,* and *Themeda-Arundinella* grass covers in sub-tropical regions; and temperate-alpine grass covers in temperate regions. Again, altitude separates *Themeda-Arundinella* cover (restricted to the northern hills) from *Phragmites-Saccharum-Imperata* (in moist to wet habitats) and *Dichanthium-Cenchrus-Lasiurus* (in relatively dry habitats) types which occurs in the plains. Investigation on grass communities and their succession trends showed that the *Sehima-Dichanthium* cover is highly productive at its climax stage, whereas *Themeda-Arundinella* yields the lowest. However, different types of grass covers have tremendous potentials in India (Table 13.1). Indian grassland soils are very poor in nutrients, which affects their productivity. Even a small dose of fertilizers can significantly improve the productivity. Shankar and Gupta (1992) observed that application of nitrogen at a rate of 40 kg $ha^{-1}$ on natural pastures dominated by *Sehima nervosum, Heteropogon contortus,* and *Iseilema laxum* increased their production from 4.1 to 7.6, 3.4 to 5.6, and 4.5 to 6.4 t $ha^{-1}$ $year^{-1}$, respectively. Similar studies conducted for different grasses such as *Cenchrus ciliaris, C. setigerus, Setaria sphacelata, Chrysopogon fulvus, Bothriochloa intermedia, Bracharia mutica, Dicanthium annulatum,* and *Panicum antidotale, Paspalum notatum* at Jhansi, India, resulted in significant increase in DM yields through application of fertilizer nitrogen. However, application of N-fertilizers on grasslands is considered as a costly affair and usually natural grasslands are not fertilized with nitrogen. But the population of livestock (>500 million heads) is increasing. Hence, in near future, there will be a need to fertilize available grasslands with nitrogen and other nutrients to increase forage production.

**Table 13.1** Production Potential (t DM $ha^{-1}$) of Indian Grass Covers

| Type of Grass Cover | Harvestable Yield | Potential Yield |
|---|---|---|
| *Sehima-Dichanthium* | 3.5 | 6.0 |
| *Dichanthium-Cenchrus-Lasiurus* | 3.3 | 5.0 |
| *Phragmitis-Saccharum-Imperata* | 5.0 | 5.0 |
| *Themeda-Arundinella* | 2.2 | 4.0 |
| Temperate-alpine | 4.0 | 6.0 |

Pathak, P.S., Dagar, J.C., 2015. Indian grasslands and their management. In: Ghosh, P.K., Mahanta, S.K., Singh, J.B., Pathak P.S. (Eds.), Grassland: A Global Resource Perspective. Range Management Society of India, Jhansi.

## Grassland Nitrogen: Transformations of N

Transformations of N in grasslands represent inputs of N that occur through the application of fertilizers and manures, the deposition of excreta from grazing animals, the fixation of $N_2$ by microorganisms, and the deposition of ammonia and nitrate from the atmosphere. Outputs of N occur through the removal of grass or animal products, leaching, volatilization of ammonia, and denitrification. Different types of grasslands vary in both the absolute and relative amounts of N undergoing transformation, with climate and grassland management practices being the important factors. Aspects of pasture management that have major effects are whether significant amounts of fertilizer, if any, are applied. Some transformations occur rapidly and some slowly. Thus, in grazed grassland, it is possible for an atom of N, initially present as nitrate or ammonium in the soil, to be taken up by the grass, consumed by a grazing animal, returned to the soil in excreta, and get transformed again to nitrate, all within a few days. As a result, some of the N in the system may be recycled several times during a growing season. On the other hand, a portion of the N returned in excreta will remain in the soil for many years, immobilized in the SOM. The intensity of management has an important influence on the outputs of N from the system: for example, the amounts of nitrate leached and ammonia volatilized are likely to be at least 50 times greater from intensively fertilized and heavily grazed grassland than from unfertilized and lightly grazed grassland. An important feature of intensively grazed grasslands is the heterogeneity in the soil that arises from the combined effects of consumption and excretion by the grazing animals (Floate, 1987; Afzal and Adams, 1992). Herbage is consumed from a wide area but the N excreted in dung and urine is concentrated into small patches within the area. The nature and intensity of N transformations in the soil may therefore show large differences even within a few centimeter distances.

## Grassland Nitrogen: Uptake of N

Grassland soils receive nitrate and/or ammonium ions through the application of fertilizer. However, the two ions differ in the reactions they undergo in the soil, and in the mechanisms by which they are taken up by plants roots. Nitrate is not adsorbed by the colloidal material (clay and organic mattes) in the soil and is therefore mobile in the soil solution, readily accessible to plant roots but also susceptible to leaching and dentrification. On the contrary, ammonium is retained by cation exchange on the clay and organic matter. Thus it is less mobile than nitrate, less accessible to roots, and less susceptible to loss. Ammonium, although, is converted slowly to nitrate by nitrification process. In the soil, nitrate moves mainly by mass flow with the movement of water and partly by diffusion, whereas ammonium moves mainly by diffusion and only slightly by mass flow (Fig. 13.1) (Whitehead, 1995). When plants were grown in solution culture it was observed that both ammonium and nitrate ions are utilized but usually plants take up nitrate rather more readily than ammonium, and show greater growth responses to

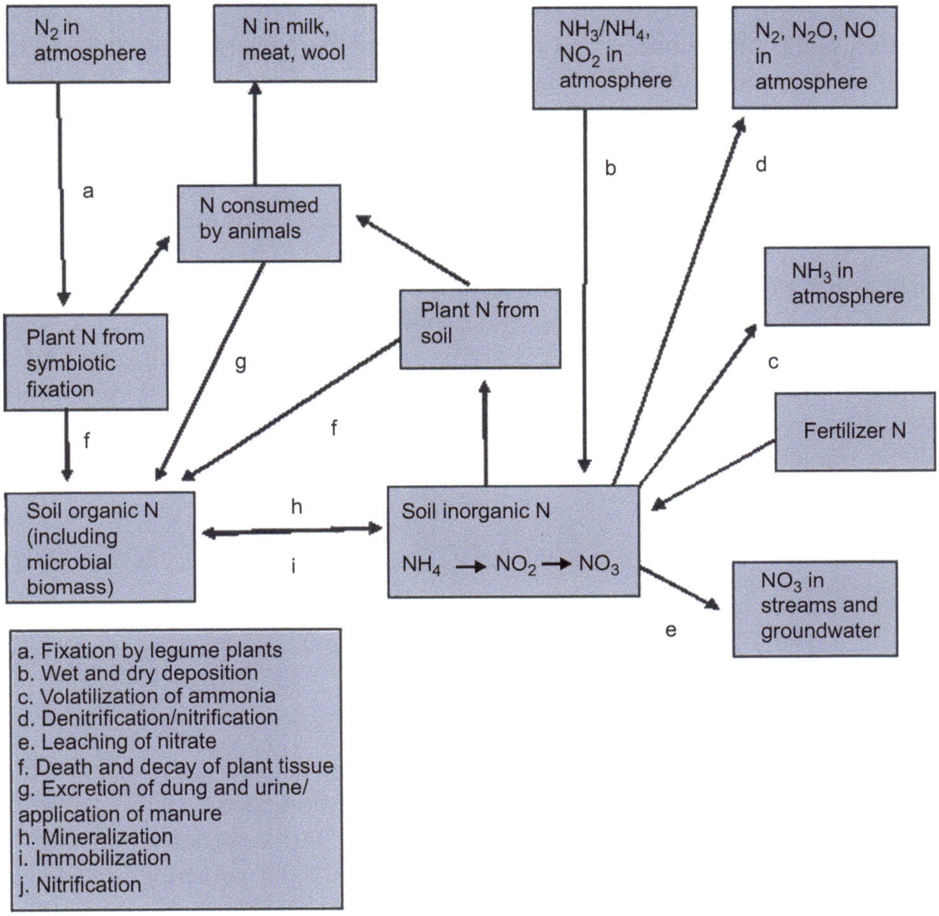


**FIGURE 13.1** Transformations of nitrogen in a grassland system (Whitehead, 1995).

nitrate. Grasses, however, often show a greater uptake of ammonium than of nitrate when the two ions are supplied in equal amounts. Again the rate of uptake of both nitrate and ammonium by roots is influenced by temperature and pH, but differently for the two ions. Studies with perennial ryegrass in solution culture indicated that absorption of nitrate increased with increasing temperature over the range of 5–35°C and was highest at pH 6.2, while absorption of ammonium was highest at a temperature of about 22°C and was only slightly influenced by pH over the range 4.0–7.5. In another study with perennial ryegrass grown in nutrient solution containing $NH_4NO_3$, the proportion taken up as ammonium was always greater than that of nitrate but was inversely related to root temperature, declining from 93% as ammonium at 3°C to 65% at 25°C. Other plant species had a smaller effect of temperature on the ratio between ammonium and nitrate uptakes. Thus grasses also absorb a considerable quantity of N in the form of nitrate, but it is reduced by the plant in order to be assimilated.

In grasslands, the annual uptake of nitrogen was reported to vary based on ecoclimatic zones, being maximum, on an average, in a dry subhumid zone (25.58 g/m$^2$) and minimum in a semiarid zone (2.93 g/m$^2$). It was observed that 53–66% uptake of nitrogen is retained in the underground net production in semiarid and humid grasslands, while 53–73% of the total uptake is associated with the above ground net production in subhumid and dry subhumid grasslands. Thus recycling of N was promoted through underground parts in semiarid and humid climatic conditions. The proportion of N absorbed by vegetation that is subsequently returned to the soil ranged from 63% to 78%, being comparatively higher in humid and semiarid grasslands than in the zones of intermediate moisture supply. Again, it was reported that the rate of N uptake by plants is highest in the rainy season in all the climatic zones except the humid zone and ranged from 56% of the annual transfer in dry subhumid grasslands to 71% in the semiarid zone. However, the rate of release back to the soil is greatest during winter, when from 47% to 64% of this transfer takes place (Coupland, 1979).

## Grassland Nitrogen: Forage Composition, Yield and Quality

In order to achieve high forage yield in intensively managed grassland, fertilizer N is applied, but it also affects the forage composition as well as quality.

### Forage Composition

Nitrogen fertilization of grasslands can have substantial impacts to overall biomass production. However, the overall impact of N application to the plant community is largely dependent on the composition of the grass field. Further, management goals and objectives of the grassland are also important drivers in the consideration of whether fertilizer N applications will be beneficial or harmful (Bauman et al., 2015). Since legumes and grasses have differential responses to N, application of N also changes pasture composition. From several trials conducted in perennial ryegrass/clover pastures in New Zealand with an application of 100 kg N ha$^{-1}$ in spring and autumn, it was concluded that grasses increase in the proportion of pasture composition, whereas clovers/legumes decrease. The decrease in proportion of clovers/legumes was due to a suppression suffered by shading of grasses since grasses were more responsive to N fertilizer. Although the proportion of clover decreases, clover yield does not necessarily reduce. However, in some cases, the clover proportion does not decline as in white clover dominant pastures. This was probably due to extra cuts stopping grasses from giving shading to white clover (Sun et al., 2008). Moreover, legumes obtain N via fixation by *Rhizobium* bacteria and, accordingly, they respond to fertilizer N differently from grasses. It was observed that there was little N response from a rate of 240 kg N ha$^{-1}$ in legumes such as red clover (*Trifolium pratense*) and white clover (*T. repens*), while perennial ryegrass increased forage DM yields by 134% (Lingorski et al., 1995). Application of 150 kg N ha$^{-1}$ resulted in 29% higher grasses DM yield and 9% in higher grasses contribution to total DM yield in comparison with zero N. As the grass content increased, the contribution of legumes and forbs declined when fertilized with 150 kg N $ha^{-1}$.

The application of fertilizer N to grasses containing non-legume dicotyledonous species usually reduced the proportion of these species, though occasionally there was little or even increased effect. Increasing the rate of ammonium nitrate over the range from 0 to 200 kg N $ha^{-1}$ reduced plant diversity in moist hay meadows of southwest England, while increasing both perennial ryegrass and *Holcus lanatus* (Mountford et al., 1993). Application of 25 kg N $ha^{-1}year^{-1}$ increased the proportion of more productive grasses within a period of 2 years; and 50 kg N $ha^{-1}year^{-1}$ significantly reduced botanical diversity within 3 years. However, in some cases, the loss of species diversity was due to the soil acidity induced by the repeated application of ammonium fertilizers.

## Forage Yield and Quality

Nitrogen is the most limiting factor for grassland productivity. Hence, application of nitrogen fertilizers severely affects grassland production systems in many ways. At low rates, nitrogen fertilization increased forage yield with little effect on forage nitrogen. It stimulated tiller development, increased leaf size, and lengthened the period of green leaves. Yield and nitrogen concentration were increased in forages at a higher level of nitrogen fertilization (Blumenthal et al., 2008). Application of fertilizer N had a significant impact on various growth parameters of Nandi grass (*Setaria sphacelata*). On the basis of the pooled analyses of variance over 3 years, the greatest plant heights and tussock diameters were recorded at the highest level of N (120 kg $ha^{-1}$); DM yield also increased significantly at the highest level of N (120 kg $ha^{-1}$) fertilization (Dwivedi et al., 1990).

In India, none of the natural grasslands in subtemperate, temperate, and alpine regions are fertilized in any form except dung and urine addition by the grazing animals. This leads to continuous depletion of plant nutrients resulting in the poor growth of plant species. Studies have revealed that there was a significant impact on the herbage productivity and quality of natural grasslands following proper fertilization (Table 13.2).

Forage yield of natural vegetation dominated by *Chrysopogon gryllus, Arundinella nepalensis*, and *Imperata cylindrica* was increased significantly and consistently with the increase in nitrogen levels. On an average 20 and 40 kg N $ha^{-1}$ registered an increase of 2.1 and 5.8 t $ha^{-1}$ green forage yields, respectively, over control. The quality of the herbage produced from nitrogen-fertilized plots was significantly better than unfertilized plots as indicated by higher in vitro DM digestibility, which was probably attributed to comparatively higher protein content as well as lower lignin and silica contents at higher level of nitrogen (Sood and Sharma, 1996). Application of different doses of nitrogen was found superior to control *Pennisetum trispecific* hybrid (*TSH*) and *Stylosanthes hamata* pastures (Table 13.3). Among the nitrogen doses, application of 90 kg N $ha^{-1}$ provided significantly higher dry biomass yield (9.61 t $ha^{-1}$) as compared to others. Nitrogen is the main constituent of the protein and its application directly influenced the crude protein

**Table 13.2** Effect of Dose and Time of Nitrogen Application on Grassland Productivity

| Treatments | Fresh Forage Yield (t $ha^{-1}$) |
|---|---|
| **N Dosage (kg $ha^{-1}$)** | |
| 0 | 10.4 |
| 40 | 15.2 |
| 80 | 19.6 |
| CD ($p < .05$) | 0.3 |
| **Time of N-Application** | |
| Basal | 14.7 |
| Two splits | 20.2 |
| CD ($p < .05$) | 0.5 |

*CD*, critical difference.

Singh, V., 1995. Technology for forage production in hills of Kumaon. In: Hazra, C.R., Misri, B. (Eds). New Vistas in Forage Production. Indian Grassland and Fodder Research Institute, Jhansi, pp. 197–202.

**Table 13.3** Forage Yield and Quality of *TSH* and *Stylosanthes hamata* Pasture and Available N as Influenced by N Levels

| Nitrogen Levels (kg $ha^{-1}$) | Dry Biomass Yield (t $ha^{-1}$) | Crude Protein Yield (kg $ha^{-1}$) | Available N (kg $ha^{-1}$) |
|---|---|---|---|
| 0 | 6.69 | 602 | 216.2 |
| 30 | 7.76 | 683 | 230.2 |
| 60 | 8.71 | 561 | 247.2 |
| 90 | 9.61 | 842 | 253.2 |
| CD ($p < .05$) | 0.26 | 23 | 1.8 |

*CD*, critical difference.

Niranjan, K.P., Burman, D., Arya, R.L., Agrawal, R.K., Shivadhar., 2004. Influence of nitrogen and phosphate levels on forage production of *Pennisetum* trispecific hybrid and *Stylosanthes hamata* under rainfed conditions. Range Management and Agroforestry 25, 98–101.

yield of grasses, which further increased significantly with successive increase in nitrogen levels. The highest ($p < .05$) crude protein yield (842.0 kg $ha^{-1}$) was obtained at 90 kg N $ha^{-1}$. Values of available nitrogen were also increased significantly in the soil with successive increase in nitrogen levels and was found to be the highest at 90 kg N $ha^{-1}$ (Niranjan et al., 2004).

Increased water concentration and decreased soluble carbohydrates were recorded after nitrogen fertilization of grasses (Blumenthal et al., 2008). Thus, the content of DM in grassland herbage is generally reduced by the fertilizer N. This is partly due to an increase in the amount of internal water in the leaves, and partly due to the fact that a larger crop, resulting from the application of fertilizer N, retains a greater quantity of superficial water from dew or rainfall. The content of water-soluble carbohydrate in grass herbage was substantially reduced by the application of nitrogen N, particularly when rate of application was more than about 200 kg N $ha^{-1}year^{-1}$. Fertilizer N generally had little effect on the amount of fiber (cell wall material) in grass herbage when the samples were taken at uniform time intervals. Similarly, there was little effect on the constituent materials of the fiber, cellulose and lignin. However, high rates of fertilizer N (>500 kg N $ha^{-1}year^{-1}$) may cause small reduction in cellulose and lignin contents. Since fertilizer N enables grass to be harvested at shorter growth interval, the herbage tends to have a lower content of fiber at the time of harvesting.

Nitrogen fertilization also had an impact on mineral composition of the herbage. If nitrogen is supplied in the ammonium form, uptake of cations such as calcium, potassium, and magnesium is reduced. By lowering magnesium concentration, ammonium fertilization may result in grass tetany. However, other effects of nitrogen fertilization on mineral composition of forages are inconsistent. Lower concentrations of macro and micronutrients were also observed, caused mainly by dilution, whereas many workers reported little effect of nitrogen fertilization on mineral composition of forage.

Nitrogen fertilization of grasses often had little impact on herbage digestibility. It was reported that as nitrogen fertilization increases lignin production, digestibility may be reduced. It was also observed that the rate of neutral detergent fiber digestion, and with it the animal feed intake, was increased upon nitrogen fertilization. Nitrogen fertilization of forages was more likely to increase digestibility of warm-season grasses because of their lower nitrogen content as compared to cool-season grasses. The main beneficial effect of nitrogen fertilization of forages, on meat production per unit land, increases in yield resulting in increased carrying capacity. In warm-season grasses, many studies found that increased nitrogen fertilization caused higher body weight gains when expressed in terms of kg/ha.

## Grassland Nitrogen: Consumption and Utilization in Ruminant Animals

When grasslands are managed extensively, the forage is consumed mainly by grazing, though some forage are harvested and conserved as hay or silage depending upon the availability. But in intensively managed grasslands, in relation to livestock farming, forage is usually consumed by grazing during the growing season and as conserved forage (hay/silage) during the lean period. Thus grasslands supply considerable amount of forage to meet the feed requirements of different categories of animals. However, many factors influence total N contents in grassland forage, which usually vary within the range of 1–5%. The factors include the species and variety of grass, the stage of maturity, the ratio of leaf and stem, application of fertilizer nitrogen and management practices, and climatic factors. The total N content, in relation to nutritional value of grassland forage for animals, is often expressed as crude protein. This is the concentration of total nitrogen multiplied by 6.25, a factor derived from the average concentration of N in plant proteins (16%). Thus crude protein includes both true protein and other nonprotein nitrogenous compounds. Normally in grass and legume forages/pastures, around 70–90% of total N is present as true protein and 10–30% is present in other forms, e.g., amino acids, peptides, and amides. Sometimes, a considerable amount of nitrate becomes available in grasses and in that case nonprotein N-compounds may even exceed 30%.

The amount of forage consumed by an animal on daily basis depends on the amount and type of forage available and also on the physiological status of animals, i.e., growing, pregnancy, lactating, etc. In a situation where animals depend exclusively on grazing to meet their forage requirements, the supply of forage restricts intake only when the average height of the sward is less than 7–10 cm in cattle and 4–6 cm in sheep. The plant/forage materials thus consumed go to the rumen for degradation and utilization. But ruminant animals lack the ability to hydrolyze the beta-1,4-glycosidic bond of plant cellulose due to the lack of the enzyme cellulase. Hence, ruminant animals completely depend on the microbial flora, present in the rumen to digest cellulose. Indeed, digestion of forage in the rumen is primarily carried out by the rumen microflora, which contains dense populations of several species of bacteria, protozoa, and sometimes yeasts and other fungi. About 1 mL of rumen was estimated to contain 10–50 billion bacteria and 1 million protozoa, as well as several numbers of yeasts and fungi. Since the environment inside a rumen is anaerobic, most of these microbial species are obligate or facultative anaerobes that can decompose complex plant material, such as carbohydrates and proteins. As bacteria carry out fermentation in the rumen, they consume about 10% of the carbon, 60% of the phosphorus, and 80% of the nitrogen that the ruminant ingests. To reclaim these nutrients, the ruminant then digests the bacteria in the abomasum. The enzyme lysozyme has adapted to facilitate digestion of bacteria in the ruminant abomasum. Pancreatic ribonuclease also degrades bacterial RNA in the ruminant small intestine as a source of nitrogen.

Enzymatic activity in the rumen converts dietary protein into amino acids, which are in turn deaminated to ammonia and various carbon skeleton compounds (organic acids). This affects the composition of dietary protein that escapes the rumen as well as the microbial protein fraction. This process appears to be wasteful since the animal requires some amino acids for its own use. However, some plant proteins may be very indigestible by the host enzymes and these proteins often have a low content of the essential amino acids. Since the microbes synthesize protein from ammonia and other suitable carbon skeletons, they contribute to the nutritional metabolism of the animal (Fig. 13.2). Nucleic acids (5–10% of dietary N) are fermented rapidly and the nitrates are converted to nitrite, which are fermented to ammonia. The most common strategy is to increase the escape of high quality dietary protein by minimizing proteolysis, peptidolysis, and deamination. It is well accepted, however, that increasing the intake of undegradable protein fraction must not be at the expense of lowering the degradable intake protein in the diet. Optimal protein supply to the animal depends on adequate degradable protein to maximize the capture of organic matter in microbial biomass. Synchronizing the rate of nitrogen hydrolysis with the rate of energy release will increase the rate of assimilation of ammonia by microbes and maximize the nitrogen use by the animal (Mahanta et al., 2009).

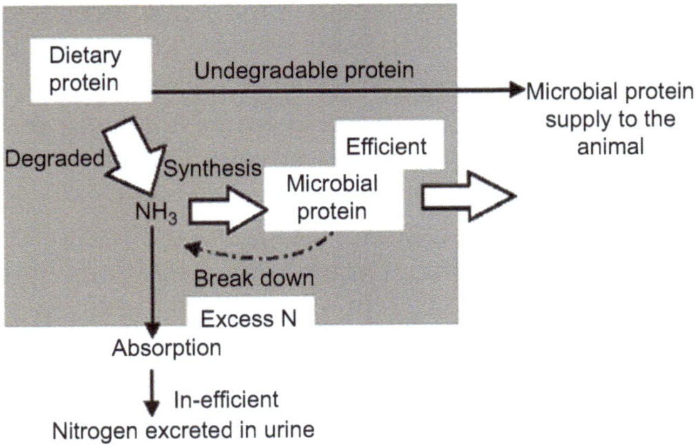


**FIGURE 13.2** Interconversion of dietary nitrogen into microbial protein (Mahanta et al., 2009).

It was reported that utilization of forage N is very high when the ratio between the available nitrogen and available energy in the forage is optimal for the amount of milk or body weight gain being produced, and the excretion of N then becomes minimum. In young/immature grass forage, where the ratio between the available N and energy is higher than optimum, the excess protein is then utilized as energy source with higher excretion of N as urea. On the contrary, mature grass forage with little or no fertilizer N, the protein content may be inadequate to ensure that the available N is utilized completely, particularly in lactating, growing, and pregnant (last trimester) animals where protein requirements are comparatively very high. The requirement for protein is about 7, 10–12, and 15–20% of DM intake for maintenance, pregnant and lactating, actively growing animals, respectively (Whitehead, 1995). However, less than 30% of dietary N consumed by the animals is actually utilized for the productive performances such as live-weight gain, milk or wool production, and the rest is excreted from the animal body. Utilization is about 15–25% of the nitrogen consumed in dairy animals and 3–10% in meat producing animals. It was also reported that grassland systems recover between 50% and 65% of the fertilizer N as harvested forage yield in the year of application. When grazed pastures fertilized with 200–250 kg N $ha^{-1}year^{-1}$ as $(NH_2)2CO$, nearly a third of the N input was retained in the SOM (25%) and root system (2–5%). The remainder was lost via ammonia volatilization from the soil following application (20%), ammonia volatilization from urine patches (20–30%), denitrification (6–12%), animal live-weight gain (7–9%), milk production (14–18%), and leaching or unaccounted losses such as excretion off pasture (20–50%) (Vendramini et al., 2007).

## Grassland Nitrogen: Animal Health and Production

In many parts of the world, livestock producers/farmers are intensifying their pasture production systems where adequate fertilization of the pastures is done to produce the

DM yields required to maintain high levels of forage and animal production. Recently, there were reports of stock mortality, bloat, reduced animal growth rates, and apparent lack of acceptability of intensive pastures etc. These adverse effects were probably due to the presence of high nitrogen levels in the forage. It was reported that nitrogen (N) fertilizer applications even above 50 kg N $ha^{-1}$ application, at certain times of the year, may result in plant N levels of 4.0–5.4% and plant nitrate levels as high as 0.9–1.0% nitrate nitrogen ($NO_3$-N) in forage. These high levels of N in forage are not only in excess of plant requirement, but are in excess of animal requirement too, which are sometimes potentially toxic to ruminant animals leading to various health problems (Eckard, 2004).

## Ammonia-Induced Bloat

Ammonia-induced bloat (unlike that of frothy-bloat) occurs in ruminant animals eating forages containing high levels of rumen-degradable nitrogen. The ingestion of excessive quantities of such nitrogen might result in a buildup of ammonia gas in the rumen, as rumen microorganisms, that convert ammonia to microbial protein, have a limit. In most grasses, the carbohydrate and nitrogen fractions are inversely related. Indeed, the energy required to metabolize the excess nitrogen increases as the nitrogen content of the forage increases. Under conditions of high nitrogen, without sufficient energy for microbial assimilation, a buildup of ammonia occurs in the rumen. This excess ammonia leads to alkalosis (pH 7.5 plus) and rumen stasis (paralysis of the rumen wall), thus preventing eructation (release of stomach gas via the mouth) and the animal suffers from bloat.

## Nitrate Toxicity

Animal health problems are a consequence when nitrate is reduced to nitrite in the rumen and absorbed in the bloodstream. Accumulation of nitrates within forages can be a problem under conditions of high nitrogen supply and impaired photosynthesis. The toxic level of nitrate in the plant was originally reported to be in the order of 0.22% $NO_3$-N. Nitrate toxicity is difficult to diagnose as affected animals are usually found dead. Symptoms in the live animal include marked anemia, bright-red (changing to brown) blood, profuse salivation and purging, and muscular spasms leading to paralysis.

## Poor Animal Performance

Poor animal performance was also reported on highly nitrogen-fertilized pastures. In fact, animals on such pastures are under an energy stress as the rumen microbes require energy to metabolize the extra nitrogen, while carbohydrate and nitrogen are inversely related in grasses/forages. The energy used to metabolize blood ammonia to blood urea was calculated to cost the equivalent of 2–3 L of milk. A blood-urea level over 19 mg $kg^{-1}$ was also found to reduce fertility in dairy animals. Thus animals on highly nitrogen-fertilized pastures may even suffer from sub-clinical nitrate and/or ammonia toxicity stress or reduced intake due to appetite repression. Hence, poor performances in animals were probably due to combined effect of these factors and need to be kept in mind.

## Grassland Nitrogen: Balances of N in Tropical Grasslands

The concept of N balance in grassland involves summing of all the inputs of N to the system and all the outputs of N from the system, the differences between the sum of inputs and the sum of outputs must then equal the change in total amount of N in the system and expressed in terms of kg N $ha^{-1}year^{-1}$. In a grassland, the total N is distributed over the three main pools: soil, plant, and animal (Fig. 13.3). The N in the soil is mainly in organic form and its concentration is generally highest in the topsoil, reflecting the fact that nearly all N inputs enter the soil through the surface. In general, the soil contains at least 98% of the total N in the system at any given time. Mineralization of organic N, a microbiological process, results in temporary accumulation of inorganic N, mainly in the nitrate and partly in the ammonium form. The mineralization coefficient, i.e., the percentage soil organic N in the topsoil that is mineralized per growing season, is generally higher in the tropics (about 5%) than in temperate climates (1–3%). However, the size of this sub-pool fluctuates strongly over time because of the plant uptake of this mineral N (Wetselaar and Ganry, 1982).

The N pool of an annual plant at sowing is usually equal to the amount of N contained in the seed. From then the pool size increases sigmoidally over time, and

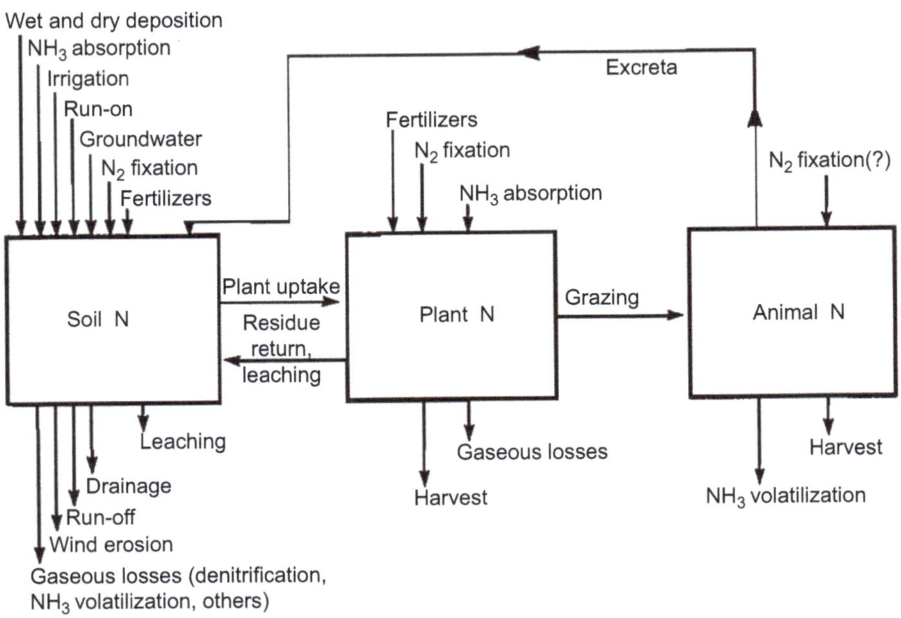


**FIGURE 13.3** Three main N pools of grassland, and their inputs and outputs (Wetselaar and Ganry, 1982).

reaches a maximum at flowering. In the case of a non-leguminous plant, the N in the plant pool is originated from mineralized soil organic N or from fertilizer N applied to the soil. In case of a leguminous plant, a substantial percentage of the total N in the plant comes from symbiotic fixation of $N_2$. When grazing occurs, the size of the animal N pool depends on the type of animal and the stocking rate.

The accurate measurements for all the inputs and outputs of N that occur in a grassland/pasture during a year are, however, very difficult and it becomes greater for the outputs than the inputs. Accurate values for fertilizer input and output in forage and/or animal production are obtained from direct measurements and analysis. The inputs from atmospheric wet deposition can also be estimated with reasonable accuracy in temperate grasslands of European and other developed countries. However, information is lacking for dry deposition. Methods are also available for measuring symbiotic N fixation, leaching, ammonia volatilization, and denitification in the pasture fields. But these methods are time-consuming and as a result the data include some interpolation for the period without actual measurements. Hence, in many cases, the accounted N balances are lacking of precision and need further systematic investigation. Moreover, the balance/equilibrium soil N content depends on factors like climate, soil type, vegetation, and management practices. Changes in any of the factors will change the equilibrium leading to a shift in soil N contents (Whitehead, 1995). Hence, there is an urgent need to generate data/information on different aspects of N budgeting particularly in tropical grasslands.

In India, few studies were conducted on grassland nitrogen cycling under different climatic situations. Indian grassland soils contain nitrogen several times than that held by plant biomass. The distribution of N in different plant compartments and in the top 30 cm soil was recorded among the protected, semi-protected and open-grazed grasslands at Kanpur, India (26° 26′ N and 80° 22′ E). The protected site had greater N pools compared to semi-protected and open-grazed sites. Nitrogen content in combined live and dead herbage varied from 26 to 535 kg $ha^{-1}$ in protected community, 16–276 kg $ha^{-1}$ in semi-protected, and 9–174 kg $ha^{-1}$ in open-grazed communities. The nitrogen content ranged in protected, semi-protected, and open-grazed communities between 10 and 177 kg $ha^{-1}$, 5 and 97 kg $ha^{-1}$, and 4 and 66 kg $ha^{-1}$ for belowground and from 2 to 11 kg $ha^{-1}$, 1 to 5 kg $ha^{-1}$, and 1 to 3 kg $ha^{-1}$ for litter compartments, respectively. The distribution pattern of nitrogen in plant/soil system indicated that the major portion of nitrogen (91% in protected, 95% in semi-protected, and 96% in open-grazed stands) in the system was retained in the soil, while a small fraction of it (9%, 5%, and 4% on protected, semi-protected, and open-grazed areas, respectively) resided in plant compartments (Mishra, 1979). In another investigation, it was observed that about 58 kg N $ha^{-1}$ was taken up annually by the grasslands, of which 29 kg N $ha^{-1}$ was released and about 29 kg $ha^{-1}$ was retained in the vegetation compartments (Billore and Mall, 1976). Greater accumulation of nutrients in the shoot was due to high nutrient concentration and greater biomass. In fact, this is the characteristic feature of tropical grasslands, whereas the greater belowground reserve is characteristic of temperate

**Table 13.4** Uptake, Transfer, and Release of Nitrogen in *Heteropogon contortus* Grazing Lands of Garhwal, Himalaya

| Components | Nitrogen (kg ha year$^{-1}$) |
|---|---|
| Uptake of N in aboveground net production | 6.25 |
| Uptake of N in belowground net production | 5.16 |
| Total uptake | 11.41 |
| Transfer from live to standing dead compartment | 6.41 |
| Transfer from standing dead to litter compartment | 5.97 |
| Release through litter decomposition | 5.71 |
| Release through root decomposition | 3.98 |
| Total release | 9.69 |

Bhandari, B.S., Mehta, J.P., Tiwari, S.C., 2000. Fire and nutrient dynamics in a *Heteropogon contortus* grazingland of Garhwal Himalaya. Tropical Ecology 41, 33–39.

grasslands (Raghubanshi, 2008). Uptake of nitrogen at 11.41 kg/ha/year was also observed in *Heteropogon contortus* grazing lands of Garhwal, Himalaya, of which 84.9% nitrogen was released through litter and root decomposition (Table 13.4).

On the contrary, higher allocation of biomass and nutrients to the belowground parts was observed in the grasslands at Cherrapunji, which was probably due to adverse climatic and edaphic conditions. In these grasslands, among different vegetation compartments, live root was the largest reservoir of N followed by live shoot, dead shoot, litter, and dead root. The total annual N uptake was 137.3 kg ha$^{-1}$. In an annual cycle, 98% N was returned to the soil through litter and belowground detritus (Fig. 13.4). Thus nitrogen cycling and its partitioning in grasslands were also influenced by climatic conditions and their topography.

## Grassland Nitrogen: Environmental Impacts

While widespread use of fertilizer N has greatly increased the agricultural production it has also contributed to environmental pollution. This is partly due to the energy requirements of fertilizer manufacture, partly due to the pollution of the wider environment by compounds derived from fertilizer N, and partly due to the loss of biodiversity. However, management practices to minimize N pollution are site-specific, which depend on soil physical and hydraulic characteristics, as well as the rainfall and irrigation facilities of a specific region (Vendramini et al., 2007).

Pollution of the wider environment occurs through the use of fertilizer N derived from the leaching of nitrate to groundwater and streams, and from the volatilization of ammonia, nitrous oxide, and nitric oxide to the atmosphere. Also, the botanical diversity of old grassland/permanent pasture is inevitably reduced by the application of large amount fertilizer N and it specifically reduces the dicotyledonous species that are visually attractive during the flowering season, and that provide food for insects such as butterflies. The rate of fertilizer N applied to grassland has a major impact on the

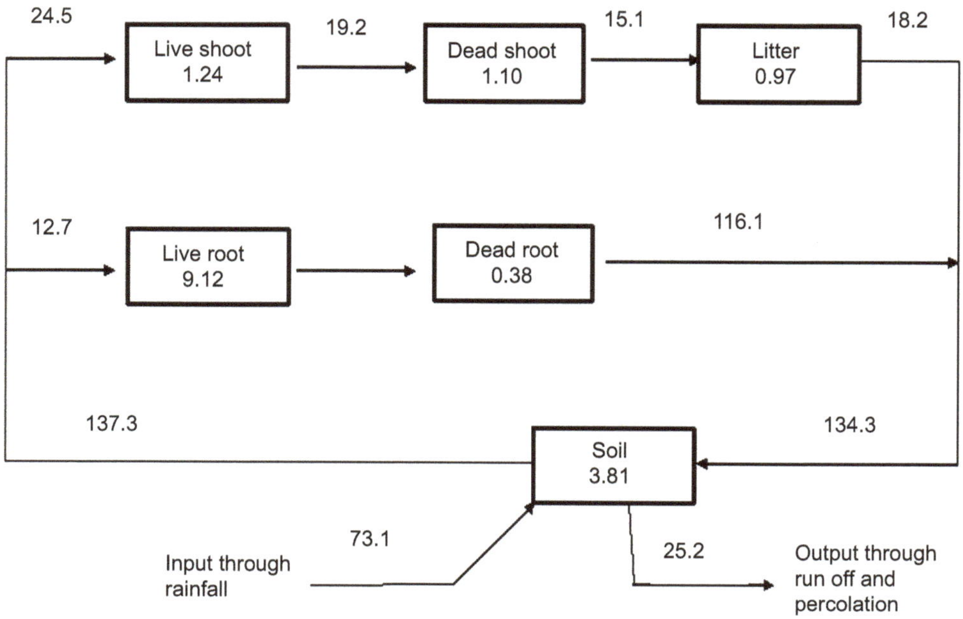


**FIGURE 13.4** Cycling of N in soil-vegetation compartments at Cherrapunji grasslands. Mean N accumulation (kg $ha^{-1}$) in different vegetation compartments and in available pool in soil down to 30 cm depth is shown boxes. The N flux rates (kg $ha^{-1}year^{-1}$) are indicated by arrows (Pandey et al., 1993).

transformations that are liable to result in pollution in addition to other factors. Thus the increase in the average rate of fertilizer N applied to grassland was responsible, either directly or indirectly, for much of the increase in amounts of N leached through the soil and lost in gaseous forms to the atmosphere, though factors such as plowing of long-term grassland and the incorporation of animal feeding stuffs were also partly responsible. Nitrogen that is leached from soil into groundwater is almost entirely in the form of nitrate and, if the groundwater is used for domestic supplies, the nitrate may occasionally constitute a health risk to consumers including animals.

In grassland, nitrous oxide is produced mainly from the soil by a combination of nitrification and denitrification, especially where the concentration of nitrate is high following the application of fertilizer or the deposition of urine. Much of the ammonia in the atmosphere is derived from the hydrolysis of livestock urine and, to a lesser extent, from the interaction of soils with fertilizers containing ammonium- or urea-N, and from the decomposition of plant material. Although some volatilization of ammonia occurs in natural ecosystems, it is greatly increased in areas of intensive livestock production. Ultimately, ammonia released to the atmosphere is deposited on the land or sea surface, where it may have potentially undesirable effects. For example, ammonia contributes to the acidifying effect of acid rain on soil. However, the effects of nitrous oxide ($N_2O$) may be more harmful in the long-term, as it acts as a "greenhouse gas" and also contributes to the destruction of ozone in the stratosphere.

## Grassland Nitrogen: Future Research

Nitrogen *albeit* an essential nutrient for plant/pasture growth is used inefficiently in most of the world's agricultural systems including pasture production, resulting in unnecessary losses to the environment with profound impacts ranging from air and water pollution to the undermining of important ecosystems (and the services and livelihoods they support). Indeed, the global nitrogen cycle has been profoundly altered by human activities over the past century. The amount of usable or "reactive" nitrogen produced by humans (about 190 million tons per year) is now greater than the amount created through natural (112 million tons per year) processes (Anonymous, 2014). In addition to inefficient application of nitrogen fertilizers, sources of excess nitrogen in the grassland environment are animal excreta, which create nitrogen oxides. Ongoing research on grassland nitrogen has improved our knowledge and understanding of the nitrogen cycle in the last few decades or so, along with our ability to quantify the sources, fate, and impacts of nitrogen on the environment. However, because of the complexity of the nitrogen cycle and the different scales involved (from local to global), scientific understanding and measures to reduce nitrogen pollution are inadequate, and need research to make further progress. Such progress will help different sources of plant-available nitrogen to be utilized more effectively in pasture as well as livestock production with minimum losses and, therefore, contribution to environmental pollution. Following are a few suggestions regarding future research directions that are particularly relevant to grassland nitrogen management:

- Instead of concentrating on individual transformations of N, there is a need for more comprehensive, integrated research and management to reduce the adverse effects of excess nitrogen in the environment while optimizing forage and livestock production from grasslands. It is important to make simultaneous assessments of the effects of soil, weather, and management factors on a range of transformations within total farming systems.
- Focusing on natural production (nitrogen "fixing") while limiting flows of reactive nitrogen. Making greater use of symbiotic fixation will benefit environment through reducing the energy costs of manufacturing, transporting, and applying fertilizer N.
- Fixing nitrogen in such a way that it does not find its way back into the environment, and, if possible, reusing of nitrogen.
- More effective utilization of fertilizer N through adjusted application to match the supply from soil and atmosphere, and the expected requirement of grasses for forage production. This approach may even include the development of computer programs to provide recommendations for fertilizer N in specific situations.
- Modifying the composition of diets to reduce the concentration of N in the excreta without adversely affecting milk production or live-weight gain in animals. This is possible by assessing the optimal concentrations of nitrogen and energy more accurately for different categories of animals in relation to different types of diets.

- Better understanding and knowledge on ammonia volatilization, nitrification and leaching of nitrates, mineralization, immobilization and availability of N, denitrification, and associated processes, specifically in tropical grasslands.
- Finding out the means and ways or technologies that convert nitrogen back into unreactive nitrogen ($N_2$) in near future.

## References

Afzal, M., Adams, W.A., 1992. Heterogeneity of soil mineral nitrogen in pasture grazing by cattle. Soil Science Society of America Journal 56, 1160–1166.

Anonymous, 2014. Excess Nitrogen in the Environment. UNEP Year Book 2014. http://www.unep.org/yearbook/2014/PDF/chapt1.pdf.

Bauman, P., Hernandez, K., Smart, S., 2015. Grassland Fertilization: Part 3: Native Pasture (Case Studies 1 & 2). http://igrow.org/livestock/beef/grassland-fertilization-part-3-native-pasture-case-studies-1-2/.

Bhandari, B.S., Mehta, J.P., Tiwari, S.C., 2000. Fire and nutrient dynamics in a *Heteropogon contortus* grazingland of Garhwal Himalaya. Tropical Ecology 41, 33–39.

Billore, S.K., Mall, L.P., 1976. Nutrient composition and inventory in a tropical grassland. Plant and Soil 45, 509–520.

Blumenthal, J.M., Baltensperger, D.D., Cassman, K.G., Mason, S.C., Pavlista, A.D., 2008. Importance and effect of nitrogen on crop quality and health. In: Hatfield, J.L., Follett, R.F. (Eds.), Nitrogen in the Environment: Sources, Problems and Management. Elsevier, Amsterdam.

Coupland, T.R., 1979. Grassland Ecosystems of the World: Analysis of Grasslands and Their Uses. Cambridge University Press, Cambridge, London.

Daly, G.T., 1990. The grasslands of New Zealand. In: Langer, R.H.M. (Ed.), Pastures, Their Ecology and Management. Oxford University Press, Auckland, pp. 1–38.

Dwivedi, G.K., Kumar, D., Tomer, P.S., 1990. Effect of cutting management and nitrogen levels on growth, seed yield attributes and seed production of *Setaria sphacelata* cv. Nandi. Tropical Grasslands 33, 146–149.

Eckard, R., 2004. Livestock Health and Production as Influenced by Nitrogen Fertilizer in the Sub-tropics. http://www.nitrogen.unimelb.edu.au/Nhealth.htm.

Floate, M.J.S., 1987. Nitrogen cycling in managed grasslands. In: Snaydon, R.W. (Ed.), Managed Grasslands: Analytical Studies. Elsevier, Amsterdam, pp. 163–172.

Galloway, J.N., Dentener, F., Burke, M., Dumont, E., Bouwman, A.F., Kohn, R.A., Mooney, H.A., Seitzinger, S., Kroeze, C., 2009. The impact of animal production systems on the nitrogen cycle. In: Steinfeld, H., Mooney, H.A., Schneider, F., Neville, L.E. (Eds.), Livestock in a Changing Landscape. Island Press, Washington, D.C, pp. 83–95.

Herrero, M., Gerber, P., Vellinga, T., Garnett, T., Leip, A., Opio, C., Westhoek, H.J., Thornton, P.K., Olesen, J., Hutchings, N., Montgomery, H., Soussana, J.F., Steinfeld, H., McAllister, T.A., 2011. Livestock and greenhouse gas emissions: the importance of getting the numbers right. Animal Feed Science and Technology 166, 779–782.

Krishnan, P., Ramakrishnan, R., Saigal, S., Nagar, S., Faizi, S., Panwar, H.S., Singh, S., Ved, N., 2012. Conservation across Landscapes: India's Approaches to Biodiversity Governance. United Nations Development Programme, New Delhi, India.

Lingorski, V., Totev, T., Mukhovski, T.S., 1995. Effect of mineral fertilizer application on the yield and botanical composition of meadow grasses and grass mixtures in the Troyan region. Rasteniev Nauk 32, 100–103.

Mahanta, S.K., Singh, K.K., Das, M.M., Misra, A.K., 2009. Microbial protein synthesis in ruminants fed forage. In: Das, N., Misra, A.K., Maity, S.B., Singh, K.K., Das, M.M. (Eds.), Forage for Sustainable Livestock Production. Satish Serial Publishing House, Delhi, pp. 295–320.

Mishra, L.C., 1979. Nitrogen cycling in grassland at Kanpur, India. Plant and Soil 53, 361–371.

Mountford, J.O., Lakhani, K.H., Kirkham, F.W., 1993. Experimental assessment of the effects of nitrogen addition under hay-cutting and aftermath grazing on the vegetation of meadows on a somerset peat moor. Journal of Applied Ecology 30, 321–332.

Niranjan, K.P., Burman, D., Arya, R.L., Agrawal, R.K., Shivadhar, 2004. Influence of nitrogen and phosphate levels on forage production of *Pennisetum* trispecific hybrid and *Stylosanthes hamata* under rainfed conditions. Range Management and Agroforestry 25, 98–101.

Pandey, H.N., Tripathi, R.S., Shankar, U., 1993. Nutrient cycling in an excessively rainfed subtropical grassland at Cherrapunji. Journal of Biosciences 18, 395–406.

Pathak, P.S., Dagar, J.C., 2015. Indian grasslands and their management. In: Ghosh, P.K., Mahanta, S.K., Singh, J.B., Pathak, P.S. (Eds.), Grassland: A Global Resource Perspective. Range Management Society of India, Jhansi.

Raghubanshi, A.S., 2008. Nitrogen cycling in Indian terrestrial natural ecosystems. Current Science 94, 1404–1412.

Shankar, V., Gupta, J.N., 1992. Restoration of degraded rangelands. In: Singh, J.S. (Ed.), Restoration of Degraded Lands-Concepts and Strategies. Rastogi Publications, Meerut, India, pp. 115–155.

Singh, V., 1995. Technology for forage production in hills of Kumaon. In: Hazra, C.R., Misri, B. (Eds.), New Vistas in Forage Production. Indian Grassland and Fodder Research Institute, Jhansi, pp. 197–202.

Sood, B.R., Sharma, V.K., 1996. Effect of nitrogen and phosphorus levels on the productivity of a natural grassland. Haryana Journal of Agronomy 12, 68–74.

Steinfeld, H., Gerber, P., Wassenaar, T., Castel, V., Rosales, M., de Haan, C., 2006. Livestock's Long Shadow: Environmental Issues and Options. FAO, Rome, Italy.

Subbarao, G.V., Sahrawat, K.L., Nakahara, K., Rao, I.M., Ishitani, M., Hash, C.T., Kishii, M., Bonnett, D.G., Berry, W.L., Lata, J.C., 2013b. A paradigm shift towards low-nitrifying production systems: the role of biological nitrification inhibition (BNI). Annals of Botany 112, 297–316.

Subbrao, G.V., Rao, I.M., Nakhara, K., Ando, Y., Sahrawat, K.L., Tesfamariam, T., Lata, J.C., Boudsocq, S., Miles, J.W., Ishitani, M., Peters, M., 2013a. Nitrogen management in grasslands and forage-based production systems: role of biological nitrification inhibition (BNI). Tropical Grassland–Forrajes Tropicales 1, 168–174.

Sun, X., Luo, N., Longhurst, B., Luo, J., 2008. Fertilizer nitrogen and factors affecting pasture responses. The Open Agriculture Journal 2, 35–42.

Vendramini, J.M.B., Silveira, M.L.A., Dubeux Jr., J.C.B., Sollenberger, L.E., 2007. Environmental impacts and nutrient recycling on pastures grazed by cattle. Revista Brasileira de Zootecnia 36, 139–149.

Wetselaar, R., Ganry, F., 1982. Nitrogen Balance in Tropical Agrosystems. http://horizon.documentation.ird.fr/exl-doc/pleins_textes/divers16-03/17371.pdf.

Whitehead, D.C., 1995. Grassland Nitrogen. CAB International, Wallingford, UK.

# 14

# Reactive Nitrogen in Agroforestry Systems of India

A. Ram[1], I. Dev[1], A.R. Uthappa[1], Dhiraj Kumar[1], N. Kumar[1], O.P. Chaturvedi[1], M.L. Dotaniya[2], B.P. Meena[2]

[1]*ICAR-CENTRAL AGROFORESTRY RESEARCH INSTITUTE, JHANSI, INDIA;* [2]*ICAR-INDIAN INSTITUTE OF SOIL SCIENCE, BHOPAL, INDIA*

## CHAPTER OUTLINE



India faces a critical imbalance in its natural resource base with about 18% human and 15% livestock population of the world being supported only on 2.4% geographical area, 1.5% forest and pasture lands, and 4.2% water resources. A serious threat is being predicted for last few years in meeting the food, fiber, fuel, and fodder requirements of the increasing population due to effects of climate change. Landuse diversification with agroforestry as a component can help to address some of these major challenges. Agroforestry provides various ecosystem services, viz. provisioning (food, fuelwood, fodder, timber, poles, etc.), regulatory (hydrological benefits, micro-climatic modifications, etc.), supporting services (nutrient cycling, agro-biodiversity conservation, etc.), and cultural services (recreation, aesthetics, etc.), besides ensuring the livelihood security of its practitioners (CAFRI Vision, 2015). It will be contributing substantially in meeting the basic essential societal needs through enhanced production and providing environmental benefits with increase in present area from 17.45 m ha (mha) (Rizvi et al., 2014) to 53.0 m ha in next 35 years (Dhyani et al., 2013). United Nations Convention to Combat Desertification acknowledges the potential of agroforestry to control desertification and rehabilitation. It is expected that by 2050 there will be substantial increase in area under tree cover, i.e., Tree Outside Forest, and a large portion of it will be devoted to agroforestry. In addition, out of the Prime minister's eight National Missions (NM) under National Action Plan on Climate Change, agroforestry is likely to play proactive role in

The Indian Nitrogen Assessment. http://dx.doi.org/10.1016/B978-0-12-811836-8.00014-8

the NM on Greening India, NM on Energy, NM on Sustainable Agriculture, and NM for Sustaining the Himalayan Ecosystem.

With the agroforestry landuse, overall (biomass) productivity, soil fertility improvement, soil conservation, nutrient cycling, microclimate modification, and carbon sequestration potential are generally greater than that of an annual system (Dupraz and Newman, 1997; Goor et al., 2000; Calfapietra et al., 2010). Nitrogen is the most limiting nutrient among all essential nutrients for enhancing agriculture production in India. Soil erosion and runoff water further aggravate the problem of nitrogen deficiency (Lehmann, 1999). Prior to green revolution, farmers were overcoming the nitrogen limitation by taking legumes in rotation, applying organic manures, and chemical fertilizers in a limited quantity. However, in the green revolution era, the natural nitrogen cycle has been altered by applying excessive quantity of synthetic N fertilizers (Singh and Singh, 2008). Industrial fixation of N fertilizer use by far represents the largest human contribution of new reactive N to N-cycle. Already, the quantity of fertilizer N used in India during 1995–2005 was equal to the total quantity used up to 1994 (FAI, 2006). At present 16.75 Tg of reactive N is added to the agroecosystem through chemical fertilizers every year in the country (Indian Fertilizer Scenario, 2013). Unbalanced and unscientific use of chemical fertilizers causes N cascade in the form of eutrophication, depletion of tropospheric ozone, increased levels of nitrates in drinking water, loss of biodiversity, and extinction of several species (Erisman et al., 2011; Rosenstock et al., 2014). A review of the best available studies shows that N recovery in various crops varies between 20–30% for rainfed and 30–40% for irrigated situations (Dotaniya and Meena, 2013). Remaining nitrogen either leaks out of agroecosystem or becomes part of the soil organic pool (Singh and Singh, 2008). Use of increasing amounts of reactive N in agroforestry system and the proper management of fertilizer N will be an important issue to improve the reactive N balance (short- and long-term).

Integration of leguminous trees into agricultural system increases the reactive N supply in rooting zone through biological nitrogen fixation (BNF), retrieval of nitrogen from the layer below to root zone, and reduces N loss through leaching and erosion. Large amounts of foliage and pruned material in agroforestry system (AFS) also release N upon decomposition and thus reduces fertilizer N need besides increasing overall productivity. For the success of agroforestry systems, selection of tree species is a decisive factor, which depends upon the economic and agronomic criteria. N-fixing trees are probably the best choice, because they actively fix N and significantly improve the N status of soil (Dommergues, 1987; Rosenstock et al., 2014).

Agroforestry also plays important role as a sink for reactive nitrogen (Nr) in N-cycling. AFS greatly reduces the N-erosion through large and frequent litter cover, checking soil and water erosion by vegetative barriers, and stabilizing soil conservation structures. It is well established by many studies that agroforestry retains greater ecosystem functions as higher biodiversity and higher N retention (Harmand et al., 2007; Tscharntke et al., 2011). There are conflicting findings about $N_2O$ emission from AFS, as some studies show that $N_2O$ emission was higher in legume-based AFS than those from non-N-fixing trees or annual cropping systems (Verchot et al., 2007; Baggs et al., 2006; Dick et al., 2006). Rosenstock et al. (2014) reported higher $N_2O$ emissions from agricultural lands because of high use of N fertilizers than AFS. This chapter attempts to do a preliminary assessment of Nr in agroforestry systems in India.

## Agroforestry and Nitrogen Cycling

N is in a constant state of change and its distribution is not uniform in all types of systems, and it is by far the most yield-limiting nutrient in agriculture system. The nitrogen use efficiency is hardly about 50% of the applied amount. Thus, if applied nitrogen is not completely taken up by plants, nor stored in organic pools, then it is bound to contribute toward emissions and associated environmental problems such as global warming ($N_2O$), groundwater contamination ($NO_3^-$), air pollution and eutrophication ($NH_3$), and biodiversity loss (Crutzen et al., 2008; Ju et al., 2006; Millar et al., 2010; Sutton et al., 2011). Tree species, climatic conditions, and age of plantation are the major governing factors for nitrogen cycling under AFS (Nair et al., 2009). The agroforestry system has the potential to reduce the emissions and nitrogen surplus from the agroecosystem as shown in (Fig. 14.1) due to its role in improving soil health as well as the deep root systems that acts as a filter net to bind the extremely mobile $NO_3^-$-N, which otherwise is lost by leaching or denitrification.

Trees may reduce leaching losses up to 47% as compared to crop monoculture (Lehmann et al., 1999). The nitrogen requirement of the short duration trees are less than the annual crops, thus the external nitrogen inputs can be reduced. Further, the agroforestry system not only contributes to BNF, but also adds a lot of litter through leaf fall,

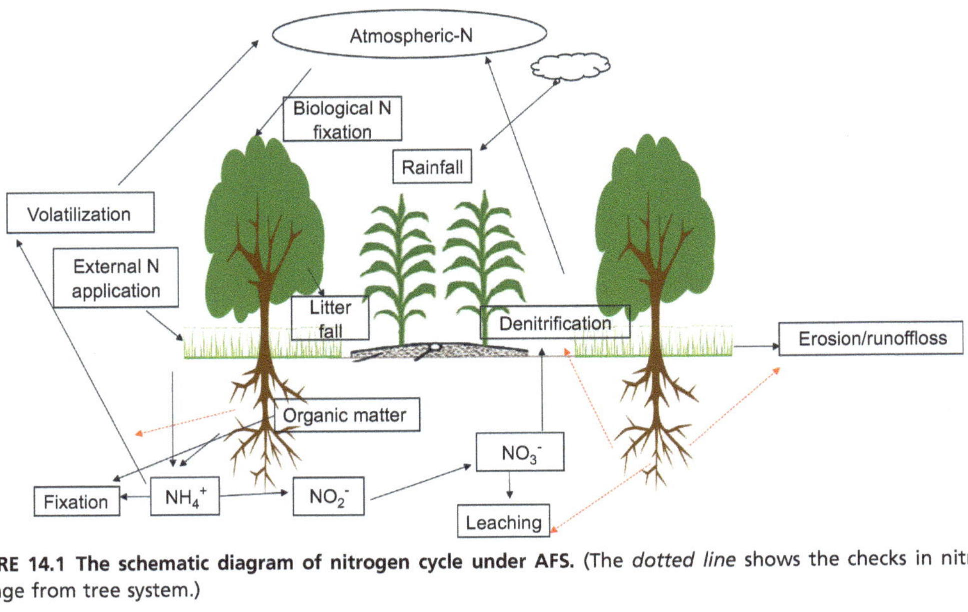


**FIGURE 14.1 The schematic diagram of nitrogen cycle under AFS.** (The *dotted line* shows the checks in nitrogen leakage from tree system.)

root decay, and pruned material. Leaf litter and organic inputs, having N (high), lignin (low), and polyphenol (low) content, release larger proportions of N at a faster rate. However, organic inputs having N (low), lignin (high), and high polyphenol (high) content release nutrients at a slower rate continuously resulting in lower potential losses through leaching. According to Gregory and Nortcliff (2013), the discrepancy between slow and quick release litters could be utilized for maximizing the synchrony between crop N demand and release. Most significantly, the added organic matter has nitrogen in the organic forms, some labile and most of the portion as organic, so it is not going to emit easily as is the case with inorganic fertilizers. Thus soil organic nitrogen in AFS will definitely make a difference as far as long-term sustainability is concerned.

## Reactive Nitrogen Addition in Agroforestry System

Nr in AFS comes from various sources, i.e., BNF through trees and associated crops, pruned materials, litter fall, manures and fertilizers, irrigation, etc.

### Through Trees

The introduction of chemical fertilizers to compensate N deficiency has created a lot of environmental problems in different ecosystems. Nature offers better solutions to tackle the environmental issues created by chemical fertilizers. BNF is one of the best sustainable alternatives to conventional N fertilizers that enhances nutrient cycling with in the soil environment. The total annual terrestrial inputs of N from BNF as given by Burns and Hardy (1975) and Paul (1988) range from 139 to 175 Tg of N, with symbiotic associations growing in arable land accounting for 25–30% (35–44 Tg of N) and permanent pasture accounting for another 30% (45 Tg of N).

Over 600 tree species have been reported as N-fixing and integration of these trees can reduce resource competition among different components in agroforestry. The contribution of leguminous trees for building up N in degraded soils through BNF is well recognized as an important component of the ecosystem service of nutrient cycling (Uthappa et al., 2015). There are significant differences in estimates of BNF in trees, ranging from high rates up to 500 kg $N_2$ $ha^{-1}$ $year^{-1}$ (*Leucaena leucocephala*) to low rates 13 kg $N_2$ $ha^{-1}$ $year^{-1}$ (*Gliricidia sepium*) as reported by Young (1989); Antangana et al. (2014); Rao (1990). Dommergues (1987) have classified the tree species into two broad categories on the basis of N-fixing potential. Species with a high N-fixing potential (in the range of 100–300 kg N $ha^{-1}$ $year^{-1}$ and more), e.g., *Acacia mangium, Casuarina equisetifolia, L. leucocephala* and species with a low N-fixing potential (<20 kg N $ha^{-1}$ $year^{-1}$), e.g., *Acacia albida, A. raddiana* and *A. senegal.* The agroforestry systems in tropical countries are designed on the basis of BNF trees which can act as fertilizers to annual components. Table 14.1 illustrates the BNF trees, with their N-fixation potential in soil, so that they can be used for designing agroforestry system on the basis of the N requirement of annual components.

**Table 14.1** Biological N-Fixing Potential of Leguminous and Actinorhizal Trees and Shrubs in Agroforestry System

| Tree Species | N-Fixation (kg $ha^{-1}$ $year^{-1}$) |
|---|---|
| Leguminous trees | |
| *Acacia mangium* | 128 |
| *Acacia mearnsii* | 200 |
| *Albizia lebbeck* | 260 |
| *Erythrina sp.* | 57–66 |
| *Faidherbia albida* | 20–50 |
| *Gliricidia sepium* | 13–50 |
| *Grevellea robusta* | 100–300 |
| *Inga edulis* | 35–40 |
| *Leucaena leucocephala* | 100–500 |
| *Prosopis* species | 25–26 |
| *Sesbania sesban* | 260–330 |
| Actinorhizal trees | |
| *Allocasuarina littoralis* | 218 |
| *Alnus acuminata* | 279 |
| *Alnus glutinosa* | 40–53 |
| *Alnus nepalensis* | 30–120 |
| *Alnus rubra* | 85–320 |
| *Casuarina equisetifolia* | 60–110 |
| *Coriaria arborea* | 192 |
| *Ceanothus velutinus* | 4–100 |

Adapted from Young, A., 1989. Agroforestry for Soil Conservation. CAB International, Wallingford; Rao, D.L.N., Gill, H.S., Abrol, I.P., 1990. Regional experience with perennial sesbania in India. In: Perennial Sesbania Species in Agroforestry, pp. 189–198; Atangana, A., Khasa, D., Chang, S., Degrande, A., 2014. Tropical Agroforestry. Springer; Russo, R.O., 2005. N-fixing trees with actinorhiza in forestry and agroforestry. In: Wenzer, D., Newton, W.E., (Eds.), Nitrogen Fixation in Agriculture, Forestry, Ecology and the Environment; pp. 143–171.

Among N-fixing trees, many legume species are nodualted by *Rhizobium*, while a few nonleguminous species are nodulated by *Frankia* (about 200 plant species covering 19 genera and 8 families). Almost all genera are nodulated by *Frankia* in the Casuarinaceae, Coriariaceae, Eleagnaceae, Datisticaceae, and Myricaceae families, whereas nodulation occurs occasionally in Betulaceae, Rhamnaceae, and Rosaceae (Benson and Clawson, 2000).

Actinorhizal-*Frankia* symbiosis can be found in a vast range of climates and have the potential to fix up to 320 kg N $ha^{-1}$ $year^{-1}$ (Table 14.1). Additionally, these species tend to be aggressive colonizers and can survive in harsh conditions. However, the integration of these trees into agroecosystems is an area of research that remains largely unexplored (Batiano, 2012). Actinorhizal trees play vital role as woody perennials in agroforestry systems. One of the most known species of actinorhizal trees is *Casuarina* which is used for firewood production as a component of a multipurpose agroforestry plantation, or when large plantations are interplanted (Taungya system) with agricultural crops. Paschke and Dawson (1992) described some of the actinorhizal tree species and their potential role in agroforestry; *Alnus* (timber and pulpwood); *Elaeagnus* spp. (nurse plant); and *Elaeagnus, Shepherdia,* and *Purshia* spp. (soil reclamation).

Management of biologically fixed N in the agroforestry system is most crucial and many studies have been carried out in order to enhance the nitrogen use efficiency. The species or provenances selected should have the highest N-fixing potential, be tolerant to adverse environmental constraints, and resistant to pests. N-fixing efficiency also varies with tree density and climatic conditions. To achieve maximum N-fixing potential, optimum pH, adequate phosphorus, calcium, and molybdenum levels need to be maintained. Secondary effects such as allelopathic interferences and development of microorganisms pathogenic to companion crops should not be overlooked.

In the recent past, various researchers have estimated the area of agroforestry in India. The estimates range from 11.15 m ha (FSI, 2013) to 25.32 m ha (Dhyani et al., 2013). Recently, Rizvi et al. (2014) have also estimated agroforestry area (17.45 m ha) for different states using Bhuvan landuse/land cover data. For calculating the number of BNF trees $ha^{-1}$, agroforestry data of eight Indian states (Punjab, Haryana, Uttar Pradesh, Bihar, West Bengal, Gujarat, Rajasthan, and Tamil Nadu) were used (NRCAF, Annual Report, 2014; CAFRI, Annual Report, 2015; Ajit et al., 2016). The average N-fixing capacity (7.46 kg N $ha^{-1}$ $year^{-1}$ $tree^{-1}$) was derived by taking the average of all BNF trees in all these eight states and their N-fixing potential (Diagne, 1996; Roskoski et al., 1986; Dreyfus et al., 1987; Sanginga et al., 1986). The average number of BNF trees was 1.5 trees $ha^{-1}$, fixing about 11.18 kg N $ha^{-1}$ $year^{-1}$. Therefore, BNF trees in an area of 17.45 m ha in India can fix N up to 0.195 Tg $year^{-1}$ (Table 14.2).

In order to estimate average litter fall from agroforestry systems, litter fall data of *Eucalyptus, Populus deltoides, Acacia tortilis, A. leucocephla, Prosopis cineraria, Dalbergia sissoo, A. nilotica, Alnus nepalensis, Leucaena leucocephala, Bambusa vulgaris, Grewia optiva, Azadirachta indica, Pongamia pinnata, Tectona grandis, Celtis australis,* and *Albizia procera* were considered. Based on literature review (Roy and Nigam, 2001; Yadav et al., 2008; Kaur, 1998; Das and Chaturvedi, 1997; Tripathi et al., 2009; Oladoye et al., 2010; Hariprasath et al., 2014; Panwar and Gupta, 2016; Singh et al., 1999; Manivasakan et al., 2007; Panwar and Gupta, 2015; Gupta et al., 2010; Isaac and Achuthan, 2006), it was found that one tree adds 18.70 kg litter $ha^{-1}$ $year^{-1}$. Considering average density of 12.44 trees $ha^{-1}$ (NRCAF, Annual Report, 2014; CAFRI, Annual Report, 2015; Ajit et al., 2016) in agroforestry, 0.055 Tg of N is added through litter fall to the system.

## Through Crops and Other External Inputs

A large amount of N is supplied through organic manure application, crop residue recycling, green manures, atmospheric deposition, and BNF; however, inorganic

**Table 14.2** Preliminary Estimates of Biological Nitrogen Fixation Through Trees in AFS in India[a]

| State/UT | Area Under Agroforestry (m ha) | N-Fixation ($t\ yr^{-1}$) |
|---|---|---|
| Andhra Pradesh | 1.67 | 18,704 |
| Arunachal Pradesh | 0.02 | 201 |
| Assam | 0.27 | 2985 |
| Bihar | 0.80 | 8888 |
| Chhattisgarh | 0.70 | 7815 |
| Delhi | 0.01 | 67 |
| Goa | 0.01 | 123 |
| Gujarat | 1.09 | 12,175 |
| Haryana | 0.35 | 3935 |
| Himachal Pradesh | 0.03 | 369 |
| Jammu and Kashmir | 0.09 | 1051 |
| Jharkhand | 0.53 | 5970 |
| Karnataka | 1.29 | 14,456 |
| Kerala | 0.09 | 1051 |
| Madhya Pradesh | 1.35 | 15,037 |
| Maharashtra | 1.92 | 21,421 |
| Meghalaya | 0.02 | 246 |
| Manipur | 0.02 | 201 |
| Mizoram | 0.00 | 45 |
| Nagaland | 0.01 | 56 |
| Odisha | 0.80 | 8989 |
| Puducherry | 0.00 | 22 |
| Punjab | 0.42 | 4696 |
| Rajasthan | 2.05 | 22,930 |
| Sikkim | 0.01 | 89 |
| Tripura | 0.03 | 291 |
| Tamil Nadu | 0.69 | 7692 |
| Uttar Pradesh | 1.97 | 22,036 |
| Uttarakhand | 0.07 | 827 |
| West Bengal | 0.41 | 4528 |
| Other states/UT | 0.73 | 8173 |
| All India | 17.45 | 195,069 |

[a]Species considered for calculating the BNF through trees: *Dalbergia sissoo, Acacia nilotica, Leucaena leucocephala, Albizia procera, Acacia auriculiformis, Prosopis cineraria, Acacia tortilis, Gliricidia sepium.*

fertilizer inputs are the major form of nitrogen supply to agriculture crops. N fixed from the air by legume crops, although the magnitude of BNF is practically difficult to enumerate, in general annual estimates, is in the order of 150 kg N $ha^{-1}$ (Giller and Wilson, 1991). Although, a large amount of the N (>97%) in the soil system is tied up in the soil organic matter and is unavailable to plants Only 2–3% is in the inorganic form of N ($NO_3^-$ and $NH_4^+$) that is available to crop plants.

**Table 14.3** Reactive Nitrogen Addition in Agriculture and Agroforestry Systems Through Various Sources

| Different Sources of N | Agriculture System (Tg year$^{-1}$) | Agroforestry System (Tg year$^{-1}$) |
|---|---|---|
| 1. Synthetic N fertilizers | 16.75 | 2.07 |
| 2. N-fixation by legume crops | 0.67 | 0.115 |
| 3. Irrigation and rainfall | 2.31 | 0.285 |
| 4. Manures and crop residues | 1.41 | 0.174 |
| 5. N-fixation by trees | – | 0.195 |
| 6. Tree litter fall | – | 0.055 |
| Total | 21.14 | 2.894 |

Data modified from Pathak, H., Mohanty, S., Jain, N., Bhatia, A., 2010. Nitrogen, phosphorus, and potassium budgets in Indian agriculture. Nutrient Cycling in Agroecosystems, 86, 287–299; Singh, K.K., Ali, M., Venkatesh, M.S., 2009. Pulses in Cropping Systems. Technical Bulletin, IIPR, Kanpur; based on agroforestry area in India.

Globally, main anthropogenic N source is fertilizer, which is continuously produced and applied at increasing rate, followed by N-fixation by agriculture crops mainly leguminous and other field crops (Swaney et al., 2015). N inputs from different sources in Indian agriculture and AFS are presented in Table 14.3. Based on the estimates of Pathak et al. (2010), it is calculated that in India about 21.14 Tg of N is added to agriculture system every year. Synthetic N fertilizers contribute maximum share (16.75 Tg) to it. In agroforestry systems (17.45 m ha), about 2.894 Tg of nitrogen is added every year through different sources, viz. synthetic N fertilizers (2.07 Tg), legume crops (0.115 Tg), irrigation and rainfall (0.285 Tg), manures and crop residues (0.174 Tg), and trees (0.250 Tg). Out of total N addition (0.250 Tg) through trees in AFS, major share of N (0.195 Tg) is added through N-fixing trees and rest (0.055 Tg) through litter fall. Trees contribute 9% of total N addition in AFS in India (Fig. 14.2). Area under pulse crops in India is around 25 m ha (Mohanty and Satyasai, 2015), which on an average fixes 37.63 kg N ha$^{-1}$ (Singh et al., 2009). Based on the assumption that of net cultivated area (141 m ha) in India, the area under pulses grown in AFS would be 17.61% (3.07 m ha), which can fix 0.115 Tg N year$^{-1}$.

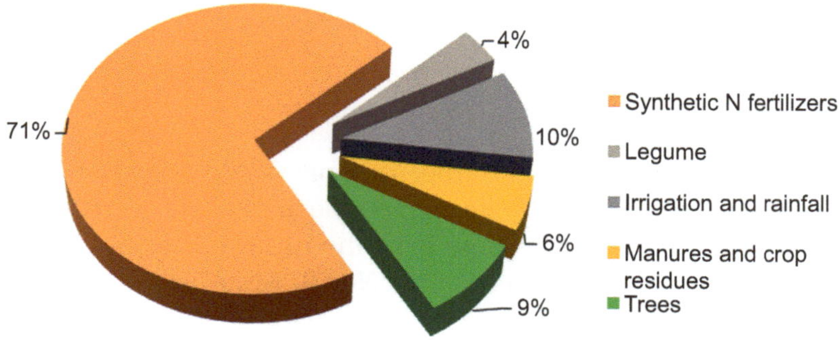


**FIGURE 14.2 Percent contribution of reactive N from different sources in agroforestry systems. Reduction in N losses through runoff in AFS.**

Removal of nitrogen through sediments and runoff will not only decline the soil fertility, but also causes environmental problems. Trees help in reducing the impacts of rain drop erosion, increases water infiltration, reduces runoff, and allows leaching nutrients to interact with natural attenuation capacity of soil (Creamer et al., 2010). Agroforestry systems act as a safety net and reduce the leaching loss of nutrients. A 6-year old poplar tree–based intercropping system has got the potential to reduce nitrate-N leaching by 227 kg N $ha^{-1}$ (Bergeron et al., 2011). In India such studies have not been conducted, therefore conclusions cannot be drawn about N leaching losses under AFS.

## Prospects of Nitrogen Focused Research in Agroforestry

Efforts are needed at national and global levels to control N loss to the environment. Use of N (synthetic fertilizer or organic source) in agriculture is essential to feed the ever growing human population, and technological changes are unlikely to provide a replacement for this as of now. A large amount of N will be required by plants to carry out their photosynthetic processes, and one of the most important factors to maintain adequate food production will be supplying plants with sufficient N. The efficiency of reactive N can be increased by recycling it within the agroecosystems, by increased use of fertilizer trees, and by redistributing reactive N from areas with high reactive N production to the areas where it is needed for food production. Field-specific corrective nitrogen management strategies for increasing productivity of different AFS and to ensure synchrony between crop N demand and supply of mineral N are essentially needed.

To precisely work out the contribution of AFS to nitrate pollution of the ground water, periodical monitoring is essential. Comprehensive models, predicting the amount of leaching of the natural water bodies are needed to be developed. Appropriate emission coefficients of N containing green house gases (GHGs) covering different seasons in the diverse AFS of the country should be necessarily developed. It is important to understand different kinds of N fluxes and to quantify variations in inputs and losses. Strategies must focus on synchronization of legume-nitrogen availability with crop demand and that may include delaying pruning until only weeks before planting; planting legume; and abandoning practices whenever N release is poorly timed. In light of growing need to produce more food and introduce N into AFS in resource-challenged regions, the integration of N-fixing trees represents a viable option in many systems. Advocacy of agroforestry with N-fixing trees would do well to orient the attention to ecosystem services (e.g., climate regulation) and locally important relevant services simultaneously.

## References

Ajit, D.S.K., Handa, A.K., Newaj, R., Chavan, S.B., Alam, B., Prasad, R., Ram, A., Rizvi, R.H., Jain, A.K., Uma, T.D., Shakhela, R.R., Patel, A.G., Dalvi, V.V., Saxena, A.K., Parihar, A.K.S., Backiyavathy, M.R., Sudhagar, R.J., Bandeswaran, C., Gunasekaran, S., 2016. Estimating carbon sequestration potential of agroforestry systems at district level in ten selected states of India. Agroforestry Systems. http://dx.doi.org/10.1007/s10457-016-9986-z.

Atangana, A., Khasa, D., Chang, S., Degrande, A., 2014. Tropical Agroforestry. Springer.

Baggs, E., Chebii, J., Ndufa, J.A., 2006. Short-term investigation of trace gas emissions following tillage and no-tillage of agroforestry residues in western Kenya. Soil and Tillage Research 90, 69–76.

Batiano, A., 2012. Fighting Poverty in Sub-saharan Africa. The Multiple Roles of Legumes in Integrated Soil Fertility Management. Dordrecht, New York. ISBN: 9789400715363, 94007153662011.

Benson, D.R., Clawson, M.L., 2000. Evolution of the actinorhizal plant symbioses. In: Triplett, E.W. (Ed.), Prokaryotic Nitrogen Fixation: A Model System for Analysis of Biological Process. Horizon Scientific Press, Wymondham, UK, pp. 207–224.

Bergeron, M., Lacombe, S., Bradley, R.L., Whalen, J., Cogliastro, A., Jutras, M.F., Arp, P., 2011. Reduced soil nutrient leaching following the establishment of tree-based intercropping systems in eastern Canada Agroforestry Systems. Agroforestry System 83, 321–330.

Burns, R.C., Hardy, R.W.F., 1975. Nitrogen Fixation in Bacteria and Higher Plants. Springer-Verlag, New York.

CAFRI Vision 2050, 2015. Jhansi: CAFRI.

CAFRI. Annual Report 2014-15, 2015. CAFRI.

Calfapietra, C., Gielen, B., Karnosky, D., Ceulemans, R., Scarascia Mugnozza, G., 2010. Response and potential of agroforestry crops under global change. Environmental Pollution 158, 1095–1104.

Creamer, R.E., Brennan, F., Fenton, O., Healy, M.G., Lalor, S.T.J., Lanigan, G.J., Regan, J.T., Griffiths, B.S., 2010. Implications of the proposed soil framework directive on agricultural systems in Atlantic Europe-a review. Soil Use and Management 26, 198–211.

Crutzen, P.J., Mosier, A.R., Smith, K.A., Winiwarter, W., 2008. $N_2O$ release from agro-biofuel production negates global warming reduction by replacing fossil fuels. Atmospheric Chemistry and Physics 8, 389–395.

Das, D.K., Chaturvedi, O.P., 1997. Litterfall and nutrient return in 4-year old plantations of different tree species in north Bihar. Journal of Tropical Forester 13 (1), 19–24.

Dhyani, S.K., Handa, A.K., Uma, 2013. Area under agroforestry in India: an assessment for present status and future perspective. Indian Journal of Agroforestry 15 (1), 1–11.

Diagne, O., 1996. Utilization and nitrogen fixation of Prosopis in Senegal. In: Felkar, P., Moss (Eds.), Prosopis: Semiarid Fuelwood and Forage Tree Building Consensus for the Disenfranchised. Center Semi-arid Forest Resources, Publications, Kingsville, TX.

Dick, J., Skiba, U., Munro, R., Deans, D., 2006. Effect of N-fixing and non N-fixing trees and crops on NO and $N_2O$ emissions from Senegalese soils. Journal of Biogeography 33, 416–423.

Dommergues, Y.R., 1987. The role of biological nitrogen fixation in agroforestry. In: Tippler, H.A., Nair, P.K.R. (Eds.), Agroforestry a Decade of Development. ICRAF, Nairobi, Kenya, pp. 245–271.

Dotaniya, M.L., Meena, V.D., 2013. Rhizosphere effect on nutrient availability in soil and its uptake by plants-a review. Proceedings of the National Academy of Sciences, India Section B: Biological Sciences 85 (1), 1–12.

Dreyfus, B.L., Diem, H.G.J., Freire, S.O., Keya, D.Y.R., 1987. Nitrogen fixation in tropical agriculture and forestry. In: Dasilva, E.J., Dommergues, Y., Nins, E.J., Ratledge, C. (Eds.), Microbial Technology in the Developing World. Oxford University Press, Oxford, pp. 7–50.

Dupraz, C., Newman, S.M., 1997. Temperate agroforestry: the European Way. In: Gordon, A.M., Newman, S.M. (Eds.), Temperate Agroforestry Systems. CAB International, Wallingford, UK.

Erisman, J.W., Galloway, J.N., Seitzinger, S.P., Bleeker, A., Butterbach, B.K., 2011. Reactive nitrogen in the environment and its effect on climate change. Current Opinion in Environmental Sustainability 3, 281–290.

Fertilizer Association of India, 2006. Annual Report 2005–2006. FAI.

Giller, K.E., Wilson, K.F., 1991. Nitrogen Fixation in Tropical Cropping Systems. CAB International, Wallingford, UK.

Goor, F., Jossart, J.M., Ledent, J.F., 2000. ECOP: an economic model to assess the willow short rotation coppice global profitability in a case of small scale gasification pathway in Belgium. Environmental Modelling & Software 15, 279–292.

Gregory, P.J., Nortcliff, S., 2013. The new challenge—sustainable production in a changing environment. In: Soil Conditions and Plant Growth. Blackwell Publishing Ltd, Oxford. http://dx.doi.org/10.1002/9781118337295.

Gupta, G., Yadav, R.S., Maurya, D., Mishra, S.V., 2010. Litter dynamics under different pruning regimes of *Albizia procera* based agroforestry system in semi-arid region. Asian Science 5 (2), 93–97.

Hariprasath, C.N., Sudarshan, A., Goroji, P.T., 2014. Quantification of litter fall and assessment of nutrient composition in bamboo (*Bambusa vulgaris* var. *Vulgaris*) plantation. International Journal of Forestry and Crop Improvement 5 (2), 54–60.

Harmand, J.M., Avila, H., Dambrine, E., Skiba, U., Miguel, S., Renderos, R.V., Oliver, R., Jimenez, F., Beer, J., 2007. Nitrogen dynamics and soil nitrate retention in a *Coffea arabica—Eucalyptus deglupta* agroforestry system in Southern Costa Rica. Biogeochemistry 85, 125–139.

Indian Fertilizer Scenario, 2013. Department of Fertilizers Ministry of Chemicals and Fertilizers Government of India. New Delhi.

Isaac, R.S., Achuthan, M.N., 2006. Litter dynamics of six multipurpose trees in a homegarden in Southern Kerala, India. Agroforestry Systems 67, 203–213.

Ju, X.T., Kou, C.L., Zhang, F.S., Christie, P., 2006. Nitrogen balance and groundwater nitrate contamination: comparison among three intensive cropping systems on the North China Plain. Environmental Pollution 143, 117–125.

Kaur, B., 1998. Organic Matter Dynamics and Nitrogen Mineralization in Agroforestry Systems. Ph.D. Thesis. Kurukshetra University, Kurukshetra, India, p. 161.

Lehmann, J., Weigl, D., Peter, I., Droppelmann, K., Gebauer, G., Goldbach, H., et al., 1999. Nutrient interactions of alley-cropped *Sorghum bicolor* and *Acacia saligna* in a runoff irrigation system in Northern Kenya. Plant and Soil 210, 249–262.

Manivasakan, S., Mani, A.K., Balasubramanian, A., Santhi, R., Karikalan, V., 2007. Litterfall and nutrient accretions to soil in teak plantation. In: Proceedings of the 12th International Forestry and Environment Symposium, Sri Lanka. http://journals.sjp.ac.lk/index.php/fesympo/article/view/459.

Millar, N., Robertson, G.P., Grace, P.R., Gehl, R.J., Hoben, J.P., 2010. Nitrogen fertilizer management for nitrous oxide ($N_2O$) mitigation in intensive corn (Maize) production: an emissions reduction protocol for US Midwest agriculture. Mitigation and Adaptation Strategies for Global Change 15, 185–204.

Mohanty, S., Satyasai, K.J., 2015. Feeling the Pulse Indian Pulses Sector. Nabard Rural Pulse. Available from: URL. https://www.nabard.org/publication/rural-pulse-july-august-2015. pdf.

Nair, P.K.R., Mohan Kumar, B., Nair, V.D., 2009. Agroforestry as a strategy for carbon sequestration. Journal of Plant Nutrition and Soil Science 172, 10–23.

NRCAF, 2014. Annual Report 2013–14. NRCAF.

Oladoye, A.O., Ola-Adams, B.A., Adedire, M.O., 2010. Litterfall dynamics in *Leuceana leucocephala* (Lam) de Wit plantation in the nigerian derived Savanna. ARPN Journal of Agricultural and Biological Sciences 5 (2), 31–38.

Panwar, V.P., Gupta, M.K., 2015. Litter fall, nutrient return and soil fertility under *Celtis australis*: an indigenous agroforestry tree species in Himachal Pradesh, India. International Journal of Environmental Sciences 6 (3), 318–328.

Panwar, V.P., Gupta, M.K., 2016. Litter fall, nutrient return and soil fertility under *Grewia optiva*: an indigenous agroforestry tree species in Himachal Pradesh. Indian Forester 142 (3), 235–244.

Paschke, M.W., Dawson, J.O., 1992. Avian dispersal of *Frankia*. Canadian Journal of Botany 71, 1128–1131.

Pathak, H., Mohanty, S., Jain, N., Bhatia, A., 2010. Nitrogen, phosphorus, and potassium budgets in Indian agriculture. Nutrient Cycling in Agroecosystems 86, 287–299.

Paul, E.A., 1988. Towards the year 2000: directions for future nitrogen research. In: Wilson, J.R. (Ed.), Advances in Nitrogen Cycling in Agricultural Ecosystems. CAB International, Wallingford, United Kingdom, pp. 417–425.

Rao, D.L.N., Gill, H.S., Abrol, I.P., 1990. Regional experience with perennial Sesbania in India. In: Perennial Sesbania Species in Agroforestry, pp. 189–98.

Rizvi, R.H., Dhyani, S.K., Newaj, R., Karmakar, P.S., Saxena, A., 2014. Mapping agroforestry area in India through remote sensing and preliminary estimates. Indian Farming 63 (11), 62–64.

Rosenstock, T.S., Tully, K.L., Arias-Navarro, C., Neufeldt, H., Butterbach, B.K., Verchot, L.V., 2014. Agroforestry with $N_2$-fixing trees: sustainable development's friend or foe? Current Opinion in Environmental Sustainability 6, 15–21.

Roskoski, J.P., Petter, I., Pardo, E., 1986. Inoculation of leguminous trees with rhizobia and VA mycorrhizal fungi. Forest Ecology Management 6, 57–68.

Roy, M.M., Nigam, G., 2001. Litter production in an *Acacia tortilis* based silvopastoral system in the dry tropics. Indian Forester 127, 457–462.

Russo, R.O., 2005. Nitrogen-fixing trees with actinorhiza in forestry and agroforestry. In: Wenzer, D., Newton, W.E. (Eds.), Nitrogen Fixation in Agriculture, Forestry, Ecology and the Environment, pp. 143–71.

Sanginga, N., Mulongoy, K., Ayanaba, A., 1986. Inoculation of *Leucaena leucocephala* (Lam.) de Wit with *Rhizobium* and its nitrogen contribution to a subsequent maize crop. Biological Agriculture & Horticulture 3, 347–352.

Singh, B., Singh, Y., 2008. Reactive nitrogen in Indian: inputs, use efficiency and leakages. Current Science 94 (11), 1382–1393.

Singh, K.K., Ali, M., Venkatesh, M.S., 2009. Pulses in Cropping Systems. Technical Bulletin, IIPR, Kanpur.

Singh, K.P., Singh, P.K., Tripathi, S.K., 1999. Litterfall, litter decomposition and nutrient release patterns in four native tree species raised on coal mine spoil at Singrauli, India. Biology and Fertility of Soils 29, 371–378.

Sutton, M.A., Oenema, O., Erisman, J.W., Leip, A., van Grinsven, H., Winiwarter, W., 2011. Too much of a good thing. Nature 472, 159–161.

Swaney, D.P., Hong, B., PaneerSelvam, A., Howarth, R.W., Ramesh, R., Purvaja, R., 2015. Net anthropogenic nitrogen inputs and nitrogen fluxes from Indian watersheds: an initial assessment. Journal of Marine Systems 141, 45–58.

Tripathi, O.P., Pandey, H.N., Tripathi, R.S., 2009. Litter production, decomposition and physico-chemical properties of soil in 3 developed agroforestry systems of Meghalaya, Northeast India. African Journal of Plant Science 3 (8), 160–167.

Tscharntke, T., Clough, Y., Bhagwat, S.A., Buchori, D., Faust, H., Hertel, D., Holscher, D., Juhrbandt, J., Kessler, M., Perfecto, I., et al., 2011. Multifunctional shade-tree management in tropical agroforestry landscapes: a review. Journal of Applied Ecology 48, 619–629.

Uthappa, A.R., Chavan, S.B., Dhyani, S.K., Handa, A.K., Newaj, R., 2015. Trees for soil health and sustainable agriculture. Indian Farming 65 (3), 2–5.

Verchot, L.V., Van Noordwijk, M., Kandji, S., Tomich, T., Ong, C., Albrecht, A., Mackensen, J., Bantilan, C., Anupama, K., Palm, C., 2007. Climate change: linking adaptation and mitigation through agroforestry. Mitigation and Adaptation Strategies for Global Change 12, 901–918.

Yadav, R.S., Yadav, B.L., Chhipa, B.R., 2008. Litter dynamics and soil properties under different tree species in a semi-arid region of Rajasthan, India. Agroforestry Systems 73, 1–12.

Young, A., 1989. Agroforestry for Soil Conservation. CAB International, Wallingford.

# 15

# Nitrogen and Soil Quality

R. Bhattacharyya[1], S. Kundu[2], H.S. Thind[3], B.N. Ghosh[4], D. Sarkar[5], K. Batabyal[5], B. Mandal[5], S. Rajendiran[2], K.S. Anil[6], M. Lalitha[6]

[1]CESCRA, ICAR-INDIAN AGRICULTURAL RESEARCH INSTITUTE, PUSA, INDIA; [2]ICAR-INDIAN INSTITUTE OF SOIL SCIENCE, BHOPAL, INDIA; [3]PUNJAB AGRICULTURAL UNIVERSITY, LUDHIANA, INDIA; [4]ICAR-NATIONAL BUREAU OF SOIL SURVEY AND LAND USE PLANNING REGIONAL STATION, KOLKATA, INDIA; [5]BIDHAN CHANDRA KRISHI VISWAVIDYALAYA, MOHANPUR, INDIA; [6]ICAR-NATIONAL BUREAU OF SOIL SURVEY AND LAND USE PLANNING REGIONAL STATION, BANGALURU

**CHAPTER OUTLINE**



## Introduction

Soil constitutes the natural resources necessary for food production. Soil quality/health is an integrated characteristic determined by biological, chemical, and physical soil properties defining the soil's capacity to function (Karlen et al., 1997), and to respond to interventions. Nitrogen being a building block of proteins, nucleic acids, and other cellular constituents is essential for all forms of life. However, it is the most deficient nutrient in Indian soils. Mineral N is applied to the soil to meet the N needs of crop plants. But sometimes N is applied indiscriminately in doses much more than the requirement of crop plants. The N fertilization can have complex interactive effects on the soil organic carbon (SOC) transformations and hence on soil quality (Bhattacharyya et al., 2008).

Excessive amounts of N fertilizers may impair soil health through long-term soil carbon degradation and negative impacts on the composition and structure of the soil biological

The Indian Nitrogen Assessment. http://dx.doi.org/10.1016/B978-0-12-811836-8.00015-X

community. Even in prevailing agricultural systems, N-fertilizer recommendations continue to follow a prescriptive approach without considering the site-specific variations in crop N demand. These recommendations have been followed by the private sectors and public institutions alike. Thus N fertilization rates, directly or indirectly subsidized in the market, often exceed crop N demands because: (1) soil N availability is not adequately accounted for, and (2) fertilizer form, placement, or timing are unlikely to synchronize with soil and crop N dynamics (Ladha et al., 2005). As a result, the fertilizer N use efficiency has been as low as 33–36% for cereal grain production (Ladha et al., 2005). The unutilized N is not only an economic loss but causes various forms of environmental hazards. Hence, this chapter will assess the Indian research results on: (1) impacts of mineral N application alone versus no N application (unfertilized control plots) on soil quality and crop productivity; (2) effects of balanced mineral fertilization (mainly, NPK application) and integrated nutrient management (INM) on soil quality and crop productivity; and (3) approaches for better use of fertilizer N and future research needs to assess impacts of N on soil quality.

## Effects of N Use on Soil Quality

Impacts of application of inorganic fertilizers, particularly effects of N fertilizers on soil physical and chemical properties have been reported earlier. For instances, from the long-term fertilizer experiments in the country, in general, it was observed that there was no drastic change in pH and electrical conductivity due to N treatments. In general, SOC tended to increase after application of N fertilizer. Available N content of soil increased marginally with continuous application of N fertilizers and manures, but the effect was nonsignificant over the initial status in most of the soils irrespective of cropping systems. However, in Palampur (Alfisol) and Pantnagar (Mollisol), where virgin soils were recently brought under cultivation, SOC and availability of N decreased from initial values even with application of N (Verma and Sharma, 2007; Singh et al., 2014).

Application of N alone tended to increase bulk density and reduce SOC content in some soils due to decreased crop productivity. The negative impact of imbalanced fertilization (only N without P and K) on soil parameters and thus on soil quality in lower Indo-Gangetic Plains (IGP) is well understood. Manna et al. (2005) found that application of only N increased surface soil bulk density over initial values and also under balanced application of 100% recommended N, P, and K (NPK) and 100% recommended N, P, and K plus farmyard manure (NPK + FYM) in rice–wheat–jute and soybean–wheat cropping systems located at Barrackpore and Ranchi, respectively (Table 15.1). Surface soil organic C and N losses were also observed under N treated plots. In the N plots, soil microbial biomass C was 20–67% lower than that under NPK + FYM treatments in these locations (Manna et al., 2005). Water-stable aggregates (<250 mm) decreased up to 50% in N treated plots compared to NPK + FYM. In Ranchi, application of fertilizer N alone decreased the SOC by 28.8% from the initial level (0.45%) in 0–15 cm soil depth, while the reduction was only 2.9% and 6.7%, respectively, in NPK + FYM and NPK + lime treatments (Hati et al., 2008). In addition to a decline in structural stability,

**Table 15.1** Long-Term Effects of Fertilizer and Manure on Some Parameters of Soil Quality in Lower Indo-Gangetic Plains

| | Barrackpore (Typic Eutrochept) | | | | Ranchi (Typic Haplustalf) | | | | Almora (Typic Haplaquept) | | |
|---|---|---|---|---|---|---|---|---|---|---|---|
| **Parameters** | **Control** | **N** | **NPK** | **NPK + FYM** | **Control** | **N** | **NPK** | **NPK + FYM** | **Control** | **NPK** | **NPK + FYM** |
| Bulk density (g $cm^{-1}$) | 1.34a | 1.36a | 1.30a | 1.27a | 1.40a | 1.40a | 1.36 ab | 1.28b | 1.30 | 1.27 | 1.24 |
| Water-stable aggregates (%) <250 mm | 19.8c | 20.8bc | 24.0b | 40.4a | 20.27 | 18 | 21.6 | 24.8 | 1.03 | | |
| AWC (0–30 cm layer) | | | | | | | | | 2.55 | 3.30 | 3.90 |
| Organic C (g $kg^{-1}$) | 5.4 d | 5.7c | 7.4b | 7.9a | 3.5 | 3.4 | 4.2 | 4.5 | 6.30 | 8.01 | 11.79 |
| Total N (mg $kg^{-1}$) | 422d | 660c | 867b | 927a | 445 | 460 | 50.5 | 570 | 552 | 637 | 835 |
| pH (1:2 soil water) | 7.04a | 7.04a | 6.9a | 7.28a | 5 | 4.2 | 4.9 | 5.31 | | | |
| Cation exchange capacity (c mol $kg^{-1}$) | 13.1c | 15.2c | 24b | 31a | 6.8 | 6.5 | 7.2 | 9.9 | | | |
| Available N (kg $ha^{-1}$) | 210c | 230b | 273a | 285a | 163 | 317 | 331 | 392.9 | 11.2 | | |
| Available P (kg $ha^{-1}$) | 20.6 d | 31.2c | 63b | 83.7a | 13.9 | 14 | 64.4 | 75.5 | 13.44 | 35.84 | 40.32 |
| Available K (kg $ha^{-1}$) | 115d | 132c | 192b | 225a | 95 | 60 | 164 | 177.8 | 103.04 | 114.24 | 197.17 |
| Soil microbial biomass C (mg $kg^{-1}$) | 190c | 189c | 398b | 576a | 154 | 185 | 210 | 231 | 97 | 155 | 253 |
| Soil microbial biomass C (mg $kg^{-1}$) | 190c | 189c | 398b | 576a | 154 | 185 | 210 | 231 | 97 | 155 | 253 |
| Soil microbial biomass N (mg $kg^{-1}$) | 7.5c | 8.9c | 14b | 18.7a | 7.78 | 6.8 | 13.6 | 14.5 | | | |
| Hot water soluble C (mg $kg^{-1}$) | 10.7 d | 18.6c | 69b | 80.4a | 13 | 14 | 18.5 | 23.5 | | | |
| Soluble carbohydrates (mg $kg^{-1}$) | 526d | 580c | 689b | 845a | 328 | 367 | 466 | 517 | 87.4 | | |
| Dehydrogenase activity (μg TPF $g^{-1}$ 24 $h^{-1}$) | 45.6c | 50.8c | 86b | 118a | 33.5 | 36 | 54.7 | 62.6 | 600 | 480 | 3600 |
| Alkaline phosphatase (*p*-nitrophenol $g^{-1}$ $h^{-1}$) | 119.3c | 120c | 191b | 223a | 160.5 | 151 | 171 | 301.5 | 12 | 12 | 14 |

*N*, 100% recommended N; *NPK*, 100% recommended N, P, and K; *NPK + FYM*, 100% NPK plus farmyard manure; control, no fertilizer. Values with similar lowercase letters within a row were not significantly different at $P < 0.05$.

soils of the N treated plots had lower moisture retention and available water capacity. In acid soils, a decrease in soil pH was also observed in N fertilized plot in *Typic Haplustalf* in Ranchi (Table 15.1). Long-term application of only N had a negative impact on soil quality as the practice significantly reduced the active fractions of SOC, dehydrogenase, and alkaline phosphatase activity (Manna et al., 2005, 2007). Application of mineral N fertilizer had little effect on fertility buildup under rice-based cropping system on an acid lateritic soil at Kharagpur, West Bengal (Sharma and Mittra, 1991).

The long-term fertilizer experiment centers in South India show that application of N fertilizers alone in the form of urea resulted in decline of soil pH in Alfisols (Bangalore) due to production of $H^+$ ions during conversion of $NH_4^-$ into $NO_2^-$ and $NO_3^-$ (AICRP, 2013). At the same time, continuous application of N alone declined the SOC status, which is obviously due to incorporation of residual biomass in less quantity due to poor yield. Thus, in acid soils, with sustained application of N, $Al^{3+}$ is released from the soil minerals, and frequently causes toxicity that induces plant nutrient disorders. A controversial report given by Sarma et al. (2013) states that the increased application of N fertilizer (at 100 kg $ha^{-1}$) improved the soil properties, viz. total porosity and water-holding capacity, but decreased the soil pH.

Although impacts of N fertilization versus NPK and NPK + FYM application on soil properties are documented, limited information is available on the effects of N fertilization versus no N fertilization on soil properties in different soil orders of the country. Moreover, information on impacts of fertilization on critical soil quality parameters, such as microbial community structure, microbial diversity, and C-, N-, and P-cycling enzymes, are lacking. In a recent review, Bijay-Singh and Ryan (2015) concluded that fertilizer use has been associated with reduction in some soil organisms but these effects are relatively short-lived and occur only at the site of the fertilizer band. Significant increases in microbial biomass have been shown by long-term application of fertilizers in nonacid soils. Transformation of ammonium-based N fertilizers in soils can adversely affect the soil health by increasing acidity. Recently, Geiseller and Scow (2014) published a meta-analysis based on 107 datasets from 64 long-term experiments from around the world and revealed that mineral fertilizer application led to a significant increase (15.1%) in the microbial biomass content above levels in the unfertilized control. Fertilizer application tended to reduce microbial biomass in soils with a pH below 5 in the fertilized treatment, but in experiments where soil pH was at least 7, the fertilization-related increase in microbial biomass averaged 48%.

## Long-Term Effect of Balanced Fertilization and Integrated Nutrient Management on Soil Quality

Bhandari et al. (2002) hypothesized that the decrease in total soil N (TSN) is likely to have played a major role in yield sustainability in a rice–wheat cropping system in Ludhiana. TSN declined markedly at 0.03, 0.02, and 0.02 g $kg^{-1}$ $year^{-1}$ (in typic Haplustalf) in treatments with NPK, green manure, and wheat straw, respectively. It was suggested that N losses were responsible for the decline in TSN despite the relatively large N inputs from fertilizers. The TSN stock, however, was maintained in FYM + NPK treated plots, signifying that organic matter application (FYM, in that case) is critical for maintaining soil fertility.

Balanced use of chemical fertilizers resulting in an increased soil organic matter (SOM) compared to plots receiving no fertilizers has been revealed by long-term studies from all over the globe (Ladha et al., 2011; Geiseller and Scow, 2014; Korschens et al., 2013) and in India (Mandal et al., 2007, Mandal et al., 2008; Majumder et al., 2008). For instances, long-term fertilizer experiment in a finger millet–maize cropping sequence revealed that continuous application of inorganic fertilizers with FYM increased SOC, and available nutrients and micronutrient content of the soil compared with the N fertilizers alone (Hemalatha and Chellamuthu, 2013). In general, balanced fertilization (NPK) and INM (NPK + FYM) treatments increased SOC and reduced soil bulk density (up to 45 cm soil layer) at all Indian sites. N fertilization significantly decreased water-soluble aggregates. However, plots with NPK or NPK + FYM, on the contrary, either maintained or increased compared to initial values. Available N, P, K, and S contents were either maintained or improved in NPK and NPK + FYM treated plots, irrespective of the soils and cropping systems in different parts of India. Also excess application of fertilizers (150% NPK) improved the SOC and other soil physicochemical and biological properties (AICRP-LTFE, 2013). Available water capacity values of 0–30 cm soil layer with the NPK + FYM and NPK + lime treated plots were greater compared with the plots with N. Increasing rates of N from 50% to 150% and use of FYM resulted in higher buildup of available N compared to control (Swarup and Wanjari, 2000).

## Long-Term Effect of Balanced Fertilization and Integrated Nutrient Management on Soil Quality and Crop Productivity

The application of organic manure alone generally sustains crop productivity (although not increases progressively). Results revealed that FYM application at 10 t $ha^{-1}$ helped to sustain higher rice, maize, and wheat yields over the years (Brar and Dheri, 2012). The application of balanced doses of mineral fertilizers significantly increased the yield of maize and wheat in a long-term fertilizer experiment at Ludhiana under maize–wheat–cowpea cropping system (Brar and Dheri, 2012). Saha et al. (2008) also reported that the balanced use of NPK produced higher rice and wheat yields at Barrackpore in comparison with N alone or NP treatments. Singh (2004) reported that balanced use of 100% NPK gave higher soybean yields than 100% N or 100% NP at Ranchi and Jabalpur. The combined use of mineral fertilizers and organic manures further improves crop productivity in all soils. For examples, the productivity of maize–wheat system could be enhanced by addition of FYM or lime along with the chemical fertilizers in Alfisols (Sharma et al., 2002). Ram (1998) also reported that the integrated use of NPK fertilizers and FYM produced highest yields of 6.14, 4.51, and 2.15 t $ha^{-1}$ of rice, wheat, and cowpea fodder, respectively, in rice–wheat–cowpea (fodder) cropping system at Pantnagar. Singh (2004) reported that integrated use of 100% NPK + 10 t $ha^{-1}$ FYM produced higher crop yields under soybean–wheat based cropping sequence in Vertisol of Jabalpur and Alfisol of Ranchi. Similar results were reported in soybean–wheat system at LTFE Centre, Jabalpur (40 years from 1972 to 2012), soybean–safflower system at LTFE Centre, Parbhani (14 years from 1996 to 2012), and rice–wheat system at LTFE Centre, Raipur (AICRP-LTFE, 2013).

Soil quality index (SQI) and sustainable yield index (SYI) for different fertility treatments under long-term rice–wheat–jute cropping system in the lower IGP (Barrackpore, West Bengal) was computed by Chaudhury et al. (2005). They considered NPK + FYM as ideal treatment (SQI = 100%, SYI = 100%). Further, they showed that the soil quality with exclusion of FYM, K, and P fertilizer (with use of only N fertilizer) would decline by 46.5%. At the same time, the encouraging yield trends were found with NPK and NPK + FYM treated plots (Manna et al., 2005, 2007). The yield trends in the various cropping systems as SYI generally followed the order: 100% NPK + FYM ≥150% NPK >100% NPK >100% NP > 50% NPK> 100% N > control. This showed the importance of P and in some cases K for sustainable crop production along with N. Further, the influence of fertilizers depends on soil and cropping system (Manna et al., 2007). For example, in rice–wheat cropping system both crop yield and SYI was recorded the highest in 150% NPK, whereas in other systems 100% NPK + FYM showed the highest yields and SYIs. The results on finger millet and maize crops grown in sequence in Alfisol, Banglore, and Verisol, Coimbatore, suggest that productivity of finger millet–maize sequence was about 10–25% higher in Vertisol than Alfisol. Finger millet and maize showed responses to graded doses of NPK, FYM, and NPK + lime but imbalanced use of either N or NP gave far low yield even less than 50% NPK. The SYI was greater for 100% NPK + FYM treatment in finger millet–maize at Coimbatore, soybean–wheat at Ranchi and Jabalpur, and maize–wheat at Ludhiana. Therefore, FYM addition along with recommended fertilizer NPK sustained yield more than other treatments in these crops (Singh et al., 2004; Brar et al., 2013).

Recently Batabyal et al. (2016) and Tamang et al. (2016) reported that adoption of INM technologies for growing vegetables such as cauliflower, broccoli, and eggplant on an Inseptisol in the lower IGP improved soil quality (137–152%) over the initial soil by improving soil organic C (10–14%), microbial biomass C (138–345%), bulk density (1–5% reduction over initial soil), and other available nutrients. Application of 125% of recommended NPK (250–55–103 kg N–P–K $ha^{-1}$) plus FYM (5 t $ha^{-1}$) or vermicompost (3 t $ha^{-1}$) for cauliflower, 125% of NPK (15–33–63 kg N–P–K $ha^{-1}$) plus FYM (6.25 t $ha^{-1}$) for broccoli, and 150% NPK (150–33–63 kg N–P–K $ha^{-1}$) plus vermicompost (3 t $ha^{-1}$) for eggplant were the best management options for maintaining soil health in addition to harvesting higher yield and quality produce, and achieving economic benefits.

## How to Use N for Improving Soil Health

Knowing and managing the N supply from soil and other indigenous sources, and maximizing the agronomic efficiency are equally important components for achieving high partial factor productivity of N (Dobermann, 2007). One important key to improve N use efficiency is achieving synchrony of N delivery from soils and fertilizers with N uptake by plant. When N application is not synchronized with crop demand, N losses from the soil–plant system are large, leading to low N-fertilizer use efficiency. The need to match N-fertilizer supply in time and space of crop demand to maximize efficiency is conceptually well known. However, achieving this ideal in practice is complicated by site-specific conditions and management decisions.

Several efficient N management strategies have been worked out to increase fertilizer N use efficiency in crops, particularly in cereals. These include optimum fertilizer N rates, improved methods and timings of application and placement, specially formulated forms of fertilization, including those with nitrification and urease inhibitors, integrated use of mineral fertilizers organic manures, and/or organic residues to attain better synchronization between the N supply and crop requirement, especially in case of cereals. As these practices would improve crop productivity in a sustainable manner, these also are believed to improve soil quality. However, much research is needed under diverse production systems of the country.

Integrated approaches of organic and inorganic nutrient management have shown an increased efficiency of applied N fertilizer in rice compared to inorganic fertilizer alone (Buresh and De Datta, 1991; Khan et al., 2008). A reference or critical point may be made if we add organic manures having narrow C:N ratio, when the N will be available for the growing plant. Some of the plants showing less yield and quality even after addition of organic manure is due to wider C:N ratio which leads to slower decomposition and loss of nitrogen (Mirchandani, 1936). The lands which are degraded due to continuous application of N alone for long years can be rejuvenated by superimposition of nutrients (P, K, and FYM) within a period of 2–3 years (Munneshwar Singh and Wanjari, 2013).

Under salt-stress conditions, the uptake of N by crop plants is generally affected (Alam, 1999). Results of a long-term experiment for 17 years in an alluvial coastal saline soil of West Bengal showed definite response to soil application of fertilizers N and P, but no such response was observed for K (Bandyopadhyay and Maji, 1996). The response to N application was observed from the initial year, while the same for P fertilizer was observed after 12 years due to high initial available P content of soil. Long-term addition or omission of fertilizers N (urea) and K (potassium chloride) showed no definite changes or negligible changes in the content of available N and K to this coastal saline soil, but available P content of the soil gradually decreased when cropping continued without P fertilizer.

If fertilizer N could be saved without a yield penalty, this sort of management practice could improve soil quality by better soil biological health and less soil and environmental pollution. Recent advances in fertilizer N management (for N saving in several production systems of the country) include using gadgets, such as leaf color chart (LCC), chlorophyll meter (SPAD meter), and optical sensors to apply different rates and distribution of mineral N within the crop growing season to complement the demand for supplemental N. These gadgets help in real time N management through the assessment of leaf N status of crops. Chlorophyll meter is a simple portable diagnostic tool used for monitoring crop N status in situ by measuring unit less SPAD value, which is then compared with a critical threshold value to arrive at a decision of whether and how much N is to be applied. LCC is another simple and easy diagnostic tool used effectively for N topdressing in rice and other cereal crops. The leaf color of standing rice is compared with the colored strips to make a decision. Maiti et al. (2004) established LCC 4 as the critical value for transplanted rice in north-eastern India and found that this could save 20–42.5 kg N $ha^{-1}$, which led to significantly large increases in N use efficiency compared to the generally recommended fixed-time split N applications. Fertilizer application for wetland rice with LCC helps in saving 30–40 kg N $ha^{-1}$ in Bihar and 42–50 kg N $ha^{-1}$ in West Bengal (Bijay-Singh and Yadvinder-Singh, 2003).

Balanced use of NPK fertilizers is indispensable to replenish the soils with nutrients removed by crops under intensive agriculture. This practice ultimately enhances the grain yield and uptake of N, P, and K by the growing crops. A sustainable soil management practice should, therefore, include integrated use of balanced NPK fertilizers and organic manure with soil reclamation measures, if needed, to maintain the soil physical quality to create suitable environment for the availability and uptake of the plant nutrients. Yadav et al. (2000) reported that available N, P, and K content showed a buildup with NPK application or INM adoption in rice–wheat cropping of lower IGP. Ghosh et al. (2012) also reported beneficial effect of INM in increasing availability of N, P, and K in rice–wheat cropping system and in improving soil quality. Chaudhary et al. (2014) reported that supplementation of 50% of recommended N with organics (poultry manure and vermicompost) increased productivity and profitability of SRI grown rice without affecting the soil health in Bihar.

Another strategy by which fertilizer N could be saved is by long-term adoption of conservation agriculture (CA). CA is a concept for resource-saving agricultural crop production that strives to achieve acceptable profits together with high and sustained production levels while concurrently improving the soil quality and conserving the environment. An appropriate CA practice generally improves soil quality through:

1. Increased agricultural productivity, profitability, and enhanced ecosystem services
2. Enhanced input use efficiency, including that of water, nutrients, energy, pesticides, land, and labor
3. Protected soil, water, and biodiversity through minimum soil disturbances
4. Judicious use of external inputs (mineral fertilizers and pesticides) and preference for alternatives (organic materials and integrated pest management)

**5** Use of the species' biodiversity to build systems' resilience to abiotic, biotic, and economic stresses with improvement of SOM for activities of soil biota coupled with reduction in greenhouse gases (GHGs).

Despite CA having many benefits, it could lead to increased $N_2O$ emission. Thus, N fertilization protocol and their impacts on soil quality under CA is a researchable issue. In CA system, broadcasting N onto the residue covered surface is an inefficient method of N application. This is because of N immobilization by surface residues and N losses, such as volatilization. Thus, placement of N fertilizers in bands below the C-enriched surface soil may be a good idea. As information on N management under CA systems is very scarce, considerable research is needed to determine how much fertilizer can be safely applied with the seed of different crops under CA of varied agro-ecological zones of the country. The addition of organic sources such as FYM, green manure, and crop residues has potential to improve soil health and increase crop productivity. However, use of organic sources also results in increased GHG emissions (Bhatia et al., 2005). Nitrification inhibitors, on the other hand, may increase N losses due to volatilization (Banerjee et al., 2002). Thus, there is a need to carry out further research on the use of organics in combination with low mineral N fertilization under CA for a better soil quality and conservation of natural resources.

The use efficiency of N fertilizers for crops in coastal areas is very low due to heavy loss of the nutrient, particularly in saline and alkaline soils by volatilization (Sen and Bandyopadhyay, 1987) and also in deep waterlogged areas by leaching and run-off. The rates of both mineralization and immobilization of N in soils considerably decreased at higher soil sodicity (Bandyopadhyay and Rao, 2001). Under such situations, one of the most efficient ways of increasing the N status and sustaining crop production in the salt-affected coastal soils is to exploit the benefit of biological N fixation, especially by cultivation of annual and perennial legumes in crop rotation, application of *Azolla* and nonsymbiotic N-fixing bacteria, etc. *Rhizobium* spp. shows considerable tolerance to high pH and can thrive well in sodic soils but these are considerably sensitive to salinity. Though *Rhizobia* could thrive and multiply in sodic soil with pH as high as 10, the effective contribution of the bacteria toward N fixation reduces because of the sensitivity of the host plants to salinity of soil. Use of green manure crops such as *Sesbania aculeata* and *Glyricidia maculata* either alone or in combination with fertilizer N has been proved to be effective for saline alkali soils (Patil et al., 1991; Rao, 1986; Evans and Rotar, 1986). Combined application of either 50% of urea N with green manure at 5 t $ha^{-1}$ or 25% of urea N with green manure at 10 t $ha^{-1}$ has been found to give best results in terms of higher rice yield and higher agronomic efficiency of applied N and could save 25% N. Likewise, application of green leaves of *Glyricidia maculata* at 10 t $ha^{-1}$ to the puddled soil before rice transplanting has been found comparable in terms of yield with inorganic fertilizer application of N and $P_2O_5$ at 100 and 50 kg $ha^{-1}$, respectively (Chavan and Dongale, 1994). In coastal saline soils in India, rice and wheat yields could be maintained even at 50% NPK used with FYM or green manure (DARE, 2003–04).

## Do We Overuse N?

Farmers, in general, have a tendency to rely primarily on N fertilizers to maximize the crop yields, rather than targeting achievable yields determined by local agronomic, economic, and environmental conditions. High levels of N fertilizers without appropriate balance with fertilizers P and K result in negative effects on crop yields and soil quality (Chaudhury et al., 2005). There are significant differences to the extent of N use in various states. This shows that states experience imbalanced fertilizer consumption. As mentioned earlier, several studies conducted on wide variety of soils and geographic locations clearly revealed that continued (even up to 40–50 years) addition of synthetic fertilizer N caused net depletion of TSN particularly its organic form and this in turn declined soil productivity and the agronomic efficiency (kg grain $kg^{-1}$ N) of fertilizer N (Manna et al., 2005; Mulvaney et al., 2009).

Indian soils mostly have low total N concentration and rice is the crop that invariably responds to the application of N fertilizer (Panda et al., 2007). Rice is also the crop, which alone consumes 37% of the chemical fertilizer produced in India followed by wheat (24%), oilseed (9%), sugarcane (5%), vegetables (4%), and cotton (3%) (FAO, 2011). Nitrogen fertilizer use with high application intensity was found in rice-growing districts of Andhra Pradesh, viz. Krishna and West Godavari as well as Karimnagar and Nizamabad districts in Telangana, and Tirunelveli district in Tamil Nadu. These districts are having N-fertilizer use intensity in the range of 151–200 kg $ha^{-1}$. Most of the districts in Andhra Pradesh, Telangana, and Tamil Nadu were found with N-fertilizer use intensity in the range of 51–150 kg $ha^{-1}$. Districts of Alapuzha, Ernakulam, and Kottayam in Kerala and districts of Belgaum, Kodagu, Hassan, Kolar, Mandya, and Shimoga in Karnataka were found with N-fertilizer use intensity in the range of 51–150 kg $ha^{-1}$(Fig. 15.1) (Chhabra et al., 2010). Nitrogen fertilization at 120 kg $ha^{-1}$ results in high $N_2O$ emission of 0.48–1.01 g $ha^{-1}$ $day^{-1}$ in rice field (Pathak, 1999). Recommended mineral fertilizers are hardly adopted by some rice–wheat farmers in the IGP. For example, farmers are using high doses of N (130–195 kg N $ha^{-1}$) to non-basmati rice in Transects 2 and Part 3 of IGP. Application of only 11–14 kg P $ha^{-1}$ is very common in these areas, but K fertilization is inadequate.

Advantages of the blanket N recommendation are its simplicity and no costs involved for soil test. However, such recommendations do not ensure economy and efficiency of fertilizer N use since the variation in soil fertility is not taken into consideration, and there will be wastage of fertilizers in some fields. In majority of the cereal growing areas of India, despite blanket recommendations for N management are practiced, it is believed that N is not overused. The mineral fertilizer N use intensity and state-wise estimated consumption of N fertilizers (kg $ha^{-1}$) in India (2010–2011 to 2012–2013) are presented in Fig. 15.1. Thus, albeit very few locations, Indian farmers do not overuse mineral N fertilizer. However, as mentioned earlier, optimum crop yields can be obtained provided that the nutrient application level is high enough to compensate for a possible low supply of the nutrients from the poorly fertile Indian soils. As efficient N use

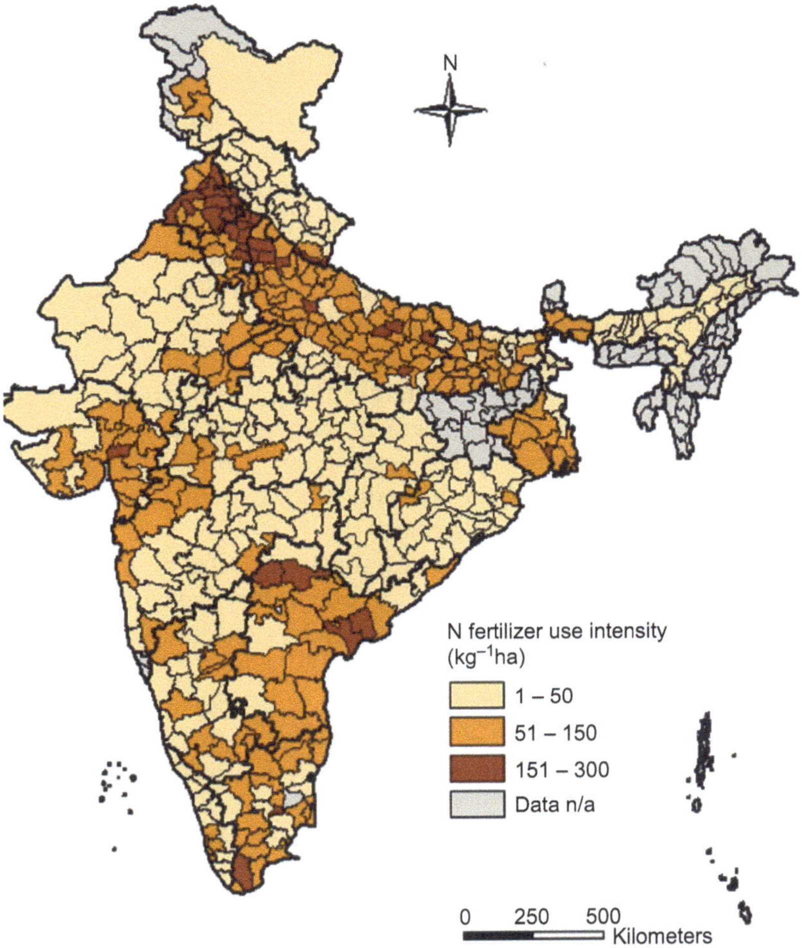


**FIGURE 15.1** District level nitrogen fertilizer use intensity during 2003–04 (Chhabra et al., 2010).

is central to eco-efficiency in agriculture, it is important to work out fertilizer N doses that will produce not only high crop yields per unit area and remain economically acceptable to farmers, but also result in minimal environmental impacts. Thus, it is imperative to identify new N-fertilizer recommended doses under novel crop management practices in the diverse agroecosystems of the country.

## Conclusion

Mineral fertilizers used in a balanced way can enhance soil quality, through increasing SOM and enhanced microbial activity. Significant increases in enzyme activities are observed by application of N fertilizers in alkaline or neutral soils. Despite application of only N is comparatively well performed than that of no use of fertilizers in some soils, in the long run it is not sustainable as N fertilization alone deteriorates soil health and declines the crop productivity, mainly in neutral to acidic soils. Therefore, adequate balanced fertilization and INM (mineral fertilization with addition of organics) should be recommended and followed.

Due to limited and sometimes contradictory reports regarding impacts of mineral fertilization on soil health (mainly on soil biological parameters such as microbial diversity and soil meta-genomics), there is a need to understand the unknown facts. These are: impacts of mineral N fertilizer alone on labile and recalcitrant C pools, soil quality, and microbial diversity under different production systems; mineral N fertilization versus organic fertilization on soil biodiversity and meta-genomics, and SOM dynamics under diverse soil types, N fertilization schedule and protocol and their impacts on soil quality under novel agronomic management practices; and so on. As enough data are not available to understand these complex interactions, there is a strong need to conduct more studies in India under novel agronomic management practices that are being promoted to face the climate change effects.

Incomplete knowledge of microbial community structure and ecological dynamics may delude estimates of fluxes of the N biogeochemical cycle, and thus may lead to improper recommendation and, in the long-run, may deteriorate the soil quality.

# References

Alam, S.M., 1999. Nutrient uptake by plants under stress conditions. In: Pessarakli, M. (Ed.), Handbook of Plant and Crop Stress, second ed. Marcel Dekker, Inc, pp. 285–313.

AICRP-LTFE (All Indian Coordinated Research Project on Long-term Fertility Experiments), 2013 http://www.iiss.nic.in/Rti%20Appl/RTI%2080.pdf.

Bandyopadhyay, B.K., Rao, D.L.N., 2001. Integrated nutrient management in saline soils. Journal of the Indian Society of Coastal Agricultural Research 19, 35–58.

Bandyopadhyay, B.K., Maji, B., 1996. Long Term Experiment to Study the Effect of Missing P and K Treatments on Yields of Crops. Annual report (1996-97). CSSRI, Karnal, pp. 80.

Banerjee, A., Sharma, A., Banerjee, U., 2002. The nitrile-degrading enzymes: current status and future prospects. Applied Microbiology and Biotechnology 60, 33–44.

Batabyal, K., Mandal, B., Sarkar, D., Murmu, S., Tamang, A., Das, I., Hazra, G.C., Chattopadhyay, P.S., 2016b. Comprehensive assessment of nutrient management technology for cauliflower production under subtropical conditions. European Journal of Agronomy 79, 1–13.

Bhandari, A.L., Ladha, J.K., Pathak, H., Padre, A.T., Dawe, D., Gupta, R.K., 2002. Yield and soil nutrient changes in a long-term rice-wheat rotation in India. Soil Science Society of America Journal 66, 162–170.

Bhatia, A., Pathak, H., Jain, N., Singh, P.K., Singh, A.K., 2005. Global warming potential of manure amended soils under rice–wheat system in the Indo-Gangetic plains. Atmospheric Environment 39, 6976–6984.

Bhattacharyya, R., Kundu, S., Ved-Prakash, Gupta, H.S., 2008. Sustainability under combined application of mineral and organic fertilizers in a rainfed soybean-wheat system of the Indian Himalayas. European Journal of Agronomy 28, 32–46.

Brar, B.S., Singh, K., Dheri, G.S., Balwinder-Kumar, 2013. Carbon sequestration and soil carbon pools in a rice-wheat cropping system: effect of long-term use of inorganic fertilizers and organic manure. Soil & Tillage Research 128, 30–36.

Brar, S., Dheri, G.S., 2012. Long-term effects of fertilizers on rice-wheat-cowpea productivity and soil properties in a Mollisols. Rice-Wheat Consortium Paper Series 6. In: Abrol, I.P., Bronson, K.F., Duxbury, J.M., Gupta, R.K. (Eds.), Long-term Soil Fertility Experiments in Rice-Wheat Cropping Systems. Rice-Wheat Consortium for the Indo-Gangetic Plains, New Delhi, India, pp. 50–55. Available from: https://www.researchgate.net/publication/237707213.

Bijay-Singh, Ryan, J., 2015. Managing Fertilizers to Enhance Soil Health. International Fertilizer Industry Association, Paris, France.

Bijay-Singh and Yadvinder-Singh, 2003. Nitrogen management in rice-wheat system in the Indo-Gangetic Plain. In: Yadvinder-Singh, et al. (Eds.), Nutrient Management for Sustainable Rice-Wheat Cropping System. National Agricultural Technology Project, Indian Council of Agricultural Research, pp. 99–114. New Delhi, and Punjab Agricultural University, Ludhiana.

Buresh, R.J., De Datta, S.K., 1991. Nitrogen dynamics and management in rice-legume cropping systems. Advances in Agronomy 4, 51–59.

Chaudhary, S.K., Singh, S.P., Singh, Y., Dharminder, 2014. Influence of integrated use of fertilizers and manures on SRI grown rice (*Oryzasativa*) and their residual effect on succeeding wheat (*Triticum aestivum*) in calcareous soil. Indian Journal of Agronomy 59, 527–533.

Chaudhury, J., Mandal, U.K., Sharma, K.L., Ghosh, H., Mandal, B., 2005. Assessing soil quality under long-term rice-based cropping system. Communications in Soil Science and Plant Analysis 36, 1141–1161.

Chavan, A.S., Dongale, J.B., 1994. Nutrient deficiencies and integrated nutrient management for optimization of agricultural production. Journal of Indian Society of Coastal Agricultural Research 12, 37–45.

Chhabra, A., Manjunath, SushmaPanigrahy, K.R., 2010. Non-point source pollution in Indian agriculture: estimation of nitrogen losses from rice crop using remote sensing and GIS. International Journal of Applied Earth Observation and Geo-information 12, 190–200.

DARE, 2003–04. Natural Resource Management. http://www.dare.gov.in/report/NATURAL_RESOURCE_MANAGEMENT.pdf.

Dobermann, A., March 7–9, 2007. Nutrient use efficiency—measurement and management. In: Fertilizer Best Management Practices: General Principles, Strategy for Their Adoption and Voluntary Initiatives Vs Regulations. Proceedings of IFA International Workshop on Fertilizer Best Management Practices. International Fertilizer Industry Association, Brussels, Belgium, pp. 1–28. Paris, France.

Evans, D.D., Rotar, P.P., 1986. Role of Sesbania in Agriculture. West View Press, Boulder, Colorado, USA.

Food and Agriculture Organization (FAO), 2011. Chapter 3, Soil Health. in, Save and Grow (Available at: http://www.fao.org/ag/save-and-grow/en/3/index.html.

Geiseller, D., Scow, K.M., 2014. Long-term effects ofmineral fertilizers on soil microorganisms- a review. Soil Biology and Biochemistry 75, 54–63.

Ghosh, S., Wilson, B., Ghoshal, S., Senapati, N., Mandal, B., 2012. Organic amendments influence soil quality and carbon sequestration in the Indo-Gangetic plains of India. Agriculture Ecosystem and Environment 156, 134–141.

Hati, K.M., Swarup, A., Mishra, B., Manna, M.C., Wanjari, R.H., Mandal, K.G., Misra, A.K., 2008. Impact of long-term application of fertilizer, manure and lime under intensive cropping on physical properties and organic carbon content of an Alfisol. Geoderma 148, 173–179.

Hemalatha, S., Chellamuthu, S., 2013. Impacts of long term fertilization on soil nutritional quality under finger millet: maize cropping sequence. Journal of Environmental Research and Development 7, 1571–1576.

Karlen, D.L., Mausbach, J.W., Doran, J.W., Cline, R.G., Harris, R.F., Schuman, G.E., 1997. Soil quality: a concept, definition and framework for evaluation. Soil Science Society America Journal 61, 4–10.

Khan, H.Z., Malik, M.A., Saleem, M.F., 2008. Effect of rate and source of organic material on the production potential of spring maize (*Zea mays* L.). Pakistan Journal of Agricultural Sciences 45, 40–43.

Korschens, M., Albert, E., Armbruster, M., Barkusky, D., Baumecker, M., Behle-Schalk, L., Bischoff, R., Cergan, Z., Ellmer, F., Herbst, F., Hoffmann, S., Hofmann, B., Kismanyoky, T., Kubat, J., Kunzova, E., Lopez-Fando, C., Merbach, I., Merbach, W., Pardor, M.T., Rogasik, J., Ruhlmann, J., Spiegel, H., Schulz, E., Tajnsek, A., Toth, Z., Wegener, H., Zorn, W., 2013. Effect of mineral andorganic fertilization on crop yield, nitrogen uptake, carbon and nitrogen balances, as well as soil organic carbon content and dynamics: results from 20 European long-term field experiments of the twenty-first century. Archives of Agronomy and Soil Science 59, 1017–1040.

Ladha, J.K., Pathak, H., Krupnik, T.J., Six, J., van Kessel, C., 2005. Efficiency of fertilizer nitrogen in cereal production: retrospect and prospects. Advances in Agronomy 87, 85–156.

Ladha, J.K., Kesava Reddy, C., Padre, A.T., van Kessel, C., 2011. Role of nitrogen fertilization in sustaining organic matter in cultivated soils. Journal of Environmental Quality 40, 1756–1766.

Mandal, B., Majumder, B., Adhya, T.K., Bandyopadhyay, P.K., Gangopadhyay, A., Kundu, S., Sarkar, D., Kundu, M.C., Gupta Choudhury, S., Hazra, G.C., Samantaray, R.N., Mishra, A.K., 2008. The potential of double-cropped rice ecology to conserve organic carbon under subtropical climate. Global Change Biology 14, 2139–2151.

Mandal, B., Majumder, B., Bandopadhyay, P.K., Hazra, G.C., Gangopadhyay, A., Smantaroy, R.N., Misra, A.K., Chowdhuri, J., Saha, M.N., Kundu, S., 2007. The potential of cropping systems and soil amendments for carbon sequestration in soils under long-term experiments in subtropical India. Global Change Biology 13, 357–369.

Majumder, B., Mandal, B., Bandyopadhyay, P.K., Gangopadhyay, A., Mani, P.K., Kundu, A.L., Mazumdar, D., 2008. Organic amendments influence soil organic carbon pools and rice–wheat productivity. Soil Science Society of America Journal 72, 775–785.

Maiti, D., Das, D.K., Karak, T., Banerjee, M., 2004. Management of nitrogen through the use of leaf color chart (LCC) and soil plant analysis development (SPAD) or chlorophyll meter in rice under irrigated ecosystem. Science World Journal 4, 838–846.

Manna, M.C., Swarup, A., Wanjar, R.H., Ravankar, H.N., Mishra, B., Saha, M.N., Singh, Y.V., Sahi, D.K., Sarap, P.A., 2005. Long-term effect of fertilizer and manure application on soil organic carbon storage, soil quality and yield sustainability under sub-humid and semi-arid tropical India. Field Crops Research 93, 264–280.

Manna, M.C., Swarup, A., Wanjari, R.H., Mishra, B., Shahi, D.K., 2007. Long-term fertilization, manure and liming effectson soil organic matter and crop yields. Soil and Tillage Research 94, 397–409.

Mirchandani, T.G., 1936. Nitrogen status of Indian soils. In: Proceedings of the National Institute of Sciences of India (August 29–31 at Bihar).

Mulvaney, R.L., Khan, S.A., Ellsworth, T.R., 2009. Synthetic nitrogen fertilizers deplete soil nitrogen: a global dilemma for sustainable cereal production. Journal of Environmental Quality 38, 2295–2314.

Munneshwar Singh, Wanjari, R.H., 2013. Measures to sustain and restore declined productivity in Alfisols under long term fertilizer experiments. Indian Journal of Fertilisers 9 (2), 24–32.

Panda, D., Samantaray, R.N., Misra, A.K., Senapati, H.K., 2007. Nutrient balance in rice. Indian Journal of Fertilisers 3, 33–38.

Pathak, H., 1999. Emissions of nitrous oxide from soil. Current Science 77, 359–369.

Patil, B.P., Jadav, S.N., Khade, V.N., 1991. Studies on integrated nitrogen management with glyricidia maculate green manuring in rice groundnut sequence. Journal of Indian Society of Coastal Agricultural Research 9, 193–196.

Ram, N., 1998. Effect of continuous fertilizer use on soil fertility and productivity of a Mollisol. In: Swarup, A., Damodar Reddy, D., Prasad, R.N. (Eds.), Long-Term Soil Fertility Management through Integrated Plant Nutrient Supply. Indian Institute of Soil Science, Bhopal, India, pp. 229–237.

Rao, D.L.N., 1986. Sesbania for Green Manuring. Central Soil Salinity Research Institute, Karnal, Haryana, India, p. 6. Brochure 7.

Saha, S., Mina, B.L., Gopinath, K.A., Kundu, S., Gupta, H.S., 2008. Organic amendments affect biochemical properties of a subtemperate soil of the Indian Himalayas. Nutrient Cycling in Agroecosystems 80, 233–242.

Sen, H.S., Bandyopadhyay, B.K., 1987. Volatilization loss of nitrogen from submerged rice soil. Soil Science 143, 34–39.

Sarma, B., Gogoi, N., Paul, S., Baroowa, B., 2013. Impact of inorganic N fertilizer on soil organic carbon and physico-chemical properties of alluvial soil of Assam. International Journal of Agriculture and Food Science Technology 4 (8), 743–744.

Sharma, A.R., Mittra, B.N., 1991. Effect of different rates of application of organic and nitrogen fertilizers in a rice-based cropping system. The Journal of Agricultural Science 117, 313–318.

Sharma, K.L., Srinivas, K., Das, S.K., Vittal, K.P.R., Kusuma-Grace, J., 2002. Conjunctive use of inorganic and organic sources of nitrogen for higher yield of sorghum in dryland Alfisol. Indian Journal of Dryland Agricultural Research & Development 17, 79–88.

Swarup, A., Wanjari, R.H., 2000. In: Swarup, A., Wanjari, R.H. (Eds.), Three Decades of All India Coordinated Research Project on Long-term Fertilizer Experiments to Study Changes in Soil Quality, Crop Productivity and Sustainability. Indian Institute of Soil Science, Bhopal, p. 59.

Singh, D.P., 2004. Vermiculture biotechnology and biocomposting. In: Singh, D.P., Dwivedi, S.K. (Eds.), Environmental Microbiology and Biotechnology. New Age International (P) Limited Publishers, New Delhi, pp. 97–112.

Singh, D.K., Pandey, P.C., Gupta, S., 2014. Long term addition of organics to sustain the system productivity of rice (*Oryza sativa* L.)–Wheat (*Triticum aestivum* L.) under Indo-Gangetic Plains of India. International Journal of Scientific Research and Innovative Technology 1, 1–10.

Tamang, A., Das, I., Batabyal, K., Sarkar, D., Murmu, S., Mandal, B., Hazra, G.C., Bhattacharyya, R., 2016. Assessment of nutrient management technologies for broccoli to improve productivity and quality and soil resources in the subtropics. International Journal of Vegetable Science. http://dx.doi.org/10.1080/19315260.2016.1216912 (in press: Published online 28 July 2016).

Verma, S., Sharma, P.K., 2007. Effect of long-term manuring and fertilizers on carbon pools, soil structure, and sustainability under different cropping systems in wet-temperate zone of northwest Himalayas. Biology and Fertility of Soils 44, 235–240.

Yadav, R.L., Dwivedi, B.S., Pandey, P.S., 2000. Rice-wheat cropping systems: assessment of sustainability under green manuring and chemical fertilisers input. Field Crops Research 65, 15–30.

16

# Reactive Nitrogen in Environment vis-à-vis Livestock Production System: Possible Remedies

C.S. Prasad[1], N.K.S. Gowda[1], S. Anandan[1], K. Sharma[2], M. Mohini[3]
[1]ICAR-NIANP, BANGALORE, INDIA; [2]ICAR, NEW DELHI, INDIA; [3]ICAR-NDRI, KARNAL, INDIA

**CHAPTER OUTLINE**



## Introduction

Impact of climate change on environment is well established and anthropogenic causes are the main reason for this phenomenon. Agriculture, including the livestock, is one of the major factors for the anthropogenic activity-led climate change. While fully acknowledging the immense benefits that intensified agriculture brought on with it, in terms of achieving food and livelihood security for the growing population in developing countries, the flip side is the associated potential health and environmental issues that need due attention. Livestock is an important component of agriculture, playing a critical role in food and nutrition security as animal-sourced foods accounts for 33% of protein and 16% of energy consumed globally. It is predicted that demand for animal-sourced foods in the developing world is expected to double over the next two

**The Indian Nitrogen Assessment. http://dx.doi.org/10.1016/B978-0-12-811836-8.00016-1**

decades mainly driven by the growing population, rising income levels, and urbanization. Enhancing productivity without considering its impact on environment and natural resources could be unsustainable unless a harmony is maintained between productivity increase and environment, which could be achieved on a sustainable basis (FAO, 2015). Impact of animal husbandry on climate change is influenced by a number of ways.

India's vast livestock wealth comprising 191 million cattle, 109 million buffaloes, 65.07 million sheep, 135.2 million goats, 10.3 million pigs, and 729.2 million poultry (DAHDF, 2014, 19th Livestock Census) has a major role to play in improving the socioeconomic status of the rural poor, particularly the small and marginal farmers who own more than 70% of livestock wealth. The country is ranked first globally in terms of livestock population as the livestock population density is among the highest in the world. Globally, India is the largest producer of milk and fish, and is ranked second in egg and third in broiler meat production (BAHS, 2015). India accounts for 2.4% of world's geographic area but supports nearly 17% of the global livestock population and around 17.5% of the human population. Further, the livestock density (7/ha) on forest land in India is one of the highest in the world. The transaction of inputs for nitrogen in terms of the nutrients—fertilizers, organic manures, and feeds to support Indian agriculture and the outputs in terms of various agriculture products and associated waste on account of the efficiency and/or inefficiency of different biological processes under various production systems are quite substantial. Livestock production is rapidly shifting to intensive production systems in developing countries to cater to the growing demand of animal products. These changes affect the livestock numbers, feed requirements, feeding and manure management practices, and related nitrogen emissions, causing acute environmental degradation (Pierre and Harald, 2006). Under intensive production systems to achieve higher production, more often the animals are fed more nutrients in terms of protein, phosphorus, and micronutrients leading to higher excretion of these nutrients causing environmental issues and nutrient wastage.

## Nitrogen Cycle and Reactive Nitrogen

Nitrogen cycle represents one of the most important nutrient cycles found in terrestrial ecosystems. Nitrogen cycle refers to the movement of nitrogen across different reservoirs, including living organisms, in different forms via chemical processes. Nitrogen as an essential constituent of DNA, RNA, and proteins is required for all organisms. Atmospheric nitrogen, though representing the largest reservoir, cannot be readily used by most organisms. Different steps of the nitrogen cycle are mostly driven by microorganisms. Most plants get the nitrogen from the soil or water in which they grow. Animals get nitrogen by consuming these plants. When the plants and animals perish, the bodies decompose bringing the nitrogen back to land and water. Humans influence the global nitrogen cycle in a major way and account for more than 75% of the reactive nitrogen with significant consequences to human and environmental health, primarily through agriculture and increased fossil fuel consumption to fulfill the demand for food and energy for the rapidly growing population (Galloway et al., 2015).

Nitrogen gas that constitutes almost the total nitrogen in air is unreactive, while other inorganic N compounds that are present in traces, such as ammonia, ammonium ion, nitrous oxide, nitric acid, nitrogen oxides, nitrogen pentaoxide, peroxyacetyl nitrate, nitrate ions, and organic compounds such as urea, amines, amides, and proteins, constitute the reactive form and play an important role in global biogeochemical cycles (Aneja et al., 2012). Over the decades, there has been a gradual buildup of reactive nitrogen due to human activities with consequences for air and water quality leading to various diseases in humans including respiratory problems (Aneja et al., 2009; Singh and Singh, 2008). Crops and animal activities are associated with spike in ammonia and $N_2O$ concentration. The $N_2O$ has approximately 300 times the global warming potential of carbon dioxide and is one of the important greenhouse gases in Earth's atmosphere (Bouwman et al., 1995; Olivier et al., 1998) besides being responsible for reduction in the ozone layer (Ravishankara et al., 2009).

Ammonia is the major alkaline component, present abundantly in the atmosphere (Aneja et al., 2008), that regulates atmospheric acidity (Brasseur et al., 1999) and is also an important source of atmospheric aerosols. Ammonia deposition contributes to soil acidification through oxidation of the deposited ammonia to acidic compounds (Roelofs et al., 1987).

## Livestock Production Systems and Nitrogen Transactions

Livestock farming systems broadly encompass four major production systems: (1) grazing, (2) mixed or semi-intensive, (3) intensive, and (4) landless systems. Depending on species, animal type, production system, and management, these systems convert feed protein into animal protein with an efficiency ranging from 5% to 45%. The inefficiency associated with this process results in loss of nitrogen in urine and dung to an extent of 5–55%. Generally, with intensification of production systems N inputs and N-losses increase, the least being in the grazing systems, followed by mixed systems, and highest in external input-based intensive system in industrialized countries. The major contribution of manure N globally is between 75 and 138 Tg N per year, the main contributors being cattle (50–60%), followed by sheep (12%) and pigs (6%). About 50% of manure N is collected from the animal sheds and barns. Out of this only half the quantity is used as organic manure for crop cultivation. The losses of N from manure are to the extent of 45–75 Tg per year (Oenema, 2006).

Inherent efficiency of nitrogen assimilation in animals is lower than crops and as a result nitrogen losses and wastage are quite large under livestock production systems. Nitrogen retention in various livestock products, such as milk, meat, eggs, and wool, as percent of nitrogen intake ranges from 5% to 20% (IPCC, 1997). Animal metabolism and fecal excretion are the major factors responsible for direct emissions (Dijkstra et al., 2011;

Tamminga, 1992). Methane and nitrous oxide emissions with global warming potentials of 25 and 298 times than that of $CO_2$ respectively are an environmental concern (IPCC, 2007). Nutrient losses from animal production are inevitable and excessive animal waste are mainly from an intensive livestock production, high stocking density, and excess or imbalanced feeding of nutrients (Tamminga, 1992). Nutrient composition of the feed and feed processing will have direct effect on animal productivity and losses through dung and urine (Van Vliet et al., 2007).

The sources of nitrogen losses in different livestock are mainly through indigestible feed and urinary losses and additionally in ruminants—nitrogen is lost in the form of ammonia from the rumen. Urinary losses of nitrogen in the form of urea accounts for over 70% in all species of livestock, except poultry where nitrogen is mainly lost as uric acid. The hydrolysis of these compounds takes place very rapidly from urine spills (IPCC, 1997). Apart from these sources of N-losses, inefficient utilization of absorbed protein for maintenance and for synthesis of milk and body proteins can lead to additional losses (Tamminga, 1992). Losses in the dung/urine are inversely related to digestibility, and the digestibility of a feed in an animal depends on its chemical composition, feed processing, particle size, dry matter intake, and feed additives (McDonald et al., 2010).

Ruminant manure from animal houses is a mixture of dung, urine, feed residue together with bedding material and water used for washing. Under intensive and semi-intensive systems, the animals are mostly kept in one place where it is possible to segregate fecal matter, leftover residue, urine, and water washings into solid and liquid manure (Sommer and Hutchings, 2001). The manure is a very good source of nutrients to crop productivity, cost-effective compared to inorganic fertilizers (Lazcano et al., 2008) and is widely used in mixed crop livestock systems. Traditionally, ruminant manure is allowed to compost with bedding and left over crop residues and normally spread in the crop land as farm-yard manure. Excessive application of manure could also lead to environmental issues, besides leading to toxicity, odor, water pollution, and increase in greenhouse gas emission and may present health risks (Dominguez and Edwards, 2011). In many countries, the manure is stored for long period before application in the field (Sommer and Hutchings, 2001), which could lead to accumulation of greenhouse gases (GHGs) and its subsequent emission to atmosphere (Kulling et al., 2001).

An estimated 48% of nitrogen excreted is lost during storage I and the remaining 52% represents the recycled nitrogen (Webb et al., 2012). Major means of nitrogen loss from manure is through ammonia, nitrous oxide, and ammonia in volatile form (Van der Meer, 2008; Ndegwa et al., 2008). Traditionally, farmers carry out composting of animal manure as it is an important input for soil enrichment. Nutrients losses during composting adds to pollution and can either be in the form of gases, liquid (leachate), or both. Nitrogen losses through composting are by ammonia volatilization, leaching and denitrification, and through chemo-denitrification (Hao et al., 2001). Depending on the manure mixture and the compost management employed, ammonia losses can account for about 92% but losses through leachate account for <0.5% (Eghball et al., 1997). Total

Nitrogen losses during composting are related to the initial nitrogen content in the manure which can be up to 42% (Gómez- Brandón et al., 2008).

During composting, a major portion of nitrous oxide is produced when the available carbon is reduced (Hao et al., 2001). Globally, including India, poultry manure or litter is used to enhance crop production by applying it to the soil. This is an effective and beneficial option when properly managed and can lead to environmental pollution when excess manure is applied beyond the utilization capacity of plants. Countries such as New Zealand use poultry manure for generating electricity. Several factors, e.g., storage of manure, exposure to environment, process of collection, transportation, and crop application, influence the extent of impact of manure on the environment.

# N Emissions From Indian Livestock Sector

Livestock production Aneja et al. (2012) reported the total emission of ammonia and $N_2O$ from livestock for the year 2003 as 1704.8 and 213.3 Gg per year, respectively, and the estimates are subjected to uncertainty due to a number of variables. Cattle and buffaloes were the largest contributors for ammonia accounting for 56.1% and 23.6%, respectively, followed by goat (10%) and sheep (6%), and the remaining 4.3% contributed by other species such as poultry, pigs, mules and asses, horses, and camels. Similarly for $N_2O$, cattle also contributed the highest proportion (42.3%), followed by buffalo (28.1%), goats (15.5%), sheep (9.5%), and the remaining 4.6% from the other species. For the country as a whole, Uttar Pradesh (12.7%), Madhya Pradesh (8.7%) and West Bengal (8.4%) were found to be the largest emitter for ammonia, while Uttar Pradesh (13.1%), Madhya Pradesh (8%) and Maharashtra (7.5%) contributed the most for the $N_2O$ emissions. In terms of the density, $NH_3$ and $N_2O$ emission per $km^2$, it was found that the states of West Bengal, Andhra Pradesh, Maharashtra, Orissa, and Bihar had the largest regions (Aneja et al., 2012).

## Manure Management

Nitrous oxide, a greenhouse gas, is produced during the nitrification–denitrification of nitrogen contained in livestock waste, and the quantity varies with the quality of diet. Nitrous oxide potential estimates in buffalo fed with paddy straw was 0.16 g $kg^{-1}$ dung, 0.04 g $kg^{-1}$ dung from buffalo calves fed with maize fodder in the ration, and 0.07 g $kg^{-1}$ dung from the buffalo calves fed with berseem fodder in the ration (Nampoothiri et al., 2016), and 0.007 g $kg^{-1}$ dung from cattle calves fed with wheat straw and concentrate mixture as per the requirement (Sarobana, 2015). Livestock dung represents one of the major sources of nitrogen and its mode of disposal has a large influence on the nitrous oxide production in India. Estimates of nitrogen loss through dung by total livestock in India indicated that around 4017.52 million tons of nitrogen is excreted through dung. Nitrous oxide is one of the important greenhouse gases and the share of nitrous oxide emissions from agriculture and its components—emission of soils, burning of crop

**Table 16.1** India's National $N_2O$ Gas Inventories in Gigagram

| | 1994 | 2000 | 2007 | 2010 |
|---|---|---|---|---|
| Total emission | 178 | 264.16 | 293.31 | 370.79 |
| Agriculture | 151 | 192.73 | 146.07 | 268.7 |
| Manure management | 1 | 0.07 | 0.07 | 0.08 |
| Agriculture crop residues | 4 | 6.17 | 6 | 7.07 |
| Emission from soils | 146 | 186.49 | 140 | 261.55 |

MOEF, 2004. Ministry of Environment and Forests, Government of India, 2004. India's Initial National Communication to the United Nations Framework Convention on Climate Change. http://unfccc.int/resource/docs/natc/indnc1.pdf; INCCA, 2010. Indian Network for Climate Change Assessment India: Greenhouse Gas Emissions 2007. Ministry of Environment and Forests Government of India. http://unfccc.int/resource/docs/natc/indnc2.pdf; MOEF, 2012. Second national Communication to the United Nations Framework Convention on climate change. Ministry of Environment & Forests Government of India. http://unfccc.int/resource/docs/natc/indnc2.pdf.

residues, and manure management documented over different years at the national level is presented in Table 16.1 (MOEF, 2004; INCCA, 2010, MOEF, 2012). Although total $N_2O$ emission increased over the years, its share from Agriculture is reduced, while soil is the major contributor and manure is the least contributor.

Disposal of dung is influenced by the species of the livestock, season, and the system of rearing. Generally, there are three major ways of disposal of dung are common (1) using dung for production of dried cake used as fuel in rural houses; (2) storing in heaps to allow for composting to be used as organic manure for crops; and (3) the leftover portion of dung which is not recovered when animals are let loose outdoors and the dung is generally allowed to decompose in the fields/pastures. Estimates of disposal pattern of dung in crossbred cattle in India indicate that 36% is used for making fuel cake, 27% by heaping for compost, and remaining 37% represents the portion that is left in the fields when animals are allowed to graze outdoors. The corresponding figures for indigenous cattle are 36, 22, and 42% while for buffaloes they are 41, 23, and 36%, respectively (Mohini et al., 2016). For other species, in the absence of reliable data, it was assumed that the disposal of dung in goat, sheep, yak, mithun, horse, mule, donkey, and camel, which are reared mostly by grazing outdoors, the dung is not recovered and is allowed to decompose 100% in pastures and fields. For pigs it was assumed that 100% of the dung is disposed through storing in heaps for compost while poultry was not included in the estimate. Based on the above assumptions, the overall distribution of dung from all livestock species combined excluding poultry, it was estimated that 36% is used for making dung cakes, 43% goes in pastures/fields, and 21% is kept as heaps for composting. Using the total quantity of dung and the emission factors for different uses, the estimates for total nitrous oxide production from livestock was found to be 80.35 (x000) tons per annum. Species-wise, buffaloes were found to be the largest contributors accounting for 43% of the total, followed by indigenous cattle (38%), crossbred cattle (11%), goats (4%), sheep (3%), and other species (1%). Based on 2012 livestock census and the IPCC default factors estimates indicate that the largest share for nitrous oxide contribution comes from the states of Uttar Pradesh (18.6%), followed by Rajasthan (9.9), Madhya Pradesh (8%), Andhra Pradesh (8.5%), and Maharashtra (6.7%) (Madhu Mohini et al., 2016).

## Implications of Other Nitrogenous Compounds

As most of the feeds for livestock are derived from plants, certain forms of nitrogen, such as nitrates, nitrites, and cyanogenic compounds, of certain plants might be toxic to livestock. Nitrates are the most readily assimilated form of nitrogen in plants. Under certain conditions, e.g., high nitrates in soil, droughts, cloudy conditions, wet weather, and wilting of certain fodder plants, such as sorghum, Sudan grass, and pearl millet, tend to accumulate higher levels of nitrates. Ruminants are more susceptible to nitrate poisoning due to the rumen micro flora which reduces nitrates into nitrites which interfere with oxygen carrying capacity of the blood leading to anoxia, abdominal pain, and diarrhea in acute cases, and retarded growth and reduced milk production in subacute cases (Susan, 2016). Management options, e.g., drying of fodders rich in nitrate do not help as nitrates persist in hay however ensiling helps in reducing the nitrates by almost 50%. A number of plants such as sorghum and cassava accumulate cyanogenic compounds, which on hydrolysis result in cyanide toxicity. Cyanogenic compounds tend to be higher in young plants or regrowth, and stressed regrowth tends to have dangerous levels. Similar to nitrates, ruminants tend to be more susceptible to cyanogens than monogastrics due to microbial hydrolysis of cyanogenic compounds in rumen. Hay-making and ensiling is known to reduce the cyanogen level by more than 50%. Apart from the agronomic factors, HCN in sorghum is also under the genetic control (Duncan, 1996). Research on reduction of antinutrients in forages in India has received less attention than other parts of the world as forage quality research is complex, expensive, and requires multidisciplinary approach and multi institutional alliances (Harinarayan et al., 2005).

Urea is a nitrogenous compound used for feeding ruminants and it is allowed legally to a maximum extent of 1% of the concentrate mixture and 15% in urea molasses block as economic ruminant feed supplement (BIS, 1992a,b). Efficiency of urea utilization is influenced by a number of factors, more prominently the level of energy in diet, and inefficiency associated with urea utilization in cattle leads to greater nitrogen losses through dung and urine, and at times when used inappropriately can lead to urea toxicity. Hence urea feeding has to be cautious and properly balanced with fermentable feed in diet.

## Mitigation Options

Mitigation options can be broadly classified under three major categories, namely enhanced livestock productivity that aims at reducing the overall nitrogen transactions; feed related interventions that aim specifically at improving the nitrogen utilization; others that are more related to manure/pasture management; and miscellaneous approaches.

## Productivity Enhancement

Livestock sector in India is gearing up to cater to the growing demand for livestock products, and the greatest option for mitigation comes through achieving a desired level of production through improvement in the productivity rather than the current number-driven production characterized by large numbers and low productivity livestock. These approaches call for maintaining an optimum number of livestock in productive phase, improving the productivity through breed, feed, and herd management on scientific lines. Breeding methods to improve the herd performance and better management can lead to reduction in unproductive animals and contribute in reducing the emissions (Gerber et al., 2013). Improvement in the average productivity of milk from 3.6 to 6 kg $d^{-1}$ in India can reduce the milch animal's numbers by 40% and feed requirement by 27% without reduction in milk yield, thus offering substantial advantages in reducing the nitrogen transactions (Blummel et al., 2009)

## Feed Related Interventions

Among the feed related mitigation approaches, improving the efficiency of nitrogen assimilation in livestock is most important and is achieved through balanced feeding, e.g., optimizing protein level, amino acids quality, and energy protein balancing. A successful example of balanced feeding is the ration balancing program implemented by National Dairy Development Board across India using locally available feed resources with improvement in farmers income, efficiency of production, reduction in methane emission by approximately 15%, better reproduction, reduced inter-calving interval, improved growth rates, longer productive life, and significant reduction in fecal and urinary nitrogen losses (FAO, 2011; Garg et al., 2013). Similar attempts have been made by NIANP, Bangalore using tools such as Feed Chart and Feed Assist. Synchronization of protein in terms of degradable and undegradable protein and fermentable energy has a greater influence on better ammonia utilization in rumen resulting in lower nitrogen excretion (Walli, 2005). Reducing the dietary level of crude protein by way of balancing the digestible amino acid requirements (precision feeding) has shown to reduce nitrogen excretion (Yulong and Bie, 2010). Exogenous use of phytase enzyme in diets of monogastric animals, e.g., pig and poultry will not only improve phosphorus utilization but also reduce nitrogen excretion. Improved feeding practices include grouping of animals based on the gender and phase of production, improving the efficiency of production through precision feeding as per the physiological requirements (FAO, 2006) where you can provide the essential amino acids and fatty acids, and reducing the total protein in the diets without compromising on targeted gains.

## Other Interventions

In spite of the best management, large amounts of nitrogen is lost through dung and during storage through volatilization as ammonia and, in an intensive system of production, the best option is to minimize this through closed tanks or alternatively maintaining a natural crust as top layer in open tank, both being equally efficient and economical. Anaerobic composting of manure and use of lime for acidification will help in minimizing the nitrogen emission (Samer, 2015). Biogas production helps in minimizing the nitrous oxide emissions, and also the methane produced is used as biogas for fuel and is not discharged to atmosphere (FAO, 2006). Manure management practices that ensure the recovery and recycling of nutrients can further contribute to mitigation. Identifying feeds and energy with low emissions and promoting their usage is another option (Gerber et al., 2013). Environmental friendly housing of farm animals to reduce volatilization of nitrogenous compounds is also an aspect which deserves due attention particularly in tropical countries like India.

Use of nitrification inhibitors along with urea and ammonium compounds in fertilizing pastures reduces nitrogen emissions (Di and Cameroon, 2003). Of late, use of neem-coated urea is being recommended in India for slow release of nitrogen in soil (Dinesh Kumar, 2015). Application of microbes that favor fixation of nitrogen into soil will reduce the negative impact on environment besides improving soil fertility and crop productivity. Legumes by virtue of their symbiotic association with nitrogen-fixing bacteria are capable of fixing atmospheric nitrogen and the recent advances in biotechnology has prompted the researchers to explore the possibility of extending this process to nonlegumes that produce bulk of food to improve productivity and reduce the fertilizer use and environmental impact mainly of nitrogen in developing countries (Christian and Giles, 2014; Mus et al., 2016). Native grasses, shrubs, or trees or a combination of these next to the water resources can serve as a natural buffer in effectively controlling the runoffs of nutrients from manure or liter applications near streams (Wenger, 1999).

For a country like India where the development objectives of livestock sector for generating income, livelihood, and food security have greater emphasis vis-à-vis the environmental concern, it is crucial that objectives of sector that serve both development and mitigation are given due importance.

## Conclusion

In the view of the changed perception about climate change, greenhouse gas emissions, and role of livestock, it is crucial to assess the impact of animal production systems and critical to understand how global warming and climate change would negatively affect animal and human systems. Mitigation measures include overall improvement in production efficiency of livestock through better breeding, feeding, and health management. Effective diet management in terms of nutrient balancing, precision nutrition, and phase feeding are essential to reduce nutrient excretion to environment. Effective measures of manure management to reduce nitrogen losses, adoption of low ammonia emission through housing designs, and precision soil fertilization are the key factors to overcome the challenges of nitrogen pollution in environment. Policy initiatives include stricter regulations, tax incentives, financial grants, targets for lower nitrogen losses, and more importantly investment on continued integrated research to innovate novel methods of mitigating the challenge of greenhouse gases.

## References

Aneja, V.P., Blunden, J., James, K., Schlesinger, W.H., Knighton, R., Gilliam, W., Jennings, G., Niyogi, D., Cole, S., 2008. Ammonia assessment from agriculture: U.S. status and needs. Journal Environmental Quality 37, 515–520.

Aneja, V.P., Schlesinger, W., Erisman, J.W., 2009. Effects of agriculture upon the air quality and climate: research, policy, and regulations. Environmental Science and Technology 43, 4234–4240.

Aneja, V.P., Schlesinger, W.H., Erisman, J.W., Behera, S.N., Sharma, M., Battye, W., 2012. Reactive nitrogen emissions from crop and livestock farming in India. Atmospheric Environment 47, 92–103.

BAHS, 2015. Basics of Animal Husbandry Statistics, Department of Animal Husbandry, Dairying and Fisheries. Annual report 2014–15. Ministry of Agriculture, Government of India. http://dahd.nic.in/sites/default/files/BAH_%26_FS_Book.pdf.

BIS, 1992a. Bureau of indian standards, indian Standard. In: Compounded Feeds for Cattle Specification IS2052-2009, New Delhi.

BIS, 1992b. Bureau of indian standards, indian Standard. In: Urea Molasses Block as Ruminant Feed Supplement-specification IS13451-1992, New Delhi.

Blümmel, M., Anandan, S., Prasad, C.S., 2009. Potential and limitations of by-product based feeding systems to mitigate greenhouse gases for improved livestock productivity. In: 13th Biennial Conference of Animal Nutrition Society of India, 17–19 December Bangalore, India, pp. 68–74, pp. 168.

Bouwman, A.F., Van der Hoeck, K.W., Olivier, J.G.J., 1995. Uncertainties in the global source distribution of nitrous oxide. Journal of Geophysical Research 100, 2785–2800.

Brasseur, G.P., Orlando, J.J., Tyndall, G.S., 1999. Atmospheric Chemistry and Global Change. Oxford University Press, New York.

Christian, R., Oldroyd, G.E., 2014. Synthetic biology approaches to engineering the nitrogen symbiosis in cereals. Journal of Experimental Botany 65 (8), 1939–1946. http://dx.doi.org/10.1093/jxb/eru098. http://jxb.oxfordjournals.org/content/65/8/1939-corresp-1.

DAHDF, 2014. 19th Livestock Census, 2012, All India Report. Department of Animal Husbandry, Dairying and Fisheries, Krishi Bhawan, New Delhi.

Dijkstra, J., France, J., Ellis, J.L., Kebreab, E., López, S., Reijs, J.W., Bannink, A., 2011. Effects of nutritional strategies on simulated nitrogen excretion and methane emission in dairy cattle. In: Sauvant, D., Milgen, J.V., Faverdin, P., Friggens, N. (Eds.), Modelling Nutrient Digestion and Utilization in Farm Animals. Wageningen Academic Publishers, The Netherlands.

Di, H.J., Cameron, 2003. Mitigation of nitrous oxide emission in spray irrigated grasslands by treating the soil with dicyandiamide, a nitrification inhibitor. Soil Use and Management 19, 284–290.

Dominguez, J., Edwards, C.A., 2011. Relationships between composting and vermicomposting. In: Edwards, C.A., Arancon, N.Q., Sherman, R. (Eds.), Vermiculture Technology Earthworms, Organic Wastes, and Environmental Management. CRC Press, Boca Raton, pp. 11–26.

Duncan, R.R., 1996. Breeding and improvement of forage sorghums for the tropics. Advances in Agronomy 57, 161–185.

Eghball, B., Power, J.F., Gilley, J.E., Doran, J.W., 1997. Nutrient, carbon, and mass loss during composting of beef cattle feedlot manure. Journal of Environment Quality 26 (1), 189–193.

FAO, 2006. Livestock Long Shadow- Environmental Issues and Options. Food and Agricultural Organization of the United Nations, Rome.

FAO, 2011. http://www.fao.org/ag/againfo/home/en/news_archive/AGA_in_action/2011_Going_to_the_basics.html.

FAO, 2015. http://www.fao.org/ag/againfo/programmes/en/lead/toolbox/Index.htm.

Galloway, J.N., Winiwarter, W., Leip, A., Leach, A.M., Bleeker, A., Erisman, J.W., 2015. Nitrogen footprints: past, present and future. Environmental Research Letters 9 (2014), 115003, 11pp.

Garg, M.R., Sherasia, P.L., Phondba, B.T., Shelke, S.K., Patel, C.T., 2013. Effect of feeding balanced ration on milk production, enteric methane emission and metabolic profile in crossbred cows under field conditions. Indian Journal of Dairy Science 66 (2), 113–119.

Gerber, P.J., Steinfeld, H., Henderson, B., Mottet, A., Opio, C., Dijkman, J., Falcucci, A., Tempio, G., 2013. Tackling Climate Change through Livestock—A Global Assessment of Emissions and Mitigation Opportunities. Food and Agriculture Organization of the United Nations (FAO), Rome.

Gómez-Brandón, M., Lazcano, C., Domínguez, J., 2008. The evaluation of stability and maturity during the composting of cattle manure. Chemosphere 70 (3), 436–444.

Hao, X., Chang, C., Larney, F.J., Travis, G.R., 2001. Greenhouse gas emissions during cattle feedlot manure composting. Journal of Environment Quality 30 (2), 376–386.

Harinarayana, G., Melkania, N.P., Reddy, B.V.S., Gupta, S.K., Rai, K.N., Kumar, P.S., 2005. Forage potential of sorghum and pearl millet. In: Sustainable Development and Management of Drylands in the Twenty-first Century Proceedings of the Seventh International Conference on the Development of Dryland, 14–17 September 2003, ICARDA Aleppo, Syria.

INCCA, 2010. Indian Network for Climate Change Assessment India: Greenhouse Gas Emissions 2007. Ministry of Environment and Forests Government of India. http://unfccc.int/resource/docs/natc/indnc2.pdf.

IPCC, 1997. Revised 1996 IPCC Guidelines for National Greenhouse Gas Inventories- Reference Manual, Vol. 3.

IPCC, 2007. Climate Change 2007: Impacts, Adaptation and Vulnerability. Summary for Policy Makers (Intergovernmental Panel on Climate Change).

Kumar, D., July, 2015. Neem coated urea: uses and benefits. Employment News 15, 63.

Kulling, D.R., Menzi, H., Krober, T.F., Neftel, A., Sutter, F., Lischer, P., Kreuzer, M., 2001. Emissions of ammonia, nitrous oxide and methane from different types of dairy manure during storage as affected by dietary protein content. Journal of Agricultural Science 137 (02), 235–250.

Lazcano, C., Gómez-Brandón, M., Domínguez, J., 2008. Comparison of the effectiveness of composting and vermicomposting for the biological stabilization of cattle manure. Chemosphere 72 (7), 1013–1019.

Mohini, M., Mondal, G., Thakur, S.S., Gupta, S., 2016. Trends in methane emission from Indian livestock. In: Proceedings of XVI Biennial Animal Nutrition Conference on "Innovative Approaches for Animal Feeding and Nutritional Research, Held at NDRI, Karnal.

McDonald, P., Edwards, R.A., Greenhalgh, J.F.D., Morgan, C.A., Sinclair, L.A., Wilkinson, R., 2010. Animal Nutrition, seventh ed. Publisher-Prentice Hall.

MOEF, 2004. Ministry of Environment and Forests, Government of India, 2004. India's Initial National Communication to the United Nations Framework Convention on Climate Change. http://unfccc.int/resource/docs/natc/indnc1.pdf.

MOEF, 2012. Second national Communication to the United Nations Framework Convention on climate change. Ministry of Environment & Forests Government of India 2012. http://unfccc.int/resource/docs/natc/indnc2.pdf.

Mus, F., Crook, M.B., Garcia, K., Garcia Costas, A., Geddes, B.A., Kouri, E.D., Paramasivan, P., Ryu, M.-H., Oldroyd, G.E.D., Poole, P.S., Udvardi, M.K., Voigt, C.A., Ané, J.-M., Peters, J.W., 2016. Symbiotic nitrogen fixation and the challenges to its extension to nonlegumes. Applied and Environmental Microbiology 82, 3698–3710.

Van der Meer, H.G., 2008. Optimising manure management for GHG outcomes. Australian Journal of Experimental Agricultrue 48 (2), 38–45.

Nampoothiri, V.M., Mohini, M., Mondal, G., Pandita, S., Sarkar, S., Rajvaidya, S., 2016. Effect of diets varying in roughage and concentrate level on manure methane and nitrous oxide emissions from buffalo calves. In: Proceedings XVI Biennial Animal Nutrition Conference on "Innovative Approaches for Animal Feeding and Nutritional Research. Held at NDRI, Karnal.

Ndegwa, P.M., Hristov, A.N., Arogo, J., Sheffield, R.E., 2008. A review of ammonia emission mitigation techniques for concentrated animal feeding operations. Biosystems Engineering 100 (4), 453–469.

Oenema, O., 2006. Nitrogen budgets and losses in livestock systems. International Congress Series 1293, 262–271.

Olivier, J.G.J., Bouwman, A.F., Van der Hoek, K.W., Berdowski, J.J.M., 1998. Global air emission inventories for anthropogenic sources of NOx, $NH_3$ and $N_2O$ in 1990. Environmental Pollution 102 (S1), 135–148.

Pierre, G., Harald, M., 2006. Greenhouse gases and animal agriculture: an update—nitrogen losses from intensive livestock farming systems in Southeast Asia: a review of current trends and mitigation options. In: Proceedings of the 2nd International Conference on Greenhouse Gases and Animal Agriculture, Held in Zurich, Switzerland, 20-24 September 2005. International Congress Series, vol. 1293, pp. 253–261.

Ravishankara, A.R., Daniel, J.S., Portmann, R.W., 2009. Nitrous oxide ($N_2O$): the dominant ozone-depleting substance emitted in the 21$^{st}$ century. Science 123–125.

Roelofs, J.G.M., Boxman, A.W., Van Dijk, H.F.G., 1987. Effect of ammonium on natural vegetation and forest. In: Asman, W.H., Diederen, H.S.M.A. (Eds.), Ammonia and Acidification, pp. 266–276.

Samer, M., 2015. GHG emission from livestock manure and its mitigation strategies. In: Sejian, et al. (Eds.), Climate Change Impact on Livestock: Adaptation and Mitigation. Springer, India, pp. 321–346, 978-81-322-2264-4.

Sarobana, S., 2015. Mitigation of Methane Emissions in Cattle Calves through Supplementation of Tree Leaves and Malic Acid. M Sc Thesis submitted to NDRI Deemed University.

Singh, B., Singh, Y., 2008. Reactive nitrogen in Indian agriculture: inputs use efficiency and leakages. Current Science 94, 1382–1393.

Sommer, S.G., Hutchings, N.J., 2001. Ammonia emission from field applied manure and its reduction. European Journal of Agronomy 15 (1), 1–15.

Aiello, S.E., 2016. The Merck Veterinary Manual, eleventh ed. Merck & Co., INC, Kenilworth, NJ, USA.

Tamminga, S., 1992. Nutrition management of dairy cows as a contribution to pollution control. Journal of Dairy Science 75 (1), 345–357.

Van Vliet, P.C.J., Reijs, J.W., Bloem, J., Dijkstra, J., de Goede, R.G.M., 2007. Effects of cow diet on the microbial community and organic matter and nitrogen content of feces. Journal of Dairy Science 90 (11), 5146–5158.

Walli, T.K., 2005. Bypass protein technology and the impact of feeding bypass protein to dairy animals in tropics: a review. Indian Journal of Animal Sciences 75 (1), 135–142.

Webb, J., Sommer, S.G., Kupper, T., Groenestein, K., Hutchings, N.J., Eurich-Menden, B., Rodhe, L., Misselbrook, T.H., Amon, B., 2012. Emissions of ammonia, nitrous oxide and methane during the management of solid manures. Sustainable Agriculture Reviews 8, 67–107.

Wenger, S., 1999. A Review of the Scientific Literature on Riparian Buffer Width, Extent and Vegetation. Institute of Ecology, University of Georgia, USA.

Yulong, Y., Bie, T., 2010. Manipulation of dietary nitrogen, amino acids and phosphorus to reduce environmental impact of swine production and enhance animal health. J.ournal of Food, Agriculture and Environment 8 (3&4), 447–462.

# 17

# Nitrogen Use Efficiency in Poultry Husbandry—Indian Perspective

A.B. Mandal, S.K. Bhanja
*ICAR-CENTRAL AVIAN RESEARCH INSTITUTE, IZATNAGAR, INDIA*

**CHAPTER OUTLINE**



## Introduction

Poultry production has undergone an enormous expansion and development during the last 50 years due to advances in genetic makeup of poultry birds, feeding and nutrition, rearing practices, and bio-security measures. Consequently, the poultry industry in India has taken a shape of the most dynamic and fast expanding segment in animal agriculture. The annual growth rate for broiler production was 6% in 1980s, 11% in 1990s, 19% from 1997 to 2002, about 6% from 2002 to 2006, and 7.79% from 2006 to 2013, while 5–7% for egg production (Mandal et al., 2005). With an annual production (BAHS, 2010, 2014) of around 69.73 billion eggs (approximately 95.2% from hen, 2.5% from duck, and

The Indian Nitrogen Assessment. http://dx.doi.org/10.1016/B978-0-12-811836-8.00017-3

2.3% from other species) and 2.682 million metric tons of poultry meat during 2012–3, India ranked 3rd and 5th in egg and meat production, respectively, in the world. The total population of fowls and ducks in 2012 was 716.2 million comprising 692.65 million fowls and 23.54 million ducks. In addition, total population of other poultry birds such as Japanese quails, turkeys, guinea fowl, emu accounted for 3.5 million (BAHS, 2010). The population growth of chicken, the major component of poultry, has been 2.42% during 2007–12. Thus the estimated population of chicken is likely to be 762.33 million in 2016, 839.02 in 2020, and 945.8 million in 2025. Similarly, with annual population growth rate of 6.45% for ducks during 2008–13, the estimated population of ducks is likely to be 21.27, 27.34, and 35.07 million in 2016, 2020, and 2025, respectively.

In addition to poultry population, the performance of birds in terms of meat and egg production and feed conversion efficiency has also improved significantly. The market age of broilers has reduced one day each year and presently 2.0 kg of body weight is being attained at 35 days of age in normal environmental temperature, but in summer it takes generally 38–40 days in an open house system of rearing which predominates in India. Similarly, the feed conversion ratio has decreased to 1.5–1.75 depending upon dietary plan of nutrition. Additionally, chicken egg production has increased by one egg per year and one commercial layer is capable of producing 315–325 eggs in a year with feed conversion of 122–125 g per egg. Egg production has also been increased substantially from 36.632 billion in 2001 to 73.89 billion in 2013–14 (http://www.gktoday.in/blog/egg-production-in-india/) at an average annual growth rate of 7.53%. Accordingly, per capita egg consumption has been increased from 36 to 58 during this period. With annual growth rate of human population at 1.2% and egg production at 7.53% in India, it is expected that the per capita availability of eggs will be 73.33 in 2016–17, 126.6 in 2025–26, 171.5 in 2030–31, and 182 in 2031–32. Thus with the present growth rate India will able to meet the requirement of 182 eggs per head as recommended by Indian Council of Medical Research in 2031–32. The average annual growth rate of poultry meat has been 10.57% during 2008–09 to 2012–13, but the per capita availability of meat is only 3.15 kg against the requirement of 10.8 kg per person per annum. Through the initiative of Indian Government's start-up program, new poultry farms are being set up by youth for self-employment, therefore, the growth of poultry will continue with the same pace. In reply of "want to end extreme poverty?" technology hyper-billionaire Bill Gates says the answer is chickens (Bill Gates: Chickens, not computers, can solve poverty, The Times of India Business, AFP | June 10, 2016, 06:52 p.m. IST).

## Total Feed and Protein Requirements for Poultry

According to United States of Department of Agriculture, the total broiler meat production was 4.2 mt in 2016, which is equivalent to sacrifice of 3465.34 million of broilers considering an average meat production of 1.212 kg per broiler (BAHS, 2014). Considering the normal mortality pattern in broiler (about 4%), the total number of broiler chicken to be raised is 3603.96 million. However, as per an estimate from BAHS

(2014) data, 2538.6 million of meat birds will be slaughtered (equivalent to 2640.12 million of broilers raised) during 2016, which is less than the USDA estimate of 3603.96 million. Thus, taking average of these two values, total number of broilers likely to be raised in 2016 is 3122.02 million. It is to add that Indian broilers generally attain a body weight of 1.836 kg (average body weight at slaughter, BAHS, 2014) at about 35 days of age, and a downtime period of 7–15 days is followed between two batches. Thus a total of seven cycles of broilers can be raised in a year. Again, the average feed conversion ratio in India is 1.65 (1.5–1.85) at present, and 3.02 kg of feed (0.5 kg pre-starter, 0.7 kg of starter, and 1.82 kg finisher) is required to attain a body weight 1.836 kg. Thus the annual (in 2016) requirement of feed for commercial broiler chickens is likely to be 9.43 mt (1.561, 2.185, and 5.682 mt of pre-starter, starter, and finisher, respectively).

Generally, one broiler breeder lays 230 eggs per year, equivalent of 184 quality chicks for commercial broiler production. Therefore, to produce 3122.02 million broiler chicks 19.587 million of broiler female and 4.897 million of male breeders are to be raised. The annual feed requirements for female (feed consumption of 68.26 kg in one cycle of 84 weeks and 42.26 kg per year) and male broiler breeders (feed consumption of 64.62 kg in one cycle of 84 weeks and 40.00 kg per year) are likely to be 0.716 and 0.169 mt, respectively. Thus the total feed requirement is likely to be 10.31 mt for broiler sector in India. Accordingly protein supply will be 1.929 mt (0.351, 0.470, and 1.108 mt in pre-starter, starter, and finisher growth phases) for commercial broilers, which is equivalent to 0.3087 mt of nitrogen. Similarly, the estimate of protein supply in the diet of male and female breeders is likely to be 0.139 mt (equivalent to 0.0223 mt of N). Thus the total protein supply in broiler sector will be 2.068 mt (or 0.331 mt of N).

The layer population has increased from 258.1 million in 2008–9 to 317.3 million in 2012–13 with an annual growth rate of 5.35%. Extrapolating this growth rate, the estimated population in 2016–17 will be 390.83 million, out of which about 77% will be commercial (300.94 million) and 23% will be native and improved native birds (89.89 million). The commercial layers are reared under intensive production system (total feeding), while major population of native birds are reared under extensive (backyard) as well as semi-intensive system (with slight supplementary feed). The commercial layer population includes 16.96 million broiler breeders, and the feed requirements of broiler breeders have already been taken care of. Thus the population of egg-type commercial chicken is likely to be 283.97 million. The average feed consumption during starting (0–8 weeks), growing (8–16 weeks) plus pre-laying (16–18 weeks) and laying (18–72 weeks of age) is 1.865, 5.688, and 41.69 kg. Thus total yearly feed intake is 36.56 kg in commercial layers and 10.67 kg for improved native or native birds (assuming 30% supplementary feed and rest from natural feed resource base) in 52 weeks. Hence, the total layer feed requirement for 283.97 million of commercial layers and 89.89 million native/improved native birds will be 10.10 mt. Considering the protein content in chick starter, grower, and layer diets as 20.0%, 16.5%, and 16% (average of three phases), respectively, the total intake of protein per layer bird is 7.98 kg during 0–72 weeks of age. Thus annual requirement is 5.765 kg per bird and total requirement of protein for commercial (1.637 mt) and native (0.155 mt) layers is 1.792 mt.

The estimated population of ducks (considering average growth rate of duck layers as 6.45% as per BAHS, 2014) in 2016 is 21.27 million. The male ducks are sacrificed for meat at about 20–24 weeks of age and are also kept for egg production. Therefore, it is likely that equal numbers of male are sacrificed. The ducks are mainly reared under range system with or without supplemental feed resources. The feed intake of ducks is about 15% higher than naïve chickens. Thus the supplementary feeds for duck layers would be 16.99 kg per cycle equivalent to 12.27 kg per year, and total requirement of feed will be 0.52 mt for layers as well as for males. Accordingly, the protein supply is likely to be 2.75 kg per cycle or 1.99 kg per year per duck, and 0.085 mt for entire duck population of India in 2016.

It becomes difficult to estimate the present population growth of turkey and other species of poultry (Japanese quails, emu, Guinea fowl, etc.) in India as the annual growth rate during 2007–12 was very high (55.46%), which may not sustain in coming years. Moreover, the emu farming has been either drastically reduced or abandoned by many farmers. Guinea fowls are reared under range feeding system. Japanese quails are reared under intensive production system but their feed intake is very less due to the lower body weight. Turkey birds are reared under semi-intensive system with supplementary feeding. Therefore, assuming feed and protein intake as 50% of duck population, the values will be 0.26 mt of feed and 0.045 mt for other poultry species in India during 2016.

In aggregate, therefore, for entire poultry sector, the requirements of feed and protein are likely to be 21.20 and 3.988 mt (equivalent to 0.638 mt of nitrogen), respectively, during the year 2016 in India. There is a big gap between the per capita production and requirement of egg (61 vs. 180 per capita per year) and meat (2.88 vs. 11.2 kg) in India. These gaps will be fulfilled in 2031–32 for eggs and 2030–31 for meat when the availability will be 182 for eggs and 10.8 kg per head per annum for meat, provided the present growth rate is sustained. However, the contributory factors for expansion of poultry industry include increase in India's share in meat and eggs export in future, population growth, urbanization, changing lifestyle, shifting of food habits, increasing per capita earning and health awareness being poultry meat healthier and environment friendly, younger population of India (average age 21 years in 2000, 26 years in 2025), and setting of organized retailing of poultry products. Therefore, there is every chance that the present growth rate may continue. With the growth rate of 8.32% in broiler sector, the requirements of feed for broiler sector will be 14.20, 21.17, and 31.57 mt in 2020, 2025, and 2030, respectively. Similarly, with growth rate of 6.5% in egg-type chicken population (egg production in India is growing at a compounded annual growth rate of over 8% according to a sector-specific analysis by Assocham), the requirements of feed for layer sector will be 12.99, 17.80, and 24.39 mt in 2020, 2025, and 2030, respectively. The feed requirement of ducks and other poultry species during respective years will be 1.0, 1.37 and 1.88 mt. Thus the total feed requirements during these years will be 28.20, 40.35, and 57.84 mt, respectively (Table 17.1). The protein (nitrogen) requirements to feed broilers will be 2.85 (0.456), 4.24 (0.679), and 6.33 (1.013); layers 2.30 (0.368), 3.16 (0.505), and 4.33 (0.693); ducks and other species 0.16 (0.026),

0.22 (0.0356), and 0.30 (0.049); and overall poultry species 5.32 (0.85), 7.63 (1.22), and 10.96 (1.75) mt by 2020, 2025, and 2030, respectively.

## Partitioning of Intake Protein into Meat, Eggs, Offal, and Excreta

The total consumption of protein for commercial broilers up to 6 weeks of age is about 615 g assuming 22.5, 21.5%, and 19.5% protein during pre-starter (0–14 days), starter (14–21 days), and finisher (>21 days) phase to attain a body weight of 1.836 kg (National average weight of slaughter, BAHS, 2014). Out of this amount, 264.4 g (assuming 72% meat yield and 20% protein on as such meat) is received back as meat with 79% biological value, 110.5 g as offal, and remaining 240.0 g in excreta. The average protein intake per female broiler breeder in one year cycle is 6.72 kg through 42.26 kg of feed, out of which 1.11 kg protein is received through settable eggs, 0.29 kg as meat, 0.12 kg as offal, and 5.20 kg in excreta. In male broiler breeders, annual protein intake is about 5.88 kg, out of which 0.33 kg is received back as meat, 0.14 kg as offal protein, and approximately 5.40 kg is excreted through excreta. However, significant quantity of protein is secreted through semen (about 5 g per 100 mL), which was not taken into consideration. In commercial layers, out of 5.76 kg annual protein intake, 1.68 kg protein with 100% biological value is received back as eggs for human consumption, 1.08 kg through meat, 0.06 kg as offal, and 3.88 kg through excreta. Accordingly, total nitrogen excretion from poultry industry is likely to be 0.415, 0.546, 0.770, and 1.089 mt in 2016, 2020, 2025, and 2030, respectively. Total excreta voided were calculated based on prediction (Mandal and Pathak, 1996) of dry matter metabolizability from dietary energy (apparent metabolizable energy-nitrogen corrected) supply to different species of poultry. The estimated values for excreta (on dry matter basis) would be 7.70, 10.18, 13.48, and 17.87 mt, equivalent to 30.79, 40.72, 53.92, and 71.48 mt of wet excreta (25% dry matter) during 2016, 2020, 2025, and 2030, respectively. However, the moisture and

**Table 17.1** Estimated Poultry Population, Feed Requirement, and Nitrogen Excretion

| | | Year | | | |
|---|---|---|---|---|---|
| **Attributes** | **Units** | **2016** | **2020** | **2025** | **2030** |
| Poultry population | Million, number | 762.3 | 839.0 | 945.8 | 1066.2 |
| Feed requirement (total) | Million tons | 21.20 | 28.20 | 40.35 | 57.84 |
| Feed requirement, broiler | Million tons | 10.31 | 14.20 | 21.17 | 31.57 |
| Layer | Million tons | 10.10 | 12.99 | 17.80 | 24.39 |
| Ducks and others | Million tons | 0.783 | 1.01 | 1.37 | 1.88 |
| Excreta (DM) | Million tons | 7.70 | 10.18 | 13.48 | 17.87 |
| Excreta (25% DM basis) | Million tons | 30.79 | 40.72 | 53.92 | 71.48 |
| N excretion | Million tons | 0.415 | 0.546 | 0.770 | 1.089 |

nutrient contents in poultry litter vary greatly due to dietary nutrient composition, rearing conditions, storage of litter, and handling methods.

## Dietary Sources of Proteins

The most commonly feedstuffs used in poultry ration are maize and soybean meal. As there is scarcity of maize and soybean meal at reasonable price in India, the locally available low-cost feed ingredients are used more and more to buffer the shortage as well as to reduce the feed cost. India, due to its highly variable landscape, rainfall, and agro-climate, produces large number of raw materials for feeding livestock and poultry. The various ingredients used in place of maize include pearl millet, sorghum, broken rice, rice bran, solvent-extracted rice bran, etc. The alternate feedstuffs to soybean meal are groundnut extraction, rapeseed meal (solvent-extracted), maize gluten meal, sunflower seed meal, high-protein roasted cluster bean meal, rice gluten meal, rice-based dried distillers' grain with solubles, sesame meal, decorticated cottonseed meal, maize germ cake, fishmeal, meat-cum bone meal, blood meal, etc. The findings on utilization of different cereals and oilseed residues in alone or in combinations have been encouraging (Mandal et al., 2002, 2004a; Elangovan et al., 2003; Ramarao et al., 2003, 2004a,b,c; Raju et al., 2003, 2004; Tyagi et al., 2003a,b, 2004a,b; Kumar et al., 2005). Combination of rapeseed meal and sunflower seed meal at 10% each replacing 67% of soybean meal resulted in similar growth and feed conversion efficiency as obtained in diet with soybean meal as sole source of vegetable protein. Solvent-extracted rapeseed meal could safely be included up to 10% in diet, while undecorticated sunflower seed meal up to 5% in diet, replacing soybean meal partially. However, combination of maize and pearl millet in a ratio of 1:1 and soybean meal alone, or with RSM or SSM, yielded apparently better performance than maize with soybean meal. Use of RSM or SSM in broiler starting diets based on sorghum or finger millet may be restricted. However, these meals can be used at 5% or 10% levels in sorghum or finger millet–based finisher diets (4–6/7 weeks of age). Majority of protein feedstuffs are generally limiting in one or more amino acids especially lysine, methionine, and/or threonine but commercial availability of these critical amino acids provides scope of improved utilization of such feedstuffs. The feed ingredients, e.g., maize gluten meal, rice gluten meal, rice-based dried distillers' grains with solubles, sesame meal, and rapeseed meal are relatively rich in methionine (Mandal et al., 2004b), thus strategic use of high methionine–containing feedstuffs along with soybean meal may reduce dietary supplementation of methionine and may improve protein utilization. Again, the digestibility of protein in some of the alternate feed resources is less. Generally, in commercial broiler diets (pre-starter, starter, and finisher), cereals and cereal by-products contribute 30.95%, protein supplements 67.30%, and limiting amino acids 1.75% of total protein. Similarly, in layer and broiler breeders 31.63%, 67.86%, and 0.51% of total dietary protein is supplied by cereals and cereal by-products, protein supplements, and amino acid supplements, respectively.

## Improving Nitrogen Utilization in Poultry

The environmental problems related to pollution of nitrogen (N) from the poultry excreta have forced the poultry nutritionists to scientifically manage the feeding strategies. Protein supplements also cost about 40% of diet in present days. Earlier ration formulation was based only on higher production performance without concerning the nutrient oversupply, especially those of protein and amino acids. Lately, judicial use of protein and amino acid supply to achieve maximum N utilization for meat and eggs, with least excretion of ingested N through excreta has gained more importance. Therefore, to achieve a sound protein feeding management different concepts and applications have been evolved for precision supply of protein and amino acids.

### Crude-Protein Concept

The proteins are composed of several amino acids, out of which some are dietary essentials, i.e., must be supplied through diets. Some of the essential amino acids (EAA) are usually deficient and are called limiting amino acids and the first, second, and third limiting amino acids in maize and soybean meal–based diets are methionine, lysine, and threonine, respectively. It is known that dietary protein level can be reduced through inclusion of synthetic amino acids or amino acid precursors (corresponding keto acids) in diet. The requirements were previously set on crude protein (CP) basis as the feed ingredients used in poultry diets are concentrates in nature and contain least amount of nonprotein nitrogen. Total CP requirement is referred to essential and non-EAA present. Generally 40% of body tissue protein is made up of non-EAA. Therefore, dietary proteins should supply all the EAA in right amount along with sufficient non-EAA to meet their requirements of other non-EAA. The requirements of EAA, expressed as a percentage of the diet, increase with the increase in dietary protein levels. However, maintaining the dietary EAA profile in relation to protein provides scope for better protein utilization, body weight gain, feed conversion efficiency, and carcass yields. To meet the requirement of all EAA at recommended levels in atypical maize and soybean meal–based broiler starter diet without supplementation of limiting amino acids, the diet should contain approximately 71% soybean meal and 14% maize with 35.6% CP, which does not seem to be economical and feasible at present day situation considering welfare and environment issues. Simple addition of synthetic DL-methionine by merely 0.23% to the feed will drastically reduce the soybean meal requirement to 34% with protein level of 21.6% to satisfy the requirements of all EAA (Pesti, 2009). A series of experiments were conducted in India on reduction of dietary protein supply through supplementation of limiting amino acids or maintaining amino acids profile in relation to protein in different poultry species with great success (Bhanja and Mandal, 2005a, 2007a; Jiang et al., 2005; Kaur et al., 2006, 2007, 2008, 2009; Kadam et al., 2007a,b; Mandal et al., 2006, 2010, 2013).

The "quality" of protein depends upon its constituting amino acids. Proteins having amino acids assortment more close to the body protein are considered of high quality. Moreover, all the amino acids required for the synthesis of a particular protein should be present at a time and in proper quantity in the body amino acid pool for an adequate supply. The purpose of maintaining calorie to protein ratio is to utilize protein for the purpose of growth and production, as excess protein are deaminated and the carbon skeleton is utilized as source of energy or synthesis of non-EAA. However, use of protein as source of energy is not economical as a significant amount of energy (25% of protein energy) is lost for synthesis of uric acid, the major excretory form of nitrogen. Uric acid excretion also adds to the soil and water pollution. Excess protein intake above the requirement enhances the heat increment, stress to the animals, and water intake for elimination of end products. Moreover, protein supplements are costlier.

## Ideal Amino Acid Concept

The requirement for a particular amino acid is proportionally linked to the requirement of other amino acids. Increasing the supply of one amino acid will improve the performance when no other amino acid is limiting (Bhanja and Mandal, 2006). In the ideal protein concept, the dietary EAA are adjusted in relation to lysine (Emmert and Baker, 1997; Morris et al., 1999; Leeson and Summers, 2001) that facilitates better utilization of protein irrespective of small changes in dietary protein or energy concentration. Expressing amino acid requirements in relation to dietary lysine overcome the variables (Han and Baker, 1994) affecting feed intake or amino acid requirements such as sex of the birds, dietary protein and energy levels, type of growth (fat or protein), floor and feeder space, diseases, stress associated with extreme climates, handling, and vaccination. Ideal dietary EAA profile play important role in the least cost feed formulation. Supplementation of methionine, lysine maize–soybean and threonine to practical maize and soybean–based poultry diets provides a means for decreasing protein supply, improving efficiency of protein utilization, and thus reduction of N excretion. Luckily these synthetic amino acids are available at affordable price. In young chicks, glycine may also often become limiting when dietary protein level is reduced (Kerr and Kidd, 1999). However, the reduction in the CP requires the knowledge of minimal indispensable amino acids need so that the maintenance and tissue accretion needs of the bird can be met. Almost all the feedstuffs are limiting in one or more amino acids. Identifying amino acids that are limiting in dietary formulation and supplementation of limiting amino acids or strategic supplementation with other feedstuff rich in that particular amino acid provides scope for use of those feedstuffs. A shortage of NEAA can be met by synthesis, while a deficit of any EAA will stop protein synthesis. There are other benefits of supplementation of limiting amino acids. Supplementation with amino acids allows reduction of protein-rich expensive feed ingredients for example soybean meal, without affecting performance. Moreover, reduction of dietary protein lowers the feed cost in addition to providing opportunities to include more energy supplements to increase feed energy value. The decrease in nitrogen excretion will decrease the nitrogen pollution of soil, particularly in regions of intensive animal production. It also improves air quality by lowering ammonia levels in farm buildings. It contributes to animal welfare as it ensures the good health of poultry.

## Digestible Amino Acid Concept

The values of total amino acids have been used for expressing amino acid requirements in various specifications (BIS, 1992; NRC, 1994). However, the availability (digestibility) of amino acids ingested through feed for maintenance, growth, and production may differ.

The digestibility of CP as such does not indicate the digestibility of amino acids, which are ultimately used for metabolic functions. Thus, supply of amino acids based on digestible amino acids in practical feed formulation becomes more precise, and thus the need for extra allowance as safety margin can be curtailed. Moreover, the digestible amino acids intake could be more accurately partitioned for maintenance and performance of birds. The terms amino acid digestibility and availability are often considered synonymous. However, certain amino acids are digested and absorbed in the forms which are not utilized. Extensive studies were conducted to evaluate the amino acid digestibility in different feedstuffs of India using caecectomized birds (Vasan et al., 2007a,b,c, 2008a,b,c,d).

There are several factors affecting protein digestibility and amino acids availability including processing. The oil-extraction process significantly affects the true amino acid digestibility. Solvent extraction also significantly affects the digestibility of lysine. Pressure processing of meat and bone meal decreases the digestibility of cysteine. Reductions in protein formulation with adoption of the digestible amino acid values greatly reduce feed cost and N excretion. The total amino acids could be adjusted downward by 10% when feeds are formulated with digestible rather total amino acids basis (Ravindran et al., 2005; Bhanja and Mandal, 2007a,b; Kaur et al., 2005, 2007; Vasan, 2006). However, the variations in amino acid digestibility as well as laboratory to laboratory variations in determined digestible amino acids values are the major constraints in using digestible amino acid values in practical feed formulations. Moreover, conventional feedstuffs have received more attention for evaluation of digestibility values. However, in India so many alternate feedstuffs are available which have received less attention (Table 17.2). Again, due to revolution in food-processing technologies, many newer feed resources are available for practical poultry feeding.

## Use of Proteases in Diet

Poultry birds have sufficient protease activities in gastrointestinal tract. However, with age the protease levels are increased and reach to a plateau at about 21–28 days of age. Presently, so many brands of proteases are available to improve digestibility of protein, and thus reduction of dietary protein supply to a tune of 0.50–0.80% resulting in efficient protein utilization and less protein excretion. Preliminary studies conducted showed that there is a scope for reduction of protein through supplementation of proteases. However, efficacy is further increased when more and more alternate feedstuffs with low digestible protein are added, thus addition of proteases provides scope for use of

**Table 17.2** True Amino Acid Digestibility Values (%) of Some of the Feed Ingredients

| Amino Acid | Maize | DORB | SBM | Sunflower | Rapeseed | Sesame | FM | MM |
|---|---|---|---|---|---|---|---|---|
| Lys | 77.8 | 78.5 | 89.5 | 91.0 | 70.0 | 88.0 | 88.3 | 87.0 |
| Met | 89.5 | 76.8 | 90.8 | 94.0 | 89.0 | 94.0 | 90.4 | 85.0 |
| Cys | 83.9 | 65.6 | 81.7 | 85.0 | 53.0 | 82.0 | 73.6 | 73.0 |
| Met + Cys | 86.0 | 69.2 | 86.5 | – | – | – | – | – |
| Arg | 85.6 | 85.6 | 90.6 | 98.0 | 91.0 | 92.0 | 89.7 | 92.0 |
| Thr | 75.5 | 71.3 | 84.7 | 90.0 | 69.0 | 87.00 | 85.97 | 91.0 |
| Ile | 80.4 | 81.7 | 88.5 | 92.0 | 80.0 | 92.0 | 89.1 | 88.0 |
| Leu | 91.4 | 77.5 | 88.4 | 93.0 | 85.0 | 91.0 | 89.8 | 89.0 |
| Val | 80.0 | 80.4 | 86.7 | 92.0 | 84.0 | 91.0 | 86.7 | 88.0 |
| Phe | 88.2 | 77.0 | 88.6 | 94.0 | 87.0 | 93.0 | 88.2 | 90.0 |
| His | 94.0 | 82.0 | 87.0 | 94.0 | 82.0 | 89.00 | 85.9 | 84.0 |
| Trp | – | – | 87.5 | – | – | – | 89.7 | – |

–, indicates no report available.

alternate protein supplements. However, protein and amino acids are present in plant cell content and the use of non-starch polysaccharidases along with proteases may be more beneficial.

## Dietary Crude Protein Content and Nitrogen Excretion

Research indicated that the performance of the bird could be maintained on reduced CP diet; however, the body fat content and feed:gain ratio increases. The reduction in dietary CP content resulted in 10–27% reduction in the total amount of N excreted during 6 weeks rearing period. However, reductions in CP level are associated with increase in N retention (Blair et al., 1999). Marginal reductions in CP content both during grower and finisher stage may result in equal body weight gain and feed conversion compared to the control diet bird. However, severe reduction in CP content significantly reduces the body weight gain and feed conversion in starting as well as finishing period (Kadam et al., 2007a,b). Nitrogen exertion dramatically reduced by 1.2% reductions in CP but not at severe reduction (2.7%). It is also reported that N excretion declined from 1.3 to 0.85 g/bird day$^{-1}$ and N retention increased from 64% to 75% for birds fed 19.4% and 18.2% CP diet, respectively.

Dietary protein level affects the growth and feed consumption, thus could change the feed utilization efficiency in birds with different genetic makeup. The carcass and abdominal fat content is influenced by dietary protein levels. There seems to be prioritization in the order of requirements (i.e., the maximum responses to CP levels) for maximum live weight, better feed utilization efficiency, and lean carcass production. Moreover, certain amino acids, especially methionine, stimulate immunity, hence the requirements for immuno-competence is also important.

The response to dietary protein level in the younger and slower growing birds is less compared to the fast growing or adult birds. The protein synthesis requires a large amount of energy and is also dependent on the development of gastrointestinal tract (GIT) (Bhanja and Mandal, 2005a,b). It is also observed that a proper balance of critical amino acid can reduce the overall amino acid requirement (Bhanja and Mandal, 2005a,b) in Indian climate as compared to that suggested by NRC (1994).

Dietary CP requirement is somewhat a misnomer as the requirement is based on the amino acids content of the protein. After digestion, proteins are broken down to amino acids and absorbed, enter into the body amino acid pool and then partition for synthesis of structural body proteins, enzymes, antibodies, hormones, and several other organic molecules associated with different functions. As body proteins are being synthesized and degraded constantly in the body, amino acid supply should be adequate to support wear and tear, growth, and egg production. The excess amino acids, not utilized for protein synthesis, are deaminated and used as a source of energy, resulting in higher nitrogenous excretions. The best way to reduce N in poultry excreta is supplementing diets with deficient amino acids in a low CP diet. Formulation based on bird amino acid requirements rather than CP can minimize N excretion by simply reducing total dietary N intake.

## Phase Feeding and N Excretion

With increase in age or body weight, the requirement of protein is decreased but that of energy is increased. Therefore, by feeding birds in different growth phases depending upon the requirements, can improve the nitrogen utilization. Conventionally, broilers were being reared in two growth phases, i.e., 0–28 days starter and 28–56 days finisher, but presently commercial broilers are fed in three (0–14 days pre-starter, 14–21 days starter, and 21 days market weight-finisher) or four (0–14 days pre-starter, 14–21 days starter, 21–32 days grower, and 32 days market weight-finisher) growth phases to minimize the cost of feeding and to improve the nutrient (including nitrogen) utilization. Reducing CP content of broiler diets by less than 2% units resulted in decrease in litter N content without changing ammonia concentration in the house. It is reported that four-phase feeding program with supplementation of critical amino acids such as lysine and methionine had similar feed conversion ratio but higher body weight of 77 g than the six-phases feeding program with supplementation of most of the EAA. However, the six-phase feeding program reduced the nitrogen intake (about 9%), nitrogen excretion (about 20%), and ammonia emission by 15.4% as compared to four-phased feeding program. In another study during finisher stage (21–63 days) the dietary changes on alternate days also reduced the N excretion by 7–13%, though the performance remained similar. Similarly, laying and breeding birds are also fed in different phases (starter, grower, pre-layer, and layer phase 1, 2, and 3) depending upon growth rate during starting, growing, and pre-laying periods and successively upon egg production, increase in egg size, and continuance of body weight gain during laying period that facilitates precise nutrient supply and thus improved nutrient utilization and least N excretion.

## Ammonia Emission in Poultry House

Ammonia is the most critical major noxious gas emitted from the poultry house due to microbial decomposition of urea and uric acid, principally the uric acid present in litter and the alkaline condition of the litter (pH above 7.0). Moreover, alkaline pH also increases release of ammonia by converting ammonium ion to ammonia, by stimulating the growth of uricase producing bacterium *Bacillus pasteurii*, as well as by increasing activity of uricase. The smell of ammonia can be detected easily when level goes beyond 20 ppm; however, it burns eyes and causes irritation leading to tearing when level reaches 50 ppm. As far as bird's performance is concerned, it affects the health at even 10 ppm and at 25 ppm it reduces broiler growth (by 4–8%) and feed conversion ratio (FCR) (by 3–6%). The ammonia levels above 25 ppm mostly affect the bird's respiratory system, liver function, and also increase incidence of *E. coli*, viral infections, and ascites. Prolong exposure to ammonia at 50 ppm leads to eye burn and blindness of broiler chicks. Even at 5 ppm concentration, it increases the incidence of chronic respiratory disease (CRD) and other respiratory diseases, as even the low concentration injures the protective lining of respiratory tract (Lacy, 2002). Of late, more research is focused on reduction of ammonia levels in litter for improvement in overall health and immunity of the bird. High moisture in excreta/litter, overcrowding, high protein and arginine in diet, and poor ventilation, especially during winter months, have been attributed to high ammonia concentrations in poultry shed. Therefore, it is necessary to reduce moisture in excreta (wet litter condition), prevent dampness of litter, and provide good ventilation in poultry house to remove moisture and humidity. The factors that responsible for wet litter conditions includes dietary factors (high salt, calcium, fiber, soluble non-starch polysaccharides, mycotoxins, etc.), environmental factors (both high temperature and humidity), certain diseases that lead to diarrhea (coccidiosis, IBD, IBH, colibacillosis, necrotic enteritis, mycotoxicoses, etc.) and vices of high water intake and urination especially in layers. Overcrowding should be avoided to keep the litter dry and maintain water quality. Proper ventilation to remove moisture from the poultry house is the most effective way to maintain air quality in shed. Addition of lime or superphosphate to the litter reduces the moisture level and decreases the pH of the litter thereby controlling the ammonia release. Treated litter is friendly to crops and the environment. In most applications, treatment is applied only once, as close to bird placement as possible and is also gaining popularity. The purpose of using agents for litter treatment is to reduce pH of the litter to 7 or below, as at this pH production of ammonia is also very less. Attempts have been made to reduce broiler house ammonia levels by using various litter treatments using phosphoric acid, sodium bisulphate, ferrous sulfate, aluminum sulfate (alum), calcium phosphate, and superphosphate. *Yucca schidigera* extract is a natural feed additive that represents one means of reducing ammonia levels via the diet. Unfortunately, the most frequently asked question "What is the best litter treatment?"

has no general answer, and the difficulties in addressing this question may be complicated and numerous. Besides the genetic potential of the birds, the other management practices, such as brooding, ambient temperature, lighting schedule, ventilation, and litter management, applied in the farm also contribute immensely to the cost of production and profit potential. However, such management practices should be cost-effective and justifiable.

## Nitrogen Elimination and Pollution Problems

Pollution from poultry farms has currently become one of the most challenging environmental problems. Large size poultry farms produce large amount of waste and their eco-friendly disposal becomes difficult. Manure causes soil, water, and air pollution. Poultry manure contains large amounts of N, P, K, and heavy metals. Excess N is converted to ammonia and nitrates, burning the fragile cells of land plants and poisoning ground and surface waters. The excretion of N is largely due to the excess protein and amino acids fed to poultry. The dietary means for reducing nitrogen excretion has already been discussed earlier. Protein and amino acids as well as the energy are the major items of poultry feed that needs special consideration to their precise supply can bring a continued and increased production of eggs and meat with an entry of improved genetic stock and less nitrogen excretion. Optimizing calorie–protein ratios for different categories of poultry birds of different ages at different climatic condition and production status is a must to improve nitrogen utilization. Application of ideal protein concept, supplementing the limiting amino acids in reduced protein diet and meeting amino acids requirements in terms of digestible in practical feed formulation, and use of proteases in diet will help in efficiently reducing nitrogen elimination significantly. Partitioning of protein and amino acids requirement for different growth and/or production rates at different body weights will help in precise supply of nitrogen. Suitable computerized models can also be developed that will benefit in estimating requirements of protein and amino acids to keep pace with the genetic improvement. Various workers have tried to predict the protein and amino acid requirement using different regression models. The linear and quadratic models are best fitted for the prediction of critical amino acids intake (gram) from the average daily gain (ADG) under hot and humid conditions (Bhanja and Mandal, 2006, 2007b). The linear regression models $Y = \mathrm{a} + b_1x_1 + b_2x_2$ or $Y = b_1x_1 + b_2x_2$, where $Y$ is the intake of a nutrient, $x_1$is the live weight or metabolic body size ($W^{0.75}$, MBW), $x_2$ is daily body weight gain in, $b_1$ and $b_2$ are the coefficients of variables $x_1$ and $x_2$, and a is a constant (Table 17.3) can act as a good predictor. Use of such models can help in formulating the balanced diets for maximizing growth in birds (Fig. 17.1).

## Poultry Litter and Dropping as Fertilizer

Poultry litter consists primarily of droppings and bedding materials (sawdust, rice husk, chaffed straw, and stovers) and also the waste feed and feathers of poultry birds reared on deep litter, i.e., on floor-raised birds. The manure or droppings are obtained from

**Table 17.3** Predicting Nutrient Intake From the Metabolic Body Size and Daily Weight Gain

| | Coefficients for 0–3 Weeks of Age | | | Coefficients for 4–6 Weeks of Age | | | |
|---|---|---|---|---|---|---|---|
| **Nutrient Intake** | **MBW (kg)** | **ADG (g)** | **$R^2$** | **MBW (kg)** | **ADG (g)** | **$R^2$** | **Sig Level (*p*)** |
| CP (g $day^{-1}$) | 5.56 | 0.288 | 0.992 | 13.025 | 0.182 | 0.995 | <.001 |
| ME (kcal $day^{-1}$) | 168.2 | 3.23 | 0.865 | 129.5 | 3.96 | 0.53 | <.001 |
| Lysine (mg $day^{-1}$) | 216.0 | 15.0 | 0.988 | 588.0 | 8.6 | 0.989 | <.001 |
| Methionine (mg $day^{-1}$) | 88.0 | 10.5 | 0.895 | 238.0 | 3.8 | 0.974 | <.001 |
| Threonine (mg $day^{-1}$) | 199 | 10.0 | 0.990 | 699.0 | 10.8 | 0.974 | <.001 |

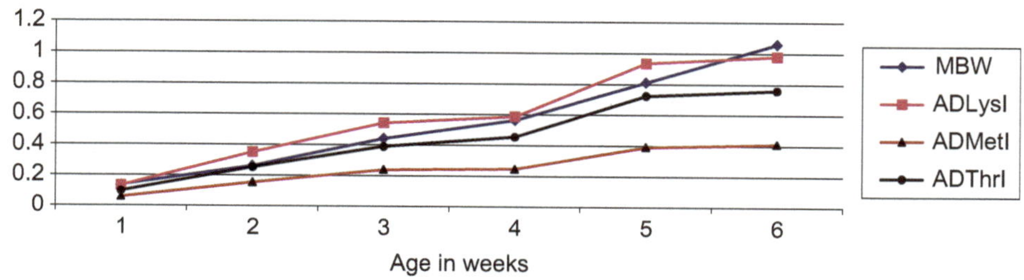


**FIGURE 17.1** Daily intake of AAs (g) in relation to MBW (kg) in slow growing broilers.

birds (layers and breeders) raised in cages. The physical and chemical composition of poultry litter and manure vary based on type of birds reared, bedding material used and amount, number of birds per unit area, feed nutrient density, time in use, management, environmental factors during production and storage, and postproduction handling methods. In order to prevent surface or groundwater contamination, it is advised that the litter should be placed in a structure. The broiler litter contains 2.6–3.0% N, 1.00–1.2% P, 0.9–1.65–1.74% K, while broiler excreta contains about 4.5–5% N, 1.80–2.61% P, and 2.4–2.8% K on as such basis. The layer droppings contain 1.5–1.6% N, 1.1–1.74% P, and 0.87–0.95% K. Phosphorus in chicken manure is predominantly present in organic form and thus becomes available much more slowly than nitrogen. Potassium is present in chicken manure as an inorganic salt and is readily available in most cases, but may be rapidly lost by leaching. Poultry manure is not only an excellent source of nutrients including minerals but it increases water infiltration rates of soil. Soil organic matter content can also be increased through application of poultry manure. Nutrient content of poultry manure is an essential element in the designing and operation of a poultry manure management system. However, for its efficient use it may be composted with cattle dung or applied in field on the basis of requirements of N, P, and K. Poultry waste being rich in P and K, its application rate should be determined based on requirements of P for soil and the crop. Nitrogen needs can be then met through application of additional N fertilizer. Excessive use of poultry litter or manure has negative impact on soil, water, and plant. Excessive use as well as continuous application of poultry excreta in soil, as fertilizer, causes leaching of nitrate nitrogen from soil to groundwater (Heathman et al., 1995). Excess ammonia level in surface water may affect the aquatic life. Ammonia accumulates in different organs, alters the metabolism, and increases the body pH of fish. However plants are more tolerant to ammoniacal nitrogen than animals. Ammoniacal nitrogen level exceeding 8–16 mg per liter water inhibits the plant growth, especially by lowering the chlorophyll concentration.

## Recycling of Poultry Slaughter House Wastes

As discussed, substantial quantity of nitrogen (about 58500 tons) is lost in the form of slaughter house waste. There is scope of recycling of such nitrogen in poultry feed. Poultry slaughter by-product meal (PSBM) a by-product produced by processing of the inedible parts of poultry carcasses including heads, feet, and viscera, with the exception of feathers. PSBM has a proper profile of available EAA and is rich in calcium, phosphorus, and vitamin $B_{12}$. This by-product is used as a protein source in diets of monogastric animals, e.g., pig (Zieret al., 2004) and poultry (Aimiuwu and Lilburn, 2006; Awachat et al., 2011a,b,c). The chemical composition, mineral contents, and protein quality of poultry by-product meal can vary greatly depending on the raw material source, storage time of raw materials prior to rendering, processing methods, processing pressure and temperature, ash content, inclusion of other poultry tissue, and difference in rendering procedure. In view of the above, attempt should also be made to standardize method of dry rendering to obtain high nutritional quality and microbiologically safe poultry by-product meal. Poultry by-product meal may be contaminated with microbial flora of intestinal origin or may be contaminated during or after the processing or handling and storage. The processed products or meat should be free from *Salmonella* organisms. Jardy and Michard (1992) tested 17 samples of poultry by-products meals from different sources and found it free from *Salmonella*. However, about 95% of birds are slaughtered for wet sale while only 5% by slaughtered by organized sector for cold chain supply. Therefore, it becomes difficult to collect and process such by-products. Experiment conducted at ICAR-CARI indicated that the poultry slaughter house by-products processed into meal can safely be used up to 5% in diet replacing soybean meal, but at higher levels (7.5 or 10%) it decreased the body weight gain of growing turkey poults (Awachat et al., 2011a,b,c). Different techniques have also been developed for utilization of feather meal but it has not yet been commercialized in India.

# References

Aimiuwu, O.C., Lilburn, M.S., 2006. Protein quality of poultry by-product meal manufactured from whole fowl co-extruded with corn or wheat. Poultry Science 85, 1193–1199.

Awachat, V.B., Majumdar, S., Mandal, A.B., Bhanja, S.K., 2011a. Effect of feeding rendered poultry slaughter byproduct meal on growth and immunocompetance of growing turkey poults. Indian Journal of Animal Sciences 82, 312–317.

Awachat, V.B., Majumdar, S., Mandal, A.B., Bhanja, S.K., 2011b. Slaughter byproducts yield of poultry, their rendering into poultry byproduct meal and evaluation of its nutrient composition and aflatoxin content. Animal Science Reporter 5, 135–140.

Awachat, V.B., Majumdar, S., Mandal, A.B., Bhanja, S.K., Yadav, A.S., 2011c. Chemical composition and microbiological quality of rendered poultry slaughter by-product meal prepared from different type of chickens. Indian Journal of Poultry Science 46, 411–414.

BAHS, 2010. Basic Animal Husbandry Statistics. Ministry of Agriculture, Govt. of India, India.

BAHS, 2014. Basic Animal Husbandry Statistics. Ministry of Agriculture, Govt. of India, India.

Bhanja, S.K., Mandal, A.B., 2005a. Effect of *in ovo* injection of critical amino acids on pre- and post-hatch growth, immuno-competence and development of digestive organs in broiler chickens. Asian–Australasian Journal of Animal Sciences 18, 524–531.

Bhanja, S.K., Mandal, A.B., 2005b. Effect of dietary energy and amino acids on meat yield and gastrointestinal tract development of naked neck broilers in hot climate. Indian Journal of Animal Sciences 75 (11), 1300–1306.

Bhanja, S.K., Mandal, A.B., 2006. Requirement of critical amino acids and their partitioning for growth and maintenance of naked neck broilers chickens in extreme summer. Animal Nutrition and Feed Technology 6, 103–113.

Bhanja, S.K., Mandal, A.B., 2007a. Optimising dietary energy and amino acid levels in the diets of naked neck broilers for growth, nutrient utilisation and immune response in a hot climate. Journal of the Science of Food and Agriculture 87, 945–950.

Bhanja, S.K., Mandal, A.B., 2007b. Validating the current amino acid and energy requirements of coloured broiler chickens. Journal of the Science of Food and Agriculture 87, 2131–2140.

BIS, 1994. Nutrient Requirements for Poultry (First revision), Bureau of Indian Standards, Manak Bhavan, 9 Bahadur Shah Zafar Marg, New Delhi 110002, 1992, IS 9863.

Blair, R., Jacob, J.P., Ibrahim, S., Wang, P., 1999. A quantitative assessment of reduced protein diets and supplements to improve nitrogen utilization. Journal of Applied Poultry Research 8, 25–47.

Elangovan, A.V., Mandal, A.B., Johri, T.S., 2003. Comparative performance of broilers fed diets containing processed meals of Bt, parental non-Bt line or commercial cottonseeds. Asian-Australasian Journal of Animal sciences 16, 57–62.

Emmert, J.L., Baker, D.H., 1997. Use of the ideal protein concept for precision formulation of amino acid levels in broiler diets. Journal of Applied Poultry Research 6, 462–470.

Han, Y., Baker, D.H., 1994. Digestible lysine requirement of male and female broiler chicks during the period three to six weeks of posthatching. Poultry Science 73, 1739–1745.

Heathman, G.C., Sharpley, A.N., Smith, S.J., Robinson, J.S., 1995. Land application of poultry litter and water quality in Oklahoma, USA. Fertilizer Research 40, 165–173.

Jardy, N., Michard, J., 1992. Salmonella contamination in raw feed components. Microbiologie Aliments Nutrition 10, 233–240.

Jiang, Q., Waldroup, P.W., Fritts, C.A., 2005. Improving the utilization of diets low in crude protein for broiler chicken 1.Evaluation of special amino acid supplementation to diets low in crude protein. International Journal of Poultry Science 4, 115–122.

Kadam, M.M., Mandal, A.B., Bhanja, S.K., Kaur, S., 2007a. Reducing protein through supplementation of limiting amino acids in maize or sorghum based diets for growth and immunocompetence of broiler chickens. Indian Journal of Animal Sciences 77, 912–917.

Kadam, M.M., Mandal, A.B., Bhanja, S.K., Vasan, P., Kaur, S., 2007b. Effect of maize and sorghum based low protein diets supplemented with limiting amino acids on nutrient utilization. Indian Veterinary Journal 84, 1159–1162.

Kaur, S., Mandal, A.B., Singh, K.B., Raj Narayan, 2005. Influence of dietary amino acids on marketing age of meat type quails. Indian Journal of Animal Nutrition 22, 1–7.

Kaur, S., Mandal, A.B., Singh, K.B., Raj Narayan, 2006. Responses of growing Japanese quails (heavy body weight line) to graded levels of essential amino acid concentrations in diets with or without fishmeal. Journal of Science of Food and Agriculture 86, 320–327.

Kaur, S., Mandal, A.B., Singh, K.B., Kadam, M.M., Elangovan, A.V., 2007. Response of laying Japanese quails to graded levels of essential amino acids profile with reduced dietary protein. Journal of the Science of Food and Agriculture 87, 751–759.

Kaur, S., Mandal, A.B., Singh, K.B., Kadam, M.M., 2008. The response of Japanese quails (heavy body weight line) to dietary energy levels and graded essential amino acid levels on growth performance and immuno-competence. Livestock Science 117, 255–262.

Kaur, S., Mandal, A.B., Tyagi, P.K., Tyagi, P.K., 2009. Effect of dietary essential amino acid profile with or without fish meal on meat production in Japanese quails. Indian Journal of Poultry Science 44, 313–318.

Kerr, B.J., Kidd, M.E., 1999. Amino acid supplementation of low-protein broiler diets: 2. formulation on an ideal amino acid basis. Journal of Applied Poultry Research 8, 310–320.

Kumar, V., Elangovan, A.V., Mandal, A.B., 2005. Response of broiler chicks fed high tannin sorghum diets supplemented with DL-methionine. Indian Journal of Animal Sciences 75, 1417–1422.

Lacy, M.P., 2002. Broiler management. In: Bell, D.D., Weaver Jr., W.D. (Eds.), Commercial Chicken and Egg Production, fifth ed. Kluwer Academic Publishers, USA, pp. 837–838.

Leeson, S., Summers, J., 2001. Scott's Nutrition of the Chicken, fourth ed. International Book Distributing Co., Charbag, Lucknow 134–138.

Mandal, A.B., Pathak, N.N., 1996. Comparative evaluation of metabolizable energy values of feedstuffs in chicken and guineafowl. Indian Journal of Animal Sciences 66, 930–934.

Mandal, A.B., Elangovan, A.V., Tyagi, P.K., Tyagi, P.K., Verma, S.V.S., 2002. Effect of mixed enzyme preparation on utilization of pearl millet (*Pennisetum typhoides*) in broiler chickens. Indian Journal of Poultry Science 37, 139–144.

Mandal, A.B., Tyagi, P.K., Elangovan, A.V., 2004a. Associative efficiency of feed resources. In: Mandal, A.B., Elangovan, A.V., Tyagi, P.K., Tyagi, P.K., Singh, R.V. (Eds.), Feedstuffs for Livestock and Poultry. Lead Centre, NATP (RNPS–16), Central Avian Research Institute, Izatnagar, UP, India, pp. 102–105.

Mandal, A.B., Yadav, A.S., Johri, T.S., Pathak, N.N., 2004b. Nutrition and Disease Management of Poultry. International Book Distributing Agency, Charbagh, Lucknow.

Mandal, A.B., Elangovan, A.V., Praveen, K., Tyagi, 2005. Poultry production for economic egg and meat production: a review. Indian Journal of Animal Sciences 75, 1215–1226.

Mandal, A.B., Kaur, S., Johri, A.K., Elangovan, A.V., Deo, C., Shrivastava, H.P., 2006. Response of growing Japanese quails to dietary concentration of L-threonine. Journal of Science of Food and Agriculture 86, 793–798.

Mandal, A.B., Panda, A.K., Elangovan, A.V., Deo Chandra, Shrivastav, H.P., 2010. Effect of different dietary protein levels supplemented with limiting amino acids on the performance of broilers in extreme summer. Indian Journal of Poultry Science 45, 283–286.

Mandal, A.B., Elangovan, A.V., Deo, C., Singh, D.P., 2013. Optimizing energy, protein and amino acid needs in diet of Kadaknath layers. Indian Journal of Poultry Science 48, 303–305.

Morris, T.R., Gous, R.M., Fisher, C., 1999. An analysis of the hypothesis that amino acid requirements for chicks should be stated as a proportion of dietary protein. Worlds Poultry Science 55, 7–22.

NRC, 1994. National Research Council, Nutrient Requirements of Poultry, ninth revised ed. National Academy Press, Washington, DC.

Pesti, G.M., 2009. Impact of dietary amino acid and crude protein levels in broiler feeds on biological performance. Journal of Applied Poultry Research 18, 477–486.

Raju, M.V.L.N., Shyam Sunder, G., Elangovan, A.V., Reddy, M.R., Sadagopan, V.R., Ramarao, S.V., 2003. Sorghum, bajra and ragi vis-a-vis maize as energy sources in broiler chicken diets. Indian Journal of Animal Nutrition 20, 185–192.

Raju, M.V.L.N., Shyam Sunder, G., Sadagopan, V.R., Elangovan, A.V., Reddy, M.R., Ramarao, S.V., 2004. Replacement of maize with jowar, bajra and ragi in broiler chicken diets. Animal Nutrition and Feed Technology 4, 53–61.

Ramarao, S.V., Raju, M.V.L.N., Panda, A.K., 2003. Replacement of yellow maize with graded levels of pearl millet (*Pennisetum typhoides*) in commercial broiler diets. Indian Journal of Poultry Science 38, 236–242.

Ramarao, S.V., Raju, M.V.L.N., Shashibindu, M.S., Murthy, O.K., 2004a. Effect of supplemental enzymes in diets containing yellow maize or pearl millet (*Pennisetum typhoides*) as the principal source of energy in broiler chicken. Animal Nutrition and Feed Technology 4, 101–111.

Ramarao, S.V., Raju, M.V.L.N., Panda, A.K., Sharma, R.P., 2004b. Replacement of maize with pearl millet (*Pennisetum typhoides*) on weight and nutrient basis in broiler chicken diet. Indian Journal of Poultry Science 39, 15–24.

Ramarao, S.V., Raju, M.V.L.N., Reddy, M.R., Panda, A.K., 2004c. Replacement of yellow maize with pearl millet (*Pennisetum typhoides*), foxtail millet (*Setaria italica*) or finger millet (*Eleusine coracana*) in broiler chicken diets containing supplemental enzymes. Asian-Australasian Journal of Animal Sciences 17, 836–842.

Ravindran, V., Hew, L.I., Ravindran, G., Bryden, W.L., 2005. Apparent ileal digestibility of amino acids in dietary ingredients for broiler chickens. Animal Science 81, 85–97.

Tyagi, P.K., Elangovan, A.V., Mandal, A.B., Tyagi, P.K., Kaur, S., Johri, A., 2003a. Effects of feeding low tannin sorghum grain to broiler chickens. Indian Journal of Animal Nutrition 20, 322–326.

Tyagi, P.K., Elangovan, A.V., Tyagi, P.K., Kaur, S., Johri, A., Mandal, A.B., 2003b. Utilizing sunflower seed meal along with maize or maize and pearl millet in the diets of broilers. Indian Journal of Poultry Science 38, 243–248.

Tyagi, P.K., Mandal, A.B., Elangovan, A.V., Tyagi, P.K., Johri, A., Kaur, S., 2004a. Growth performance and carcass traits of broiler chickens as influenced by feeding of finger millet. Indian Journal of Poultry Science 39, 131–135.

Tyagi, P.K., Mandal, A.B., Elangovan, A.V., Tyagi, P.K., Johri, A., Kaur, S., 2004b. Inclusion of rapeseed meal in the broiler diets based on maize-soybean or maize-pearl millet –soybean meal. Animal Nutrition and Feed Technology 4, 153–160.

Vasan, P., Dutta, N., Mandal, A.B., Sharma, K., 2007a. Apparent and true digestibility of amino acids of meat cum bone meal and fish meal in caecectomized roosters. Journal of Poultry Science 44, 396–400.

Vasan, P., Dutta, N., Mandal, A.B., Sharma, K., 2007b. Protein Digestibility and energy metabolizability of finger millet and pearl millet grains in cockerels and quails. Indian J. Poultry Science 42, 248–250.

Vasan, P., Dutta, N., Mandal, A.B., Sharma, K., 2007c. Amino acid digestibilities of protein supplements using caecectomized roosters. Indian Journal of Animal Nutrition 24, 205–208.

Vasan, P., Dutta, N., Mandal, A.B., Sharma, K., 2008a. Influence of caecectomy on the Bioavailability of minerals from vegetable protein supplements in adult roosters. Asian-Australasian Journal of Animal Science 21, 1178.

Vasan, P., Mandal, A.B., Dutta, N., Maiti, S.K., Sharma, K., 2008b. Digestibility of amino acids of maize, low tannin sorghum, pearl millet and finger millet in caecectomized roosters. Asian-Australasian Journal of Animal Science 21, 701–706.

Vasan, P., Dutta, N., Mandal, A.B., Sharma, K., 2008c. Effect of grinding on apparent and true digestibility of amino acids of pearl millet (*Pennisetum typhoides*) in adult roosters. Animal Nutrition and Feed Technology 8, 111–114.

Vasan, P., Dutta, N., Mandal, A.B., Sharma, K., 2008d. Determining digestibility of amino acids in Japanese quails. Indian Journal of Poultry Science 43, 75–78.

Vasan, P., 2006. Assessment of the Bioavailability of Nutrients in Poultry Feedstuffs. Ph. D. thesis. Deemed University, Indian Veterinary Research Institute, Uttar Pradesh, India.

Zier, C.E., Jones, R.D., Azain, M.J., 2004. Use of pet food grade poultry by-product meal as an alternate protein source in weanling pig diets. Journal of Animal Science 82, 3049–3057.

18

# Assessment of Nitrogen in Freshwater Aquaculture in India

K.K. Vass[1], G.N. Chattopadhyay[2], S. Adhikari[3], S. Datta[4], S. Samanta[5]

[1]NATIONAL ACADEMY OF AGRICULTURAL SCIENCES, NEW DELHI, INDIA; [2]ASHIRBAD, SEVAPALLI, SANTINIKETAN, WEST BENGAL, INDIA; [3]ICAR-CENTRAL INSTITUTE OF FRESHWATER AQUACULTURE, BHUBANESWAR, ODISHA, INDIA; [4]ICAR-CENTRAL INSTITUTE OF FISHERIES EDUCATION, KOLKATA, WEST BENGAL, INDIA; [5]ICAR-CENTRAL INLAND FISHERIES RESEARCH INSTITUTE, BARRACKPORE, WEST BENGAL, INDIA

## CHAPTER OUTLINE



## Introduction

India is an important producer of fish, contributing 5.7% of global fish production as of 2013–14 (Thakur, 2014), and about 50% of this production mainly comes from freshwater aquaculture sector. With the growing thrust toward increasing the aquatic production of the country to 12.71 mt by 2020 (Ayyappan et al., 2011), this sector is being entrusted to assume larger responsibility with a production target of 7.5 mt (Vass et al., 2015). With potential areas for this kind of aquaculture being used up for various other purposes, such a high target of production can only be achieved by bringing more water

The Indian Nitrogen Assessment. http://dx.doi.org/10.1016/B978-0-12-811836-8.00018-5

areas under optimum management for various productivity attributes. One very important factor to be taken into account in this respect is the availability of different nutrients present in the pond soil and water environment. A good numbers of studies have shown that these nutrients are essential for autotrophic and subsequently heterotrophic production of different fish food organisms and, thereby, growth and production of fishes (Banerjea, 1967; Kundu and Jana, 1994). Among all these nutrients, again, nitrogen plays a significant role in influencing the productivity of various aquaculture systems. Fish being a protein-rich harvest, requires high amount of N for their growth and production. Ayyappan et al. (1976) estimated average concentrations of N in the fishes commonly occurring in India to be 3.00% as against 0.82% and 0.25% for other two primary nutrients namely P and K. Based on this estimate it can be perceived that huge quantity of N is harvested from fish ponds of the country through aquaculture operations. A trend of gross annual removal of N through freshwater aquaculture since 1950s has been presented in Fig. 18.1. This quantum of N removal will increase with increment in fish-yield levels and any shortage in the supply of N in the pond environment will act as a major barrier for achieving the targeted production in fish culture.

In addition, the rate of conversion of N from natural foods to fish flesh is very low owing to the occurrence of long food chains and large-scale losses of this nutrient through various mechanisms such as leaching, denitrification, and volatilization, under aquatic environment (Chattopadhyay, 2004). This necessitates supply of substantially higher amount of N to the culture system for providing adequate N nutrition to fishes after compensating for these losses. However, such practice is also likely to induce further losses of N through various pathways, apart from increasing the cost of aquaculture operations. While the problem relating to leaching losses of $NO_3^-$ ions and their adverse effects on ground water is well known (Lenka et al., 2016), denitrification of this form of N in the anaerobic zones of the pond bottom and subsequent gaseous losses as

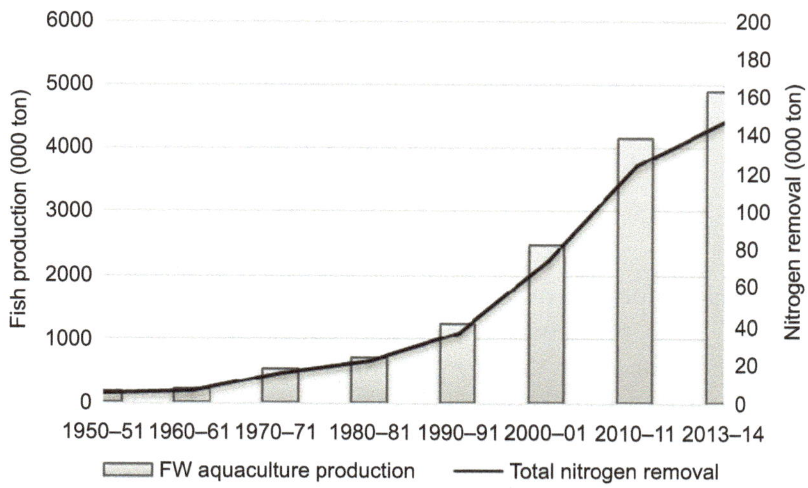


**FIGURE 18.1** Trend of estimated N removal through freshwater aquaculture in India.

$N_2$ and $N_2O$ adds to the quantum of greenhouse gases, a component of global warming (Pathak, 2015). Transformation of $NH_4^+$–N to $NH_3$ through volatilization is also considered as a common pathway of N escape from pond environment (Mandal and Chattopadhyay, 1992). In addition, in some situations, increased concentrations of this nutrient in the pond environment may lead to a change in the existing trophic state to a higher one leading to eutrophication, especially in weakly managed ponds.

In view of the significant influence of N in providing nutrition to the cultured fishes in freshwater fish ponds, and at the same time accelerating environmental pollution through release of several reactive as well as nonreactive forms of this nutrient, the present chapter is aimed to assess the nature of nitrogen transformation in freshwater aquaculture system with major focus on the processes, magnitudes, and dynamics of various losses of N from such culture system pertaining to Indian condition and also to suggest possible remedial measures. Since, all the pathways of N transformation in fish ponds have not been studied in depth in India, relevant observations from investigations carried out in other sectors as well as countries have also been used in the discussion for explaining different mechanisms.

# Nitrogen in Pond Environment

## Sources of N

Importance of N in providing nutrition to the fish food organisms and, thereby, the fishes, under Indian context is discussed by Saha (1969), Manissery (1995), Jhingran (1995), and others. In fish ponds, native N occurs mostly in various organic forms in the bottom soils. Singh and Ram (1971), while working on nature and properties of some fish pond soils with relation to their nearby upland ones, reported pond soils to exhibit higher amount of organic matter. Chattopadhyay (2004) attributed this behavior to the continued deposition of planktonic biomass in the pond bottom and also accumulation of some allochthonous organic matter in the form of leaves, plant debris, etc., and considered this high organic matter content to be the major source of N in such ponds. A general trend of using organic manures in most of the freshwater fish ponds of India also helps to supply substantial amount of organic N to pond environment. In wastewater aquaculture also, good amount of organic N is added to the culture system through sewage effluents. On the other hand, while supplementary feeds provide N nutrition through direct uptake by the fishes, its unutilized and half-digested residues also contribute to the organic sources of N in the pond environment. This organic N is mineralized and subsequently transformed to various reactive forms, e.g., $NH_4^+$, $NO_2^-$ and $NO_3^-$, through which N becomes available to primary producers.

To maintain adequate supply of N to the primary producers in readily available forms, N fertilizers are commonly used at periodic intervals which constitute a major source of available N in freshwater aquaculture (Saha, 1979). Runoff waters from nearby fertilized agricultural fields may also contribute some amount of N to pond environment in several situations. In addition, fixation of atmospheric N by blue-green algae, an important component of primary producers in aquatic environments, and also by some $N_2$-fixing bacteria constitute important sources of N in a tropical country like India (Jana and Roy, 1986). Additionally, dissolution of atmospheric N deposits into water is known to be an important source of N in any aquatic body. A good amount of water-soluble nitrogen in fish ponds also comes from phytoplankton mineralization (Schroeder, 1987). However, studies on these aspects are limited. All these sources together contribute to N nutrition in a freshwater aquaculture system. A gross picture of the N requirements and the suggested applications of N in freshwater fish culture system through some common nutrient inputs is presented in Table 18.1. As observed from the table, 50% of the N requirements of the fish ponds are provided through chemical fertilizers and the rest need to be supplied from organic sources. In addition, with intensification of aquaculture practices, varying doses of supplementary feeds containing appreciable amount of organic N are also being used. Mechanisms of transformations of these major sources of N in fish pond environment are discussed in Section 2.2.

## Transformation of N in Freshwater Ponds

Nitrogen becomes available to primary fish food organisms as $NH_4^+$, $NO_3^-$ and $NO_2^-$. Of these, $NH_4^+$ is generally considered preferable than the other two forms because the uptake of the other forms requires a supplementary energy expenditure to be transformed to ammonium form (Thompson et al., 1989). On the other hand, significant

**Table 18.1** Nitrogen Requirements and Use of Inputs in Freshwater Fish Ponds of India Under Varying Productivity Levels

| | Soil Condition | | | | | |
|---|---|---|---|---|---|---|
| **Pond Types** | **pH** | **Organic Carbon (%)** | **Available Nutrients (mg/100g)** | **N Requirements (kg/ha)** | **Fertilizer or Manure (kg/ha)** | |
| Low productive | <5.5 | <0.5 | N below 25 $P_2O_5$ below 3 | N 200–250 | CAN or urea<br>cow dung<br>or<br>Gobar gas slurry | 500–600<br>225–290<br>10,000–12,000<br>20,000–30,000 |
| Medium productive | 5.5–6.5 | 0.5–1.5 | N 25–30 $P_2O_5$ 3–6 | N 150–200 | CAN or urea<br>cow dung or<br>gobar gas slurry | 350–500<br>156–225<br>8,000–10,000<br>16,000–20,000 |
| High productive | 6.5–7.5 | 1.5–2.0 | N above 50 $P_2O_5$ above 6 | N 100–150 | Urea cow dung<br>or gobar gas<br>slurry | 112–156<br>5000–8,000<br>10,000–16,000 |

*In addition, supplementary feeds consisting about 120–170 kg N $ha^{-1}$ $year^{-1}$ are also generally used in freshwater aquaculture operations.

Based on Anon, 1985. Package of Practices for Increasing Production in Carp Culture Ponds (Aquaculture Extension Manual, New Series No. 2). ICAR-Central Inland fisheries Research Institute, Barrackpore, 30 pp.; *Estimated based on information provided by Jena, J.K., and Dash, P.C., 2011. Carp culture. In: Handbook of Fisheries and Aquaculture. ICAR, New Delhi, pp. 380–400.

occurrence of $NO_2^-$—N in the pond environment is toxic and undesirable. While chemical fertilizers release N directly to the available phases, organic forms of N are mineralized gradually to ammonium ($NH_4^+$) state by a wide spectrum of microorganisms (Barat and Jana, 1991). A good part of this $NH_4^+$—N subsequently undergoes nitrification to produce nitrate ($NO_3^-$) by nitrifying bacteria which are strictly aerobic in nature (Thakur, 2006).

An important condition that influences the transformation and availability of N in the aquatic ecosystem is the presence of a thick layer of water overlying the bottom soils which restricts the entry of oxygen to the deeper depths of these soils. This divides the pond soil surface into two distinct zones: (1) an oxidized or aerobic layer at the soil—water interface exhibiting the presence of molecular oxygen, and (2) a reduced or anaerobic zone below the aerobic layer where molecular oxygen is not available (Vass et al., 2015). While the aerobic layer includes the presence of various oxidized forms, such as ferric, nitrate, and sulfate, and also aerobic groups of various microorganisms, the anaerobic zone beneath this layer remains practically devoid of molecular oxygen, contains anaerobic and/or facultative bacterial population, and is characterized by the presence of reduced components e.g., ferrous iron, manganous, sulfide. Redox potential (Eh) is the electro-chemical property that serves to differentiate the condition of an upland soil from that of a pond soil. The oxidized state of a soil is reflected by its high Eh value. The depletion of oxygen and subsequent decline in redox potential are accompanied by sequential developments of various reduction reactions as summarized in Table 18.2.

It has been discussed earlier that N becomes available to primary producers mainly in the form of $NH_4^+$ and $NO_3^-$ and also as $NO_2^-$ in some cases. Due to limited availability of molecular oxygen in the bottom soils, mineralization of organic N in fish pond soils is largely carried out by some obligate and facultative anaerobic microbes (Mandal and Chattopadhyay, 1992). Since the efficiency levels of these organisms are lower than aerobic microorganisms, mineralization of N in fish pond environment occurs at comparatively slower pace than in the upland soils. However, the N requirement of these

**Table 18.2** Sequential Formation of Some Reduction Products Under Submergence

| Chemical Transformation | Eh (mv at pH 7.0) |
|---|---|
| Disappearance of mol. $O_2$ | 660—500 |
| Disappearance of $NO_3^-$ | 660—500 |
| Formation of $Mn^{2+}$ | 600—400 |
| Formation of $Fe^{2+}$ | 500—300 |
| Formation of sulfide | 000-(—) 190 |
| Formation of $H_2$ gas | (—)150—(—)220 |
| Formation of methane | (—)150—(—)190 |

Adopted from Ghildyal, B.P., 1976. Fertility and properties of submerged rice soils. In: Kanwar, J.S. (Ed.), Soil Fertility—Theory and Practice. pp. 301—317.

anaerobic microbes is less than the aerobic ones, and hence they can release relatively higher amount of N in mineralized form per unit of organic matter after their own nitrogen requirements are satisfied (Chattopadhyay, 2004). Apart from the nature of the microorganisms, the C:N ratio of the soil also contributes largely to the release of N in pond soils (Jana et al., 2001). It is well known that an organic matter with narrow C:N ratio mineralizes faster than the one having wider C:N ratio and also transfers higher amount of N.

A portion of the N in $NH_4^+$ form gets adsorbed in the negatively charged exchange complex of the bottom soil (Mandal and Chattopadhyay, 1990) and remains available to the primary producers, especially the benthic flora (Mandal, 1962). The rest amount may either be utilized by the primary producers in the water phase or undergo nitrification to form $NO_3^-$−N in the presence of dissolved oxygen in the ponds. A part of the adsorbed $NH_4^+$−N may also be subsequently released to water phase through cation-exchange reactions and undergo similar transformation. Some of this $NH_4^+$ may also be lost as free ammonia ($NH_3$) through volatilization. This process is accelerated by high pH and temperature of pond water, a condition which is commonly observed in the afternoon hours for productive fish ponds of tropical countries like India (Chattopadhyay, 2004). On the other hand, the nitrification process occurs only in the aerobic zones of the pond soil−water system forming nitrite ($NO_2^-$−N) by *Nitrosomonas* sp. in the intermediate phase of the reaction and nitrate ($NO_3^-$) by *Nitrobacter* sp. during the second step (Mandal and Chattopadhyay, 1990). Since complete oxidation of 1 mg of $NH_4^+$−N requires 4.6 mg of oxygen (Delince, 1992), occurrence of fairly aerobic condition is necessary to complete this process. Otherwise, incomplete nitrification may result in the production of high amount of intermediate product $NO_2^-$. Although small amount of $NO_2^-$−N can be utilized by the primary producers, larger concentration of it will affect the pond health seriously. The $NO_3^-$ ions are readily utilized by the floating primary producers. However, this form of N is highly soluble in water and, being negatively charged, has high mobility in the soil. A considerable part of this $NO_3^-$−N may be leached down to the anaerobic layer of the pond bottom from where it may be lost from the aquatic environment either by further leaching or through denitrification by denitrifying microbes prevailing under the anaerobic condition (Thakur, 2006). This denitrification process, where a good amount of $NO_3^-$−N is used by some anaerobic denitrifying bacteria as an electron acceptor in the absence of molecular oxygen producing gaseous nitrogen as a byproduct, constitutes an important process of N transformation in fish ponds. As shown in Table 18.2, denitrification of N may occur during the initial phase of decline in redox potential values, along with the disappearance of molecular oxygen. This indicates that only marginally reduced state of the pond soil−water system may initiate denitrification process and result in substantial loss of N from the pond environment. Such condition is likely to be observed in a large number of properly cultured pond bottoms which generally exhibit good occurrence of easily degradable organic matter from organic manuring and also accumulation of residues of supplementary feeds. In addition, wastewater fish ponds are also susceptible to oxygen depletion and consequently to such reduction reactions owing to the frequent intake of organic matter–rich effluents (Chattopadhyay et al., 1988). The magnitudes of various losses of N from pond environment are discussed later on.

## Nitrogen Cycle

It has been discussed earlier that in fish ponds N occurs mostly as $NH_4^+$, $NO_3^-$, $NO_2^-$, $N_2$ and $NH_3$, and also in various organic forms. All these forms are closely associated with the supply, availability, and loss of N in fish ponds through various mechanisms which can be explained by N transformation cycle in fish ponds (Jana and Roy, 1986; Mandal and Chattopadhyay, 1990; Saraswathy et al., 2013). However, N cycle in a fish pond system appears to be much more complicated than under upland conditions owing to the varying nature of its transformation under submerged condition and also occurrence of complex food chains. A simplified form of N cycle for freshwater fish culture system has been presented in Fig. 18.2. Since the transformation of N in fish ponds has already been discussed previously, only the salient mechanisms related to N cycle are being summarized here.

- Organic components of N come from soil organic matter, pond biomass residues, organic manures, animal excretion, unutilized supplementary feeds, etc., and form a major source of N in fish ponds, apart from fertilizers, along with dissolution and biological fixation of atmospheric $N_2$ in freshwater fish ponds.
- Inorganic N sources from fertilizers are mostly in $NH_4^+$ form, largely converted from applied urea, which undergoes hydrolysis in the pond environment to form $NH_4^+$–N.

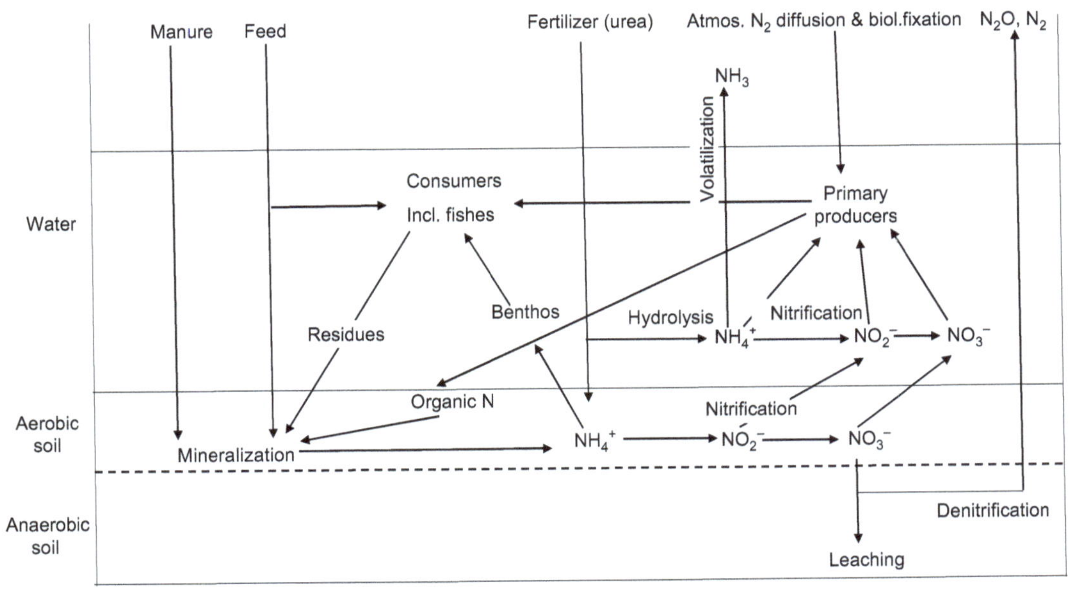


**FIGURE 18.2** Nitrogen cycle in freshwater aquaculture.

- This $NH_4^+$—N constitutes the primary form of available N in fish pond system. This form of N is also produced by mineralization of organic matter with the help of various heterotrophic microorganisms (Jana and Roy, 1986). $NH_4^+$—N may remain adsorbed in the exchangeable phase of the bottom soils to provide N nutrition to the benthic flora or may be released to water and be available to floating primary producers directly.
- In addition, a good part of this $NH_4^+$—N can be oxidized through nitrification and give rise to production of $NO_3^-$—N through an intermediate step involving production of $NO_2^-$. While both of these forms are available to primary producers, presence of appreciable amount of $NO_2^-$ is observed only under low oxygen condition and can be toxic to pond health.
- A portion of the $NH_4^+$ can also be transformed to unionized $NH_3$ under high alkaline condition and be lost from the pond environment through volatilization.
- The N occurring in $NO_3^-$ form becomes directly available to the floating phytoplankton population, while a good share of this highly soluble form of N may enter the bottom soil and undergo leaching loss, and is also subjected to escape through denitrification after entering the anaerobic zone of the soil.

## Availability Under Different Culture Situations

The soil—nitrogen system being a dynamic one has varying contents widely depending on the environment (Kanwar, 1976). In accordance with these observations, fish pond soils situated under different agro-ecological regions also exhibit considerable variations in total as well as available N contents. Jhingran (1995) discussed the variations in the properties of different soil groups of India in the context of aquatic productivity but was largely confined to upland soils. Thakur (2006) studied the availability of N in bottom soils of fish ponds in three major soil zones, i.e., red and laterite, hilly, and alluvial. Considerable variations in easily mineralizable and mineralized ($NH_4^+$, $NO_3^-$) forms of N in such soils were reported (Table 18.3).

Pond soils under red and laterite zones, being poor in organic carbon showed the lowest status of available N among the three groups. However, the high temperature

**Table 18.3** Variations in Availability of Nitrogen in Freshwater Fish Pond Soils (mg $kg^{-1}$) Under Different Soil Zones

| Soil Groups/Availability of N | Red and Laterite | Hilly | Alluvial |
|---|---|---|---|
| Easily mineralizable | 78.4—224.6<br>(124.6) | 55.2—375.2<br>(211.9) | 72.7—299.5<br>(196.3) |
| Mineralized<br>($NH_4^+ + NO_3^-$) | 47.6—184.6<br>(120.4) | 36.0—141.6<br>(73.9) | 66.0—216.4<br>(152.32) |

Figures in parenthesis indicate average values.

Thakur, A. 2006. Nitrogen Nutrition to Primary Producers in Fish Pond Soil Water System (Ph.D. thesis). Visva-Bharati University, Santiniketan, India. p. 133.

ranges prevailing in such areas and also large-scale development of almost dry pond conditions during summer season helped to accelerate the mineralization of organic matter and maintained large share of the easily available N in mineralized form. On the other hand, the occurrence of easily mineralizable N was the highest in the hilly soils, owing to the maintenance of higher content of organic matter in such soils. However, the amount of mineralized form of N was the lowest in such ponds owing to the high acidic condition and the prevailing low temperature regimes which affected the mineralization process. Similar observations, on the occurrence of easily mineralizable N in freshwater fish pond soils situated under two different agro-ecological regions in Nadia and Purulia districts of West Bengal, India, representing hot moist subhumid and hot dry subhumid agro-ecological subregions, respectively, were also made by Bhowmick (2008) while surveying the variations in nature and properties of fish pond soil and water under two contrasting agro-ecological regions.

## Critical Levels of N

Banerjea (1967), while studying the soil and water properties of large numbers of freshwater fish ponds of India, suggested that water-soluble inorganic N below 0.1 mg $L^{-1}$ can be considered as indicator of poor productivity of fish ponds while the values above 0.2 mg $L^{-1}$ indicated favorable conditions. As regards to soil phase, available N status (easily mineralizable form) below 250 mg $kg^{-1}$ was considered to be the index of poor productivity and 250–750 mg $kg^{-1}$ appeared to exhibit medium to high production potential. This study was carried out about 50 years ago and the N categorization values established through this investigation are followed till date for scheduling the fertilizer application doses in freshwater fish ponds of India. Since the fish-yield levels in the country have increased substantially since 1967 (Fig. 18.1), it is essential that the ratings of available N status be revised accordingly. Recently, Banerjee et al. (2009) carried out a systematic study on determination of lower critical levels of different nutrients in fish pond soils situated in red and lateritic soil zones. They used the classical method of Cate and Nelson (1965) on determination of the critical levels of plant nutrients in upland soils after modifying it to fit fish pond condition. This study showed that presence of less than 200 mg $kg^{-1}$ easily mineralizable N to be critical for such pond soils, below which the primary productivity declined sharply. They suggested that for assuring a tolerable level of natural productivity in these ponds, this level of easily mineralizable N needs to be maintained in the pond soils. Although inorganic forms of N , e.g., $NH_4^+$ and $NO_3^-$ constitute the readily available sources of N for the primary producers and exhibit closer relationship with growth of primary fish food organisms than the easily mineralizable form (Banerjee and Chattopadhyay, 2011), the information pertaining to occurrence of these two readily available forms of N in various freshwater fish pond soil–water systems and their relative contributions to primary production is still meager in India.

## Addition of Nitrogen Through External Sources

The major objective of using fertilizers in fish ponds is to enhance the growth of fish food organisms by improving the nutrient status of the pond soil–water environment and, thereby, the growth of fishes (Mandal and Chattopadhyay, 1992). The beneficial effects of fertilizers and manures in increasing fish yields under Indian condition were discussed earlier (Saha, 1979; Verma and Prakash, 1991; Tripathi et al., 1991; Adhikari, 2003; Jena and Dash, 2011). Schedules for using N inputs in freshwater fish ponds of varying productivity levels are discussed in Table 18.1. Detailed information about the variations in productivity as well as nutrient requirement of numerous fish ponds occupying about 2.36 mha area in the country (Ayyappan et al., 2011) is, however, not available. A gross estimate of the total requirement and supply of N in the freshwater fish ponds of India has been made by using the following calculations:

Assuming all the fish ponds of the country are having medium productivity, the average N requirement of these ponds will be @150–200 kg $ha^{-1}$. Since 50% of this requirement is to be provided through chemical fertilizers and rest 50% as organic sources, 75–100 kg N $ha^{-1}$ is to be applied to these ponds from each of these sources. This will lead to requirement of total 0.18–0.24 mt of fertilizer N for meeting the requirement of 2.36 mha freshwater aquaculture ponds of the country. On the other hand, Ayyappan and Ahmad Ali (2007), while assessing the quantum of mineral fertilizers being used in aquaculture, calculated that 0.53% of the total N fertilizer consumption of the country is used in aquaculture. Taking into consideration this assessment and assuming that bulk of these applications went to freshwater aquaculture sector, it can be estimated that out of total 16.7 mt of fertilizer N used in this country during 2013–14, about 0.08 mt N fertilizer was used in freshwater aquaculture. This is a substantially low figure in the context of total fertilizer N requirement for freshwater aquaculture, as discussed. However, when we consider that only 40% of the available fish culture areas of the country are presently being properly utilized (Ayyappan et al., 2011) and the rest areas use only small amount of fertilizers, the actual requirement of N fertilizer is likely to come down to around 0.07–0.10 mt which has a close proximity with the stated amount of N fertilizer consumption in aquaculture. On the other hand, while assessing the quantum of organic manure application in India, Ayyappan and Ali (2007) stated that 99.73 mt of cattle dung was used for aquaculture in the country. Since uses of traditional organic manures are not increasing in this country (FAI, 2015), the values mentioned by Ayyappan and Ali (2007) may be expected to remain almost unchanged for current organic manure application scenario in India. Considering these manures to contain 0.5% N (FCO, 2006), on an average, the contribution of this 99.73 mt organic manures to total N use in freshwater aquaculture comes to around 0.5 mt.

By putting these two values together, it may be surmised that fish ponds of India are adequately supplied with N when uses of fertilizers and manures are taken together, and major share of this supply comes from organic manures. This is in agreement with an age old observation of Prowse (1966), who reported organic manure to be by far the most common nutrient input in Asian countries.

Along with these inputs, bio-fertilizing microorganisms constitute a steady source of N in fish ponds through fixation of atmospheric $N_2$ (Jana and Roy, 1986). Ayyappan and Pandey (1989) stated the magnitude of such biological fixation in freshwater fish ponds to be around 5–8 kg $N_2$ $ha^{-1}$ $year^{-1}$. It can be mentioned here that under some intensively cultured high N input systems the pond water is likely to be well supplied with N. This may result in a consequent decline in the rate of biological fixation of the nutrient. However, there is no doubt that biological fixation of atmospheric $N_2$ acts as a steady source of N nutrition in freshwater aquaculture, especially in the ponds which do not receive adequate N inputs.

In addition to common organic manures, increasing use of supplementary feeds is also making important contribution to organic N nutrition to fishes in various aquaculture systems of the country. Based on the estimates of Jena and Dash (2011), it has been calculated that around 3000 kg supplementary feed is usually used in a fish pond producing 4000 kg fish $ha^{-1}$ in 10 months. Since commonly used supplementary feeds generally contain 25–35% protein, addition of N through such feeds is expected to be around 120–170 kg $ha^{-1}$.

However, as discussed earlier, use of these N inputs are not uniformly distributed in all the fish ponds of the country and more intensively cultured ponds receive larger shares of these fertilizers, manures, and feeds. Since only 40% of the existing fish ponds of India are being properly utilized for fish production, it may be stated that a large share of all these N inputs remains concentrated within this core aquaculture areas. This is likely to result in gross imbalance in N nutrition in fish ponds of the country and also to lead to larger magnitude of losses of this nutrient from these intensively cultured ponds.

## Magnitudes of Nitrogen Loss

Large-scale losses of N through various pathways pose a major problem for maintaining the use efficiency of N fertilizers at satisfactory levels in freshwater aquaculture. However, studies in this respect are limited in India. For the sake of better understanding of this problem, therefore, salient features of some important studies carried out in other countries are considered here.

Schroeder (1987), while working on N budget of channel cat fish ponds, observed that only 18% of the total N added as manure and fertilizer was actually incorporated in the fish flesh while 52% was lost from pond environment. Among different forms of N, again, maximum loss was encountered through $NO_3^-$ form. Similarly, Gross et al. (2000) also showed that the loss of N from fish pond through denitrification and leaching was about 40.5% while that through $NH_3$ volatilization was around 12.5%. Although no such study on quantification of N loss in different forms are available in Indian context, yet volatilization loss of N is likely to be more in a tropical country like India than any temperate country because such process is aggravated under high temperature and pH ranges

(Boyd, 1984), which are very common features for productive tropical ponds during post noon period (Chattopadhyay, 2004).

## N Budget

To optimize the use of N inputs in freshwater aquaculture it is essential to develop gross N budgets in various culture systems pertaining to the use of N inputs in different forms, utilization by the fishes, various losses of the nutrient from the pond environment, and accumulation of residual amount in water and bottom soil. Unfortunately, this kind of study was lacking from Indian perspective for a long time. In recent years, studies on this important field have been initiated in ICAR-Central Institute of Freshwater Aquaculture, Bhubaneswar (Adhikari et al., 2012). Adhikari et al. (2014a) reported a decline in fish yield with an increase in the stocking density from 7500 to 12,500 $ha^{-1}$ of Indian Major Carps. This resulted in substantial reductions in the N harvest from total N addition as organic manure, fertilizer, and supplementary feed under higher stocking levels. Since manure and fertilizer were applied at uniform rates in all the treatments while the dose of supplementary feeding increased with stocking density, the lowering in N harvest under higher stocking rates would result in accumulation of higher amount of unutilized N largely from the supplementary feeds. The study further reported that out of the total N added, 24.1–43.2% was harvested through fish flesh, 41.2–50.0% accumulated in soil, 1.8–2.5% was detected in effluent water, and 13.8–23.3% could not be accounted for. Since about half of the total N inputs in the culture practice was supplied in organic form through manure and supplementary feeds, the amount of mineralized N was comparatively less in these ponds and most of the unutilized N was expected to be retained in the bottom soil as organic form, limiting larger losses of N from the pond environment during the culture period. Similarly, while working on N budget in freshwater prawn culture using organic sources of N, through manure and supplementary feeds, Adhikari et al. (2014b) found that out of 55.7 kg $ha^{-1}$ of N added to the system, 20.5 kg was harvested through prawns, 30.4 kg was retained in soil and water phase, and 4.8 kg $ha^{-1}$ remained unaccounted. In this case also, organic form constituted almost the sole source of added N in the culture system which probably reduced the loss of N from the pond environment. Almost similar result was reported by Sahu et al. (2015) while working on polyculture with major carps and shrimp in several village ponds of Orissa, India. In these ponds also, supplementary feeds formed the dominant source of N and on an average, about 45% of the added N was harvested through fish and prawns, 47% could be detected in bottom sediment, while 8% remained unaccounted.

All these studies indicate better conversion of N in fish ponds with supplementary feeds. Such result is expected since these feeds are prepared and used for providing direct nutrition to the fishes. However, irrational uses of such feeds may result in the accumulation of a good share of these unutilized organic residues in the bottom sediments affecting the pond health as well as the N transformation in the culture system. Anaerobic degradation of the unutilized protein-rich supplementary feeds having narrow C:N ratio will not only result in greater release of reactive N through denitrification but also trigger off the decomposition process of dormant organic matters in the bottom sediment through lowering their C:N ratios. With the growing use of supplementary feed in freshwater aquaculture systems of the country, there is an urgent need for a detailed study on the relative contributions of these N inputs on N budgeting and also on pond health so that optimum combinations of N fertilization, manuring, and supplementary feeding can be developed for the aquaculture systems leading to better utilization and consequently lesser losses of this nutrient.

In the light of the above discussion, an attempt was made to prepare an N budget for freshwater aquaculture sector of the country under the proposed target of 7.5 mt fish production from freshwater aquaculture during 2020, recording a 50% jump over the current (2013–14) production of 4.9 mt. To reach this targeted production over the present status, we have estimated that the area under properly managed aquaculture will also be expanded from present 40% of the total pond area of the country (Ayyappan et al., 2011) to at least 60%, bringing a total area of about 1.4 mha under regular aquaculture. Making a generalized assessment that bulk of the targeted production will come from this 1.4 mha area and all the ponds under this area are medium in productivity, this 1.4 mha pond area will have the N requirement of 0.21–0.28 mt which need to be met through use of fertilizers and organic manures (Table 18.4). In addition, as per the report of Jena and Dash (2011), about 0.17–0.24 mt of N is expected to be added to this 1.4 mha pond area through use of supplementary feeds at 3 ton $ha^{-1}$. Biological fixation of atmospheric $N_2$ is also likely to add 0.007–0.012 mt of N to the total freshwater aquaculture area of the country (Ayyappan and Pandey, 1989). On an average, these will lead to a total addition of about 0.46 mt of N to the 1.4 mha of freshwater aquaculture ponds which are to be effectively explored for achieving the projected production from freshwater aquaculture in 2020. However, a large share of this added N will expectedly be lost from the pond

**Table 18.4** Gross N Budget for Production of 7.5 Million Ton Fish from Freshwater Aquaculture in India in 2020

| Source | Amount of N (Million Ton) |
|---|---|
| **Addition**[a] | |
| 1. Fertilizers and manures | 0.25 |
| 2. Supplementary feeds | 0.20 |
| 3. Biological fixation | 0.01 |
| Total | 0.46 |
| **Transformation** | |
| 1. Conversion to fish flesh | 0.16 |
| 2. Denitrification and leaching | 0.18 |
| 3. $NH_3$ volatilization | 0.06 |
| 4. Others | 0.06 |
| Total | 0.46 |

[a]Since nutrient requirements of the ponds (Table 18.1) take care of inherent fertility of pond soil and water also, contributions from native fertility have not been included here.

environment through different mechanisms, as discussed earlier. Taking the observations of Schroeder (1987) and Adhikari et al. (2014b) into consideration, it has been approximated that about 30% of the total N added through fertilizers and manures amounting to 0.05–0.07 mt N is likely to be ultimately utilized into fish flesh. On the other hand, based on the report of Sahu et al. (2015), supplementary feeds are expected to contribute 0.08–0.11 mt of N to formation of fish flesh. Thus, on an average, a total amount of about 0.16 mt of N will ultimately be utilized for fish production. A large part of the rest of the added N is likely to escape from the pond environment. Since information on pathways of N escape in aquatic ecosystems is limited in this country, the study of Gross et al. (2000) was considered for approximating the magnitudes of loss of N through various forms (Table 18.4). The exercise shows that the major route for escape of the N added to the culture system will be in $NO_3^-$–N through denitrification and leaching. Along with volatilization of $NH_3$, total loss of N through these three major pathways is likely to be 150% of what will be ultimately transformed to fish flesh. Moreover, a substantial share of this loss from the pond environment will be as reactive N. This, once again, emphasizes the imperative need for developing some efficient N management programs for minimizing the loss of N from such aquaculture system.

## Options for Better Nitrogen Management in the Culture Systems

All the studies on N use efficiency have shown that maximum loss of added N in a fish pond environment occurs from $NO_3^-$–N through denitrification and leaching. Under this context, Mandal and Chattopadhyay (1992) suggested that maintaining a higher amount of $NH_4^+$-N than $NO_3^-$-N in the pond environment may increase the efficiency of nitrogenous fertilizers. Since $NH_4^+$ ions can be adsorbed by bottom soil colloids in easily exchangeable phase, loss of nitrogen in this form will be less and the nutrient will become consistently available to the primary fish food organisms. However, some amount of these $NH_4^+$ ions may also be lost through volatilization under high temperature and alkaline condition, a typical situation encountered in tropical productive fish ponds—especially during intense sunshine periods.

In rice soils, use of different nitrification inhibitors is gaining popularity for increasing the use efficiency of nitrogenous fertilizers by restricting the release of $NO_3^-$–N. In view of the similarity of fish pond condition to that of submerged rice soils (Hickling, 1971), Thakur et al. (2004) carried out a mesocosm study to assess the effects of three nitrification inhibitors, neem (*Azadirachta indica*) extract, karanj (*Pongamia glabra*) extract, and sodium azide ($NaN_3$), used in conjunction with urea, on primary productivity of water under simulated fish pond condition. The study revealed use of the nitrification inhibitors to result in substantial increments in $NH_4^+/NO_3^-$ ratios in soil and water and also to significantly improve the primary production in water, as compared to the treatment without any nitrification inhibitor.

In India, high doses of organic manures are commonly used in fish culture ponds (Table 18.1). As a general norm, these manures and mineral fertilizers are recommended to be used separately keeping an interval of 15 days (Anon, 1985). During the course of decomposition of these organic manures, a good part of the dissolved oxygen of the water bodies can be utilized by the decomposing microbes. As a result, a semi-aerobic or even anaerobic condition develops near these decomposing organic materials. It was thought that the use of organic matter along with urea will help to develop a semi-aerobic environment around the added fertilizer, thus restricting the rapid transformation of the nutrient into $NO_3^-$–N. Thakur et al. (2012) reported the results of such a study, carried out to assess the effect of using urea along with starch as organic matter, on use efficiency of added N in a simulated fish pond condition. They showed that use of organic matter treated urea maintained higher level of $NH_4^+ + NO_3^-$–N with wider $NH_4^+/NO_3^-$ ratio in both soil and water phases. This behavior also helped to increase the gross primary production of water to 45.04–65.66% over the treatment without organic matter. However, development of anaerobic condition around the N fertilizer due to presence of organic matter may also lead to some loss of N through denitrification. In this context, use of carbonaceous organic matter with high C:N ratio can be a good option. This technology was evolved and successfully used by Avnimelech et al. (1994) for immobilizing excess inorganic N, especially $NO_3^-$–N, thus restricting their escape from the pond environment. The N thus immobilized was entrapped in the culture system and recycled to achieve higher amount of biological production from unit application of N. This approach may be validated under Indian conditions.

In fish culture practices in India, fertilizers are generally used in monthly doses. However, in view of the large-scale loss of nitrogenous fertilizers from the fish ponds, small split application of such fertilizers is likely to provide a steady source of N to the primary fish food organisms. This will prevent larger accumulation of this nutrient in the soil–water system at any point of time and thus help in reducing the loss of unutilized nitrogen from the culture system. To assess the efficiency of this concept, use of 100 kg N $ha^{-1}$ $year^{-1}$ was split to monthly, fortnightly, and weekly intervals keeping the total N application rate same during each of the treatment. The study revealed more frequent application of urea to result in higher production of primary fish food organisms as compared to monthly use of the fertilizer (Thakur et al., 2012). Field trial on beneficial effects of integrated use of these few simple N management practices showed 35.05% and 29.64% increase in gross and net primary production of fish food organisms, respectively.

## Way Forward

Earlier discussions have shown that the large-scale losses of N through various reactive forms pose a serious issue in the assessment of N under freshwater aquaculture systems of India. This not only contributes to substantial environmental threat but also affects the use efficiency of the inputs, and consequently the economy of the culture systems. Since yield levels of fishes are directly linked with consumption of N, there cannot be any compromise with adequate supply of N in the culture system. Under this situation, concerted efforts need to be made for development of some effective N management strategies for reducing various losses of N inputs through all possible means. Since availability of information in this regard is limited in India, some suggestions on researchable issues and also possible modifications in aquaculture technologies have been made in this chapter.

- Studies pertaining to transformation of N and also to magnitudes of loss of this nutrient from pond environment through different pathways, especially the behaviors of the reactive forms of N, are limited in India. Such studies need to be carried out under various pond management conditions in order to improve the rate of N utilization. Good amount of studies are carried out on increasing use efficiency of N fertilizers in submerged rice soils. Since conditions of rice soils have a close proximity to those of fish pond soils, some of these studies may be tailored suitably to fit fish pond environment for developing more effective N fertilization programs for freshwater aquaculture. Thakur et al. (2012) reported some encouraging results on this aspect. However, most of their investigations were carried out under mesocosm environment. On-farm studies under varying agro-ecological conditions are required for assessing the efficiency levels of these technologies under field condition.
- Oxidation-reduction reactions play a very important role in controlling the transformation of N and, thereby, the uptake and loss of N in pond environment (Muralidhar et al., 2014). These behaviors need to be studied in depth under different management conditions, especially in high N input systems. High doses of different kinds of organic manures often encourage the development of anoxic condition in the pond environment during the courses of their decomposition adding to the quantum of loss of N. In this context, use of humic acid, which is resistant to further degradation and also used in much smaller doses, is likely to appear as a viable option. Initial studies carried out in India showed a possibility of this organic input in replacing the large uses of bulky organic manures (Chattopadhyay and Dutta, 1993). More studies need to be carried out on this potent substitute of organic manures in order to maintain a healthy redox condition in the pond environment and, thereby, to restrict escape of reactive N.
- Development of actual need based N fertilization programs is essential for optimizing N budget of fish ponds and also for reducing the unnecessary uses of N fertilizers. For this, determination of critical levels of N in pond soil and water under different agro-ecological regions and also varying culture systems are necessary. Unfortunately, such kind of study is lacking globally, barring a few like Banerjee et al. (2009). Some research efforts need to be diverted in this field for avoiding overuse of N fertilizers.
- With intensification of aquaculture practices, use of supplementary feeds is increasing steadily. However, indiscriminate and excess use of such feeds may

contribute to high rate of accumulation of N under fish pond environment resulting in significant loss of the nutrient. While application of feeds with high food conversion ratio is likely to be a major strategy for reducing the loss of N, some other practices such as preparation of fish specific feeds, use of slow-sinking feeds, installation of demand feeding mechanisms are some options to be investigated in this respect. In addition, successful use of carbonaceous organic matter for immobilizing excess inorganic N, especially $NO_3^-$−N and re-using this N for productive purpose (Avnimelech et al., 1994), is discussed in previous section. However, in this case also, indiscriminate or overuse may be harmful to N nutrition. This approach needs to be assessed under Indian condition for recycling the excess N and also reducing N application rates.

- Little information is available on N dynamics under sewage-fed as well as animal-linked integrated fish culture systems of the country where organic source of N plays a major role. Since these kinds of fish farming may create oxygen-stressed situation conducive to larger escape of reactive N through denitrification, more information need to be collected on this field and suitable management strategies be developed.
- Aquaponics is emerging as a new concept for simultaneous production of fishes and hydroponics from aquatic ecosystems leading to better utilization of the available nutrients (Datta, 2015). This technology may be refined further and standardized suitably in order to use it with common fish culture systems for increasing the magnitude of N utilization reducing, thereby, the loss of N from fish ponds.

## References


Adhikari, S., Sahu, B.C., Dey, L., 2012. Nutrients budget and effluents characteristics in polyculture of scampi (*Machrobrachium rosenbergii*) and Indian major carps ponds using organic inputs. Water Science and Technology 66 (7), 1540–1548.

Adhikari, S., Pani, K.C., Mishra, B., Jena, J.K., Jayasankar, P., 2014a. Carbon, nitrogen and phosphorus budget for Indian major carps with different stocking density. Hydrology: Current Research 5, 176. http://dx.doi.org/10.4172/2157-7587.1000176.

Adhikari, S., Sahu, B.C., Mahapatra, A.S., Dey, L., 2014b. Nutrient budgets and effluent characteristics in giant freshwater prawn (*Machrobrachium rosenbergii*) culture ponds. Bulletin of Environmental Contamination and Toxicology 92, 509–513.

Adhikari, S., 2003. Fertilization, soil and water quality management in small scale ponds. Aquaculture Asia VIII (4), 6–8.

Anon, 1985. Package of Practices for Increasing Production in Carp Culture Ponds. Aquaculture Extension Manual, New Series No. 2. ICAR-Central Inland Fisheries Research Institute, Barrackpore, 30 pp.

Avnimelech, Y., Kochva, M., Diab, S., 1994. Development of controlled intensive aquaculture systems with a limited water exchange and adjusted carbon to nitrogen ratio. The Israeli Journal of Aquaculture—Bamidgeh 46 (3), 119–131.

Ayyappan, S., Ahmad Ali, S., 2007. Analysis of feeds and fertilizers for sustainable aquaculture development in India. In: Hasan, M.R., Hecht, T., De Silva, S.S., Tacon, A.G.J. (Eds.), Study and Analysis of Feeds and Fertilizers for Sustainable Aquaculture Development. FAO, Rome, pp. 191–219. Fisheries Technical Paper. No. 497.

Ayyappan, S., Pandey, B.K., 1989. Biological nitrogen fixation in fish ponds under different management practices. In: Proc. Nat. Sem on Forty Years of Freshwater Aquaculture in India. CIFA, Bhubaneswar, India.

Ayyappan, S., Kutty, M.P., Shenog, A.V., Gopakumar, K., 1976. Proximate composition of 17 species of Indian fishes. Fishery Technology 2, 153–155.

Ayyappan, S., Sugunan, V., Jena, J.K., Gopalkrishnan, A., 2011. Indian fisheries. In: Handbook of Fisheries and Aquaculture. ICAR, pp. 1–31.

Banerjea, S.M., 1967. Water quality and soil conditions of fish ponds in some states of India in relation to fish production. Indian Journal of Fisheries 14, 115–144.

Banerjee, A., Chattopadhyay, G.N., 2011. Soil System Based Fertility Management of Fish Ponds with Reference to Red and Lateritic Zones. Lambert Acad. Publ, Saarbrucken, Germany.114.

Banerjee, A., Chattopadhyay, G.N., Boyd, C.E., 2009. Determination of critical limits of soil nutrients for use in optimizing fertilizer rates for fish ponds in red and lateritic soil zones. Journal of Aquaculture Engineering 40, 144–148.

Barat, S., Jana, B., 1991. Distribution of nitrifying bacteria in fish culture tanks: influence of management factors. Acta Microbiologica Hungarica 38 (1), 7–16.

Bhowmick, A.K., 2008. Utilization of Soil Resources for Assessment of Pond Productivity under Two Contrasting Agro-Ecological Regions of West Bengal (Ph.D. thesis). Visva-Bharati University, p. 158.

Boyd, C.E., 1984. Water Quality in Warm-Water Fish Ponds. Auburn University, AL, USA, p. 369.

Cate Jr., R.B., Nelson, L.A., 1965. A Rapid Method for Correlation of Soil Test Analysis with Crop Response Data. International Soil Testing Series Technical Bulletin No. 1. North Carolina State University. Agricultural Experiment Station, Releigh.

Chattopadhyay, G.N., Dutta, J.K., 1993. Possibility of using humic acid in increasing primary productivity in aquaculture. In: Joseph, M., Mohan, C.V. (Eds.), Proc. 3rd. Indian Fish Forum, pp. 155–158.

Chattopadhyay, G.N., Saha, P.K., Ghosh, A., Karmakar, H.C., 1988. A study on optimum BOD levels for fish culture in wastewater ponds. Biological Wastes 25, 79–85.

Chattopadhyay, G.N., 2004. Balanced fertilization in aquaculture. Fertilizer News 49 (4), 87–93.

Datta, S., 2015. Aquaponics: present status and its potential. Fishing Chimes 34 (11), 44–48.

Delince, G., 1992. The Ecology of the Fish Pond Ecosystem. Kluwer Academic Publishers, pp. 1–230.

FCO, 2006. Fertiliser Control Order. Fertiliser Association of India, New Delhi.

Gross, A., Boyd, C.E., Wood, C.E., 2000. Nitrogen transformations and balance in channel catfish ponds. Aquaculture Engineering 24, 1–14.

Hickling, C.F., 1971. Fish Culture. Faber and Faber, London, p. 225.

Jana, B., Roy, S.K., 1986. Seasonal and spatial distribution pattern of nitrogen fixing bacteria in fish ponds under different management systems. Hydrobiologia 137 (1), 45–54.

Jana, B.B., Chakraborty, P., Biswas, J.K., Ganguly, S., 2001. Biogeochemical cycling bacteria as indices of pond fertilization: importance of CNP ratios of input fertilizers. Journal of Applied Microbiology 90 (5), 733–740.

Jena, J.K., Dash, P.C., 2011. Carp culture. In: Handbook of Fisheries and Aquaculture. ICAR, New Delhi, pp. 380–400.

Jhingran, V.G., 1995. Fish and Fisheries in India. Hindusthan Publications, New Delhi, p. 727.

Kanwar, J.S., 1976. Soil fertility-problems and prospects. In: Kanwar, J.S. (Ed.), Soil Fertility—Theory and Practice. ICAR, New Delhi.

Kundu, G., Jana, B.B., 1994. Influence of an atypical summer on the primary productivity of phytoplankton in two tropical ponds. Internatonal Revue der gesamten Hydrobiologie 79 (2), 249–257.

Lenka, S., Rajendiran, S., Coumar, V.S., Dotaniya, M.L., Saha, J.K., February 26, 2016. Impact of fertilizer use on environmental quality. In: Souvenir, Seminar on Environmental Concern of Fertilizer Use in Future, Society for Fertilizers and Environment, Kolkata, pp. 19–26.

Mandal, L.N., Chattopadhyay, G.N., 1990. Chemical and electrochemical environment in fish ponds and its impact on aquaculture. In: Manna, G.K., Jana, B.B. (Eds.), Impact of Environment on Animals and Aquaculture. University of Kalyani, India, pp. 79–84.

Mandal, L.N., Chattopadhyay, G.N., 1992. Nutrient management in aquaculture. In: Tandon, H.L.S. (Ed.), Non-Traditional Sectors in Fertilizer Use. FDCO, New Delhi, pp. 1–17.

Mandal, L.N., 1962. Effect of salinity on the transformation of nitrogen in the brackishwater fish farm soils. Journal of the Indian Society of Soil Science 10 (4), 255–261.

Manissery, J.K., 1995. Use of nitrogenous fertilisers in aquaculture. In: Chattopadhyay, G.N. (Ed.), Nutrient Management in Aquaculture. Institute of Agriculture, Visva-Bharati University, Sriniketan-731236, India, pp. 32–33.

Muralidhar, M., Saraswathy, R., Lalitha, N., Jithendran, K.P., Ponniah, A.G., 2014. Redox Potential as an Indicator of Pond Bottom Sediment Sondition. CIBA, pp. 1–6. e-Publ. Series. No.26.

Pathak, H., 2015. Greenhouse gas emission from Indian agriculture: trends, drivers and mitigation strategies. Proceedings of the Indian National Science Academy 81, 1133–1149.

Prowse, G.A., 1966. A review of the methods of fertilizing warm-water fish ponds in Asia and Far East. In: Proceedings of the FAO, World Symposium on Warm-Water Pond Fish Culture, Rome. http://www.fao.org/docrep/005/66573t/66573T02.htm.

Saha, G.N., 1969. Relative efficiency of different nitrogenous fertilizers in fish pond soils. Journal of the Indian Society of Soil Science 17 (1), 51–57.

Saha, G.N., 1979. Technologies of pond fertilization and use of fertilizers in freshwater aquaculture for increased fish production. Fertilizer News 24 (2), 3–6.

Sahu, B.C., Adhikari, S., Mahapatra, A.S., Dey, L., 2015. Nitrogen, phosphorus and carbon budgets in polyculture ponds of Indian major carps and giant freshwater prawns in Orissa state, India. Journal of Applied Aquaculture 27 (4), 365–376.

Saraswathy, R., Muralidhar, M., Syma Dayal, J., Ambasankar, K., Arulmozhiselvan, K., Lalitha, N., Kumararaja, P., Nagavel, A., Ravichandran, P., Ponnia, A.G., 2013. Methodology to Trace the Nitrogen Pathway in Shrimp Culture. CIBA, pp. 1–5. e-Publ. Series No. 22.

Schroeder, G.L., 1987. Carbon and nitrogen budgets in manured fish ponds on Israel's coastal plain. Aquaculture 62, 259–279.

Singh, S., Ram, H., 1971. A comparative study of pond mud and adjoining cultivated soils. Indian Journal of Agricultural Chemistry 4, 13–20.

Thakur, A., Banerjee, A., Chattopadhyay, G.N., 2004. Use of nitrification inhibitors to increase efficiency of nitrogenous fertilizers in aquaculture. The Israeli Journal of Aquaculture—Bamidgeh 56 (4), 256–263.

Thakur, A., Banerjee, A., Chattopadhyay, G.N., 2012. Increasing use efficiency of nitrogenous fertilisers in fish ponds. Better Crops with Plant Food 96 (1), 24–26.

Thakur, A., 2006. Nitrogen Nutrition to Primary Producers in Fish Pond Soil Water System (Ph.D. thesis). Visva-Bharati University, Santiniketan, India, p. 133.

Thakur, A.K., 2014. Preface. In: Handbook of Fisheries Statistics. Dept. of Animal Husbandry, Dairying and Fisheries. Govt. of India, New Delhi, p. 166.

Thompson, P.A., Levasseur, M.E., Harrison, P.J., 1989. Light-limited growth on ammonium vs. nitrate: what is the advantage for marine phytoplankton. Limnology and Oceanography 34, 1048–1061.

Tripathi, S.D., Bhandari, M., Ayyappan, S., 1991. Evaluation of manurial application levels of biogas slurry in carp farming. In: Proc. Nat. Symp. New Horizons in Freshwater Aquaculture. Assoc. of Aquaculturists and CIFA, Bhubaneswar, India, pp. 157–160.

Vass, K.K., Chattopadhyay, G.N., Jana, B.B., 2015. Soil, fishery and aquaculture. In: Pathak, H., Sanyal, S. K., Takkar, P.N. (Eds.), State of Indian Agriculture-Soil. National Academy of Agricultural Sciences, New Delhi, pp. 165–182.

Verma, D.K., Prakash, C., 1991. Certain observations on nitrogen profile in ponds under controlled conditions. Agricultural and Biological Research 7 (1), 38–46.

19

# Nitrogen Assessment and Management in Brackish-Water Aquaculture of India

M. Muralidhar[1], R. Saraswathy[1], J. Syama Dayal[1], K.K. Vass[2]

[1]ICAR-CENTRAL INSTITUTE OF BRACKISHWATER AQUACULTURE, CHENNAI, INDIA; [2]NATIONAL ACADEMY OF AGRICULTURAL SCIENCES, NEW DELHI, INDIA

## CHAPTER OUTLINE



The Indian Nitrogen Assessment. http://dx.doi.org/10.1016/B978-0-12-811836-8.00019-7

# Introduction

## Brackish-Water Aquaculture and Culture Systems

Brackish-water aquaculture (BWA) requires natural resources such as land, water, and biological resources such as seed and feed. Shrimp farming, a biology-based production activity, is synonymous with BWA in India. The quality and quantity of available resources and their management therefore plays a key role in the economic success and long-term sustainability of coastal aquaculture. Shrimp culture technology and the system of culture to be undertaken depend entirely on the site characteristics. The natural resources, namely land and water, if poorly planned and managed, can result in irreversible environmental damage and may lead to ultimate loss in economic gains. Shrimp farming in India till now has three distinct phases. First, the "rising phase" with *Penaeus monodon* culture from late 80s to 1995, second, the "falling phase" during 1995–2001 affected with viral diseases, predominantly the "white spot syndrome virus" (WSSV) and monodon baculo virus (MBV), and from 2002 onward, the third phase during which the farming is "sustainable" due to involvement of research and development (R&D) institutions in the country (Kumaran et al., 2003). Since 2009 after the introduction of *Penaeus vannamei*, India witnessed rapid expansion of shrimp culture activity. The inland production of marine shrimp provides an alternative to traditional coastal aquaculture where land costs and user conflicts can inhibit commercial development.

The culture systems practiced vary depending on the availability of inputs and farmers capacity to invest in the farming. In low input farming systems, due to adoption of scientific principles of farming, 500–1500 kg crop $ha^{-1}$ production is predictable. Surendran et al. (1991) demonstrated the technology of semi-intensive shrimp farming with 4–6 tons $ha^{-1}$ production.

## Source Waters (Creek/Sea Water)

India has a long coastline of 8129 km, including the mainland, islands (Andaman and Nicobar, and Lakshadweep), 14 large river systems, and innumerable small seasonal coastal rivers draining into the seas. The main source waters for BWA are estuaries, tidal creeks, lagoons and backwaters, and mangrove swamps to a large extent. India has a rich resource of underground saline water that is not commonly used for agriculture and could be used to raise marine shrimp. Water quality is critical for survival, health, and growth of cultured organisms, especially in semi-intensive and intensive culture systems and for the production of good quality seeds in the hatchery. The physical, chemical, and biological properties of water should be kept within the safe levels to maintain good water quality.

### N Flows in Coastal Water Ecosystems

Atmospheric deposition of nitrogen (N) onto land, freshwaters, and estuary, use of N fertilizers in agriculture and sewage wastewater are the predominant N sources into coastal ecosystems. As these sources are in close proximity to the coastal waters compared to open seas, N atmospheric deposition rates are higher into coastal systems. Rivers and groundwater interfering complicates the assessment of the impacts of atmospheric deposition of N into coastal water systems (Jickells, 1998; Paerl et al., 2002; Jickells, 2006). The coastal water ecosystems such as estuaries and lagoons that act as a link between land and sea receive substantial nutrients inputs, particularly N, phosphorus (P), and organic matter. Data from the Central Pollution Control Board (www.cpcb.nic.in) as well as the Central Water Commission (cwc.nic.in) indicate that most of the rivers of India carry dissolved N in the form of $NO_3$-N. Reactive nitrogen (Nr) has a profound impact in the coastal areas, though its residence time is little compared to terrestrial ecosystems.

### N Flows in Brackish-Water Aquaculture Ponds

In shrimp aquaculture ponds, the application of external inputs into the pond such as fertilizer, feed, chemicals, probiotics, etc. along with good water quality not only results in increased shrimp production but also leads to changes in water quality. Nitrogen added into aqua ecosystem from the atmosphere through biological and industrial N fixation comes out of the system through aquatic produce harvest. The accumulated feed on the pond bottom is expected to increase the suspended solids, organic load, and nutrients concentration in pond water and changes the bottom soil condition (Boyd and Fast, 1992). The release of soluble forms of nutrients (N and P) from aquaculture sites has the potential of eutrophication of water bodies (GESAMP, 1991).

Nutrient budgeting studies in shrimp farming (Phillips et al., 1993; Csavas, 1994) indicated that N and P from feed are major contributors to the deterioration of coastal waters quality. However, nutrient enrichment of coastal waters is improbable with the present level of coastal aqua farming in India. But accumulation of such nutrients is possible in semi-enclosed coastal water bodies with poor dilution rate and flushing capacity.

## Role of N in Brackish-Water Ecosystems

### Role as a Nutrient

N may limit the development of autotrophic food webs in ponds on which fish/shrimp yields are dependent. Productivity of BWA system depends on the nature and properties of bottom soils (Hickling, 1971; Mandal and Chattopadhyay, 1979). The main source of nutrient elements is the reserves in the bottom soil, which through the activity of several groups of microorganisms are transferred into the soluble forms (Mandal, 1962a). $NH_4^+$ and $NO_3^-$ forms are primary available N sources to food organisms for aquatic species in ponds. Boyd and Munsiri (1997) predicted the availability of nutrients in aquaculture ponds from bottom soil properties.

### N as Nutrient Through Manures and Fertilizers

Manuring and fertilization increase primary productivity by phytoplankton (Rao et al., 1982; Singh and Sharma, 1999) and ultimately increase the productivity of aquaculture ponds (Bhimachar and Tripathi, 1966; Chen, 1972; Apud et al., 1981). The availability of N for phytoplankton has to be increased through organic and inorganic fertilization programs (Schroeder et al., 1990; Knud-Hansen et al., 1991). Phytoplankton abundance of more than 300,000 to 400,000 cells $mL^{-1}$ is usually considered sufficient.

Nitrogen fertilization is considered more important in brackish-water (BW) ponds than in freshwater (FW) ponds and application of nitrogen fertilizers is a common practice to increase the N:P ratio (Boyd and Daniels, 1994). An N:P ratio of 20:l in ponds favors diatoms, which is more preferable in BW ponds (Das and Jana, 1996). A diatom abundance of 20–30% of the total phytoplankton cells is considered adequate. Urea is more effective in stimulating diatoms compared to ammonium and nitrate fertilizers. Cyanobacteria typically dominate aquatic ecosystems in N-limited conditions to satisfy phytoplankton nutritional needs (Beversdorf et al., 2013). However, many non-diazotrophic cyanobacteria species, such as *M. aeruginosa*, also tend to dominate the phytoplankton community during periods of N-limitation (Paerl et al., 2011; Monchamp et al., 2014). Ayyappan et al. (1991) reported low primary production as the ponds did not receive any external nutrient input. Saraswathy et al. (2012) reported a positive correlation of phytoplankton density with nitrate ($r = 0.87$), phosphate ($r = 0.79$), silicate ($r = 0.97$), and chlorophyll *a* content ($r = 0.98$) in shrimp culture ponds.

### N as Nutrient Through Feed

Feed alone accounted for almost 95% of the total N input in the form of crude protein in shrimp aquaculture. Protein levels in feed range from 10% to 50% depending on the nature of protein source. Native protein seems better hydrolyzed than processed ones (Zwilling et al., 1981). The recommended dietary protein level for *P. vannamei* varies from 20 to 48% (Kureshy and Davis, 2002; Cuzon et al., 2004). This wide variation in the reported values might be due to the protein quality, protein:energy ratios and a lower content of dietary protein could be compensated for by higher consumption. Changes in particular species biotic factors (weight gain, feed, and protein conversion efficiencies) and abiotic factors (temperature and salinity) determine the feed protein requirement (Guillaume, 1997).

Effect of dietary protein level variation on feed conversion and growth of aquatic species was evaluated. Aranyakananda and Lawrence (1993) reported the maximum dietary protein level of 15% and energy:protein ratio of 28.57 kcal $g^{-1}$ protein for juvenile *P. vannamei.* Kureshy and Davis (2002) reported the dietary protein requirements per kg body weight per day (g DP $kg^{-1}$ BWd) of 1.8–3.8 g for juvenile shrimp and 1.5–2.1 g for subadult shrimp. In addition to the above, the recycling of protein by natural productivity should also be considered for optimization of protein based on the stocking densities and farming systems.

## N as Toxicant

Nitrogenous compounds such as ammonia and nitrite were high during extensive shrimp culture, which was due to the shrimp feed protein catabolism and subsequent nitrification. Ammonia is the major end product of protein metabolism in most aquatic animals (Hartenstein, 1970) and is also produced upon decomposition of organic waste by bacteria. Free or unionized form of ammonia ($NH_3$) is toxic, while the ionized form ($NH_4^+$) has little toxicity (Burkhalter and Kaya, 1977; Armstrong, 1978). Concentrations of unionized ammonia above 1 mg/L are potentially lethal for shrimps, while concentrations greater than 0.1 mg/L may adversely affect growth. At pH 9 and 20 ppt salinity, about 25% of total ammonia is unionized. Therefore, total ammonia concentrations above 0.4 mg/L could negatively affect the growth when pH is high (Boyd and Fast, 1992).

Timmons et al. (2002), based on the quantity of feed applied, calculated the amount of ammonia-nitrogen generated per day in aquaculture ponds.

$$\text{PTAN} = \text{F} * \text{PC} * 0.092$$

where PTAN (kg $\text{day}^{-1}$) = total ammonia nitrogen production rate; F (kg $\text{day}^{-1}$) = feed rate; and PC = protein concentration in the feed.

In the above equation, the constant is arrived based on the assumptions of 16% protein N, 80% N assimilated by aquatic species, 80% of assimilated N as excreta, 90% excreted N as TAN, and 10% as urea. It is also assumed that N in unutilized feed and feces is removed quickly from the system by sedimentation and sludge removal. In zero water exchange–based aquaculture production systems with dominant heterotrophic bacteria, it is not possible to remove the solids from the system and hence there is a requirement of a modified formula.

Ammonia ($NH_3$) volatilization from aquatic system can be a source of atmospheric $NH_3$. Only 25 to 30% of the N applied in feed is recovered in the cultured animals at harvest and the remaining is converted to $NH_3$ that enters the water (Boyd, 1990; Boyd and Tucker, 1995) that can be lost by volatilization. The dominant factors controlling volatilization are $NH_3$ concentration in solution, pH, and atmospheric conditions such as temperature, wind speed, and solar radiation (Jayaweera and Mikkelsen, 1991). Fox et al. (1996) reported an average loss of 40 kg $ha^{-1}$ $NH_3$ in 25 day in fish ponds with onetime application of fertilizer urea @ 134 kg $ha^{-1}$ during one growing cycle.

Nitrifying bacteria oxidize ammonia nitrogen and utilize it as energy source for their metabolism and growth resulting in the formation of nitrite-nitrogen ($NO_2$-N), which upon accumulation is toxic to fish. Another byproduct nitrate-nitrogen ($NO_3^- - N$), produced by nitrifying oxidizing bacteria that utilize nitrite as an energy source, is less toxic. In shrimp ponds nitrite is seldom at concentrations great enough to kill them, but growth may be adversely affected by concentrations above 4 or 5 mg L (Boyd and Fast, 1992). Nitrite is toxic to fish and causes brown blood disease, a condition in which the hemi ion in blood is oxidized from the ferrous to ferric state resulting in the formation of methemoglobin which gives blood a brown color and kills fish by hypoxia. The toxicity of nitrite is significantly reduced in seawater due to high concentration of chloride and calcium (Lewis and Morris, 1986).

### N Role in Eutrophication

Gowen and Bradbury (1987) termed the increase in dissolved nutrients concentration from shrimp farms into receiving water body as "hyper-nutrification" and the increase is mainly due to feed-related N and P (Phillips et al., 1993; Csavas, 1994). Increased accumulation of Nr inputs leads to increased growth of algae in tropical coastal systems (Corredor et al., 1999). The consequences of hyper-nutrification, such as increase in phytoplankton production, and turbidity, and decrease in oxygen levels and biodiversity are becoming frequently reported (Nixon, 1995; Cloern, 2001; Rabalais, 2002).

## Nitrogen Transformation/Dynamics and Budgeting

### N Forms, Sources and Emissions in Brackish-Water Aquaculture Ponds

Major portions of N remain locked in the atmosphere as gaseous form of N ($N_2$) and only a fraction of N as Nr goes to plants (plankton in aquaculture ponds), animal, and ultimately to human beings (Velmurugan et al., 2008). N compounds in nature are both nonreactive ($N_2$) and reactive. Photochemically, biologically, and radiatively active N compounds in the earth's atmosphere and biosphere (reduced forms—$NH_3$ and $NH_4^+$; oxidized forms—nitrogen oxides [$NO_x$], $HNO_3$, $N_2O$, and $NO_3$; organic compounds—urea, proteins, nucleic acids, and amines) fall under Nr group (Galloway, 1998; Galloway et al., 2002).

Fractions of soil nitrogen, namely amino acid, peptide, and easily decomposable proteins are known as easily available forms of nitrogen. This form of nitrogen is lower in BW ponds than FW ponds. However, the productivity of BW ponds (Pillay et al., 1962) was directly related to the amount of soil nitrogen in this form. The amount in the $NH_4$ form was in general higher than that in the $NO_3$ form due to the prevailing reducing condition in the soils (Moritimer, 1941). This form of N is mainly adsorbed on the soil exchangeable complex and is beneficial for benthic algae in BWA (Mandal, 1962a), which obtain the nutrients for their growth directly from the pond bottom soil.

N loss can take place through several ways in aquaculture ponds. During the periods of high pond water pH and heavy aeration, ammonia escapes into atmosphere through volatilization from the surface of water column (Reeves, 1972). Biological oxidation of ammonia N to nitrite and nitrate forms does not result in N loss from the ponds. Denitrification of $NO_2$ to $N_2$ and $N_2O$ production are possibly more significant pathways for N loss (Seitzinger et al., 1984). Nishio et al. (1983) reported that $N_2$ production by denitrification in pond sediments quickly moves to the overlying water due to biological and physical turbulences near sediment–water interface. Nitrate can be converted back to ammonia through denitrification, but the quantity of ammonia produced by this process is mostly small compared to other N compounds (Vanderborght et al., 1977). Generally, active sites for both nitrification and denirification initiated from ammonia are limited mainly to shallow depth pond sediment. Loss of N through water seepage is an important pathway in sandy porous soils (Krom et al., 1985).

Nitrous oxide ($N_2O$), an important greenhouse gas has the largest concentration in the atmosphere among the trace nitrogenous species, and contributes to global warming potential. $N_2O$ emission into the atmosphere is conspicuous from natural water bodies and soils, and also from aquaculture—an anthropogenic activity. To increase the aquaculture production globally, use of aquaculture feed has to be increased, which may result in the increased $N_2O$ emission. Schlesinger et al. (2012) reported that N use efficiency reduces $N_2O$ emissions, which indirectly reduces greenhouse gas emissions from N fertilizer use. N use efficiency can also be increased by minimizing the N loss through volatilization and leaching, which indirectly reduces $N_2O$ emissions. Adjustments in feed application rates based on the actual requirements through check tray observations in aquaculture ponds may reduce surpluses in N balance.

## Factors Affecting the Mineralization of Organic N in Brackish-Water Fish/Shrimp Pond Soils

Transformation of nutrients in pond soil depends considerably on the widely fluctuating salinities of water in ponds during different seasons of the year (Johnson and Guenzi, 1963). Muralidhar et al. (1999) reported maximum available nitrogen at 20 ppt and the higher salinities had some adverse effect on the microbial population responsible for the process of mineralization (Mandal, 1962b). A decrease in available nitrogen content was observed with increase in salinity level as compared to initial values. The increase of available nitrogen in the water with a corresponding decrease of nitrogen content in the soil at the initial stages of increasing salinity can be explained due to the replacement of $NH_3$ from the soil complex by sodium.

## Role of Manuring, Fertilization, and Feeding on N Dynamics

Accumulation of the inorganic N forms ($NH_4^+ - N$ and $NO_2^- - N$) in intensive aquaculture pond waters is responsible for the deterioration of water quality. Nutrient concentration ($NO_3$-N) has been observed more with fertilizer treatment which might be because of immediate release of nutrients from fertilizer. A steady increase in nutrient concentration has been recorded with progress of time. Nitrate-nitrogen concentration was significantly more in treatments whenever chemical fertilizer was applied (Muralidhar et al., 2001). Average nitrate *N* values of 0.008–0.009 ppm were observed with chemical fertilizers compared to 0.001 ppm with manure treatments. Nitrogen in manure occurs largely in organic forms and the C:N ratio in FYM is 15:30. Primary production and plankton numbers were significantly high in manure and fertilizer combination compared to other treatments. Ammonia *N* values did not vary among the manure and fertilizer treatments, and nitrite-N concentration was more in poultry manure and fertilizer tanks compared to cattle manure and manure plus fertilizer treatments.

## N Budgeting

Nitrogen budgeting is a prerequisite to achieve nitrogen reduction in waste and increase nitrogen recovery in the aquatic animal. N budgeting is usually expressed as a mass-balance approach, where total N inputs plus generation is equal to total N outputs plus consumption. N input into the pond (quantity and quality), aquatic species under culture, water-exchange frequency, and other environmental parameters determine the magnitude and rates of N fluxes.

Many studies on N budget revealed that feed nitrogen contributes 90–98% of the total nitrogen input. Other minor contributors are intake water (1.54%–3.04%) and post-larvae stages. Nitrogen recovery in shrimp was high (34.2–43.6%) under zero water–exchange *Penaeus monodon* culture systems (Fig. 19.1) (Xia et al., 2004; Saraswathy et al., 2013) compared to water-exchange system (22–31%). Under zero water–exchange system, nitrogen is accumulated in the sediment whereas under water-exchange system, before fixing in the soil, nitrogen is removed through discharge water. The nitrogen in discharge water varied between 11.2% and 14% and the lowest was recorded under zero water–exchange system. Unaccounted loss (volatilization of ammonia and denitrification) varied between 15% and 57% and this variation is due to stocking density, aeration intensity, and culture practices. Under water-exchange system, the discharge water contained 14–35% of N and only 3% of N was unaccounted and assumed to be lost through denitrification and voltalization into atmosphere (Briggs and Funge-smith, 1994; Jackson et al., 2003; Thakur and Lin, 2003). Generally, unaccounted loss is high under zero water–exchange and aided by aeration (Reeves, 1972) under high stocking density. Accumulation of organic sludge may sequester some of the unaccounted nitrogen in the pond sediment (Briggs and Funge-Smith, 1994).

In ponds where aquatic species are fed with commercial feeds, N fixation was minor but an important contributor to the N budget. In tropical fish ponds, N fixation in mg N

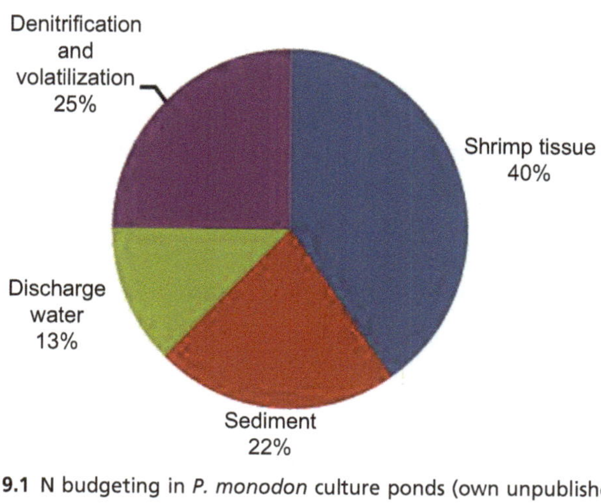


**FIGURE 19.1** N budgeting in *P. monodon* culture ponds (own unpublished data).

$m^{-2}$ $day^{-1}$ was 6–23 and 21–57 during dry and rainy seasons, respectively (Lin et al., 1988). $NH_4$-N concentration in the pond water and phytoplankton species composition largely determine the N fixation and ammonia concentration is inversely related to inhibition of N fixation (Lin et al., 1988).

As nitrification rate measurements are not available in aquaculture ponds, 15–25 mg N $m^{-2}$ $day^{-1}$ of nitrification rate in estuarine sediments (Henriksen and Kemp, 1988) is likely to be representative for pond sediment. Assuming the sediment oxygen demand of 5–10% for nitrification, about 25–50 mg N $m^{-2}$ $day^{-1}$ is oxidized, which is equivalent to 5–10% of N input per day. In order to get energy from the reduced forms of N, oxygen is required for nitrifying bacteria. The oxygen half-saturation concentration ($K$) values were 0.3–0.9 mg $L^{-1}$ and are directly proportional to temperature (Painter, 1970).

# Nitrogen Management

## Better Management Practices to Maintain Nitrogen Metabolites Levels

The best method for preventing nitrogen metabolites accumulation in aquaculture ponds lies in the selection of good site with availability of high quality water, following optimum stocking densities and aiming at modest levels of shrimp production. Providing aeration, liming, and pond bottom treatments to improve the decomposition of organic matter helps in the maintenance of optimum soil and water parameters in aquaculture ponds. Aquaculture sustainability depends on the adoption of better management practices, which in turn reduce the nutrients concentration and pollution potential of pond discharge water.

In case of high levels of ammonia detected in pond water, feeding should be reduced or stopped immediately followed by flushing with the new water, aerating the water, and reducing the stocking density in the pond. The other pathways to reduce the metabolites load are:

- Use of photoautotrophic algae–based systems (green water systems)
- Autotrophic bacterial conversion: In intensive recirculating aquaculture production systems, bioreactors are routinely used that rely on the nitrification of ammonia nitrogen to nitrate-nitrogen by ammonia oxidizing bacteria (AOB) and nitrite oxidizing bacteria (NOB) (Timmons et al., 2002).
- Heterotrophic bacterial conversion: Controlling inorganic nitrogen by manipulating carbon/nitrogen ratio is a potential method for aquaculture systems. This approach offers a practical and inexpensive means to reduce the accumulation of inorganic nitrogen in culture ponds. In case of increased ammonium concentration, addition of cheap sources of carbonaceous substrate such as cassava meal and flour can control the N metabolites concentration. Bacteria and other microorganisms use carbohydrates (sugar, starch, and cellulose) as a food to generate energy and produce proteins and new cells (Avnimelech, 1999). The resulting heterotrophic

bacterial production (single cell protein) may be utilized as a food source for fish and shrimp (Beveridge et al., 1989; Rahmathulla and Beveridge, 1993; Burford et al., 2004), thus lowering the demand for supplemental feed protein (Avnimelech, 1999).

- Periphyton-based aquaculture: The whole complex of aquatic biota attached to the substratum including the associated detritus and microorganisms is termed as Periphyton. Thus, the periphyton community comprises bacteria, fungi, protozoa, phytoplankton, zooplankton, benthic organisms, and a range of other invertebrates and their larvae. Any material with high surface area such as tree branches, bamboo, higher aquatic plants, PVC pipes can be used for the production of periphyton.
- Use of chemicals: Fast ammonia removal by zeolites is a well-known process in freshwater aquaculture (Sand and Mumpton, 1977) and shrimp ponds in South East Asia (Boyd, 1990). Joseph et al. (2002b) reported that zeolite reduced the ammonia concentration only to a maximum of 10.5% by 5th day. Poly allyl amine hydrochloride (PAA·HCl) polymer hydrogels efficiently remove nitrate ($NO_3$), nitrite ($NO_2$), and orthophosphate ($PO_4$) anions from the aquaculture wastewater.
- The feed protein levels are to be adjusted to avoid the inorganic N forms load in the pond water. In intensive ponds with high water exchange and aeration, this method of decreasing the protein levels without compromising the production was successful. Avnimelech (1998) reported that intensive culture of aquatic species in these ponds is like biotechnological reactors.

After the harvest, pond sludge, if not removed regularly during culture period, comprises about 15–22% of the total N inputs, and about 42% of the total N inputs may be lost through denitrification. Re-aeration of pond sludge generally minimizes the denitrification. Regular removal of pond sludge during culture period significantly reduces the concentration of inorganic N forms in the pond water column. The disposal of harvested pond sludge on high lands not only reduces the impact of discharge water on the receiving water bodies, but also improves soil fertility and in turn increases the yield of terrestrial crops.

## Characteristics of Brackish-Water Aquafarms Discharge Water with Reference to Nitrogen

N in discharge waters from aquaculture ponds may deteriorate the quality of receiving waters. The intrinsic N use efficiency determines the quantity of N released into the environment. The N discharge to the environment from intensive raceways and cages is more than the fishponds.

Increased levels of $NO_3$-N, $NH_3$-N, and $NO_2$-N were reported by Joseph et al. (2002a) in pond and discharge waters as compared to intake water. However, discharge water quality during culture as well as harvest time is within the permitted limits of Ministry of Environment and Forests (MOEF, 1993) and guidelines of Ministry of Agriculture/

Department of Agriculture and Cooperation (MOA/DAC, 1995). Comparison of sea and creek (Kandaleru) water–based shrimp farms indicated high organic and nutrient loading in creek water–based ponds, and the likely impact of discharge water from these ponds is comparatively higher than seawater-based farms (Joseph et al., 2002a). As the farmers practiced extensive system of shrimp culture, discharge water from these ponds could not pollute the natural creek, though same source water serves as source and sink.

Adverse changes in environmental parameters of receiving water body will occur only when the pollution load in pond discharge water exceeds the assimilative/flushing capacity of water body. Site location, tidal amplitude variation, flushing/dilution rates, volume of water body during high and low tides, creek morphology, type of culture system, and the management practices determine the environmental impact of aquaculture farm's discharge water. There is a likely risk of eutrophication in the water body when semi-intensive shrimp farming is practiced (Fig. 19.2) and Joseph et al. (2003) suggested the guidelines to prevent this problem.

In most of the shrimp farming countries, use of settlement ponds is recommended as a treatment option to remove the suspended solids and particulate material effectively from discharge water, but it is not effective for nutrients removal (Jackson et al., 2003).

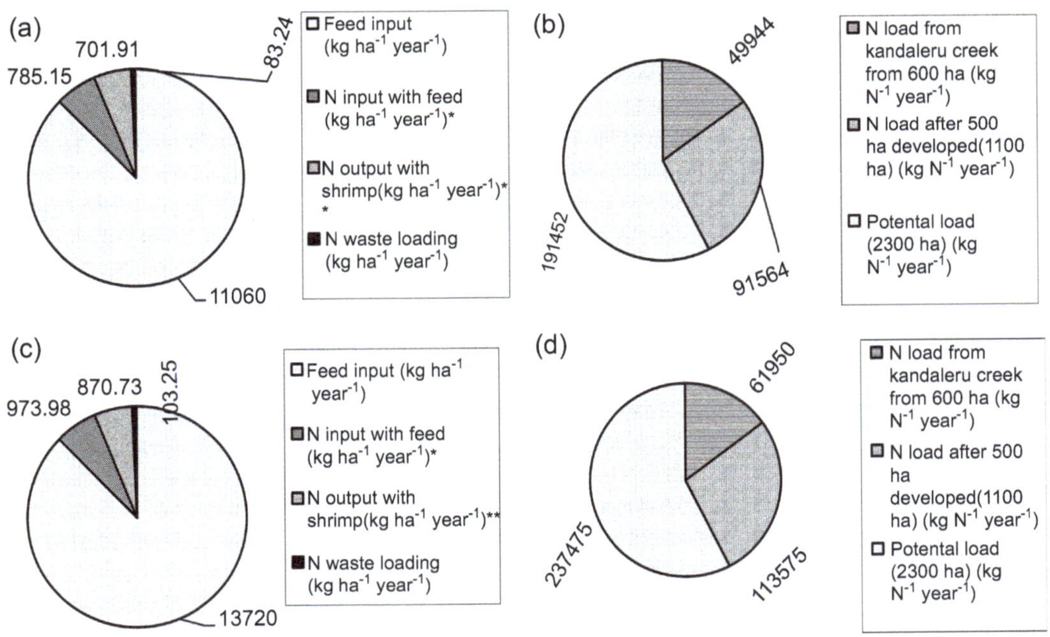


**FIGURE 19.2** Nitrogen budget (A, B low range and C, D high range) for shrimp production (7900–9800 Kg ha Year$^{-1}$) in the Kandaleru creek, Nellore District of Andhra Pradesh. *, Assumes feed quality of 7.099% Nitrogen; **, N content of shrimp 8.885%; FCR 1.4:1. *Joseph, K.O., Gupta, B.P., Thirranavakkarrassu, A.R., Muralidhar, M., Krishnani, K.K., 2003. Effect of shrimp farming on nitrogen levels in the waters of Kandaleru creek, Andhra Pradesh. Indian Journal of Fisheries 50, 291–296.*

In shrimp hatcheries, though there was an increase in N and its metabolites concentration due to biological activity, practically there was no difference in the quality of water at intake and outfall area due to the mixing of discharge water with large volume of seawater (Muralidhar and Gupta, 2007). Hatchery discharges depend on the size of the facility and its water utilization efficiency. Treatment methods for the water discharge of the hatcheries mostly include sedimentation/oxidation/percolation of the ponds before the water is released back to sea/backwaters.

## Research Needs/Gaps

The synthesis of existing knowledge on N dynamics in aquaculture ponds and coastal ecosystems indicated a lack of understanding in many aspects. Biome-level integrated studies on N cycling inputs, collection of reliable data on the large scale, and processes of Nr uptake by plankton and fish/shrimp are needed to fulfill this gap. Biotic and abiotic factors governing the availability of N in aquaculture ponds need to be studied. Aquatic species that are assimilating the N need to be effectively identified to increase the feed N use efficiency and also for the modeling studies on N dynamics for particular species. There are gaps in our knowledge of atmospheric N effects on coastal ecosystems. The spatial and temporal variations in emission of the dominant Nr trace gases from aquaculture ponds and coastal ecosystems of India have to be documented. Currently, there are no sector-wise and regional N budgets for the country. Quantitative estimations of various Nr emissions are required to clearly understand the nitrogen cycle at country level.

N exchange across the sediment–water interphase in aquaculture ponds plays an impotant role in N cycling of coastal waters because of lesser volume of water compared to higher surface area of the sediment. New pathways and transformations in N cycle are to be discovered. At the broader scale, more attention is required on specific functional groups of organisms that play key roles in overall N biogeochemistry coupling them with ecosystem modeling. Research over the next decade will likely concentrate more intensively on such groups particularly with respect to their role in the response of the aquaculture ponds to climate change.

The following research gaps requires to be addressed with the latest modern tools such as molecular techniques, remote sensing and GIS, $^{15}N$ tracer techniques, etc.

- Accurate measurement of inputs of Nr to the coastal ecosystems.
- Dependable data on larger scale and pathways of Nr uptake from aquaculture ponds and coastal ecosystems.
- Aquatic organisms effectively assimilating N are to be identified.
- Complete accounting of the N cycle in aquaculture ponds and coastal ecosystems to understand the overall changes in N cycling and productivity.
- Quantification of spatial and temporal variation in the emission of Nr gases from aquaculture ponds and coastal ecosystems.

# Conclusion

Proper management of feed N will remain at the forefront of issues to improve the global Nr balance over both short and long terms. Collaborative multidisciplinary effort can help the aquaculture sector's contribution to reduce the global Nr load. Research on improvements in nitrogen utilization and simultaneous reduction of nitrogen loss is highly desirable for sustainable aquaculture development. In this context, it is essential to identify the optimum protein requirements of a species that will assure higher growth and lower nitrogen waste to the environment. Quantitative estimations of Nr emissions not only give the database and the related impacts, but also provide strategies to formulate and evaluate mitigation options. In spite of impressive and rapid advances in our understanding of many important aspects of the marine nitrogen cycle, there are a number of fundamental issues that remain unresolved. In order to understand the ecosystem response to changes in N cycle, investigations have to be carried out for the sustainable management of the vulnerable coastal ecosystems for future generations. Interactions and feedbacks between atmospheric, terrestrial, and aquatic systems are important aspects to be studied. To decrease N export to coastal systems, an array of innovative methodologies is required. Modeling and synthesis of data at a greater magnitude along with policy and economic studies in the country will be required to minimize environmental deterioration of coastal ecosystems and to meet the food and energy demands of society through sustainable aquaculture simultaneously.

# References

Apud, F.D., Gonzales, K., Deatras, N., 1981. Survival, growth and production of Penaeus monodon fabricus at different stocking densities in earthen ponds with flow – through system and supplemental feeding. Fisheries Research Journal of the Philippiness 6, 1–9.

Aranyakananda, P., Lawrence, A.L., 1993. Dietary protein and energy requirements of the white-legged shrimp, Penaeusvannamei and the optimal P/E ratio. In: Carrillo, M., Dahle, L., Morales, J., Sorgeloos, P., Svennevig, N., Wyban, J. (Eds.), Aquat. Resour. Res. Inst., ChulalongkornUniv, Bangkok Thailand WAS'93 Intl. Conf., Torremolinos (Spain), 26–28 May 1993 from Discovery to Commercialization. Spec. Publ.-Eur. Aquac. Soc, vol. 19, p. 21. Oostende (Belgium).

Armstrong, D.A., 1978. Interaction of ionized and un-ionized ammonia on short-term survival and growth of prawn larvae, macrobrachiumrosenbergh. The Biological Bulletin 154 (1), 15–31.

Avnimelech, Y., 1998. Minimal discharge from intensive fish ponds. World Aquaculture 29, 32–37.

Avnimelech, Y., 1999. C/N ratio as a control element in aquaculture systems. Aquaculture 176, 227–235.

Ayyappan, S., Purushothaman, C.S., Saha, P.K., Pani, K.C., 1991. Heterotrophic bacterial populations and primary production in fish ponds under different management practices. Journal of Aquaculture 1, 11–29.

Beveridge, M.C.M., Begum, M., Frerichs, G.N., Millar, S., 1989. The ingestion of bacteria in suspension by the tilapia (*Oreochromis niloticus*). Aquaculture 81, 373–378.

Beversdorf, L.J., Miller, T.R., McMahon, K.D., 2013. The role of nitrogen fixation in cyanobacterial bloom toxicity in a temperate, eutrophic lake. PLoS One 8 (2), e56103. http://dx.doi.org/10.1371/journal.pone.0056103.

Bhimachar, B.S., Tripathi, S.D., 1966. A review of culture fisheries activities in India. In: F.A.O. World Symp. On Warm Water Pond Fish Culture, Rome.

Boyd, C.E., Tucker, C.S., 1995. Sustainability of channel catfish culture. World Aquaculture 26, 45–53.

Boyd, C.E., Munsiri, P., 1997. Water quality in laboratory soil water microcosms with soils from different areas of Thailand. Journal of the World Aquaculture Society 28 (2), 165–170.

Boyd, C.E., Fast, A.W., 1992. Pond monitoring and management. In: Fast, A.W., Lester, L.J. (Eds.), Marine Shrimp Culture Principals and Practices. Elsevier, The Netherlands, pp. 497–513, 862 pp.

Boyd, C.E., 1990. Water Quality in Ponds for Aquaculture. Ala. Agric. Exp. Sta., Auburn University, AL, USA, 482 pp.

Boyd, C.E., Daniels, H.V., 1994. Liming and fertilization of brackish water shrimp ponds. Journal of Applied Aquaculture 2 (3–4), 221–234.

Briggs, M.R.P., Funge-Smith, S.J., 1994. A nutrient budget of some intensive marine ponds in Thailand. Aquaculture and Fisheries Management 25, 789–811.

Burford, M.A., Thompson, P.J., Mc Intosh, P.R., Bauman, R.H., Pearson, D.C., 2004. The contribution of flocculated material to shrimp, *Litopenaeus vannamei* nutrition in a high-intensity, zero-exchange system. Aquaculture 232, 525–537.

Burkhalter, D.E., Kaya, C.M., 1977. Effects of prolonged exposure to ammonia on fertilized eggs and sac fry of rainbow trout (Salmo gairdneri). Transactions of the American Fisheries Society 106 (5), 470–475.

Chen, T.P., 1972. Fertilisation and feeding in coastal fish farms of Taiwan. In: Coastal Aquaculture in the Indo-Pacific Region. Fishing News (Books) Ltd, London.

Cloern, J.E., 2001. Our evolving conceptual model of the coastal eutrophication problem. Marine Ecology Progress Series 210, 223–253.

Corredor, J.E., Morell, J., Bauza, J., 1999. Atmospheric nitrous oxide fluxes from mangrove sediments. Marine Pollution Bulletin 38, 473–478.

Csavas, I., 1994. Important factors in the success of shrimp farming. World Aquaculture 25 (1), 34–56.

Cuzon, G., Lawrence, A., Gaxiola, G., Rosas, C., Guillaume, J., 2004. Nutrition of Litopenaeus vannamei reared in tanks or in ponds. Aquaculture 235 (2004), 513–551.

Das, S.K., Jana, B.B., 1996. Pond fertilisation through inorganic sources: an overview. Indian Journal of Fisheries 43 (2), 137–155.

Fox, R.H., Piekielek, W.P., Macneal, K.E., 1996. Estimating ammonia volatilization losses from urea fertilizers using simplified micro-metrological sampler. Soil Science of American Journal 60, 596–601.

Galloway, J.N., 1998. The global nitrogen cycle: changes and consequences. Environmental Pollution 102 (1), 15–24.

Galloway, J.N., Cowling, E.B., Seizinger, S.P., Socolow, R.H., 2002. Reactive nitrogen: too much of a good thing? Ambio 31, 60–63.

GESAMP, 1991. Joint group of Experts on the scientific aspects of marine pollution. Reducing Environmental Impacts of Aquaculture 47, 35.

Gowen, R.J., Bradbury, N.B., 1987. The ecological impact of salmon farming in coastal waters: a review. Oceanography Marine Biology Annuals Review 25, 563–575.

Guillaume, J., 1997. Protein and amino acids. In: D'Abramo, L.R., Conklin, D.E., Akiyama, D.M. (Eds.), Crustacean Nutrition. World Aquaculture Society, Baton Rouge, LA, pp. 26–50.

Hartenstein, R., 1970. Nitrogen metabolism in non-insect arthropods. Comparative Biochemistry of Nitrogen Metabolism 1, 299–385.

Henriksen, K., Kemp, W.M., 1988. Nitrification in estuarine and coastal marine sediments. In: Blackburn, T.H., Sorensen, J. (Eds.), Nitrogen Cycling in Coastal Marine Environments. Wiley, London, pp. 207–249.

Hickling, C.F., 1971. Fish Culture. Faber and Faber, London.

Jackson, C., et al., 2003. Nitrogen budget and effluent nitrogen components at an intensive shrimp farm. Aquaculture 218 (1), 397–411.

Jayaweera, G.R., Mikkelsen, D.S., 1991. Assesment of ammonia volatilization from flooded soil system. Advances in Agronomy 45, 303–356.

Jickells, T.D., 1998. Nutrient biogeochemistry of the coastal zone. Science 281 (5374), 217–222.

Jickells, T., 2006. The role of air-sea exchange in the marine nitrogen cycle. Biogeosciences Discussions 3 (1), 183–210.

Johnson, D.D., Guenzi, W.D., 1963. Influence of salts on ammonium oxidation and carbon dioxide from soils. Proceedings of Soil Science Society of America 27, 663–666.

Joseph, K.O., Muralidhar, M., Gupta, B.P., Krishnani, K.K., 2002a. Characterisation of intake, pond and outlet water of creek and seawater based shrimp farms in Nellore district of Andhra Pradesh. Aquaculture 3 (1), 51–59.

Joseph, K.O., Muralidhar, M., Gupta, B.P., Krishnani, K.K., 2002b. Effect of commercially available bio-augmentors on the removal of ammonia and sulphide from brackishwater. Indian Journal of Fisheries 49 (3), 229–233.

Joseph, K.O., Gupta, B.P., Thirranavakkarrassu, A.R., Muralidhar, M., Krishnani, K.K., 2003. Effect of shrimp farming on nitrogen levels in the waters of Kandaleru creek, Andhra Pradesh. Indian Journal of Fisheries 50, 291–296.

Knud-Hansen, C.F., Batterson, T.R., McNabb, C.D., Harahat, I.S., Sumantadinata, K., Eidman, H.M., 1991. Nitrogen input, primary productivity and fish yield in fertilized freshwater ponds in Indonesia. Aquaculture 94, 49–63.

Krom, M.D., Porter, C., Gordin, H., 1985. Nutrient budget of a marine fish pond in Eilat, Israel. Aquaculture 51 (1), 65–80.

Kumaran, M., et al., 2003. Shrimp farming practices and its socio-economic consequences in East Godavari district, Andhra Pradesh, India-A case study. Aquaculture Asia 8 (3), 48–52.

Kureshy, N., Davis, D.A., 2002. Protein requirement for maintenance and maximum weight gain for the Pacific white shrimp, Litopenaeusvannamei. Aquaculture 204 (1–2), 125–143.

Lewis Jr., W.M., Morris, D.P., 1986. Toxicity of nitrite to fish: a review. Transactions of the American Fisheries Society 115 (2), 183–195.

Lin, C.K., Apinhapath, C., Tansakul, V., 1988. Biological nitrogen fixation as a source of nitrogen input in fish ponds. In: Pullin, R.S.V., Bhukaswan, T., Tonguthai, K., Maclean, J.L. (Eds.), The Second International Symposium on Tilapia in Aquaculture. ICLARM Conference Proceedings 15. International Center for Living Aquatic Resources Management, Manila, Philippines, pp. 53–58.

Mandal, L.N., 1962a. Nitrogenous feirtilizers for brackishwater fish ponds – ammonium or nitrate form. Indian Journal of Fisheries 9 (1.A), 123–134.

Mandal, L.N., 1962b. Effect of salinity on the transformation of nitrogen in the brackishwater fish farm soils. Journal of Indian Society Soil Sciences 10 (4), 255–261.

Mandal, L.N., Chattopadhyay, G.N., 1979. Effect of organic manuring on the availability of phosphorus in brackishwater fish pond soils under different levels of water salinity. In: Symp. Inland Aquaculture, Central Inland Fisheries Research Institute, Barrackpore, India.

MOA/DAC, September 11, 1995. Notification-guidelines for Sustainable Development and Management of Brackish Water Aquaculture. Ministry of Agriculture, Department of Agriculture and Cooperation, Govt. of India.

MOEF, May 19, 1993. Notification – General Standards for Discharge of Environmental Pollutants, Ministry of Environment and Forests. Part a – Effluents.

Monchamp, M.-E., Pick, F.R., Beisner, B.E., Maranger, R., 2014. Nitrogen forms Influence Microcystin concentration and composition via changes in cyanobacterial community Structure. PLoS One 9 (1), e85573. http://dx.doi.org/10.1371/journal.pone.0085573.

Moritimer, C.H., 1941. The exchange of dissolved substances between mud and water in takes. Journal of Ecology 30, 147–201.

Muralidhar, M., Gupta, B.P., Joseph, K.O., Krishnani, K.K., 1999. Effect of salinity level on the release of native soil nutrients from brackishwater soil. Journal of Indian Society Coast Agricultural Research 17 (1&2), 182–187.

Muralidhar, M., Gupta, B.P., Joseph, K.O., Krishnani, K.K., Sudeesh, P.S., 2001. Effect of manures and Fertilizers singly and in combination on shrimp production in environment,. In: Menon, N.G., Pillai, P.P. (Eds.), Perspectives in Mariculture. The Marine Biological Association of India, Cochin, pp. 193–206.

Muralidhar, M., Gupta, B.P., 2007. Quality of water discharged from shrimp hatcheries and its impact on surrounding coastal environment. Indian Journal of Fisheries 54 (2), 189–194.

Nishio, T., Koike, I., Hattori, A., 1983. Estimates of denitrification and nitrification in coastal and estuarine sediments. Applied and Environmental Microbiology 45, 444–450.

Nixon, S.W., 1995. Coastal marine eutrophication: a definition, social causes, and future concerns. Ophelia 41 (1), 199–219.

Paerl, H.W., Xu, H., McCarthy, M.J., Zhu, G., Qin, B., Li, Y., 2011. Controlling harmful cyanobacterial blooms in a hyper-eutrophic lake (lake taihu, China): the need for a dual nutrient (N & P) management strategy. Water Research 45 (5), 1973–1983.

Paerl, H.W., Dennis, R.L., Whitall, D.R., 2002. Atmospheric deposition of nitrogen: implications for nutrient over-enrichment of coastal waters. Estuaries 25 (4), 677–693.

Painter, H.A., 1970. A review of literature on inorganic nitrogen metabolism in microorganisms. Water Research 4 (6), 393–450.

Phillips, M.J., Lin, C.K., Beverigde, M.C.M., 1993. Shrimp culture and environment: lessons from the most rapidly expanding warm water aquaculture sector. In: Pullin, R.S.V., Rosenthal, H., Maclean, J.L. (Eds.), Environment and Aquaculture in Developing Countries, ICLARM Conf. Proc., pp. 171–197.

Pillay, T.V.R., Vijayraohavan, M.R., Thakurta, S.C., 1962. An ecological study of benthic flora in brackish waterbheries. Proceedings of National Institute of Science India 28B (5), 416–431.

Rabalais, N.N., 2002. Nitrogen in aquatic ecosystems. AMBIO 31 (2), 102–112.

Rahmathulla, S.M., Beveridge, M.C.M., 1993. Ingestion of bacteria in suspension by Indian major carps (*Catla catla, Labeo rohita*) and Chinese carps (*Hypophthalmichthys molotrix, Aristichthys nobilis*). Hydrobiology 264, 79–84.

Rao, P.B., Sharmaand, A.P., Singh, J.S., 1982. Limnology and phytoplankton production in high attitude lake. International Journal of Ecological and Environmental Science 8, 39–51.

Reeves, J.G., 1972. Nitrogen removal: a literature review. Journal of Water Pollution Control Federation 44, 1895–1908.

Sand, L.B., Mumpton, F.A., 1977. Natural Zeolites. Perganon Press, New York, USA, 463 pp.

Saraswathy, R., Muralidhar, M., Kailasam, M., Ravichandran, P., Gupta, B.P., Krishnani, K.K., Ponniah, A. G., Sundaray, J.K., Panigrahi, A., Nagavel, A., 2013. Effect of stocking density on soil, water quality and nitrogen budget in *Penaeus monodon* (Fabricius, 1798) culture under zero water exchange system. Aquaculture Research 44, 1578–1588.

Saraswathy, R., Muralidhar, M., Ravichandran, P., Ponniah, A.G., Panigrahi, A., Kailasam, M., Nagavel, A., 2012. Studies on the effect of nutrient level on phytoplankton population in zero water exchange shrimp culture farms. Indian Journal of Fisheries 59 (2), 115–120.

Schlesinger, W.H., Abrahams, A.D.A., Parsons, J., Wainwright, J.C., 2012. Nutrient losses in runoff from grassland and shrubland habitats in southern New Mexico: I. Rainfall simulation experiments. Biogeochemistry 45, 21–34.

Schroeder, G.L., et al., 1990. The dominance of algal-based food webs in fish ponds receiving chemical fertilizers plus organic manures. Aquaculture 86 (2), 219–229.

Seitzinger, S.P., Nixon, S.W., Pilson, M.E.Q., 1984. Denitrification and nitrous oxide production in a coastal marine ecosystem. Limnology and Oceanography 29 (1), 73–83.

Singh, V.K., Sharma, A.P., 1999. Hydrological characteristics and primary production in fish ponds manured with different organic manures. Indian Journal of Fisheries 46 (1), 79–85.

Surendran, V., Madhusudhan Reddy, K., Subba Rao, V., February 1991. Semi-intensive shrimp farming. TASPARC's experience at Nellore. Journal of Fisheries Chimes 23–29.

Thakur, D.P., Lin, C.K., 2003. Water quality and nutrient budget in closed shrimp *(Penaeus monodon)* culture systems. Aquaculture Engineering 27, 159–176.

Timmons, M.B., Ebeling, J.M., Wheaton, F.W., Summerfelt, S.T., Vinci, B.J., 2002. Recirculating Aquaculture Systems, second ed. 769. Cayuga Aqua Ventures, New York.

Vanderborght, J.P., Wollast, R., Billen, G., 1977. Kinetic models of diagenesis in disturbed sediments. Part 2. Nitrogen diagenesis. Limnol. Oceanogr. 22, 794–803.

Velmurugan, A., Dadhwal, V.K., Abrol, Y.P., 2008. Regional nitrogen cycle: an Indian perspective. Current Science 94, 1455–1468.

Xia, L.Z., Yang, L.Z., Yan, M.C., 2004. Nitrogen and phosphorus cycling in shrimp ponds and the measures for sustainable management. Environmental Geochemistry and Health 26, 245–251.

Zwilling, R., et al., 1981. Low molecular mass protease: evidence for a new family of proteolytic enzymes. FEBS Letters 127 (1), 75–78.

# 20

# Reactive Nitrogen in Coastal and Marine Waters of India and Its Relationship With Marine Aquaculture

D. Prema, V.V. Singh, R. Jeyabaskaran, V. Kripa

*CENTRAL MARINE FISHERIES RESEARCH INSTITUTE, KOCHI, INDIA*

**CHAPTER OUTLINE**



## Introduction

India is bordered in the south, south-west, and south-east with Indian Ocean, Arabian Sea (AS), and the Bay of Bengal (BOB), respectively. Indian coast is 7517 km long comprising 5423 km in the peninsular India and 2094 km in Andaman and Nicobar, and Lakshadweep Islands. The Indian exclusive economic zone (EEZ) is spread in 2.02 million sq km (0.86 million sq km in west coast, 0.56 million sq km in east coast and 0.6 million sq km in Andaman and Nicobar Islands). The Indian marine environment

The Indian Nitrogen Assessment. http://dx.doi.org/10.1016/B978-0-12-811836-8.00020-3

consisting of adjoining coastal areas and EEZ directly sustains useful habitats and supports the livelihood of 3.9 million fishers. Nearly 25% of the country's population resides in these areas and about 340 communities are primarily occupied in marine and coastal fisheries (MoEF, 2009; SACEP, 2014).

Nitrogen (N) exists in various chemical forms, produced by marine biota through several chemical transformations during their growth and metabolism in the marine environment. Nitrogen as $N_2$ is generally unavailable in marine conditions and thereby, the equilibrium of the processes of $N_2$ fixation (conversion of atmospheric $N_2$ to organic nitrogen) and denitrification (conversion of nitrate to $N_2$) decides the bioavailable nitrogen supply and productivity (Gruber, 2008).

Potential for transfer of reactive nitrogen (Nr) to the coastal and marine waters is relevant, as it contributes to *new* production (primary production supported by external supply of nutrients) and forms the basis of energy transfer toward the higher pelagic trophic levels. Nitrogen assessment in the surface waters, as concentration of nitrite and nitrate was carried out in the Indian EEZ and the adjoining areas during the years 1998–2007 (Nair, 2010). The study revealed that the annual concentrations (μM) of $NO_2$ ranged from 0 to 0.4, 0 to 0.6, and 0 to 0.7 in the AS, BOB, and Andaman Sea, respectively. The corresponding values for $NO_3$ (μM) were 0–2.5, 0–3 and 0–3.5, respectively. The average marine fisheries production of the country during the period 2010–2015 has been estimated at 3.36 million tons consisting of approximately 667 different fish species inhabiting the neritic zone that depend on the productivity of marine ecosystem and in turn is influenced by the Nr available in the system.

## Nitrogen Dynamics in Coastal and Marine Environment

Open ocean N-cycle seems to be unharmed, in comparison with that of coastal ecosystems although recent observations have indicated increasing anthropogenic influences (Purvaja et al., 2008a). Most marine organisms make use of nitrogen as nitrate $(NO_3^-)$ and ammonium $(NH_4^+)$. Combined nitrogen occurs in oxidation states ranging from −3 to +5. Its speciation in ocean is dependent on the redox potential of seawater influenced mostly by the ambient oxygen ($O_2$) levels. The bulk of oceanic water column is well oxygenated making $NO_3^-$ the predominant species of N (Naqvi et al., 1998). In productive areas of the ocean surface, nitrification in the photic zone provides recycled nitrate that is used in primary production (Newell et al., 2011).

### Export of Reactive Nitrogen Toward the Sea

Nr accumulation possibility is less in coastal ecosystems, being subject to rapid turnover. Potential for Nr transfer to continental shelf regions is large. However, the transport is only partial by means of high denitrification rates and the Nr transferred is generally denitrified to $N_2$ prior to its transport to the open ocean (Purvaja et al., 2008a). Transport of N to the ocean mainly occurs through river runoff, atmospheric deposition, and

N-fixation, while its loss is principally through conversion of the combined nitrogen to $N_2$. Oceanic currents supply enough $O_2$ at all depths to prevent the development of anoxic conditions in most oceanic areas (Naqvi and Jayakumar, 2000).

Most of the dissolved inorganic nitrogen (DIN) brought from rivers to the northern part of Indian Ocean (~81% in the case of AS and 96% in the case of BOB) is not transported to the coastal ocean and is consumed on the course of the rivers or in the estuaries; however, a considerable contribution (~13%, geometric mean) of this nitrogen to coastal new production is observed (Singh and Ramesh, 2011). Export of DIN by Indian rivers to coastal waters was estimated to be $0.22 \pm 0.05$ Tg year$^{-1}$ (Krishna et al., 2016), which is almost similar to BOB ($0.12 \pm 0.03$ Tg year$^{-1}$) and AS ($0.10 \pm 0.02$ Tg year$^{-1}$).

In the south-west coast of India, in the Cochin estuary, export fluxes of nitrate were high during spring tide, whereas for ammonium during neap tide, because nitrate is mainly a riverine source and ammonium has other external sources in the estuary. During the spring tide, the Cochin estuary received $516 \times 10^3$ kg day$^{-1}$ of DIN and exported $592 \times 10^3$ kg day$^{-1}$ of DIN into the sea. Similarly, the neap tide input and export fluxes of the estuary varied from $383 \times 10^3$ to $468 \times 10^3$ kg day$^{-1}$ DIN. Between successive spring and neap tides, there was an increase in the net fluxes of ammonium ($25.8 \times 10^3$ to $43.1 \times 10^3$ kg day$^{-1}$) and DIN ($76.4 \times 10^3$ to $85.8 \times 10^3$ kg day$^{-1}$) whereas nitrate flux showed a reduction ($41.7 \times 10^3$–$35.8 \times 10^3$ kg day$^{-1}$). Transports of nutrients were directed toward the sea during monsoon and post-monsoon months, and were directed toward river during pre-monsoon in the Chaliyar estuary in the south-west coast of India (Lallu et al., 2014; Xavier et al., 2005). It is mainly the benthic $NH_4^+$ efflux that sustains high estuarine productivity in the $NO_3^-$ depleted dry season (Pratihary et al., 2009). The DIN transport to coastal areas can also occur through aerosol deposition (Srinivas et al., 2011) and eddies (Prasannakumar et al., 2007).

## Nitrogen Dynamics in the Regional Seas

The northern Indian Ocean is separated into BOB and AS, by the Indian landmass. During the oligotrophic inter-monsoon period, surface waters are depleted of combined nitrogen thus restricting the production of phytoplankton (Bhattathiri et al., 1996). In April–May and September–November, *Trichodesmium* spp. blooms occur in the eastern and central AS when bioavailable nitrogen is relatively scarce at the surface due to denitrification. In the AS, conditions favor denitrification as the dominant N-loss pathway (Ward et al., 2009) and may be predominantly stimulated by dissolved organic matter. The rate of denitrification (Naqvi et al., 2009) in the coastal hypoxic zone is less than that in the open ocean oxygen-minimum zones (OMZ). The hypoxia along the Western Indian continental shelf might have increased by augmented nitrogen input from runoff and atmospheric deposition through anthropogenic activities (Naqvi et al., 2009).

The $NO_3^-$ concentrations within the OMZ are about two times higher in the BOB in contrast to that of AS (Naqvi et al., 2006a). The DIN from rivers exported to the BOB is less (<0.5 Tg N a$^{-1}$) compared to AS (Naqvi et al., 2010). In BOB, upwelling is very weak and does not rise close to the surface. In the AS, oxygen-depleted waters ascend to very shallow depths (Naqvi et al., 2006a).

The $N_2$ fixation rates in the eastern AS varied from ~0.1–34 mmol $Nm^{-2}day^{-1}$. These higher $N_2$ fixation rates were consistent with higher chlorophyll *a*. $N_2$ fixation by *Trichodesmium* spp. takes place mostly in the upper ocean (within 10 m from the surface) whereas carbon uptake happens throughout the euphotic zone. $N_2$ fixation is the most important process among all the nitrogen gain processes and the AS gains ~92% of its "new" nitrogen through this process. Lower values of $d^{15}$ N of particulate organic nitrogen (PON) associated with higher fixation rates confirm the presence of $N_2$ fixing bacteria (Gandhi et al., 2011). Gandhi et al. (2010) reported that around ~79% of the nitrogen in suspended particles of surface waters in the northeastern AS was contributed from recently fixed nitrogen by *Trichodesmium*. Denitrification and recurrent occurrence of $N_2$ fixing microorganisms in the AS makes it a suitable global basin for testing marine nitrogen equilibrium (Naqvi, 1987).

### Productivity of the Regional Seas in Terms of Nitrogen Availability

In the south-west coast of India, high concentrations of urea and ammonia-N were observed prior to the formation of mudbanks. The variability of nitrite concentration here is brought about by the nitrification–denitrification process. The marked increase in the concentration of organic nitrogen as compared to a non-mudbank area is indicative of the presence of organically webbed forms of nitrogen. This organic form acts as a "reserved nutrient" (or inducer) for upcoming utilization following the assimilation of inorganic nitrogen forms. The sediments play an active role in nitrate uptake and its regeneration, which is related to forms of nitrate in the overlying water (Nair et al., 2013).

## Eutrophication in Coastal/Marine Scenario

In South Asia, the estuarine and coastal systems were recognized for their N limitation. Any kind of Nr input into these systems can cause algal blooms and eutrophication. Mandovi–Zuari and Cochin estuaries on the west coast, and Godavari and Hoogly estuaries on the east coast were reported to be nitrogen limiting (SACEP, 2014). The rate of application of synthetic N-fertilizers in the agricultural fields in India is around 17.66 Tg per annum (Jaga and Patel, 2012) and the rate of transport of DIN to the northern Indian Ocean is $1.62 \pm 0.32$ Tg $year^{-1}$, which results in the retention and/or elimination of around 91% of anthropogenic N before its transport to the sea. In the watershed of the west-flowing rivers, both natural and anthropogenic processes direct the export of DIN to the AS (Krishna et al., 2016).

As a typical case study, Cochin backwaters (CBW) were reported to change from a river-dominant system during summer monsoon to a tide-dominant system during pre-monsoon season affecting N and P stoichiometry (Martin et al., 2010). Increase was well marked in the levels of $NO_3$ and $PO_4$ in the CBW (Martin et al., 2013) after 1975 (av. 15 μM and 3.5 μM, respectively) compared to the period before (av.2 μM and 0.9 μM, respectively).

Algal bloom predominates along west coast of India especially in southern part. Major causal organism was reported to be dinoflagellate. Along the east coast, diatom blooms were reported to prevail. The Centre for Marine Living Resources and Ecology (CMLRE) worked on algal blooms formation, their spread and ecological consequences on marine ecosystems (Padmakumar et al., 2012) and documented the records of 80 algal blooms (of which 27 were by Cyanobacteria) in the Indian EEZ during 1998–2010. A comprehensive overview of historical harmful algal blooms in the AS, their causes and impact, and seasonal variability was presented by Al Shehhi et al. (2014). Blooms mostly occurred during withdrawal of south-west monsoon and/or in pre-monsoon period (D'Silva et al., 2012; Raghukumar and Anil, 2003; Patil and Anil, 2008). Enhanced phytoplankton growth and blooms usually occur just after the south-west to north-east monsoon transition in the BOB (Prasannakumar et al., 2007). Secondary production is highest from October–January and lowest from June–September (Desai and Bharghava, 1998). In BOB, $NO_3^-$ entrained in water column during January contributed to the development of blooms during late February–early March (Kumar et al., 2010).

Increase in the nitrogen species in the coastal surface waters (SACEP, 2014) was indicated from the assessment by Coastal Ocean Monitoring and Prediction Systems (COMAPS). Prema et al. (2014) observed that between 2002 and 2012, there was an increase in the annual mean DIN in the inshore waters, off Kochi, but the concentrations remained in the "good" grading, well within the water quality indexing criteria for DIN ($<0.1$ mg $L^{-1}$) by USEPA (2006) without reaching the excess limits.

## Nitrogen in Coastal/Marine Habitats and Food Webs

In the northern Indian Ocean, primary production in the bright upper ocean is mainly limited by the availability of DIN (Prasannakumar et al., 2002; Madhupratap et al., 2003). Since the contribution of DIN by atmospheric deposition is relatively less in the northern Indian Ocean (Srinivas et al., 2011; Singh et al., 2012), the riverine DIN input to this area is important for primary production. Export of $1.84 \pm 46$ Tg year$^{-1}$ of DIN from rivers would support $\sim 12.2 \pm 3.1$ Tg year$^{-1}$ of new carbon production ($11.5 \pm 2.9$ and $0.66 \pm 0.17$ Tg C year$^{-1}$ in BOB and AS, respectively). The N/P ratios in the Central and West BOB were lower than 16, indicating nitrate deficiency, whereas N:Si ratio was $<1$ (Paul et al., 2008).

In the surface ocean, N is present mostly as dissolved organic nitrogen (DON), but only a little fraction of it is chemically characterized in surface ocean waters (Purvaja et al.,2008a). It is estimated that the Indian monsoonal estuaries transfer $\sim 0.41 \pm 0.08$ Tg of DON during wet season to the northern Indian Ocean. The BOB receives around three times higher DON (0.30 Tg) than AS (0.11 Tg). DON in the estuaries located in the southwest was found to be higher, and this may be due to higher soil organic carbon, biomass carbon, and heavy rainfall (Krishna et al., 2015).

Increased Nr in coastal waters harms seagrass, macroalgae, and coral reefs affecting ecosystem, habitat structure, trophic dynamics, and human health (Purvaja et al., 2008b). In the seagrass ecosystem, variation of nitrate ($0.62 \pm 0.45–5.47 \pm 0.25$ μM) and nitrite ($0.14 + 0.03–0.69 \pm 0.08$ μM) was noted in different islands waters of Lakshadweep (Nobi et al., 2016). Higher nitrite–nitrate concentration in the seagrass environment is common as N-fixation is generally more in seagrass bed sediments than the carbonate sediments of the adjacent coral reef environment (Qasim, 1970).

In oceanic Lakshadweep waters, DIN accounted for less than 10% and DON for more than 90% of the total dissolved N in the euphotic zone. Urea formed 17–18% of total dissolved N and 19–20% of DON. The phytoplankton in these waters is nitrogen limited. Production associated with $NO_3$ assimilation varied from 11% to 60% (average 37%) of the observed primary production. Zooplankton regenerated $NH_4$ at an average rate of 0.59 $\mu$g-at N (mg dry wt)$^{-1}$day$^{-1}$. Assimilation of this quantity of $NH_4$ would account for 9–50% (average 23%) of the measured primary production (Wafar et al., 1986). $NO_3$ production rates were equal to $NO_3$ uptake rates by the zooxanthellae, suggesting a close coupling between these processes. Wafar et al. (1993) suggested that the algae would be limited by N if there is no regenerative N supplied.

Rajkumar (1997) reported that in coral reef ecosystems in Lakshadweep, there was sufficient *new* nitrogen input ($NO_3$) in monsoon; while in post-monsoon, regenerative flux (Ammonium and Urea) were dominated. There was no indication of either *new* or regenerative fluxes in pre-monsoon. Heterotrophic biomass buildup was not interfered even at low concentrations of available nitrogen. Nitrogen uptake was dependent not only on the substrate availability but also on the prevailing weather conditions. Production of N and its transformation proceeded at appreciable rates. Reefs export particulate organic matter and dissolved inorganics in measurable quantities. There is a predominance of heterotrophy over autotrophy in nitrogen flux in reef waters.

Prema et al. (2015) assessed biogeochemistry and ecosystem processes in the Ashtamudi Lake in the south-west coast of India, in clam beds as well as in no clam regions. Total ammonia-N in water in the clam bed was three times that of no clam region, whereas the nitrate and nitrite concentrations were similar in both. The benthic region of clam beds, due to bioturbation by clams showed higher amount of nitrite and nitrate, compared to the no clam region. Ammoniacal N in the benthic region in the clam bed was around twice of that of no clam region.

The environmental features including nitrogenous nutrients of seawater in the mudbank region of Purakkad, Alleppy, Kerala indicated considerable reduction in the more oxidized forms of nitrogenous nutrients in the surface and bottom waters in the mudbank after a gap of 30 years (CMFRI, 1984, 2014). This is in conformity with the earlier reports that the organic nitrogenous forms act as reserved nutrients and control the periodic variability of nitrate in the mudbank area (Nair et al., 2013). During the post- and pre-monsoon seasons, flocks of sea birds migrate to the mudbank region of Purakkad, Alleppy. Analysis of the leachable nutrients from this bird area beach sediment indicated that the leachable DIN was four times that of non-bird area sediment (CMFRI, 2014).

The covariation of nitrate in surface layers in the northern AS, chlorophyll *a* and primary production in its euphotic zone with mixed-layer depth, and wind are all indicative of carbon fixation mainly controlled by physical forcing (Prasannakumar et al., 2001). A swarm of pelagic tunicate (*Pyrosomas spinosum*) was found in surface open waters of AS (Gauns et al., 2015) during late south-west monsoon, associated with surface concentrations of 2.5 μM nitrate, 0.3 μM phosphate, 0.9 μM silicate, and 5.0 μM ammonium. Lower molar C/N ratio (5) in *P. spinosum* suggests that the growth of these organisms is carbon-limited. In CBW, rotifers showed significant positive correlations with nitrite, chlorophyll *a*, total suspended solids, phosphate, and biochemical oxygen demand (Varghese and Krishnan, 2011). *Brachionus rotundiformis* was found to dominate (85.76%) among the 13 species of *Brachionus* and showed significant positive correlations with nitrite, biochemical oxygen demand, chlorophyll *a*, and total suspended solids (Varghese and Krishnan, 2013).

In BOB, the subsurface maximum (2.0–3.51 $\mu mol\ dm^{-3}$) of ammonium in the euphotic zone is attributed to the regeneration of ammonium by zooplankton excretion associated with stratification of water column during summer. The spatial distribution of recycled nitrogen (ammonium) showed an increasing trend from inshore to offshore regions in the northern Bay (Satyanarayana et al., 1991).

In the seagrass beds of Gulf of Mannar, high positive correlation was noticed (CMFRI, 2014) between chlorophyll, nitrate, and silicate concentrations in seawater ($p < .001$). Sulochanan et al. (2011) observed highly significant influence of nitrite, ammonia ($p < .001$) seasonally in seagrass beds in Palk Bay and Gulf of Mannar. Inorganic phosphate and total organic nitrogen were considered as most important limiting factors for the growth and reproduction of seagrass (Govindasamy et al., 2012, 2013).

## Nitrogen in Coastal Aquaculture/Mariculture

There are reports of pollution due to shrimp aquaculture when the ponds are emptied and the problem of algal overgrowth in receiving waters is indicative of nutrients sourced from aquaculture wastes (Aquaculture Authority, 2001; Rajitha et al., 2006). Prema et al. (2012) reported high nutrient imbalance (N/P) in the water, with excess phosphate in the traditional shrimp farms at Vypin, Kochi due to direct release of domestic wastewater containing detergents.

Marine fisheries are an important traditional livelihood along coastal India and also contribute to the protein component of food. The marine fisheries sector has high population along the Indian coast. Traditional fishermen have their habitations very close to the shoreline. Information regarding the marine fishing sector along the east and west coast is provided in Table 20.1. Coastal aquaculture/mariculture activities are presently limited to farming of fishes, crustaceans, bivalves, and seaweed, but adverse effects are still possible on coastal water quality, by enrichment of nitrogen from uneaten feed or waste products, other inputs for aquaculture, and suspended solids (SACEP, 2014).

**Table 20.1** Marine Fisheries Related Statistics in the East and West Coast of India

| State/Union Territory | | Average Landings (Lakh tonnes) | Landing Centers | Fishing Villages | Fisher Families | Traditional Fisher Families | Fisher Population |
|---|---|---|---|---|---|---|---|
| West coast | Kerala | 5.76 | 187 | 222 | 118,937 | 116,321 | 610,165 |
| | Karnataka | 4.74 | 96 | 144 | 30,713 | 28,533 | 167,429 |
| | Goa | 1.53 | 33 | 39 | 2189 | 2,147 | 10,545 |
| | Maharashtra | 3.45 | 152 | 456 | 81,492 | 74,203 | 386,259 |
| | Gujarat | 7.12 | 121 | 247 | 62,231 | 59,469 | 336,181 |
| | Daman & Diu | 0.46 | 5 | 11 | 7374 | 7,181 | 40,016 |
| East coast | West Bengal | 0.77 | 59 | 188 | 76,981 | 53,532 | 380,138 |
| | Odisha | 1.39 | 73 | 813 | 114,238 | 87,541 | 605,514 |
| | Andhra Pradesh | 3.42 | 353 | 555 | 163,427 | 161,039 | 605,428 |
| | Tamil Nadu | 6.65 | 407 | 573 | 192,697 | 185,465 | 802,912 |
| | Puducherry | 0.65 | 25 | 40 | 14,271 | 14,248 | 54,627 |

Central Marine Fisheries Research Institute [CMFRI], 2014. Central Marine Fisheries Research Institute Annual Report, 2013–14. pp. 174–178; MoA, CMFRI, 2012. Marine Fisheries Census 2010 Part I India. Ministry of Agriculture, KrishiBhavan, New Delhi and Central Marine Fisheries Research Institute, Kochi. Available from: http://eprints.cmfri.org.in/id/eprint/8998.

Live feeds are mainly used as diet for cultured fish larvae. Microalgae form the basic requirement for live feed culture. Using marine grow-out effluents as culture media for the live feed *Nannochloropsis salina* it was found that more than 95% of total $NH_3$-N and $NO_3$-N, and more than 90% of $NO_2$-N of the effluents were used within 50 days (Prema et al., 2006).

No significant alterations in the levels of nitrate, phosphate, and silicate in seawater were observed by cage culture fisheries at Munambam, off Kochi (Prema et al., 2010), nor the cage culture activity influenced the plankton population (Varghese et al., 2010). No significant difference was observed between the total ammonia-N of the cage and reference sites ($p > 0.05$), but they differed significantly between culture periods in cage site ($p < 0.05$) at Karwar, west coast of India (Philipose et al., 2012). No significant difference was observed in the nitrite and nitrate levels of the cage and reference sites and also between the two harvests.

Oysters in fish–oyster integrated farming stimulate the bacterial process of nitrification and denitrification, helping the escape of nitrogen gas, thus, lowering the ammonia content in culture water. The lower eutrophication index (E) value evidenced that oysters can control eutrophication in an integrated aquaculture system. The filter-feeding oysters process the DIN and DIP pool, improve water quality and the growth of fish (Viji et al., 2014a,b). In oyster farm sites, lower ammonia-N levels and chlorophyll *a* values were recorded by Ramalinga (2006). The ammonia-N generated by oysters was taken up by rapidly regenerating phytoplankton in the water column which in turn accounted for increased chlorophyll *a*.

In the outdoor culture of agar-yielding seaweed *Gracilaria edulis*, crop growth rate increased with proportionate supply of nutrients including nitrogen (Kaliaperumal et al., 2003). Higher phosphate and nitrate contents in the water accelerated the growth of

*G. edulis* in nearshore waters of Gulf of Mannar during November to March. The green seaweed *Ulva reticulata* cultured along with shrimp was found to be an effective eliminator of ammonical nitrogen from 249.5 to 17.39 $\mu$mol N $L^{-1}$ (94%). The nitrate nitrogen came down from 28.39 to 24.21 $\mu$mol N $L^{-1}$ (5%) and nitrite nitrogen from 14.51 to 9.03 $\mu$mol N $L^{-1}$ (22%). The total N removal from the aquaculture system was 45%. The toxic nitrogenous wastes were always found to come down in the integrated system in comparison to the monoculture system (Seema and Jayasankar, 2005). The removal of nitrogenous species showed a gradation in the effluent treated with *U. lactuca* > *G. corticata* > control (Seema et al., 2005).

## Atmospheric Nitrogen Contribution and Sea to Air Exchanges in the Indian Seas

Based on the chemical analysis of ambient aerosols, collected from the marine atmospheric boundary layer during the continental outflow (January–April), Srinivas and Sarin (2013) quantified that dry-deposition fluxes ($\mu$mol $m^{-2}$ $day^{-1}$) of N (2–167) to the BOB are significantly higher compared to those over the AS (N: 0.2–18.6).

The AS is an active site for N-fixation (Devassy et al., 1978). The predicted $N_2$ fixation rate for the AS is 3.3 Tg N $year^{-1}$ (Bange et al., 2005). Significant amount of nitrogen is fixed ($\sim$11% of the global $N_2$ fixation) during a *Trichodesmium* bloom in the AS. Later, Parab and Matondkar (2012) observed that in the AS, *Trichodesmium* species fixed a total of 0.2976 Tg N $year^{-1}$. Higher consumption ratios suggest possible underestimation of $N_2$ fixation or presence of additional microbes that are incapable of fixing $N_2$ (Gandhi et al., 2011). *Trichodesmium erythraeum* is abundant in high ($\geq$300 $\mu$atm) $pCO_2$ concentrations. The N: P ratios almost doubled ($\sim$10) at high $pCO_2$ regions of BOB leading to enhanced productivity in this N-limited region and increased biological carbon sequestration (Shetye et al., 2013). Estimates of $N_2$ fixation are helpful in lessening the ambiguity in the marine N budget.

The OMZ of AS is responsible for a loss of up to 40 Tg $year^{-1}$ (30–50%) of fixed N (Naqvi, 2008) by denitrification. The AS seems to explain for at least one-third of the global oceanic water column denitrification (Naqvi, 2002). When extrapolated to the entire AS, the overall denitrification rate was $41 \pm 18$ Tg N $year^{-1}$ (Devol et al., 2006). Removal of ammonia and methylamines was mostly observed to happen through wet deposition, contributing less than 1% to the new production (Naqvi, 2002). Emissions of $N_2$ and $N_2O$ to the atmosphere from the AS are globally significant. The AS appears as a net sink for combined nitrogen of approx. $3 \times 10^{13}$ g N $year^{-1}$ and a significant source for atmospheric $N_2O$. The distribution of $N_2O$ indicates considerable losses to reducing zones, pointing to a very rapid turnover of $N_2O$ within the region (Naqvi et al., 1993). Isotope enrichment in the combined $NO_3$ and $NO_2$ pool and in $N_2O$ in reducing waters seems to be noticeably lesser in the coastal region than in the open ocean, implying more varied sources/sinks and/or different isotopic fractionation factors (Naqvi et al., 2006a). The isotopic data are indicative of multiple pathways of $N_2O$ production, i.e., nitrification, denitrification, and the coupling of the two processes (Naqvi et al., 1998). Out of the total amount of denitrified nitrogen ~7% was lost to the atmosphere (Somasundar et al., 1990).

Upwelled waters from the AS to Cochin backwaters leads to anoxia due to weak flushing along with high organic input from anthropogenic loadings, causing denitrification and formation of $N_2O$ in the bottom waters (Martin et al., 2010). Annual emissions of $N_2O$ in the AS ranged from 0.33 to 0.70 Tg $N_2O$ and are dominated by fluxes from coastal regions during the south-west and north-east monsoons (Bange et al., 2001). During the north-east monsoon, $N_2O$ concentrations are low and less variable (<25 nM to a depth of 100 m). But the onset and amplification of $O_2$ deficiency during south-west monsoon causes changes in the $N_2O$ field (Naqvi et al., 2006b, 2009).

The total emission of $N_2O$ from the Indian shelf is calculated to be 0.05–0.38 Tg for the upwelling season. Surface saturations (89–214%) and atmospheric fluxes (−0.10–10.67 $\mu mol\ m^{-2}\ day^{-1}$) from BOB are much lower. The upwelled water does not rise to very shallow depths (Naqvi et al., 2006a), and maximal $N_2O$ concentrations over the shelf are much lower compared to AS (Narvekar et al., 1998; Naqvi et al., 2006a). The AS N-cycle can have an impact on climate change by means of its $N_2O$ emission and denitrification modules (Bange et al., 2005; Naqvi et al., 2010). Nutrient ratios in the AS suggest that nitrogen is limiting and denitrification dominates N-fixation (JGOFS, 2002).

# Conclusion

The nitrogen dynamics, especially the oceanic nitrogen transformations, in the surface water of coastal and marine environment in India has been studied. However, the productivity of the coastal and marine ecosystem in relation to nitrogen is in its infancy, as the link among nutrients, primary, secondary, and territory production is yet to be established through modeling. Studies on nitrogen and/or nutrient transfer into the coastal and marine habitats and food web are rare in India and the explorations already done in this line are location specific, while an ecosystem-based approach needs to be followed. Research on nitrogen and/or nutrient management/environment feasibility/ environmental impacts in coastal aquaculture and mariculture also needed to be strengthened. The nitrogen budget in the coastal and marine ecosystem including sea to air exchanges and vice versa, needs updating due to its relevance in the present scenario of climate change.

## Specific Research Needs/Gaps From the Works Reviewed

- It is not yet obvious how the suboxic ecosystem of AS will respond to anthropogenic changes. Future work must therefore focus on prediction of the magnitude and direction of these impacts (Naqvi and Jayakumar, 2000).
- A meaningful quantitative prediction of increase in $N_2O$ emission in coastal and marine environment is not possible at present because of continuing uncertainties

concerning the formative pathways to $N_2O$ as well as insufficient data from key coastal regions (Naqvi et al., 2010).

- The role of N-fixation in the AS is still complex to appraise because of less database availability (Bange et al., 2005).
- Reports on pollution of coastal waters with effluents from the fertilizer industries are many. But, whether quantification of such pollution loads from material managing in ports or from the industries is yet to be established (SACEP, 2014).
- Further investigations are needed for estimation of the impact of riverine DIN fluxes on primary production in the Indian coastal waters (Krishna et al., 2016).
- Total input of DON fluxes will provide better insight on the contribution of riverine nitrogen inputs toward coastal productivity in the BOB and the AS (Singh and Ramesh, 2011) and alterations in the freshwater discharge due to natural or anthropogenic events and their impact on microbial food web dynamics needs further appraisal (Krishna et al., 2015).

## Policy Issues

Effective ecosystem-based policy actions and recommendations integrating social, economic, and environmental concerns, coastal and marine ecosystem sustainability to abate anthropogenic nutrient pollution are suggested by SACEP (2014) and CPCB (2009) also strongly advocates waste water treatment, before its disposal, both of which still needs implementation.

# References

Al Shehhi, M.R., Gherboudj, I., Ghedira, H., 2014. An overview of historical harmful algae blooms outbreaks in the Arabian Seas. Marine Pollution Bulletin 86 (1), 314–324.

Aquaculture Authority, 2001. Shrimp Aquaculture and the Environment – an Environment Impact Assessment Report. Aquaculture Authority, Chennai, India, p. 114.

Bange, H.W., Andreae, M.O., Lal, S., Law, C.S., Naqvi, S.W.A., Patra, P.K., et al., 2001. Nitrous oxide emissions from the Arabian Sea: a synthesis. Atmospheric Chemistry and Physics 1, 61–71. http://dx.doi.org/10.5194/acp-1-61-2001.

Bange, H.W., Naqvi, S.W.A., Codispoti, L.A., 2005. The nitrogen cycle in the Arabian Sea. Progress in Oceanography 65, 145–158. http://dx.doi.org/10.1016/j. pocean.2005.03.002.

Bhattathiri, P.M.A., Pant, A., Sawant, S., Gauns, M., Matondkar, S.G.P., Mohanraju, R., 1996. Phytoplankton production and chlorophyll distribution in the eastern and central Arabian Sea in 1994–95. Current Science 71, 857–862.

Central Marine Fisheries Research Institute [CMFRI], 1984. Ecology of Mudbanks – hydrography in: coastal zone management: Mudbanks of Kerala coast. Central Marine Fisheries Research Institute Bulletin 31, 25–27.

Central Marine Fisheries Research Institute [CMFRI], 2014. Central Marine Fisheries Research Institute Annual Report, 2013–14, pp. 174–178.

Central Pollution Control Board [CPCB], 2009. Status of Water Supply, Wastewater Generation and Treatment in Class-i Cities and Class-ii Towns of India. Control of Urban Pollution Series, CUPS/70/2009 – 10.

Desai, B.N., Bharghava, R.M.S., 1998. Biological production and fishery potential of the exclusive economic zone of India. In: Sherman, K., Okemwa, E., Ntiba, M. (Eds.), Large Marine Ecosystems of the Indian Ocean: Assessment, Sustainability and Management. Blackwell Science Inc, Cambridge, MA, pp. 297–309.

Devassy, V.P., Bhattathiri, P.M.A., Qasim, S.Z., 1978. *Trichodesmium* phenomenon. Indian Journal of Marine Science 7, 168–186.

Devol, A.H., Uhlenhopp, A.G., Naqvi, S.W.A., Brandes, J.A., Jayakumar, A., Codispoti, L.A., et al., 2006. Denitrification rates and excess nitrogen gas concentrations in the Arabian Sea oxygen deficient zone. Deep Sea Research Part I 53, 1533–1547. http://dx.doi.org/10.1016/j.dsr.2006.07.005.

D'Silva, M.S., Anil, A.C., Naik, R.K., D'Costa, P.M., 2012. Algal blooms: a perspective from the coasts of India. Natural Hazards 63, 1225–1253.

Gandhi, N., Ramesh, R., Srivastava, R., Sheshshayee, M.S., Dwivedi, R.M., Raman, M., 2010. Nitrogen uptake rates during spring in the NE Arabian Sea. International Journal of Oceanography 2010, 127493. http://dx.doi.org/10.1155/2010/127493.

Gandhi, N., Singh, A., Prakash, S., Ramesh, R., Raman, M., Sheshshayee, M.S., 2011. First direct measurements of $N_2$ fixation during a *Trichodesmium* bloom in the eastern Arabian Sea. Global Biogeochemical Cycles 25, GB4014. http://dx.doi.org/10.1029/2010GB003970.

Gauns, M., Mochemadkar, S., Pratihary, A.K., Roy, R., Naqvi, S.W.A., 2015. Biogeochemistry and ecology of *Pyrosoma spinosum* from the Central Arabian Sea. Zoological Studies 54 (3). http://dx.doi.org/10.1186/s40555-014-0075-6.

Govindasamy, C., Arulpriya, M., Ruban, P., Meenakshi, V.R., 2012. Hydro-chemical evolution of Palk strait region, Bay of Bengal. Journal of Tropical Life Sciences 2 (1), 1–5.

Govindasamy, C., Arulpriya, M., Anantharaj, K., Ruban, P., Srinivasan, R., 2013. Seasonal variations in seagrass biomass and productivity in Palk Bay, Bay of Bengal, India. International Journal of Biodiversity and Conservation 5 (7), 408–417. http://dx.doi.org/10.5897/IJBC12.132.

Gruber, N., 2008. The marine nitrogen cycle: overview of distributions and processes. In: Capone, D.G., Bronk, D.A., Mulholland, M.R., Carpenter, E.J. (Eds.), Nitrogen in the Marine Environment, second ed. Elsevier, Amsterdam, pp. 1–50.

Jaga, P.K., Patel, Y., 2012. An overview of fertilizers consumption in India: determinants and outlook for 2020-a review. International Journal of Scientific Engineering and Technology 1, 285–291.

JGOFS, January 2002. Report of the Indian Ocean Synthesis Group on the Arabian Sea Process Study. JGOFS REPORT No. 35.

Kaliaperumal, N., Jayasankar, R., Ramalingam, J.R., 2003. Outdoor culture of agar yielding red alga *Gracilaria edulis* (Gmelin) Silva. Seaweed Research and Utilisation 25 (1 & 2), 159–162.

Krishna, M.S., Prasad, M.H.K., Rao, D.B., Viswanadham, R., Sarma, V.V.S.S., Reddy, N.P.C., 2016. Export of dissolved inorganic nutrients to the northern Indian Ocean from the Indian monsoonal rivers during discharge period. Geochimica et Cosmochimica Acta 172, 430–443.

Krishna, M.S., Prasad, V.R., Sarma, V.V.S.S., Reddy, N.P.C., Hemalatha, K.P.J., Rao, Y.V., 2015. Fluxes of dissolved organic carbon and nitrogen to the northern Indian Ocean from the Indian monsoonal rivers. Journal of Geophysical Research (Biogeosciences) 120 (10), 2067–2080.

Kumar, S., Ramesh, R., Dwivedi, R.M., Raman, M., Sheshshayee, M.S., D'Souza, W., 2010. Nitrogen uptake in the northeastern Arabian Sea during winter cooling. International Journal of Oceanography. http://dx.doi.org/10.1155/2010/819029, 819029.

Lallu, K.R., Fausia, K.H., Vinita, J., Balachandran, K.K., Naveen Kumar, K.R., Rehitha, T.V., 2014. Transport of dissolved nutrients and chlorophyll a in a tropical estuary, south west coast of India. Environmental and Monitoring Assessment 186 (8), 4829–4839.

Martin, G.D., Muraleedharan, K.R., Vijay, J.G., Rejomon, G., Madhu, N.V., Shivaprasad, A., et al., 2010. Formation of anoxia and denitrification in the bottom waters of a tropical estuary, southwest coast of India. Biogeosciences Discussions 7, 1751–1782. http://dx.doi.org/10.5194/bgd-7-1751-2010.

Martin, G.D., Jyothibabu, R., Madhu, N.V., Balachandran, K.K., Nair, M., Muraleedharan, K.R., et al., 2013. Impact of eutrophication on the occurrence of *Trichodesmium* in the Cochin backwaters, the largest estuary along the west coast of India. Environmental Monitoring and Assessment 185, 1237–1253.

Madhupratap, M., Gauns, M., Ramaiah, N., Kumar, S.P., Muraleedharan, P.M., deSousa, S.N., et al., 2003. Biogeochemistry of the Bay of Bengal: physical,chemical and primary productivity characteristics of the central and western Bay of Bengal duringsummer monsoon 2001. Deep Sea Research Part II 50 (5), 881–896.

MoA, CMFRI, 2012. Marine Fisheries Census 2010 Part I India. Ministry of Agriculture, KrishiBhavan, New Delhi and Central Marine Fisheries Research Institute, Kochi. Available from: http://eprints.cmfri.org.in/id/eprint/8998.

MoEF, 2009. State of Environment Report India-2009. Environmental Information System (ENVIS). Ministry of Environment and Forests, Government of India. Available from: http://www.moef.gov.in.

Nair, K.K.C., 2010. Atlas on Environmental Parameters and Productivity (Primary and Secondary) on the Indian EEZ and Adjoining Areas. A technical publication of the Centre for Marine living Resources and Ecology, Kochi. CMLRE, Ministry of Earth Sciences< Government. of India, 296 pp.

Nair, S.M., Balchand, A.N., Prashob, P.K.J., Anu, G., 2013. Phenomenal changes in the forms of nitrogen: results from a coastal zone of south west India. Chemical Speciation & Bioavailability 25 (2), 89–96.

Naqvi, S.W.A., 1987. Some aspects of the oxygen deficient conditions and denitrification in the Arabian Sea. Journal of Marine Research 45, 1049–1072. http://dx.doi.org/10.1357/002224087788327118.

Naqvi, S.W.A., 2002. Air-sea interactions and exchanges. In: Watts, L., Burkill, P., Smith, S. (Eds.), Arabian Sea Process Study. JGOFS International Planning Office, Bergen, Norway. JGOFS Report No. 35.

Naqvi, S.W.A., 2008. The Indian Ocean (chapter 14). In: Capone, D.G., Bronk, D.A., Mulholland, M.R., Carpenter, E.J. (Eds.), Nitrogen in the Marine Environment, second ed. Elsevier Press.

Naqvi, S.W.A., Jayakumar, D.A., 2000. Ocean biogeochemistry and atmospheric composition: significance of the Arabian sea. Current Science 78, 289–298.

Naqvi, S.W.A., Sengupta, R., Dileepkumar, M., 1993. Carbon dioxide and nitrous oxide in the Arabian Sea. In: McBean, G.A., Hantel, M. (Eds.), Interactions Between Global Climate Subsystems: The Legacy of Hann, pp. 85–92.

Naqvi, S.W.A., Yoshinari, T., Jayakumar, D.A., Altabet, M.A., Narvekar, P.V., Devol, A.H., et al., 1998. Budgetary and biogeochemical implications of $N_2O$ isotope signatures in the Arabian Sea. Nature 394, 462–464.

Naqvi, S.W.A., Naik, H., Pratihary, A., D'Souza, W., Narvekar, P.V., Jayakumar, D.A., Devol, A.H., Yoshinari, T., Sainoet, T., 2006a. Coastal versus open-ocean denitrification in the Arabian Sea. Biogeosciences 3, 621–633. http://dx.doi.org/10.5194/bg-3-621-2006.

Naqvi, S.W.A., Narvekar, P.V., Desa, E., 2006b. Coastal biogeochemical processes in the north Indian Ocean. In: Robinson, A., Brink, K. (Eds.), The Sea, vol. 14. Harvard University Press, pp. 723–780.

Naqvi, S.W.A., Naik, H., Jayakumar, D.A., Pratihary, A., Narvenkar, G., Kurian, S., 2009. Seasonal anoxia over the western Indian continental shelf. In: Wiggert, J.D., Hood, R.R., Naqvi, S.W.A., Brink, K.H., Smith, S.L. (Eds.), Indian Ocean: Biogeochemical Processes and Ecological Variability, Geophys. Monogr. Ser, vol. 185. AGU, Washington, D.C, pp. 333–345.

Naqvi, S.W.A., Naik, H., D' Souza, W., Narvekar, P.V., Paropkari, A.L., Bange, H.W., 2010. Carbon and nitrogen fluxes in the north Indian Ocean. In: Liu, K.K., Atkinson, L., Quinones, R., Talaue-McManus, L. (Eds.), Carbon and Nutrient Fluxes in Continental Margins: A Global Synthesis. Springer-Verlag, NewYork, pp. 180–192.

Narvekar, P.V., Naqvi, S.W.A., Dileepkumar, M., 1998. $CO_2$ and $N_2O$ fluxes from the northern Indian Ocean. In: Parashar, D.C., Sharma, C., Mitra, A.P. (Eds.), Global Environment Chemistry, pp. 128–140.

Newell, S.E., Babbin, A.R., Jayakumar, A., Ward, B.B., 2011. Ammonia oxidation rates and nitrification in the Arabian Sea. Global Biogeochemical Cycles 25, GB4016. http://dx.doi.org/10.1029/2010GB003940.

Nobi, E.P., Dilipan, E., Thangaradjou, T., 2016. Ecosystem perspective of seagrass environments of Lakshadweep islands, India. International Journal of Applied Science Innovation 2016 (1), 42–51. ISSN: 2458-6501.

Padmakumar, K.B., Menon, N.R., Sanjeevan, V.N., 2012. Is occurrence of harmful algal blooms in the exclusive economic zone of India on the rise? International Journal of Oceanography 2012. Available from: http://dx.doi.org/10.1155/2012/263946.

Parab, S.G., Matondkar, S.G.P., 2012. Primary productivity and nitrogen fixation by *Trichodesmium* spp. in the Arabian Sea. Journal of Marine Systems 105-108, 82–95.

Patil, J.S., Anil, A.C., 2008. Temporal variation of diatom benthic propagules in a monsoon influenced tropical estuary. Continental Shelf Research 28, 2404–2416.

Paul, J.T., Ramaiah, N., Sardessai, S., 2008. Nutrient regimes and their effect on distribution of phytoplankton in the Bay of Bengal. Marine Environmental Research 66 (3), 337–344. http://dx.doi.org/10.1016/j.marenvres.2008.05.007. Elsevier.

Philipose, K.K., Sharma, S.R.K., Loka, J., Damodaran, D., Rao, G.S., Vaidya, N.G., et al., 2012. Observations on variations in physico-chemical water parameters of marine fish cage farm off Karwar. Indian Journal of Fisheries 59 (1), 83–88.

Prasannakumar, S., Ramaiah, N., Gauns, M.U., Sarma, V.V.S.S., Muraleedharan, P.M., Raghukumar, S., et al., 2001. Physical forcing of biological productivity in the northern Arabian Sea during the northeast monsoon. Deep Sea Research Part II: Topical Studies in Oceanography 48 (6), 1115–1126.

Prasannakumar, S., Muraleedharan, P.M., Prasad, T.G., Gauns, M.U., Ramaiah, N., de Souza, S.N., et al., 2002. Why is the Bay of Bengal less productive during summer monsoon compared to the Arabian Sea? Geophysical Research Letters 9 (24), 2235. http://dx.doi.org/10.1029/2002GL016013.

Prasannakumar, S., Nuncio, M., Ramaiah, N., Sardesai, S., Narvekar, J., Fernandes, V., et al., 2007. Eddy-mediated biological productivity in the Bay of Bengal during fall and spring inter-monsoons. Deep-Sea Research I 54, 1619–1640.

Pratihary, A.K., Naqvi, S.W.A., Naik, H., Thorat, B.R., Narvenkar, G., Manjunatha, B.R., et al., 2009. Benthic flux in a tropical estuary and its role in the ecosystem. Estuarine, Coastal and Shelf Science 85 (3), 387–398.

Prema, D., Kuriakose, S., Thomas, V.J., 2006. Culture of *Nannochoropsissalina* in different marine grow-out effluents. In: Pandey, P.K., Vasudevappa, C., Ayyappan, S., Kumar, D., Mukherjee, S.C. (Eds.), The Sixth Indian Fisheries Forum, Proceedings. AFSIB, Mangalore and IFA, Mumbai, India, pp. 1–5.

Prema, D., Sobhana, K.S., Laxminarayana, A., Joseph, I., Joseph, S., Ignatius, B., et al., 2010. Observations on selected characteristics of water and sediment at the open sea cage culture site of Asian seabass *Lates calcarifer* (Bloch) off Cochin, south-west coast of India. Indian Journal of Fisheries 57 (4), 53–59.

Prema, D., Khambadkar, L.R., Anilkumar, P.S., Remya, R., Kripa, V., 2012. Scientific Advisory on Water Quality Management for Traditional Fish Farms. Marine Fisheries Information Service, pp. 7–8. T&E Ser. No. 212.

Prema, D., Jeyabaskaran, R., Kaladharan, P., Khambadkar, L.R., Anilkumar, P.S., Nandakumar, A., et al., 2014. Water Quality Indexing of Coastal Waters off Cochin. Technical and Extension Series 219. Marine Fisheries Information Service, pp. 15–16.

Prema, D., Mohamed, K.S., Kripa, V., Jenni, B., Alloycious, P.S., Sajikumar, K., et al., 2015. Assessment of Biogeochemical Processes of Ashtamudi Lake Ecosystem in Relation to Clam Fishery. Technical and Extension Series 225. Marine Fisheries Information Service, pp. 8–12. ISSN 0254-380 X.

Purvaja, R., Ramesh, R., Shalini, A., Rixen, T., 2008a. Nitrogen cycling: a review of the processes, transformations and fluxes in coastal ecosystems. Current Science 94, 1419–1438.

Purvaja, R., Ramesh, R., Shalini, A., Rixen, T., 2008b. Biogeochemistry of nitrogen in seagrass and oceanic systems. Memoirs of the Geological Society of India (73), 435–460.

Qasim, S.Z., 1970. Some characteristics of a *Trichodesmium* bloom in the Laccadives. Deep-Sea Research 17 (3), 655–660.

Raghukumar, S., Anil, A.C., 2003. Marine biodiversity and ecosystem functioning: a perspective. Current Science 84, 884–892.

Rajkumar, R., 1997. Aspects of Nitrogen Recycling in a Coral Reef Ecosystem. PhD Thesis. Goa University.

Rajitha, K., Mukherjee, C.K., Vinu, C.R., 2006. Application of Remote Sensing and GIS for Sustainable Management of Shrimp Culture in India. Aquacultural Engineering. http://dx.doi.org/10.1016/j.aquaeng.2006.05.003.

Ramalinga, 2006. Environmental Impact Assessment of Suspended Oyster, Crassostreama Drasensis (Preston) Culture. PhD Thesis. Central Institute of Fisheries Education, Mumbai.

SACEP, 2014. Nutrient Loading and Eutrophication of Coastal Waters of the South Asian Seas – a Scoping Study. Available from: www.sacep.org/pdf/Scoping_study_ on_Nutrient_loading_in_SAS_Region.pdfpp56.

Satyanarayana, D., Sahu, S.D., Panigrahy, P.K., Sarma, V.V., Suguna, C., 1991. Subsurface ammonium maxima in northern Bay of Bengal. Marine Environmental Research 31, 123–136.

Seema, C., Jayasankar, R., 2005. Removal of nitrogen load in the experimental culture system of seaweed and shrimp. Journal of the Marine Biological Association of India 47 (2), 150–153.

Seema, C., Jayasankar, R., Mathew, G., 2005. Influence of physicochemical parameters on the removal of nitrogenous compounds during treatment of fish effluent with seaweeds. Seaweed Research and Utilisation 27 (1 & 2), 99–104.

Shetye, S.S., Sudhakar, M., Jena, B., Mohan, R., 2013. Occurrence of nitrogen fixing cyanobacterium *Trichodesmium* under elevated $pCO_2$ conditions in the western Bay of Bengal. International Journal of Oceanography. Available from: http://dx.doi.org/10.1155/2013/350465.

Singh, A., Ramesh, R., 2011. Contribution of riverine dissolved inorganic nitrogen flux to new production in the coastal northern Indian Ocean: an assessment. International Journal of Oceanography. http://dx.doi.org/10.1155/2011/983561.

Singh, A., Gandhi, N., Ramesh, R., 2012. Contribution of atmospheric nitrogen deposition to newproduction in the nitrogen limited photic zone of the northern Indian Ocean. Journal of Geophysical Research 117, C06004. http://dx.doi.org/10.1029/2011JC007737.

Somasundar, K., Rajendran, A., Dileepkumar, M., Sengupta, R., 1990. Carbon and nitrogen budgets of the Arabian Sea. Marine Chemistry 30, 363–377.

Srinivas, B., Sarin, M.M., 2013. Atmospheric deposition of N, P and Fe to the northern Indian Ocean: implications to C- and N-fixation. Science of the Total Environment 456–457, 104–114.

Srinivas, B., Sarin, M.M., Sarma, V.V.S.S., 2011. Atmospheric dry deposition of inorganic and organic nitrogen to the Bay of Bengal: impact of continental outflow. Marine Chemistry 127, 170–179.

Sulochanan, B., Kumaraguru, A.K., Korabu, L.S., 2011. Hydrological conditions in seagrass beds in Palk Bay and Gulf of Mannar, southeast coast of India. Journal of Marine Biological Association of India 53 (1), 108–115.

USEPA, 2006. National Coastal Assessment, Alabama 2000–2004. Final Report. Alabama Department of Environment Management and United States Environmental Protection Agency, Office of the Research and Development, Gulf Ecology Division, Gulf Breez, p. 37. F1 32561.

Varghese, M., Joseph, S., Joseph, I., Boby, I., Manisseri, M.K., Thomas, V.J., et al., 2010. Preliminary studies on the impact of open sea cage culture of *Lates calcarifer* (Bloch) on the planktonic and benthic fauna off Cochin, Kerala. Indian Journal of Fisheries 57 (3), 75–77.

Varghese, M., Krishnan, L., 2011. Ecology of rotifers in Cochin backwaters, Kerala, India. Indian Journal of Fisheries 58 (3), 109–115.

Varghese, M., Krishnan, L., 2013. *Brachionus* species distribution in relation to environmental characteristics in Cochin backwaters, Kerala, South India. Indian Journal of Fisheries 60 (1), 133–138.

Viji, C.S., Chadha, N.K., Kripa, V., Prema, D., Prakash, C., Sharma, R., et al., 2014a. Can oysters control eutrophication in an integrated fish-oyster aquaculture system? Journal of the Marine Biological Association of India 56 (2), 67–73.

Viji, C.S., Mohamed, K.S., Kripa, V., Prema, D., Sharma, R., Jenni, B., 2014b. Oysters Improve Growth of Fish in an Integrated aquaculture System in a Tropical Estuary. Marine Fisheries Information Service, pp. 10–14. Technical and Extension Series 220.

Wafar, M.V.M., Wafar, S., Devassy, V.P., 1986. Nitrogenous nutrients and primary production in a tropical oceanic environment. Bulletin of Marine Science 38, 273–284.

Wafar, M.V.M., Wafar, S., Rajkumar, R., 1993. Nitrogen uptake kinetics of freshly isolated zooxanthellae. Indian Journal of Marine Science 22, 83–88.

Ward, B.B., Devol, A.H., Rich, J.J., Chang, B.X., Bulow, S.E., Naik, H., et al., 2009. Denitrification as the dominant nitrogen loss process in the Arabian Sea. Nature 461, 78–81.

Xavier, J.K., Joseph, T., Paimpillii, J.S., 2005. Fluxes of nitrogen in Chaliyar river estuary India. International Journal of Ecological and Environmental Science 31 (3), 223–229.

SECTION C

# Nitrogen Flows (Air, Soil, Water)

# 21

# Assessment of Nitrate Threat to Water Quality in India

E.V.S. Prakasa Rao[1], K. Puttanna[2], K.R. Sooryanarayana[3], A.K. Biswas[4], J.S. Arunkumar[5]

[1]CSIR-CENTRE FOR MATHEMATICAL MODELLING AND COMPUTER SIMULATION, BENGALURU, INDIA; [2]CENTRAL INSTITUTE OF MEDICINAL AND AROMATIC PLANTS, BENGALURU, INDIA; [3]CENTRAL GROUND WATER BOARD, MINISTRY OF WATER RESOURCES, RIVER DEVELOPMENT & GANGA REJUVENATION, BENGALURU, INDIA; [4]ICAR-INDIAN INSTITUTE OF SOIL SCIENCE, BHOPAL, INDIA; [5]UNIVERSITY OF AGRICULTURAL SCIENCES, BENGALURU, INDIA

**CHAPTER OUTLINE**



## Introduction

Human activities are responsible for consistent reactive nitrogen (Nr) export to the environment (Prakasa Rao and Puttanna, 2000). Nitrogen enrichment of aquatic systems has deleterious effects on the water quality of rivers, lakes, aquifers, and coastal and marine waters, and contributes to the phenomenon of eutrophication. Nitrate concentration of more than 45 mg $L^{-1}$ in drinking water can lead to human and animal health problems (Gupta et al., 2010). Water-soluble nitrogen is fundamental to the global food production. Enhanced-nitrogen mobilization through crop cultivation, animal husbandry, industrial and municipal waste water discharges, and fossil fuel combustion harms the aquatic environment and affects human, animal, and ecosystem health. Since aquatic ecosystems have a limited capacity to remove incoming nitrogen load, they need

The Indian Nitrogen Assessment. http://dx.doi.org/10.1016/B978-0-12-811836-8.00021-5

intervention to limit nitrate threat. In India, groundwater quality is increasingly threatened by agricultural, urban, and industrial wastes, and this phenomenon is location specific with high-spatial variability. Rapid urban growth and increasing standard of living have resulted in generation of large quantities of waste and poor waste management has resulted in groundwater pollution exposing people to polluted drinking water.

The trend of nitrate pollution of aquifers, rivers, river deltas, and agro-ecosystems needs an assessment in India. Nitrate pollution due to land-use changes, economic growth, and climate changes need greater understanding. Such a knowledge base will facilitate science–policy interactions and help in devising adaptation and mitigation strategies.

# Nitrogen Enrichment in Indian Waters

## Groundwaters

Human activities have threatened the groundwater quality and infrastructure, for collection, treatment, and disposal of solid and liquid wastes is inadequate in India (CPCB, 2007).Monitoring groundwater quality is an aspect in planning for safe drinking water, and was started in the year 1969 by Central Ground Water Board (CGWB, 2012, 2014a,b, 2016). At present CGWB has a network of 22,339 groundwater observation wells, out of which 16,190 are dug wells and 6149 are piezometers (i.e., bore wells) (CGWB, 2016). Central Ground Water Board monitors groundwater quality annually through a network of about 15,000 observation wells located throughout the country (CGWB, 2014a,b); however, data on decadal changes in nitrate concentration in waters are not available. Apart from these observation wells, the quality is also monitored through various studies such as groundwater management studies, groundwater exploration, water supply investigations, which includes nitrate in groundwater on a regional scale. A study done by the Ministry of Drinking Water and Sanitation in 2011 indicated that out of 16,64,186 habitations of the country, 121,046 (7.27%) are quality-affected habitations, out of which 3866 (0.232%) are nitrate-affected.

It was observed that in coarse textured soils, movement of nitrate is quicker compared to fine textured soils. Indiscriminate disposal of wastes, absence of sewage systems accompanied by high density of septic systems, and excessive N fertilizer applications in agriculture resulted in leaching of the residual nitrate causing high nitrate concentration in groundwater in India.

Though nonagricultural sources cannot affect the aquifers regionally, they can cause nitrate pollution locally. Nitrate pollution of groundwater has been reported from different hydrogeological terrain in India (Somasundaram et. al., 1993, Namade and Srivastava, 1997; Datta et. al., 1997). Nitrate contamination has been a concern in the groundwater of different parts of India; such as the Gangetic plains (Handa, 1983, Pandey and Mukherjee, 1994; Saha et al.,2011), Rajasthan (Yadav and Dey, 2016; Saxena and Saxena, 2015), east coastal region (Umaran and Ramu, 2015), Maharashtra (Sachin et al., 2015), rural areas of Lucknow (Verma et al., 2014) and Meerut (Veena, 2011),

Andhra Pradesh and Telangana (Kuity, 2015'; Brindha and Elango, 2013), Assam (Das et al., 2003), and Karnataka (Divya and Belagali, 2012; Prakash and Somashekar 2006).

In India, high concentration of nitrate (>45 mg/L) was found in many districts of (Table 21.1) Andhra Pradesh, Telangana, Bihar, Chhattisgarh, Delhi, Gujarat, Haryana, Himachal Pradesh, Jammu and Kashmir, Jharkhand, Karnataka, Kerala, Madhya Pradesh,

**Table 21.1** States/Districts Reported high Nitrate Concentration in Groundwater Beyond BIS (Bureau of Indian Standards) Norms

| Sl No | State | Districts |
|---|---|---|
| 1 | Andhra Pradesh | Srikakulam, Vizianagaram, Visakhapatnam, East Godavari, West Godavari, Krishna, Guntur, Prakasam, Nellore, Chittoor, Kadapa, Kurnool, Ananthapur |
| 2 | Telangana | Adilabad, Nizamabad, Karimnagar, Warangal, Khammam, Hyderabad, Mahabubnagar, Medak, Nalgonda, Ranga Reddy |
| 3 | Bihar | Aurangabad, Banka, Bhagalpur, Bhojpur, Darbhanga, Kaimur (Bhabua), Patna, Rohtas, Saran, Siwan |
| 4 | Chhattisgarh | Bastar, Bilaspur, Dantewada, Dhamtari, Jashpur, Kanker, Kawardha, Korba, Mahasamund, Raigarh, Raipur, Rajnandgaon |
| 5 | Delhi | East Delhi, Central Delhi, New Delhi, North Delhi, North West Delhi, South Delhi, South West Delhi, West Delhi |
| 6 | Gujarat | Ahmedabad, Amreli, Anand, Banaskantha, Bharuch, Bhavnagar, Dahod, Jamnagar, Junagadh, Kachchh, Kheda, Mehesana, Narmada, Navsari, Panchmahals, Patan, Porbandar, Rajkot, Sabarkantha, Surat, Surendranagar, Vadodara |
| 7 | Haryana | Ambala, Bhiwani, Faridabad, Fatehabad, Gurgaon, Hisar, Jhajjar, Jind, Kaithal, Karnal, Kurukshetra, Mahendragarh, Panchkula, Panipat, Rewari, Rohtak, Sirsa, Sonipat, Yamuna Nagar |
| 8 | Himachal Pradesh | Himachal Pradesh |
| 9 | Jammu and Kashmir | Jammu, Kathua, Anantnag, Kupwara |
| 10 | Jharkhand | Chatra, Garhwa, Godda, Gumla, Lohardaga, Pakur, Palamu, Paschimi Singhbhum, Purbi Singhbhum, Ranchi, Sahibganj |
| 11 | Karnataka | Bagalkot, Bangalore, Belgaum, Bellary, Bidar, Bijapur, Chikmagalur, Chitradurga, Davanagere, Dharwad, Gadag, Gulbarga, Hassan, Haveri, Kodagu, Kolar, Koppal, Mandya, Raichur, Shimoga, Udupi, Uttar Kannada |
| 12 | Kerala | Alappuzha, Idukki, Kollam, Kottayam, Kozhikode, Malappuram, Palakkad, Pathanamthitta, Thiruvananthapuram, Thrissur, Wayanad |
| 13 | Madhya Pradesh | Alirajpur, Anuppur, Ashoknagar, Balaghat, Barwani, Betul, Bhind, Bhopal, Burhanpur, Chhatarpur, Chhindwara, Damoh, Datia, Dewas, Dhar, Dindori, Guna, Gwalior, Harda, Hoshangabad, Indore, Jabalpur, Jhabua, Khandwa, Khargone, Katni, Mandla, Mandsaur, Morena, Narsinghpur, Neemuch, Panna, Raisen, Rajgarh, Ratlam, Rewa, Sagar, Satna, Sehore, Seoni, Shahdol, Shajapur, Sheopur, Shivpuri, Sidhi, Singrauli, Tikamgarh, Ujjain, Umaria, Vidisha |
| 14 | Maharashtra | Ahmednagar, Akola, Amravati, Aurangabad, Beed, Bhandara, Buldhana, Chandrapur, Dhule, Gadchiroli, Gondia, Hingoli, Jalgaon, Jalna, Kolhapur, Latur, Mumbai, Nagpur, Nanded, Nandurbar, Nashik, Osmanabad, Parbhani, Pune, Sangli, Satara, Solapur, Wardha, Washim, Yavatmal |

*Continued*

**Table 21.1** States/Districts Reported high Nitrate Concentration in Groundwater Beyond BIS (Bureau of Indian Standards) Norms—cont'd

| Sl No | State | Districts |
|---|---|---|
| 15 | Orissa | Angul, Balasore, Bargarh, Bhadrak, Bolangir, Baudh, Cuttack, Deogarh, Dhenkanal, Gajapati, Ganjam, J. Singhpur, Jajpur, Jharsuguda, Kalahandi, Kendrapara, Keonjhar, Khurda, Koraput, Malkangiri, Mayurbhanj, Nawapada, Nayagarh, Phulbani, Puri, Sambalpur, Sundergarh, Sonpur |
| 16 | Punjab | Amritsar, Barnala, Bhathinda, Faridkot, Fatehgarh Sahib, Firozpur, Gurdaspur, Hoshiarpur, Jalandhar, Kapurthala, Ludhiana, Mansa, Moga, Muktsar, NawanShahr, Patiala, Ropar, Rupnagar, Sangrur, Tarn-Taran |
| 17 | Rajasthan | Ajmer, Alwar, Banswara, Baran, Barmer, Bundi, Bharatpur, Bhilwara, Bikaner, Chittaurgarh, Churu, Dausa, Dhaulpur, Dungarpur, Ganganagar, Hanumangarh, Jaipur, Jaisalmer, Jalore, Jhalawar, Jhunjhunu, Jodhpur, Karauli, Kota, Nagaur, Pali, Partapgarh, Rajasamand, Sirohi, Sikar, Sawai Madhopur, Tonk, Udaipur |
| 18 | Tamil Nadu | Chennai, Coimbatore, Cuddalore, Dharmapuri, Dindigul, Erode, Kancheepuram, Kanyakumari, Karur, Madurai, Namakkal, Nilgiris, Perambalur, Pudukkottai, Ramanathanpuram, Salem, Sivagangai, Theni, Thiruvannamalai, Thanjavur, Tirunelveli, Thiruvallur, Trichy, Tuticorin, Vellore, Villupuram, Virudhunagar |
| 19 | Uttar Pradesh | Agra, Aligarh, Allahabad, Ambedkar Nagar, Auraiya, Azamgarh, Badaun, Baghpat, Balrampur, Banda, Barabanki, Bareilly, Basti, Bijnor, Bulandshahr, Chitrakoot, Etah, Fatehpur, Firozabad, GB Nagar, Ghaziabad, Ghazipur, Hamirpur, Hardoi, Hathras, Jaunpur, Jhansi, Kannauj, Kanpur Dehat, Lakhimpur, Mahoba, Mathura, Meerut, Mau, Moradabad, Muzaffarnagar, Mirzapur, Raebarelli, SantRavidas Nagar, Shajahanpur, Sitapur, Sonbhadra, Sultanpur, Shravasti, Siddarthnagar, Unnao |
| 20 | Uttara khand | Dehradun, Haridwar, Udham Singh Nagar |
| 21 | West Bengal | Bankura, Bardhaman |
| Total districts | | 387 in 21 states |

CGWB, Overview of Ground Water Quality (http://www.cgwb.gov.in/wqoverview.html).

Maharashtra, Orissa, Punjab, Rajasthan, Tamil Nadu, Uttar Pradesh, Uttarakhand, and West Bengal, the highest concentration of 3080 mg/L was found in Bikaner, Rajasthan. These reports suggest that nitrate pollution of groundwater is reaching unacceptable levels in different parts of India. Central Pollution Control Board (CPCB) reported high nitrate concentrations in the groundwaters of 23 metropolitan cities (Fig. 21.1), indicating that improper treatment of urban wastes threaten groundwater quality.

## Rivers

Monitoring of Indian rivers for nitrates is scanty and some data are available for rivers in developed and less-developed regions of the country. Jamwal et al. (2015) observed high levels of nitrates in the inflow water samples of Vrishabhavathi river valley treatment plant, Bengaluru, Karnataka (35–40.2 mg $L^{-1}$). Nitrate content in the upstream was 23.9 mg $L^{-1}$ and in the downstream was 26.1 mg $L^{-1}$. Mallika et al. (2015) studied the water quality of Vaigai River, Tamil Nadu and reported high nitrate content in Thenur at

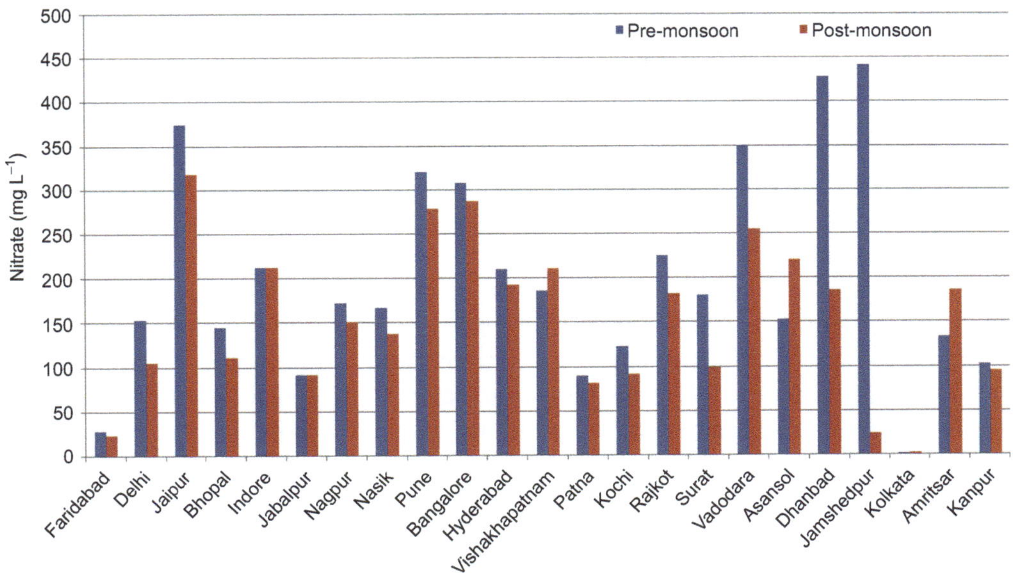


**FIGURE 21.1** Nitrate content in groundwater of metropolitan areas. *CGWB, 2014a. Concept Note on Geogenic Contamination of Ground Water in India with a Special Note on Nitrate; CGWB, 2014b. Ground Water Year Book, 2013–14.*

1.5 mg L$^{-1}$. Nitrate content in Cauvery River around Krishna Raja Sagara (KRS) Dam, Karnataka ranged from 0.037 to 4.04 mg L$^{-1}$ during monsoon, from 0.02 to 0.04 mg L$^{-1}$ during post-monsoon, and from 0.0035 to 0.088 mg L$^{-1}$ in pre-monsoon season (Susheela et al., 2014). Bhagat et al. (2013) reported nitrate content of river Sutlej in Nangal area of Punjab in the range of 95–120.0 mg L$^{-1}$ at effluent drain of National Fertilizers Limited indicating eutrophic conditions throughout the year. Jadhav et al. (2013) reported that nitrate concentration of Mula-Mutha River in Pune, Maharashtra was within the prescribed limits (13.1–18.5 mg L$^{-1}$). In Chandlodia Lake in Ahmedabad, Gujarat, the amount of nitrate recorded ranges between 5.8 ± 0.18 and 7.4 ± 0.22 (Pradeep et al., 2012). Nitrate levels in River Yamuna, Delhi, ranged from 1.38–2.9 mg L$^{-1}$ in summer to 1.51–3.1 mg L$^{-1}$ in winter seasons (Kaur and Indu Singh, 2012). Nitrate content in Chambal river water in National Chambal Sanctuary (Madhya Pradesh) ranged from 0.008 to 0.025 mg L$^{-1}$ (Saksena et al, 2008). These data suggest that river waters are mostly safe for direct consumption, except in the areas where urban and industrial wastes are released.

## River Deltas

Vetrimurugan et al. (2013) carried out water quality studies in the Cauvery deltaic region, India. Spatial and vertical distributions of ions reveal that the shallow wells and wells in coastal parts have high nitrate content ranging from below detectable limit to 120 mg L$^{-1}$. Studies of the nitrate concentrations in the groundwater samples of Krishna delta, Andhra Pradesh, have shown a large variation of nitrate from 10 to 135 mg $L^{-1}$ (Mondal et al., 2008). About 39% of 79 groundwater samples contained high nitrate contents (>50 $mgL^{-1}$). Nitrate pollution level was found more in dug wells (49%) compared to hand pumps/bore wells (31%). Nitrate concentrations in groundwater of Vamsadhara River basin, Andhra Pradesh varied from 2 to 146 mg $L^{-1}$ during the pre-monsoon period (April) and 2–324 mg $L^{-1}$ during the post-monsoon period (November), and more than 75% of the groundwater samples contain nitrate concentration within the safe limit of 45 mg $L^{-1}$ during the post-monsoon period (Srinivasa Rao, 1998). Vinod et al. (2015) estimated the nitrate leaching to groundwater at six different places around the village of Srirangapatna taluk near Cauvery River, where organic manure and inorganic fertilizers, together with irrigation water, were the major sources of nitrate in the area. Kundu and Mandal (2008) reported that nitrate content in drinking groundwater of an intensively cultivated district of Bengal basin in India is a function of its agricultural activities. Habitational areas contributed more to nitrate pollution of groundwater than agriculture; even in agricultural areas, the nitrate threat is more in shallow than deep aquifers.

## Nitrates in Agro-Ecosystems

Intensification of agricultural production involving the application of N fertilizers and organic wastes in the past 50 years led to the problems of nitrate leaching and the contamination of ground and surface waters in India. Nitrate leaching can occur in many agricultural production systems, and the amounts leached may typically vary in the following order: forest < cut grassland < grazed pastures < arable cropping < plowing of pastures < vegetables (Biswas et al., 2010).

In light-textured soils of Punjab, nitrate leaching to groundwater has been reported due to irrigation and high N application rates (>200 kg $ha^{-1}y^{-1}$) (Bijay Singh et al., 2007). Bijay Singh and Sekhon (1976) reported that nitrate concentration of water decreased significantly with depth of water table and correlated positively with amount of N applied on the farms located in the vicinity of the wells (4–10 m deep). Monitoring of the nitrate concentrations in the shallow-well waters in 1982 and 1988 revealed that the increase in N fertilizer applications was associated with an increase of nitrate-N by almost 2 mg $L^{-1}$ (Bijay Singh et al., 1991). However, when water samples were drawn from deep wells (21–38 m), the nitrate concentrations were not related to N application, and the nitrates were higher (>5 mg/l $NO_3$-N) in vegetable growing areas (17%) in comparison with rice–wheat (3%) and potato–wheat (6%) areas in Punjab (Bajwa et al., 1993). These results suggest that in similar agro-ecosystems in India where high N fertilizer use is in practice, nitrate pollution of shallow wells is a threat to human health and deeper wells may be safer.

Biswas et al. (2010) have mapped some diverse agro-ecosystems (West Godavari, Ferozepur, Hooghly, Coimbatore, and Hoshangabad) (Fig. 21.2). There were wide and significant variations in the levels and extent of nitrate in groundwaters. High fertilizer usage areas accompanied by high soil organic matter and light soil texture recorded

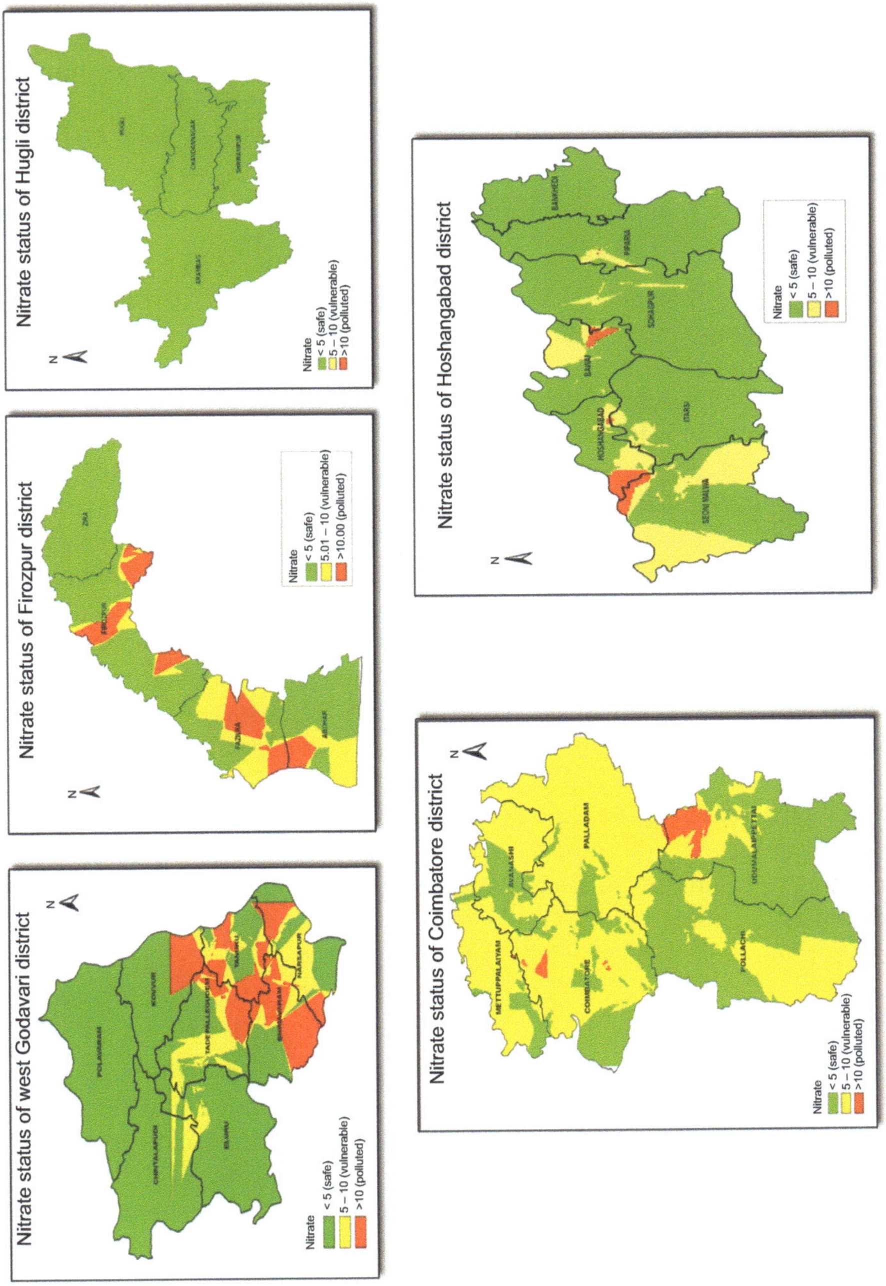


**FIGURE 21.2** Delineation of nitrate contaminated areas in diverse regions in India (Biswas et al., 2010).

higher levels of nitrate pollution as compared to medium and low areas. A significant negative correlation was observed between nitrate content in water and depth of sampling aquifers in all the districts. The sources from habitation areas were much more polluted (17.8%) than those from crop fields (5.5%) and dairy (3.7%). Intensive systems such as rice (double cropping), vegetables, and shallow root crops have contributed to more nitrate pollution of aquifers. Thus strategies for efficient N management in agriculture, urban waste treatment, and use of drinking water from proper depth need to evolve depending on local conditions.

### Nitrate and Climate Change

Climate change brought about by global warming is expected to bring about many changes including temperature, rain fall distribution, frequency of extreme events, pests and diseases, and possible human responses to these by way of alternate cropping systems, altered irrigation and fertilizer application practices, etc. All these are going to affect nitrate concentrations in soil, plants, and water. Predicted physical and biophysical impacts of climate change in India include heat extremes, water and groundwater scarcity, changes in monsoon pattern, droughts, glacial loss, and rising sea-level. Increased temperatures and moisture availability will lead to rapid mineralization of both organic and inorganic N and release nitrates to the environment. Climate change may affect nitrate movement to aquifers in a complex way. Urban, industrial, and agricultural sources will continue to be the major factors for high nitrate concentrations in groundwaters. Climate change accompanied by changes in temperature, quantity and distribution of rainfall, and higher $CO_2$ concentration in atmosphere will affect the nitrate source through changes in biomass productivity, and soil and groundwater processes. Nonagricultural sources, urbanization, and industries are also expected to be affected. Climate change will affect the hydrological cycle with changes to recharge, groundwater levels, and resources and flow processes which will likely result in changes in nitrate leaching ranging from limited increases to a possible doubling of aquifer concentrations by 2100 (Stuart et al., 2011).

## Conclusion

Nitrate pollution of waters is poised to increase with rapid economic, urban, and agricultural development in India; however, the phenomenon could be spatially and temporally variable. Monitoring groundwater is costly and time-consuming; however, dissemination of available data though web portals such as that of CGWB will help to take up corrective measures by regional agencies. Mitigation strategies are warranted to combat nitrate pollution (Prakasa Rao and Puttanna, 2006). Habitations affected by nitrate contamination are being covered by multivillage water supply schemes by surface water source as a long-term measure. Installations of reverse osmosis plants are the short-term options to address the issue. The key to reducing nitrate pollution is by preventing the accumulation of mineral N in the soil profile before the leaching season starts which can be achieved by proper treatment of urban wastes and sewage and efficient N management methods in agriculture. It is important to adopt an integrated approach in regional planning, to balance the mix of different land uses and optimize their geographical locations. Decisions on land-use changes, switching from less intensive to more intensive systems, should take possible impacts into account that this change may have on nitrate leaching and other environmental implications. Nitrate pollution in ecosystem compartments has to be considered in both science and policy perspectives. Improved N fertilizer management in different agro-ecosystems in India will help to reduce nitrate load on the environment. Suitable policy measures for N fertilizer subsidy and initiatives such as mandatory coating of all urea with neem which reduces N fertilizer use will help in reduced nitrate pollution from agriculture. Some recent government initiatives for river water cleaning will help in arresting nitrate pollution of water bodies. More comprehensive monitoring of nitrate dynamics in surface and groundwaters accompanied by the implementation of water policy regulations will be needed to combat the threat of nitrate pollution of waters. Positive synergies among interdisciplinary sciences, sectoral policy implementing agencies, and public awareness must be encouraged to address the threat of nitrate pollution in India.

## References

Bajwa, M.S., Bijay Singh, Singh, P., 1993. Nitrate pollution of groundwater under different systems of land management in the Punjab. In: 1st Agriculture Science Congress., 1992 Proceedings of National Academy of Agriculture Science, New Delhi, pp. 223–230.

Bhagat Chauhan, S., Sagar, S.K., 2013. Impact of pollutants on water quality of river Sutlej in Nangal area of Punjab, India. Biological Forum – An International Journal 5 (1), 113–123.

Bijay Singh, Prakasa Rao, E.V.S., Singh, Y., Puttanna, K., 2007. Nitrate Pollution of Ground Water vis-à-vis Nitrogen Fertilizer Use in India. Agricultural Nitrogen Use and its Environmental Implications. In: Abrol, Y.P., Raghuram, N., S.Sachdev, M. (Eds.). I. K. International Publishing House Pvt. Ltd., New Delhi, pp. 439–458.

Bijay Singh, Sekhon, G.S., 1976. Nitrate pollution of ground water from nitrogen fertilizers and animal wastes in the Punjab, India. Agriculture Environment 3, 57–67.

Bijay Singh, Singh, Y., Khind, C.S., Meelu, O.P., 1991. Leaching losses of urea-N applied to permeable soils under wetland rice. Fertilizer Research 28, 179–184.

Biswas, A.K., Srivastava, S., Rao, S.A., Lakshmi, G.V., Mandal, B., Singh, B., Katkar, R.N., Boopathi, M., 2010. Nitrate Contamination of Groundwater in Some Heavily Fertilized and Intensively Cultivated Districts of India. IISS Research Bulletin No. 1, Indian Institute of Soil Science (ICAR) Nabibagh, Bhopal, pp. 1–89.

Brindha, K., Elango, L., 2013. Soil and groundwater quality with reference to nitrate in a semiarid agricultural region. Arabian Journal of Geoscience 4683–4695.

CGWB, 2012. Aquifer Systems of India.

CGWB, 2014a. Concept Note on Geogenic Contamination of Ground Water in India with a Special Note on Nitrate.

CGWB, 2014b. Ground Water Year Book, 2013–14.

CGWB, January, 2016. Ground Water Scenario in India.

CGWB, Overview of Ground Water Quality (http://www.cgwb.gov.in/wqoverview.html).

CPCB, 2007. Status of Ground Water Quality in India – Part – 1, Ground Water Quality Series: GWQS/09/2006–2007.

Das, B., Talukdar, J., Sarma, S., Gohain, B., Dutta, R.K., Das, H.B., Das, S.C., 2003. Fluoride and other inorganic constituents in groundwater of Guwahati, Assam, India. Current Science 85 (5), 657–661.

Datta, P.S., Deb, D.L., Tyagi, S.K., 1997. Assessment of groundwater contamination from fertilizers in the Delhi area based on $NO_3$ and $K^+$ composition. Journal of Contaminant Hydrology 27, 249–262.

Divya, J., Belagali, S.L., 2012. Impact of chemical fertilizers on water quality in selected agricultural areas of Mysore district, Karnataka, India. International Journal of Environmental Sciences 2 (3), 1449–1458.

Gupta, S., Gupta, R.C., Gupta, A.B., Eskiocak, S., Prakasa Rao, E.V.S., Puttanna, K., Aditi, S., 2010. Pathophysiology of nitrate toxicity in human and its mitigation measures. In: Raghuram, N. (Ed.), ING Bulletins on Regional Assessment of Reactive Nitrogen, Bulletin No. 20. SCON-ING, New Delhi, 1–4 and 1–78.

Handa, B.K., 1983. The effect of fertilizers use on groundwater quality in the phreatic zone with special reference to Uttar Pradesh. In: Proceedings of the Seminar on Assessment Development and Management of Groundwater Research, 2. Central Groundwater Board, New Delhi, India, pp. 415–425.

Jadhav, S.D., Jadhav, M.S., Jawale, R.W., 2013. Study of chloride and nitrate concentration of MulaMutha River in Pune city (Maharashtra). International Journal of Chemical and Life Sciences 2 (4), 1140–1142.

Jamwal, P., Zuhail, T.M., Raje, P., Srinivasan, V., Lele, S., 2015. Contribution of sewage treatment to pollution abatement of urban streams. Current Science 108 (4), 677–685.

Kaur, S., Indu Singh, 2012. Accelerated phosphate & nitrate level: factors to blame for eutrophication in Yamuna river, Delhi, India. International Journal of Plant Animal and Environmental Science 2 (3), 183–187.

Kuity, D.P., 2015. Groundwater quality: Indian scenario. Vistas in Geological Research 13, 91–100.

Kundu, M.C., Mandal, B., 2008. Agricultural activities influence nitrate and fluoride contamination in drinking groundwater of an intensively cultivated district in India. Water Air and Soil Pollution 198 (1), 243–252.

Mallika, S., Umamaheswari, R., Krishnamoorthy, S., 2015. Physico-chemical analysis of Vaigai River water quality in Madurai district. Tamil Nadu. International Journal of Applied Research 1 (8), 597–600.

Mondal, N.C., Saxena, V.K., Singh, V.S., 2008. Occurrence of elevated nitrate in ground waters of Krishna delta, India. African Journal of Environmental Science and Technology 2 (9), 265–271.

Namade, P.N., Srivastava, N.S., 1997. Nitrate ($NO_3$-N) status of groundwater in tribal belt of Satpura valley. Indian Journal of Environmental Protection 17, 430–433.

Pandey, D.S., Mukherjee, S., 1994. Nitrate contamination in groundwater in Varanasi city and its environs, Uttar Pradesh, India. In: Proceedings of the Regional Workshop on Environmental Aspects of Groundwater Development, 17–19 October (123–129). IAH Roorkee: Kurukshetra University, India.

Pradeep Verma, U., Anshita Purohit, R., Naimesh Patel, J., 2012. Pollution status of Chandlodia lake located in Ahmedabad-Gujarat. International Journal of Engineering Research and Applications 2 (4), 1600–1606.

Prakasa Rao, E.V.S., Puttanna, K., 2000. Nitrates, agriculture and environment. Current Science 79 (9), 1163–1168.

Prakasa Rao, E.V.S., Puttanna, K., 2006. Strategies for combating nitrate pollution. Current Science 91 (10), 1335–1339.

Prakash, K.L., Somashekar, R.K., 2006. Ground water quality – assessment on Anekal Taluk, Bangalore urban district, India. Journal of Environmental Biology 27 (4), 633–637.

Sachin, A.A., Vijay Yeul, S., Bhagat, P., 2015. Integration of ground water quality: identifying potential hazards in Yavatmal District, India. Journal of Chemical and Pharmaceutical Research 7 (4), 512–517.

Saha, D., Sinha, U.K., Dwivedi, S.N., 2011. Characterization of recharge processes in shallow and deeper aquifers using isotopic signatures and geochemical behavior of groundwater in an arsenic-enriched part of the Ganga Plain. Applied Geochemistry 10–16.

Saksena, D.N., Garg, R.K., Rao, R.J., 2008. Water quality and pollution status of Chambal river in National Chambal sanctuary, Madhya Pradesh. Journal of Environmental Biology 29 (5), 701–710.

Saxena, U., Saxena, S., 2015. The statistical assesment of fluoride and nitrate contamination status of ground water in various tehsils of district Jaipur, Rajasthan, India. International Journal of Research Studies in Biosciences 3 (3), 107–131.

Somasundaram, M.V., Ravindran, G., Tellam, J.H., 1993. Ground-water pollution of Madras urban aquifer, India. Groundwater 31 (1), 4–11.

Srinivasa Rao, N., 1998. Impact of clayey soils on nitrate pollution in the groundwater of the lower Vamsadhara River basin, India. Journal of Hydrological Sciences 43 (5), 701–714.

Stuart, M.E., Gooddy, D.C., Bloomfield, J.P., Williams, A.T., 2011. A review of the impact of climate change on future nitrate concentrations in groundwater of the UK. Science of the Total Environment 409, 2859–2873.

Susheela, S., Swamy, S.S., Kumar, S.D., Gowda, A., Jagadish, K., 2014. Study of Cauvery river water pollution and its impact on socio-economic status around KRS Dam, Karnataka, India. Journal of Earth Sciences and Geotechnical Engineering 4 (2), 91–109.

Umaran, P., Ramu, A., 2015. Assessment of groundwater quality of east coastal region in Tamilnadu, India. International Journal of Research in Chemistry and Environment 5 (4), 80–90.

Veena, C., 2011. Nitrate toxicity and possible contaminant sources in groundwater of Western U.P., India. Vegetos 24 (2), 55–61.

Verma, A., Rawat, A.K., More, N., 2014. Extent of nitrate and nitrite pollution in ground water of rural areas of Lucknow, U.P. India. Current World Environment 9 (1), 114–122.

Vetrimurugan, E., Elango, L., Rajmohan, N., 2003. Sources of contaminants and groundwater quality in the coastal part of a river delta. International Journal of Environment Science and Technology 10 (3), 473–486. http://dx.doi.org/10.1007/s13762-012-0138-3.

Vinod, P.N., Chandramouli, P.N., Koch, M., 2015. Estimation of nitrate leaching in groundwater in an agriculturally used area in the state Karnataka, India, using existing Model and GIS. Aquatic Procedia 4, 1047–1053.

Yadav, S.S., Dey, A., 2016. Ground Water Year Book 2014–2015-Rajasthan–Western Region. Central Ground Water Board, pp. 1–108.

22

# Reactive Nitrogen Dynamics in the Mangroves of India

N. Priya, P. Ranjan, S.M. Sappal, AL. Ramanathan
*JAWAHARLAL NEHRU UNIVERSITY, NEW DELHI, INDIA*

**CHAPTER OUTLINE**



## Mangrove and Its Distribution in India

Mangroves occupy less than 1% of the world's surface (Singh et al., 2012). Globally, mangroves occupy an area of 1,37,760 km$^2$ covering about 118 countries and territories in the tropical and subtropical regions of the world, with three-fourth of its area in only 15 countries. Indian mangroves share 3.1% of the global mangrove distribution covering an area of 4461 km$^2$ of the 7500-km long Indian coastal line. They are unevenly distributed along the coast with their dominance along the east coast (60%), followed by

**The Indian Nitrogen Assessment. http://dx.doi.org/10.1016/B978-0-12-811836-8.00022-7**

west coast (27%), and the remaining 13% are distributed along Andaman and Nicobar Islands. The major Indian mangroves are as follows:

Sundarbans (West Bengal): It is a tide-dominated estuarine wetland, located at the deltaic plains of the Ganges–Brahmaputra rivers and extended over an area of 9600 $km^2$ (Banerjee et al., 2012). Indian Sundarbans mangrove, covers an area of 2106 $km^2$ (FSI, 2015).

Bhitarkanika (Odisha): It is located in Kendrapara district. It was recognized as a wildlife sanctuary with total area of 672 $km^2$ in 1975 (Sarangi et al., 2002) out of which mangrove cover is 231 $km^2$ (FSI, 2015).

Coringa and Gaderu (Andhra Pradesh): It lies to the north of Gautami–Godavari rivers. The area between Gautami–Godavari and Kakinada bay has dense vegetation of mangrove forests associated to Coringa wildlife sanctuary (Ray et al., 2006). Total mangrove forest of Andhra Pradesh is extended over an area of 367 $km^2$ (FSI, 2015).

Pichavaram (Tamil Nadu): Pichavaram mangrove forest is located between the Vellar and Coleroon estuaries. The mangrove vegetation is mainly dominated in southern part, whereas northern part is dominated by mudflats (Ranjan et al., 2013; Ramanathan et al., 1999). Mangrove cover of Tamil Nadu occupies an area of 47 $km^2$ (FSI, 2015).

Kumarakam Mangrove (Kerala): It is extended along the south-eastern border of the Vembanad estuary; mangrove forest covers an area of 11 $km^2$ (FSI, 2015). Kumarakam mangrove is severely affected by Thannirmukham barrier and shell-mining activities in Kumarakam mangroves which lead to severe regeneration problem of mangrove (Badarudeen et al., 1996).

Goa mangrove (Goa): It is the most prominent and extensive backwaters with mangrove located along the east of the capital city Panaji. It extends over Mandovi–Zuari estuarine complex (Attri and Kerkar, 2011) covering an area of 26 $km^2$ (FSI, 2015).

Gulf of Kachchh (Gujarat): It is one of the most diversified and productive habitats along the western coast of India. It covers an area of 1107 $km^2$ (FSI, 2015) of mangrove, out of which Gulf contributes approximately 96% of the total.

Andaman and Nicobar (Andaman and Nicobar Islands): It is located at a distance of 1200 km from the mainland of India, extended over an area of about 8249 $km^2$ covered by luxuriant tropical rain forest out of which 617 $km^2$ (FSI, 2015) is mangrove forest. Coral reefs form the dominant ecosystem of Andaman Islands and it creates home ground for sea grasses and mangrove in the lagoons and creeks (Nobi et al., 2010).

## Reactive Nitrogen Species and Its Transformation Processes

All forms of nitrogen species which are biologically, photochemically, and radiatively active come under reactive nitrogen species such as nitric oxide (NO), nitrous oxide ($N_2O$), ammonia ($NH_3$), and ammonium ion $\left(NH_4^+\right)$. These reactive nitrogen species play very important roles in the chemistry and climate of the present-day earth. Reactive nitrogen species are mainly the intermediate or by-product of different nitrogen cycles.

Nitric oxide (NO) is chemically very reactive; hence, it is a short-lived compound in aquatic systems. NO is produced and consumed as an intermediate during nitrification and denitrification, respectively (Bange, 2008). It is also produced photochemically in nitrite-rich surface water via reduction of nitrite.

Nitrous oxide ($N_2O$) is also produced mainly as a by-product of nitrification and as an intermediate product in denitrification (Bange, 2006). $N_2O$ production via nitrification or denitrification is mainly dependent on $O_2$ availability.

Ammonia ($NH_3$) and ammonium $(NH_4^+)$ distribution is strongly coupled with each other by $NH_3/NH_4^+$ equilibrium. $NH_3$ prevail in the ocean water as a dissolved, non-protonated gas; hence, it is accessible for gas exchange across the ocean–atmosphere interface; whereas, $NH_4^+$ is the by-product of nitrification, assimilation by phytoplankton, and excretion of zooplankton, and can also be produced by photochemical decomposition of dissolved organic nitrogen (DON).

Estuary and wetland receive nutrient from many sources such as organic matter, river input, runoff, atmospheric deposition, onshore transport of nutrient from sea, and to a small scale from $N_2$ fixation. Transformation process of nitrogenous compound is mainly mediated by bacteria. Nitrogen transformation processes which most likely occur in coastal ecosystem include ammonification, nitrification, denitrification, anaerobic ammonia oxidation (anammox), chemodenitrification, dissimilatory nitrate reduction to ammonia (DNRA), and oxygen-limited autotrophic nitrification–denitrification (OLAND).

Ammonification is carried out by bacteria as well as by actinomycetes and fungi, which help in converting organic-N to inorganic-N (Herbert, 1999). Nitrification is a major source of $N_2O$ production under aerobic conditions (William et al., 1998). Nitrification is a part of bacterial respiratory process carried by ammonia-oxidizing and nitrite-oxidizing bacteria (Inamori et al., 2008). It is a two-step process where in the first step ammonia is oxidized to nitrite and in the second step nitrite is converted into nitrate. Ammonia-oxidizing bacteria produces nitrite $(NO_2^-)$ catalyzed by ammonia monooxygenase (Amo) and hydroxylamine oxireductase (Hao), whereas nitrite-oxidizing bacteria convert nitrite $(NO_2^-)$ to nitrate $(NO_3^-)$ catalyzed by oxidoreductase (Nor). $N_2O$ is formed during the incomplete oxidation of $NH_2OH$ to $NO_2^-$. As a by-product of this process, $N_2O$ is formed by oxidation, mediated by hydroxylamine oxireductase (Hao) at low oxygen concentration (below 10 $\mu$mol $L^{-1}$) (Hynes and Knowles, 1984; Jørgensen et al., 1984). Nitrification process takes place in the water column as well as in oxic-sediment layers, and produces nitrate which is either assimilated by phytoplankton and algae, or produces $N_2$ and $N_2O$ during denitrification.

Denitrification is the process which permanently removes reactive nitrogen from coastal ecosystem; under hypoxic (below 10 $\mu$mol $L^{-1}$) or anaerobic condition, facultative aerobic bacteria and aerobic microorganisms convert $NO_3^-$ to $N_2$, producing $N_2O$ as an intermediate product. Conversion of $NO_3^-$ to $NO_2^-$ is catalyzed by nitrate reductase (Nar); $NO_2^-$ to NO is catalyzed by copper nitrite reductase (Nir); NO to $NO_2^-$ is catalyzed by nitrous reductase (Nr); whereas, $N_2O$ to $N_2$ is catalyzed by nitrous oxide reductase

(Nos) (Chon et al., 2011). Denitrification process is strongly governed by factors such as temperature, nitrate concentrations, and organic carbon availability (Hulth et al., 2005); these factors make process seasonally variable (Tuominen et al., 1998). There is a strong coupling between nitrification and denitrification processes in sediment below oxic layer where nitrite is produced by nitrification in oxic layer leach in hypoxic region and is denitrified (Herbert, 1999). $N_2O$ is formed by nitrification and denitrification processes at suboptimal condition, i.e., low oxygen during nitrification or low C/N ratio during denitrification (Fuchs et al., 2011).

Anaerobic ammonia oxidation (anammox) process produces $N_2$ by anaerobic chemoautotrophic bacteria which fixes $CO_2$ and reduces $NH_4^+$ to $NO_2^-$, finally producing $N_2$ (Thamdrup and Dalsgaard, 2002). In chemodenitrification, $N_2$ is produced by reaction of manganese species with nitrate or ammonium (Brandes et al., 2007). DNRA is an anaerobic process by fermentative bacteria (Herbert, 1999); it mainly occurs under high reducing conditions (Lam et al., 2009; Christensen et al., 2000).

## Significance of Nitrogen in Mangrove Ecosystem

Carbon (C), nitrogen (N), and phosphorus (P) are very important nutrients in the wetland ecosystem (Zhang et al., 2012), and may cause nutrient limitation, generally with N and sometimes with P (Boto, 1982) in mangrove forests. Study of nitrogen dynamics along the coastal area becomes significant due to low biologically available nitrogen in the system (Purvaraja et al., 2008). It directly influences the carbon cycle in the system. High N can cause more biomass production of phytoplankton and mangroves, and hence affects the removal and addition of $CO_2$ from atmosphere. Globally, estuarine nutrient loads have steadily increased due to rising human population and its perturbations through many activities such as industries. Such increases may enhance primary production and provide either an additional sink and/or source for carbon as well as for phosphorus, nitrate, and sulfate in coastal waters. Higher anthropogenic nitrogen input in the system will make it more acidic or basic, which may, in turn, affect many other biogeochemical processes that are pH dependent. Nitrogen may also affect the biodiversity in the system, as higher nitrogen in the system may encourage only certain groups of organisms and decrease its biodiversity. Also, N excess can cause vegetation loss and loss of diversity. The phytoplankton composition changes with nutrient fluxes because individual taxa have different requirements (Tilman, 1977). Enclosure experiments have confirmed that nutrient enrichment alters the community composition of phytoplankton in such dynamic conditions (Goldman and Stanley, 1974). Various studies have identified mangrove ecosystem as sink for N and P (Nedwell et al., 2002; Tam et al., 1995) and have proposed it to be used as a record of nutrient loading (Khan and Brush, 1994). Entrapped N and P are concerned as the nutrients which most likely limit the mangrove productivity (Kraus et al., 2008). Nutrient status in the sediment and water also indicates the pollution status in the area. High nutrient level recorded in Pichavaram mangrove water was linked with the expansion of aquaculture ponds (75%)

in the area (Bala Krishna Prasad et al., 2006). High nutrient in the region can lead to sulfate oxidation and release of sulfide, which is toxic to the biomes of the area. Indian estuaries and mangroves are heavily affected by anthropogenic activities which include agricultural runoff, sewage discharge, aquaculture ponds runoff, proximity of fertilizer industries, and establishment and expansion of port and cities in the coastal area. India with the second largest population in the world and being an agriculture-dependent country has possibility of increase in these anthropogenic influences to its coastal system making it more vulnerable to eutrophication and deterioration. Mangroves act as sink for atmospheric carbon dioxide and riverine nutrients including nitrogen. Most of the nitrogen gets recycled within the system, e.g., in the Sundarban mangrove ecosystem $N_2$-fixation and sediment–water exchange led to the storage of $5 \times 109$ mol year$^{-1}$ and $8 \times 106$ mol year$^{-1}$ of N. This serves 74% of nitrogen demand for mangrove net production (Ray et al., 2014). Mangroves are, however, being destructed and deteriorated at an alarming rate due to numerous human perturbations, including cutting of trees for fuel woods and aquaculture expansion, polluting them though sewage and industrial discharge, and damming the rivers causing higher salinity in mangrove water. This may affect various nitrogen-related biogeochemical processes and composition of its large reservoirs (the biosphere, the hydrosphere, the atmosphere, and the geosphere). Also, nitrogen is the limiting nutrient for primary productivity in some coastal systems (Howarth and Marino, 2006); thus it becomes critical to understand the present status of reactive nitrogen in Indian mangrove ecosystem along with various environmental and biological factors effecting its cycling.

## Dissolved Nitrogen Dynamics in Mangrove Water

### Sources and Sinks

Dissolved nitrogen in mangrove water consists of organic and inorganic nitrogen. The dissolved inorganic nitrogen (DIN) includes nitrate, nitrite, and ammonium species. The possible sources of DIN in mangrove ecosystem include (1) river carrying runoff from agricultural fields, (2) regeneration of nutrients from allochthonous (particulate matter from forest and sewage) and autochthonous (mangrove litter, dead and decaying organisms), (3) domestic sewage discharge and effluents from aquaculture ponds, (4) benthic nutrient regeneration, (5) groundwater flux from mangrove and adjoining regions, (6) atmospheric deposition, and (7) nitrogen fixation. During the south-west Indian monsoon (September–October), concentration of nitrate increases considerably in almost all estuaries and its mangrove ecosystem, ascribed to land drainage and precipitation (Sarma et al., 2009). Anthropogenic inputs have also been identified in almost all estuaries, the common sources being runoff from agricultural fields, sewage and industrial discharge, and aquaculture pond effluents.

Mangrove could provide important ecosystem service in recycling different forms of atmospheric and aquatic nitrogen, thereby protecting the coastal ecosystems from negative impacts of nutrient enrichment and atmospheric pollution. The major sink of DIN in many estuaries includes uptake by primary producers including mangrove trees, recycling through mineralization, and grazing within system and sedimentation. Lower amounts of nitrate during post-monsoon season in most of the Indian estuaries are due to higher primary productivity causing more utilization of nutrients during this period (Sarma et al., 2009). More autotrophic uptake is seen during low water levels and low flushing rates (Harrison et al., 1997). A net removal of nitrate (up to 14–32%) was been reported in Mandovi–Zurai estuaries (De Sousa et al., 1981), and also in Mahanadi estuary.

## Status of Dissolved Nitrogen in Indian Mangroves and Its Control

DIN species are in low concentrations in tropical mangrove water (Alongi et al., 1993) and Indian mangrove water is no exception with concentrations within μM range (Table 22.1).Very few measurements of DON have been done in the Indian mangroves. The DON along Hooghly–Matla estuarine mangrove ecosystem was reported with the maximum value of DIN during the monsoon season ranging from 0.72 to 0.87 mg $L^{-1}$ (Mandal et al., 2009). This was explained as leaching from mangrove soil organic nitrogen by rainfall. DIN variation among Indian mangrove ecosystems (Table 22.1) can be caused by differences in the freshwater and groundwater input plus productivity in the area. Ammonium is the most dominant DIN pool in mangrove water (Alongi et al., 1993). This holds true for most of the Indian mangroves; however, some mangroves have nitrate as the major DIN pool (Table 22.1). The dominant DIN pool varies depending on source, processing, and other environmental factors, redox potential being the major

**Table 22.1** Status of the Dissolved Nitrogen in Indian Mangrove Water (Range and Average Values, Unit μM)

| Study Area | Period | $NO_3$ | $NO_2$ | $NH_3$ | References |
|---|---|---|---|---|---|
| Mandovi estuary | Monsoon | 7–12 | 0–1 | 1.5–5.5 | Maya et al. (2011) |
| | Post-monsoon | 0–4.15 | 0.24–1.3 | 0.08–0.5 | |
| Sundarban mangrove | Monsoon | 17.74–22.74 | – | – | Mandal et al. (2009) |
| | Pre-monsoon | 6.12–14.68 | – | – | |
| | Annual average | 14.5 ± 1.5 | 1.6 ± 0.5 | 5.3 ± 0.6 | Ray et al. (2014) |
| Coringa mangrove | monsoon | 37.42 | – | – | Unpublished data, 2014 |
| | Pre-monsoon | 17.42 | – | – | |
| Pichavaram mangrove | Monsoon | 260 ± 29 | 4.4 ± 1 | 5.3 ± 1.1 | Bala Krishna Prasad et al. (2006) |
| | Post-monsoon | 149 ± 35 | 5.3 ± 0.9 | 12.6 ± 2.9 | |
| | Pre-monsoon | 107 ± 52 | 6 ± 1.2 | 8.1 ± 1.2 | |
| Kerala coast mangrove | Monsoon | 13.62 ± 6.70 | 0.50 ± 0.28 | 42.15 ± 34.67 | Manju et al. (2012) |
| | Post-monsoon | 3.63 ± 2.82 | 0.28 ± 0.16 | 18.20 ± 10.13 | |
| | Pre-monsoon | 1.82 ± 1.59 | 0.63 ± 0.25 | 22.38 ± 42.45 | |
| Bhitarkanika mangrove | Post-monsoon | 16.12 ± 3.2 | – | 9.4 ± 9.98 | Chauhan and Ramanathan (2008) |
| Muthupet mangrove | Annual average | 4.97 ± 1.7 | 2.02 ± 1.00 | 52.37 ± 22.84 | Krithika et al. (2008) |

one. Under anaerobic conditions of mangrove soil, ammonium is the end product of organic nitrogen nitrification, while the reaction proceeds with nitrate production under oxic conditions (Alongi et al., 1993).

Pore water DIN concentration from Indian mangrove ecosystem is not widely published; however, a few published data indicate lower concentration of nitrate in pore water than in the tidal water, signifying its utilization by the biota communities or reduction to ammonium (Singh, 2009; Ray et al., 2014). In the Sundarban mangrove, ammonium dominated the DIN pool of pore water, while nitrate remained the major pool for the tidal water (Ray et al., 2014). In the Mangalavanam mangrove ecosystem (Bava and Seralathan, 1999) located on south-west coast of India, similar finding was reported with ammonium concentration in interstitial water of mangrove region being threefold higher than the mangrove tidal water, while nitrate concentration being less than the tidal water. The ammonium concentration further increased consistently with increasing depth and reached up to 202 μM. Also, ammonium showed tidal variation with higher values at low tide, indicating the pore water input (Bava and Seralathan, 1999; Singh, 2009). Ammonium is primarily produced from ammonification of organic nitrogen in anaerobic mangrove soil, but it shows poor cation-exchange capacity in the presence of abundant $Na^+$ ion in saline mangrove sediment, and hence is mainly present in the interstitial water of mangrove sediment, being highly susceptible to leaching by heavy rain or draining out during tidal inundation (Alongi et al., 1993).This is interesting to note that most of the mangrove sites had higher DIN concentration during monsoon periods compared to non-monsoon periods, while lower DIN during the post-monsoon period. This is attributed to higher river discharge, land drainage, and anthropogenic factor during the monsoon period, while higher phytoplankton productivity and dilution with sea water during the post-monsoon period. Salinity controls the nitrogen transformation processes in the saline aquatic environments (Dittmar and Lara, 2001) and was identified as the major regulator of DIN pool in Pichavaram mangrove as average ammonium concentration increased with an increase in salinity while nitrate showed the reverse trend. Further, higher DIN values in Pichavaram mangrove creek during monsoon season have been linked to higher discharge and higher litter fall and degradation during this period. However, during most part of the year, nutrients get produced through in situ regeneration process (Senthilkumar et al., 2008). Hence, the system is highly productive and rich in phytoplankton abundance. Rajkumar et al. (2009) added from the decadal increase in DIN concentration in this mangrove has been linked with rapid degradation and conversion of mangrove to aquaculture ponds along with several years of below average rainfall. Bhitarkanika mangrove ecosystem has been reported with $16.12 \pm 3.2$ μM of dissolved $NO_3^-$ and $9.4 \pm 9.98$ μM of dissolved $NH_4^+$ during the post-monsoon season. Higher values in the adjoining estuary and bay were attributed to agricultural runoff (fertilizers), while lower values in the mangroves have been attributed to microbial consumption (Chauhan and Ramanathan, 2008). Riverine influence, land use pattern (less disturbed vs. anthropogenically disturbed), and redox potential were identified as the major regulators of DIN in the Sundarban mangrove ecosystem

(Singh, 2009). This includes higher $NO_3^-$ values in anthropogenically stressed areas as well as areas with high redox potential. The diurnal variation in DIN was also observed especially in pre-monsoon season with distinct tidal activities, lowering its value at high tide, and vice versa. Sundarban mangrove ecosystem has also been reported as a sink for DIN in its water column (Ray et al., 2014). DIN concentration was higher in tidal water than the pore water indicating removal of nitrogen during tidal recirculation. Sundarban receives anthropogenic inputs from nearby aquaculture farms, effluents from industries adjacent to Haldia port, and domestic sewage from the Kolkata mega city (Ray et al., 2014), which add nutrients including reactive nitrogen to the system. Besides the anthropogenic input, high litter fall and its degradation in the Sundarban also acts as an additional source (Ray et al., 2014). In fact, in a recent study along the Hoogly–Matla estuary, mangrove litter was identified as the major contributor to the DIN pool in the system (Mandal et al., 2009). Actually, mangrove litter is the major contributor to total nitrogen of mangrove soil of Hoogly–Matla estuarine system (little from anthropogenic input through agricultural activities), part of which converts to SIN (sediment inorganic nitrogen), which mainly contributes to DIN pool upon leaching. The seven-dimension conceptual model used for this study identifies redox potential and temperature (regulates microbial activity) as important factors for conversion of SON (sediment organic nitrogen) to SIN. Water DON and PON (particulate organic nitrogen) are primarily contributed by SON, while DIN is being added from the conversion of DON, PON, and SIN. DON gets converted to DIN through hydrolysis process which is dependent on water pH, temperature, and dissolved oxygen (DO).

The sources of nutrients in the mangrove of Gulf of Kachchh include anthropogenic input and waste water discharge. Along the Gulf of Kachchh, impact of terrestrial runoff was identified with $NO_3^-$ concentration elevating to an unusual value of 743.16 μM. The population of nitrate-forming bacteria was reported highest as $4.0 \times 10^3$ cfu $g^{-1}$ soil at few stations indicating dominance of nitrogen-fixing process which perhaps adds $NO_3^-$ in the overlying water (Kumar and Ramanathan. 2013). Along the Zurai estuary, results significantly revealed physical forcing (freshwater discharge and tidal circulation) facilitating sedimentary release as more important than the biogeochemical process in determining the intra- and interannual variability of nutrients including DIN (Aanand et al., 2014). Higher nitrate during low tide of Nava mangrove, Mumbai, was linked with domestic wastes, sewage, and effluents from container freight stations, oil tanking systems, and other maritime activities (Pawar, 2013). Along the mangrove regions of Kerala coast, higher DIN values in monsoon were explained by higher land drainage and anthropogenic input. Anaerobic environment induced by domestic load and crab holes resulted in ammonification and, hence, higher $NH_3$ values (Manju et al., 2012). Our unpublished data of DIN along Gautami–Godavari estuarine mangrove ecosystem (Coringa mangrove ecosystem) reveal average concentrations of 37.42 μM and 17.42 μM nitrate during wet and dry season, respectively, signifying the role of high river discharge and runoff in bringing these nutrients to this system during monsoon season. This is also reflected in distribution pattern as these nutrients are highest in the estuary followed by mangroves and Kakinada Bay. The mixing model (wet season) identified nonconservative behavior of DIN with a net removal of 1.06 ppm (24.60%) $NO_3^-$ in the mangrove creek and 2.44 ppm (40.7%) $NO_3^-$ in the lower estuary.

It is to be noted that majority of the Indian mangrove waters have N:P ratio ranging from 4.64 to 11.43 (Prasad, 2012) and based on the Redfield ratio (N:P = 16:1), it should be considered potentially N-limiting.

# Reactive Nitrogen in the Mangrove Sediments

The coastal ecosystems such as mangroves are highly dynamic and spatially heterogeneous compartments at the land–ocean interface. Due to their strategic location, they receive relatively large amounts of nutrients and particles from land via rivers, and exchange matter with the open ocean and atmosphere. Nitrogen is an essential nutrient for a variety of biological and chemical processes; both at micro (organism) as well as macro (ecosystem) levels. It is present in both inorganic (namely ammonium, nitrate, nitrite) as well as organic forms. Major N-sources in coastal ecosystems include atmospheric deposition on land, freshwater and estuary, fertilizers from agriculture, and wastewater from sewage. The increasing anthropogenic forcing in the recent decades is known to introduce greater amount of nitrogen in the coastal ecosystems. Some studies have reported the input nitrogen in the Indian mangroves. For example, Ray et al. (2014) stated that anthropogenic inputs from human inhabitation, aquaculture farms, and waste discharge from industries in the Indian Sundarbans apart from high mangrove litter fall and its degradation and remineralization (Ramos e Silva et al., 2007) contribute to the increased nitrogen and has the potential to retain N via sediment absorption during the riverine transport to the coastal water. The mangrove sediments in general are rich in nitrogenous organic matter (Khan et al., 2007; Purvaja et al., 2008). The sediment N pool in mangroves is composed largely of the organic-N form which is not readily available for plant uptake (Clarke, 1985). But a variety of processes come to play in the mangrove sediments which significantly affect the nitrogen transformations and cycling in the marine environment.

## Processes and Factors Governing Nitrogen Dynamics in Mangrove Sediments

The N cycling in mangrove is mediated predominantly by microbial rather than chemical processes (Alongi et al., 1993). The transformations of N species in the mangrove sediments are governed by a series of microbial processes which largely determine the N availability and its fluxes (Fernandes et al., 2012b,c). The nitrogen balance in the mangrove sediments is mainly controlled by two processes, namely nitrogen fixation and its subsequent denitrification. The fixation of nitrogen (performed by diazotrophs) is also a source of $NH_4^+$. Bacterial activity regulates most of the available $NH_4^+$ pool, particularly in deeper sediments, devoid of other biota (Trimmer et al., 2003). Ammonia immobilization and assimilation by microbes, plants, etc., always accompany and counteract the mineralization process (Alongi et al., 1993). The extent to which these processes balance each other depends upon the C:N ratio of the decomposing OM. Substances rich in N, favor net mineralization, whereas those poor in N, favor net immobilization. Further, the availability of sedimentary nutrients to microbes and plants is complicated by geochemical processes, such as the involvement of some nutrients in adsorption reactions to clay minerals. It has also been found that the quality of organic carbon/matter and quantity of iron govern the availability of N in mangroves via the process of nitrification (Krishnan and Loka Bharathi, 2009). The $NH_4^+$ adsorption is low in mangrove sediments compared to temperate salt marsh sediments, probably due to the presence of higher concentrations of competitive cations such as iron (Holmboe and Kristensen, 2002). Microbial involvement in N transformations (ammonification, nitrification, and denitrification) in Indian mangroves is more in soils with plants than soils without plants (Routray et al., 1996). Denitrification (performed by heterotrophic anaerobic facultative bacteria) is favorable under oxygen-deficient conditions, and it utilizes the fixed and converted nitrogen in the form of $NO_2^-$ and $NO_3^-$ to reduce it to $N_2O$ or $N_2$ (Desnues et al., 2007). However, denitrification activity varies in different ecosystems and could be affected by prevailing environmental conditions. Some of the factors influencing the activity of denitrifiers are oxygen concentration (Bonin and Raymond, 1990), sediment pH (Rust et al., 2000; Simek et al., 2002), redox conditions (Lee and Joye, 2006), sediment nitrification (Stockenberg and Johnstone, 1997), nitrate availability (Seitzinger, 1990; Corredor and Morell, 1994; Kunu et al., 1998), and organic carbon (McCarty and Bremner, 1993; Pfenning and McMahon, 1997).

Apart from the nitrification and denitrification processes, anaerobic ammonium oxidation (anammox), another important process, has gained prominence in the context of marine nitrogen cycle in recent times (Devol, 2003; Dalsgaard et al., 2005; Nakajima et al., 2008). In this process, an alternate pathway of N loss from the marine sediments is observed in which $NH_4^+$ is directly oxidized to $N_2$ by using up the available $NO_2^-$ (Meyer et al., 2005) which is produced either by the reduction of $NO_3^-$ (Dalsgaard et al., 2003) or nitrification by aerobic ammonium oxidation (Francis et al., 2005). According to a study by Chiu et al. (2004), denitrification is the main source of N loss (in the form of NO, $N_2$, or $N_2O$) in mangrove systems, accounting to as much as 55% of the total loss. However, recent evidences suggest that anammox can account for up to 67% N removal (Thamdrup and Dalsgaard, 2002) in marine sediments. In sediments rich in organic matter, anammox is prominent under low concentrations of $NO_2^-$ (Trimmer et al., 2003) and such conditions are frequently present in mangrove sediments since they are largely anaerobic where nitrate/nitrite availability is controlled by denitrification rates (Seitzinger, 1990). Another factor which can affect the availability of nitrate/nitrite in the benthic environment is the presence of bioturbating crabs which can play with the anaerobic conditions. The $NO_3^-$ or $NO_2^-$ abundance in the mangrove sediments is governed by many factors, namely inherent nitrification (Nielsen, 1992; Lohse et al., 1993), external supply through terrestrial runoff (Naqvi et al., 2000), or marine inputs via tidal activity (Ji et al., 2008). It is seen that denitrification and anammox operate concurrently leading to N removal from mangrove sediments, but anammox process is restrained only in organic-rich mangrove sediments, whereas denitrification becomes the major mechanism of N removal in mangroves in general. Another process that is known to exist in mangrove sediments is DNRA which can account for up to 99% of nitrate removal in mangrove sediments (Fernandes et al., 2012a,b,c). In a study conducted in Goa mangrove, the DNRA process was found to be two times higher in the relatively pristine areas compared to the anthropogenically influenced areas (Fernandes et al., 2012a,b,c).

Mangrove sediments provide conditions conducive to bacterially mediated removal of anthropogenically derived inorganic nitrogen (Dong et al., 2009). In mangroves which have low $N_2$-fixation but high inorganic-N inputs through anthropogenic interferences, the predominance of N-removal processes holds a significant role in moderating the elevated N levels by eliminating them in various forms (Fig. 22.1). However, in turn they also pose a threat by increasing greenhouse gas production ($N_2O$). A study of Goa mangrove shows that in mangroves prone to elevated nutrient levels, benthic $N_2O$ production was three orders higher than natural production rates in mangrove sediments anywhere else (Fernandes et al., 2010). However, in mangrove systems receiving low nutrient inputs, DNRA mechanism effectively conserves and recirculates N minimizing nutrient loss that would otherwise occur through denitrification and helps in buffering the climate by modulating the production of the greenhouse gas $N_2O$ (Fernandes et al., 2012a,b,c).

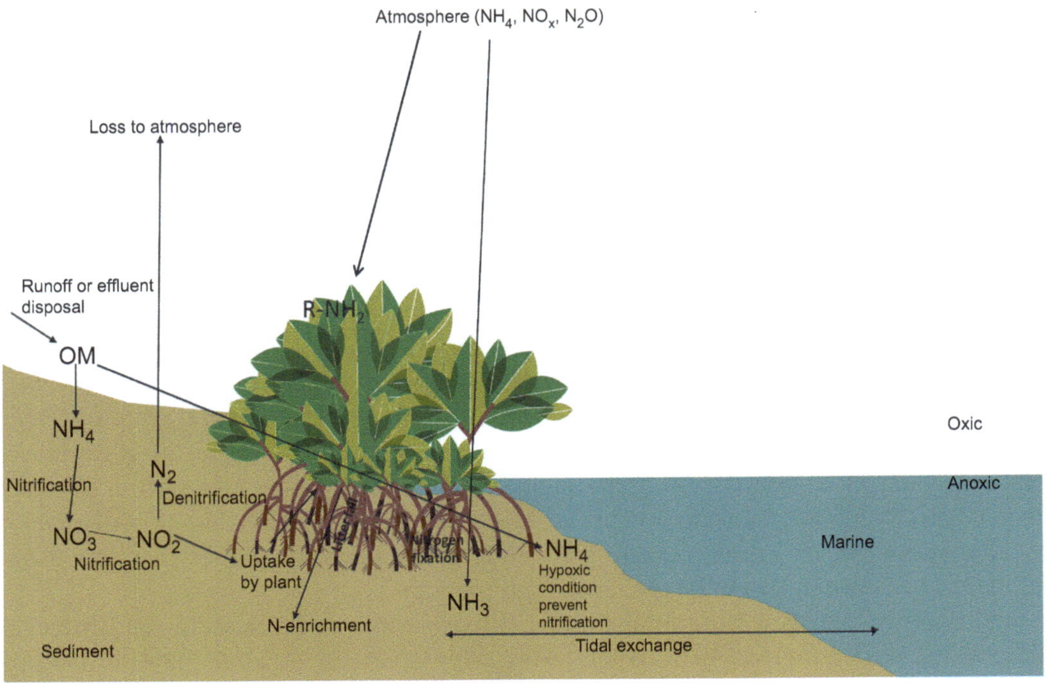


**FIGURE 22.1** A conceptual diagram depicting the mechanism of the N biogeochemical cycle in the Indian mangrove ecosystem.

In addition to the knowledge of various processes of N removal, an important factor still understudied is how the redox sensitive metals affect the denitrification activity. Presence of redox sensitive metals such as Fe and Mn could lead to several secondary reactions of diagenesis creating new pathways of oxidation/reduction (Fernandes et al., 2013). These pathways could contribute to the production and subsequent emission of the greenhouse gas $N_2O$. Thus, lowering of the anthropogenic inputs (N compounds, metals, and sewage discharge) into estuarine systems would help minimize the flux of the radiative gas to the atmosphere.

In addition to the sediment–atmospheric nitrogen flux, sediment–water (sediment to water) and water–sediment nitrogen fluxes (water to sediment) are important processes in the coastal system which are affected by porosity, sedimentation rate, and crab holes–dependent advective, and diffusive flux (Ray et al., 2014). Tidal pumping is an important mechanism in the sediment water flux in which nutrient-enriched mangrove pore water seeps into tidal water due to hydrostatic pressure (Purvaraja et al., 2008). Tidal pumping was identified to be a major process in the Mangalavanam mangrove forest adding $NH_4^+$ to the creek water from the mangrove sediment pore water (Bava and Seralathan, 1999). This DIN directly effects phytoplankton productivity in coastal areas including mangroves. The significant decrease in the DIN concentration during the post-monsoon season compared to monsoon season in most of the Indian estuaries has been linked to its utilization by these primary producers (Sarma et al., 2009). Phytoplankton occupies the base position of the trophic structure in the food chain. This is being grazed by zooplanktons which are the primary food for fish, a very important economic source for majority of population along the Indian coast. However, the increase in nutrient load may cause eutrophication and sometimes even harmful algal blooms. Eutrophication and presence of toxic Dinoflagellates and Cyanophyceae have been reported in Sundarban mangrove ecosystem, nitrogen being an important regulating factor (Manna et al., 2010).

## Use of Stable Isotopes of Nitrogen to Study Sources and Behavior of Organic Matter in Mangroves

Since N exists in several oxidation states and mainly in gaseous and liquid forms, N isotopes are very useful for studying biogeochemical processes in natural systems (Hoefs and Hoefs 1997). Three processes of importance for the $\delta^{15}N$-signal in organic-rich sediments are (1) N-fixation, (2) assimilation, and (3) diagenesis. All reaction products are isotopically lighter than the precursor substrates (Owens, 1988). The overall effect of OM diagenesis causes enrichment of $\delta^{15}N$ and, thus, nitrogen isotopes can be a useful indication of OM diagenesis in mangrove sediments.

Additionally, nitrogen isotope also finds extensive use in delineating the sources of nitrogen in the mangrove ecosystems. Mangrove trees, in general, have N isotopic compositions that reflect the overall nutrient status of the ecosystem (Fry et al., 2000; Muzuka and Shunula, 2006; Bouillon et al., 2008). Thus, trees growing in regions associated with human population will have $\delta^{15}N$ similar to the isotopic composition of the N released from sewage treatment plants (Fogel et al., 2008). A study by Wooller et al. (2003) reported that a frigate bird rookery found in mangrove ecosystem, only a few kilometers from the site, had $\delta^{15}N$ as positive as $+17‰$, which was identical to the N in the sediments and in the bird droppings (Wooller et al., 2003). Thus, stable isotopes can be used for the identification of N-sources in mangrove ecosystems apart from studying the diagenetic behavior.

In spite of the large application of stable isotopes in mangrove ecosystems, very few studies exist in context of the Indian mangroves and use of stable isotopes of N (Prasad and Ramanathan, 2009; Ranjan et al., 2011; Shynu et al., 2015). Prasad and Ramanathan (2009) observed that runoff of nutrients enriched in $\delta^{15}N$ from agricultural fields, effluents from aquaculture ponds, and drainage of domestic wastes from human settlements are responsible for the change of nutrient load in the Pichavaram. Ranjan et al. (2011) found that the $\delta^{15}N$ signatures in estuarine sediments are heavier ($\sim 1‰$) compared to mangrove sediments. The lighter $\delta^{15}N$ values observed in mangrove sediments were attributed to either atmospheric N-fixation or utilization of inorganic-N depleted in $\delta^{15}N$. Shynu et al. (2015) studied Mandovi and Zuari mangroves for their $\delta^{15}N$ content in order to delineate the controls on their sediment biogeochemistry. They found that $\delta^{15}N$ of OM is modified by biogeochemical process in both the systems and there is also a seasonal trend. The $\delta^{15}N$ content for Mandovi system varied from 4 to 10‰ whereas for Zuari it varied from $-5.1$ to $6.0‰$.

# Nitrogen Dynamics in Mangrove Atmosphere

## Release of Greenhouse Gases ($N_2O$) and Processes Controlling Them

Estuarine and coastal areas are major contributor of atmospheric nitrous oxide ($N_2O$) production in marine environment (Law et al., 1991; Middleburg et al., 1995; Bange et al., 1996; Seitzinger and Kroenze, 1998). Marine ecosystems are estimated to contribute 14% of world $N_2O$ emission (Seiler and Conrad, 1987) whereas estuarine and coastal areas comprise approximately 60% of total oceanic $N_2O$ flux (Bange et al., 1996).

Fig. 22.1 depicts a conceptual diagram of nitrogen biogeochemical process in Indian mangrove ecosystem. Nitrous oxide is produced in natural system mainly as a byproduct during several microbiological processes such as nitrification, denitrification (Yoshinari, 1990; De Wilde and De Bie, 2000; Wrag et al., 2001), and nitrate ammonification; other than these process, nitrate assimilation (Satoh et al., 1981), dissimilatory reduction of $NO_3^-$ to $NH_4^+$ (Bleaky and Tiedje, 1982), fungal denitrification (Schoun et al., 1992), and production by green algae (Weathers, 1984) also add $N_2O$ to atmosphere. However, nitrification and denitrification appears to be the major sources of $N_2O$ emission in natural system (Bremner and Blackmer, 1978; Firestone and Davidson, 1989; Bonin et al., 2002).

$N_2O$ emission affected by many environmental condition such as pH, dissolved oxygen, organic carbon, oxidation–reduction potential, water-retention time, $NO_3^-$, $NO_2^-$, temperature (Wu et al., 2009; Zhou et al., 2008).

Nitrate ammonifying bacteria form $N_2O$ by reducing nitrite $\left(NO_3^-\right)$ to ammonium ion $\left(NH_4^+\right)$ and in the process also form $N_2O$ and $NH_4^+$ (Smith, 1982); this process occurs under low oxidation–reduction potential (Wu et al., 2009). Oxidation–reduction potential shows status sediment from aerobic (supporting nitrification) to anaerobic (supporting denitrification). Inamori et al., 2007 found that ORP values ranging from 50 to 200 mV are favorable for $N_2O$ production by microbial ecology.

The supply of oxygen can also increase $N_2O$ production as a positive effect on nitrification and coupled nitrification and denitrification process. Study done by Zheng et al. (1994) showed that maximum microbial conversion and production of $N_2O$ is observed at a DO concentration of 0.2 mg $L^{-1}$. Schuthess et al. (1994) found increase in $N_2O$ production from 0 to 6% by change in DO from 0 to 4 mg $L^{-1}$. Thus, high $NO_2^-$ as well as oxygen supply favor the production of $N_2O$, and also $N_2O$ reduction is slower under low oxygen pressure than the reduction of $NO_3^-$ and $NO_2^-$; whereas highest emission of $N_2O$ is observed under complete anoxic condition.

Naqvi et al. (2000) and Bange et al. (2001) reported high $N_2O$ production in the oxygen-minimum zones of Arabian sea; similar high value of $N_2O$ flux of 0.41–0.77 $\mu mol\ m^{-2}h^{-1}$ was found in a South Indian mangrove study done by Krithika et al., 2008.

Changes in the season and temperature alter the microbial activity and thus affect the $N_2O$ production (Inamori et al., 2007). Study done on the effect of seasonal change on $N_2O$ production by Inamori et al. (2007) reported more emission (average 67%) in summer and lower emission (average 2%) in winter. Study done by Usi et al. (2001) showed that an increase in the seasonal temperature increases the rate of nitrification as well as denitrification. The increased seasonal temperature enhances the faunal activity resulting in higher $N_2O$ production.

$N_2O$ reduction by denitrification in the sediment is enhanced by an increase in the temperature and supply of organic carbon (Kieskamp et al., 1991; van Raaphorst et al., 1992).

Availability of ammonium $\left(NH_4^+\right)$ is one of the most important factors controlling the nitrification rate and $N_2O$ emission (Firestone and Davidson, 1989). Diurnal change in emission of $N_2O$ was observed with highest emission during night due to increase in production and availability of ammonium at night. Even during the dark periods of daytime, higher emission of nitrous oxide were observed, which can be correlated with the increase in the nocturnal activity and $NH_4^+$ production by microbes and different fauna. Ammonium concentration also increases by the process of N-fixation by non-phototropic bacteria (Potts, 1984). In a study done by Boto and Robertson (1990), it was found that the nitrogen-fixation activity in prop roots is much higher at nighttime than daytime; thus ammonium absorption decreases in nighttime which increases the nutrient availability for nitrifying bacteria to produce $N_2O$.

Inamori et al. (2007) observed that the oxygen released from the roots of aquatic plant, stimulates the nitrification process. The oxygen released to rhizosphere by macrophytes can affect $N_2O$ emission. Different plant species affect $N_2O$ production differently, e.g., higher $N_2O$ flux is observed in the cells of *Z. latifolia*, whereas lower $N_2O$ production is seen in the cell of *P. australis*.

Denitrification and $N_2O$ emission is also affected by C:N ratio as organic carbon provides the electron donor required in the denitrification process. Park et al. (2000) observed a decrease in $N_2O$ emission followed by increase in C:N ratio in anaerobic conditions.

Anthropogenic action also affects $N_2O$ production as an increased nitrogen input from wastewater, sludge, and sewage to sediment enhances the microbial activity which further enhances the rate of nitrous flux.

Thus, temperature, oxygen supply, $NH_4^+$, temperature, organic carbon are the major factors controlling $N_2O$ production (Henriksen, 1988; Koike and Sørensen, 1988; Seitzinger, 1988; Kieskamp et al., 1991).

## Net $N_2O$ Fluxes From Mangroves of India

$N_2O$ emission data from major Indian mangrove are gathered from published literature; existing literature shows the studies were done only on the east coast of India whereas west coast of India was not studied so well. The $N_2O$ emission from mangrove ecosystem can be inventoried mainly by three kinds of fluxes (1) air–water flux, (2) sediment–air flux, and (3) canopy–air/biosphere–atmosphere fluxes.

There are very few studies on air–water $N_2O$ flux of east coast of India and not much work is done regarding $N_2O$ flux of west coast; the range of $N_2O$ flux for east coast is 1.1–11.4 mmol N $m^{-2}s^{-1}$ with mean value for Indian coast as 5.36 mmol N $m^{-2}$ $year^{-1}$.

Sediment–air $N_2O$ flux, for the entire east coast of India, were estimated to be an average value of $5.8 \times 103$ g $N_2O$ $year^{-1}$ (Selvam, 2003). The $N_2O$ flux from Bhitarkanika mangrove ranges from 0.22 to 4.99 $mgm^{-2}$ $day^{-1}$ having an average annual emission of $1.4 \times 106$ $gN_2O$ $year^{-1}$ (Chauhan et al., 2008); in Pichavaram mangrove it varies from 0.94 to 1.90 $mgm^{-2}$ $day^{-1}$ with an average value of $5.6 \times 104$ $gN_2O$ $year^{-1}$ (Senthilkumar et al., 2008); in Muthupet it varies from 0.43 to 0.81 $mgm^{-2}$ $day^{-1}$ having an average value of 0.63 $mgm^{-2}$ $day^{-1}$ (Krithika et al., 2008); Fernandes et al. (2010) reported the $N_2O$ emission rate from Goa coast and the flux values was as high as 1.95 nmol $N_2O$-N $g^{-1}h^{-1}$ at Divar mangrove of Goa.

Among all other mangrove ecosystems studied done so far for $N_2O$ emission, Bhitarkanika shows the highest $N_2O$ emission a 100 times higher $N_2O$ emission than Pichavaram mangrove (Chauhan et al., 2008).

The $N_2O$ flux from Bhitarkanika mangrove showed the highest emission during post-monsoon period, which could be due to very high bird population in this season. These water birds excrete nitrogenous compounds, enriching sediment with substrate for nitrification–denitrification processes to emit $N_2O$ (Chauhan et al., 2008; Munoz-Hincapie et al., 2002). The high value of $N_2O$ emission from Bhitarkanika mangrove in post-monsoon period can be reasoned with mobilization of surface sediments due to high rainfall and release of nutrients, which could be available to microbes for mineralization. Also, in Bhitarkanika mangrove, building of dams on river reduced the freshwater flow, thus even the episodic rain has a profound effect on $N_2O$ emission.

Most of the mangroves are located at the other end of the city, town, or places influenced with rapid urbanization and industrialization. Thus, mangrove ecosystems receive high urban effluent, industrial waste, sewage, and agricultural as well as aquaculture wastes (Krithika et al., 2008). These effluents especially from agriculture and aquaculture wastes are rich in nitrogenous waste, which further increases the $N_2O$ emission. Similar is the case of Goa mangroves receiving a considerable amount of mining waste (De Souza, 1983, 1999), runoff (Sardessai and Sundar, 2007; Divya et al., 2009), and sewage effluent (Ansari et al., 1986) etc. through the Mandovi river, and also in Divar mangroves, this higher inorganic nitrogen content accelerates the process of nitrification–denitrification, thus enhancing the $N_2O$ production. Fernandes et al., 2010 in their study of Goa mangrove found that $N_2O$ production indicates 93% incomplete denitrification at the surface of mangrove sediment which further enhances the emission of $N_2O$. Similarly, effluent and agricultural runoff is found to be the reason for $N_2O$ emission in the study of Pichavam mangrove done by Senthilkumar et al. (2008).

Mangrove sediments were considered as a minor source of greenhouse gas emission and many early studies done on different mangrove ecosystems of the world supported this idea, e.g., study done on Kenyan mangroves (Middelburg et al., 1996), Thailand mangroves (Alongi et al., 2001), and Mexican mangroves (Giani et al., 1996). However, recent advancement and interest in the study on greenhouse gas emission from mangrove, all over the world and also in India such as study done on Bhitakanika (Chauhan et al., 2008), Muthupet (Krithika et al., 2008), and Pichavaram (Senthilkumar et al., 2008), etc., confirmed that mangroves are a significant source of $N_2O$.

## Conclusion

DIN in Indian mangrove ecosystems is comparable to other tropical mangroves with low values within μM range. Most of the ecosystems show higher concentration during wet season due to higher river discharge, anthropogenic input, and leaching of nitrate along with low productivity during this period. The lower concentration during the non-monsoon periods has been attributed to limited sources and more consumption (higher productivity) along with dilution by sea water during this period. Mangrove pore water has higher ammonium due to the reducing condition. Other factors such as salinity, temperature, pH, and tidal flushing regulating DIN dynamics in mangrove water have also been identified. Recent studies identify mangrove litter as the major source of DIN in Sundarbans mangrove ecosystem; however, we need more such researches along other Indian mangroves to make a clearer picture. DON is also limited to few mangroves and requires more studies. Few of the studies have identified Indian mangroves acting as sink of DIN.

The input and removal of N from the mangrove sediments is mainly controlled by processes of nitrification–denitrification in general. However, studies of certain Indian mangrove ecosystems have also revealed two alternate pathways of N removal, namely anammox and DNRA. The occurrence of these processes are governed by factors such as C:N ratio of the OM, availability of oxygen, sediment pH, redox conditions, nutrient availability, and trace metals. Denitrification/N-removal processes majorly prevail in the mangrove sediments and help in moderating the nutrient levels in the mangrove ecosystems. However, in turn, these may also lead to the emission of potential greenhouse gases. Use of stable nitrogen isotopes show promising role in delineating the diagenetic behavior and sources of nitrogen in the mangrove ecosystems, but there is still paucity of such studies in Indian mangroves. Mangroves are important sites in view of their strategic location at the interface of land and sea which makes them prone to nutrient enrichment/inputs from both, the anthropogenic forcing in the hinterland and contributions from the oceans. There is a need to quantify various sources of N into the mangrove sediments and their diagenetic alteration so as to predict the fate of N.

$N_2O$ is produced in natural systems mainly as a by-product during several microbiological processes. $N_2O$ emission is affected by many environmental factors, e.g., pH, dissolved oxygen, organic carbon, oxidation–reduction potential, water-retention time, $NO_3^-$, $NO_2^-$, temperature, etc. Anoxic condition supports denitrification process in which $N_2O$ is produced as an intermediate species, whereas under oxic condition it is produced as a by-product of nitrification. The $N_2O$ flux for eastern coast varies from 1.1 to 11.4 $mmolNm^{-2}s^{-1}$. Bhitarkanika shows the highest $N_2O$ flux ranging from 0.22 to 4.99 $mgm^{-2}\ day^{-1}$, the lowest $N_2O$ flux is seen from Muthupet having average value of 0.63 $mgm^{-2}\ day^{-1}$.

$N_2O$ flux showed the highest emission during post-monsoon that may be due to high nutrient supply by runoff from urban as well agricultural waste; which act as a substrate for microbial mediated processes for $N_2O$ production. The highest $N_2O$ emission from Bhitarkanika can be due to high population of birds which produce nitrogenous waste as excretion and provide substrate for $N_2O$ production by transformation processes. Increased $N_2O$ production in Goa mangrove is due to mining waste, sewage, industrial waste, and runoff. Various studies done on Indian mangroves (Pichavaram, Muthupet, Bhitarkanika, and Goa) for $N_2O$ emission showed it as a source of $N_2O$ emission.

Mangrove ecosystems are at the boundary of terrestrial and marine environments, and receive nitrogen inputs from both marine and terrestrial environments which get deposited and reworked over the course of time. These ecosystems witness intense biogeochemical activities and any reactive nitrogen introduced to these wetlands is prone to remineralization leading to a possible efflux back to the atmosphere. $N_2O$ emission data from the Indian mangroves establish that these ecosystems are the significant emitters of this potent greenhouse gas. Furthermore, studies have also suggested an increased human input into these ecosystems over the time. It is observed that the runoff of nutrients from agricultural fields, effluents from aquaculture ponds, and drainage of domestic wastes from human settlements, etc., have changed the nutrient load to these ecosystems significantly. As the input of reactive nitrogen to these ecosystems increases, so does the production of $N_2O$. There is an urgent need to manage the nitrogen inputs into the Indian mangrove ecosystems. In future, it is of utmost importance to ensure an adaptive transboundary management plan for overall nitrogen reduction, especially managed at the sources such as river catchments, domestic waste, and sewage sludge which provide nitrogenous waste. Also, we need to manage the use of different nitrogenous fertilizers in agricultural activities which contribute to a major part in runoff to coastal ecosystem for N-mediated processes. Further, with several policies proposing reduced nutrient transport to rivers, coastal system needs to be seriously looked into and followed.

## Acknowledgments

The authors are thankful to UGC, BSR-UGC, and CSIR for providing fellowship for the research. The School of Environmental Sciences and Jawaharlal Nehru University are also acknowledged for promoting research works.

## References

Alongi, D.M., et al., 2001. Organic carbon accumulation and metabolic pathways in sediments of mangrove forests in southern Thailand. Marine Geology 179 (1), 85–103.

Alongi, D.M., Boto, K.G., Robertson, A.I., 1993. Nitrogen and Phosphorus Cycles. American Geophysical Union.

Anand, S.S., Sardessai, S., Muthukumar, C., Mangalaa, K.R., Sundar, D., Parab, S.G., Kumar, M.D., 2014. Intra-and inter-seasonal variability of nutrients in a tropical monsoonal estuary (Zuari, India). Continental Shelf Research 82, 9–30.

Ansari, Z.A., Ingole, B.S., Parulekar, A.H., 1986. Effect of high organic enrichment of benthic polychaete population in an estuary. Marine Pollution Bulletin 17 (8), 361–365.

Attri, K., Kerkar, S., 2011. Seasonal assessment of heavy metal pollution in tropical mangrove sediments (Goa, India). Journal of Ecobiotechnology 3 (8).

Badarudeen, A., et al., 1996. Texture and geochemistry of the sediments of a tropical mangrove ecosystem, southwest coast of India. Environmental Geology 27 (3), 164–169.

Bala Krishna Prasad, M., et al., 2006. Seasonal variations and decadal trends in concentrations of dissolved inorganic nutrients in Pichavaram mangrove waters, southeast India. Bulletin of Marine Science 79 (2), 287–300.

Banerjee, K., et al., 2012. Sedimentation and trace metal distribution in selected locations of Sundarbans mangroves and Hooghly estuary, northeast coast of India. Environmental Geochemistry and Health 34 (1), 27–42.

Bange, H.W., 2008. Gaseous nitrogen compounds (NO, $N_2O$, $N_2$, $NH_3$) in the ocean. In: Nitrogen in the Marine Environment, pp. 51–94.

Bange, H.W., 2006. Nitrous oxide and methane in European coastal waters. Estuarine, Coastal and Shelf Science 70.3, 361–374.

Bange, H.W., et al., 2001. Nitrous oxide emissions from the Arabian Sea: a synthesis. Atmospheric Chemistry and Physics 1 (1), 61–71.

Bange, H.W., Rapsomanikis, S., Andreae, M.O., 1996. Nitrous oxide in coastal waters. Global Biogeochemical Cycles 10 (1), 197–207.

Bava, K.A., Seralathan, P., 1999. Interstitial water and hydrochemistry of a mangrove forest and adjoining water system, south west coast of India. Environmental Geology 38 (1), 47–52.

Bleakley, B.H., Tiedje, J.M., 1982. Nitrous oxide production by organisms other than nitrifiers or denitrifiers. Applied and Environmental Microbiology 44 (6), 1342–1348.

Bonin, P., Raymond, N., 1990. Effects of oxygen on denitrification in marine sediments. Hydrobiologia 207 (1), 115–122.

Bonin, P., Tamburini, C., Michotey, V., 2002. Determination of the bacterial processes which are sources of nitrous oxide production in marine samples. Water Research 36.3, 722–732.

Boto, K.K.G., 1982. Nutrient and organic fluxes in mangroves. In: Mangrove Ecosystems Australia: Structure, Function and Management, pp. 239–258.

Boto, K.G., Robertson, A.I., 1990. The relationship between nitrogen fixation and tidal exports of nitrogen in a tropical mangrove system. Estuarine, Coastal and Shelf Science 31 (5), 531–540.

Bouillon, S., Connolly, R.M., Lee, S.Y., 2008. Organic matter exchange and cycling in mangrove ecosystems: recent insights from stable isotope studies. Journal of Sea Research 59 (1), 44–58.

Brandes, J.A., Allan Devol, H., Deutsch, C., 2007. New developments in the marine nitrogen cycle. Chemical Reviews 107 (2), 577–589.

Bremner, J.M., Blackmer, A.M., 1978. Nitrous oxide: emission from soils during nitrification of fertilizer nitrogen. Science 199 (4326), 295–296.

Chauhan, R., Ramanathan, A.L., Adhya, T.K., 2008. Assessment of methane and nitrous oxide flux from mangroves along Eastern coast of India. Geofluids 8.4, 321–332.

Chauhan, R., Ramanathan, A.L., 2008. Evaluation of water quality of Bhitarkanika mangrove system, Orissa, east coast of India. Indian Journal of Marine Sciences 37.2, 153.

Chiu, C.-Y., et al., 2004. Denitrification associated N loss in mangrove soil. Nutrient Cycling in Agroecosystems 69 (3), 185–189.

Chon, K., et al., 2011. Abundance of denitrifying genes coding for nitrate (*narG*), nitrite (*nirS*), and nitrous oxide (*nosZ*) reductases in estuarine versus wastewater effluent-fed constructed wetlands. Ecological Engineering 37 (1), 64–69.

Christensen, Bondo, P., et al., 2000. Sediment mineralization, nutrient fluxes, denitrification and dissimilatory nitrate reduction to ammonium in an estuarine fjord with sea cage trout farms. Aquatic Microbial Ecology 21 (1), 73–84.

Clarke, P.J., 1985. Nitrogen pools and soil characteristics of a temperate estuarine wetland in eastern Australia. Aquatic Botany 23 (3), 275–290.

Corredor, J.E., Morell, J.M., 1994. Nitrate depuration of secondary sewage effluents in mangrove sediments. Estuaries 17 (1), 295–300.

Dalsgaard, T., et al., 2003. $N_2$ production by the anammox reaction in the anoxic water column of Golfo Dulce, Costa Rica. Nature 422 (6932), 606–608.

Dalsgaard, T., Thamdrup, B., Canfield, D.E., 2005. Anaerobic ammonium oxidation (anammox) in the marine environment. Research in Microbiology 156 (4), 457–464.

De Sousa, S.N., 1983. Studies on the behaviour of nutrients in the Mandovi Estuary during premonsoon. Estuarine, Coastal and Shelf Science 16 (3), 299–308.

de Wilde, H.P.J., de Bie, M.J.M., 2000. Nitrous oxide in the Schelde estuary: production by nitrification and emission to the atmosphere. Marine Chemistry 69 (3), 203–216.

Desnues, C., et al., 2007. Seasonal and diel distributions of denitrifying and bacterial communities in a hypersaline microbial mat (Camargue, France). Water Research 41 (15), 3407–3419.

De Sousa, S.N., 1999. Effect of mining rejects on the nutrient chemistry of Mandovi estuary, Goa. Indian Journal of Marine Sciences.

De Sousa, S.N., Sengupta, R., Sanzigiri, S., Rajgopal, M.D., 1981. Studies on nutrients of Mandovi and Zuari river systems. Indian Journal of Marine Sciences 10, 314–321. http://dx.doi.org/10.1007/s10653-005-9001-5.

Devol, A.H., 2003. Nitrogen cycle: solution to a marine mystery. Nature 422 (6932), 575–576.

Dittmar, T., Lara, R.J., 2001. Do mangroves rather than rivers provide nutrients to coastal environments south of the Amazon River? Evidence from long-term flux measurements. Marine Ecology Progress Series 213, 67–77.

Divya, B., et al., 2009. Limno-tolerant bacteria govern nitrate concentration in Mandovi Estuary, India. Estuarine, Coastal and Shelf Science 82 (1), 29–34.

Dong, L.F., et al., 2009. Changes in benthic denitrification, nitrate ammonification, and anammox process rates and nitrate and nitrite reductase gene abundances along an estuarine nutrient gradient (the Colne Estuary, United Kingdom). Applied and Environmental Microbiology 75 (10), 3171–3179.

Fernandes, L., Nayak, G.N., Ilangovan, D., 2012a. Geochemical assessment of metal concentrations in mangrove sediments along Mumbai coast, India. World Academy of Science, Engineering & Technology 61, 258–263.

Fernandes, S.O., et al., 2013. Denitrification activity is closely linked to the total ambient Fe concentration in mangrove sediments of Goa, India. Estuarine, Coastal and Shelf Science 131, 64–74.

Fernandes, S.O., et al., 2012b. Denitrification prevails over anammox in tropical mangrove sediments (Goa, India). Marine Environmental Research 74, 9–19.

Fernandes, S.O., et al., 2010. Denitrification: an important pathway for nitrous oxide production in tropical mangrove sediments (Goa, India). Journal of Environmental Quality 39 (4), 1507–1516.

Fernandes, S.O., et al., 2012c. Nitrogen-limited mangrove ecosystems conserve N through dissimilatory nitrate reduction to ammonium. Scientific Reports 2, 419.

Firestone, M.K., Davidson, E.A., 1989. Microbiological basis of NO and $N_2O$ production and consumption in soil. In: Exchange of Trace Gases Between Terrestrial Ecosystems and the Atmosphere, 47, pp. 7–21.

Fogel, M.L., et al., 2008. Unusually negative nitrogen isotopic compositions ($\delta^{15}N$) of mangroves and lichens in an oligotrophic, microbially-influenced ecosystem. Biogeosciences Discussions 5 (1), 937–969.

Forest survey of India report, 2015. http://fsi.nic.in/isfr-2015/isfr-2015-mangrove-cover.pdf.

Francis, C.A., et al., 2005. Ubiquity and diversity of ammonia-oxidizing archaea in water columns and sediments of the ocean. Proceedings of the National Academy of Sciences of the United States of America 102 (41), 14683–14688.

Fry, B., et al., 2000. $\delta^{15}$ N studies of nitrogen use by the red mangrove, *Rhizophora mangle* L. in South Florida. Estuarine, Coastal and Shelf Science 50 (2), 291–296.

Fuchs, V.J., James Mihelcic, R., Gierke, J.S., 2011. Life cycle assessment of vertical and horizontal flow constructed wetlands for wastewater treatment considering nitrogen and carbon greenhouse gas emissions. Water Research 45 (5), 2073–2081.

Giani, L., et al., 1996. Characteristics and methanogenesis of the Balandra lagoon mangrove soils, Baja California Sur, Mexico. Geoderma 72 (1), 149–160.

Goldman, J.C., Stanley, H.I., 1974. Relative growth of different species of marine algae in wasterwater-seawater mixtures. Marine Biology 28 (1), 17–25.

Harrison, P.J., et al., 1997. Nutrient and phytoplankton dynamics in two mangrove tidal creeks of the Indus River delta, Pakistan. Marine Ecology Progress Series 157, 13–19.

Henriksen, K., 1988. Nitrification in estuarine and coastal marine sediments. Nitrogen Cycling in Coastal Marine Environments 207–249.

Herbert, R.A., 1999. Nitrogen cycling in coastal marine ecosystems. FEMS Microbiology Reviews 23 (5), 563–590.

Hoefs, J., Hoefs, J., 1997. Stable Isotope Geochemistry, vol. 201. Springer, Berlin.

Holmboe, N., Kristensen, E., 2002. Ammonium adsorption in sediments of a tropical mangrove forest (Thailand) and a temperate Wadden Sea area (Denmark). Wetlands Ecology and Management 10 (6), 453–460.

Howarth, R.W., Marino, R., 2006. Nitrogen as the limiting nutrient for eutrophication in coastal marine ecosystems: evolving views over three decades. Limnology and Oceanography 51 (1 Part 2), 364–376.

Hulth, S., et al., 2005. Nitrogen removal in marine environments: recent findings and future research challenges. Marine Chemistry 94 (1), 125–145.

Hynes, R.K., Knowles, R., 1984. Production of nitrous oxide by *Nitrosomonas europaea*: effects of acetylene, pH, and oxygen. Canadian Journal of Microbiology 30 (11), 1397–1404.

Inamori, R., et al., 2007. Investigating $CH_4$ and $N_2O$ emissions from eco-engineering wastewater treatment processes using constructed wetland microcosms. Process Biochemistry 42 (3), 363–373.

Inamori, R., et al., 2008. Seasonal effect on $N_2O$ formation in nitrification in constructed wetlands. Chemosphere 73 (7), 1071–1077.

Ji, R., et al., 2008. Influence of local and external processes on the annual nitrogen cycle and primary productivity on Georges Bank: a 3-D biological-physical modeling study. Journal of Marine Systems 73 (1), 31–47.

Jørgensen, K.S., Beck Jensen, H., Sørensen, J., 1984. Nitrous oxide production from nitrification and denitrification in marine sediment at low oxygen concentrations. Canadian Journal of Microbiology 30 (8), 1073–1078.

Khan, H., Brush, G.S., 1994. Nutrient and metal accumulation in a freshwater tidal marsh. Estuaries 17 (2), 345–360.

Khan, Md N.I., Suwa, R., Hagihara, A., 2007. Carbon and nitrogen pools in a mangrove stand of *Kandelia obovata* (S., L.) Yong: vertical distribution in the soil–vegetation system. Wetlands Ecology and Management 15 (2), 141–153.

Kieskamp, W.M., et al., 1991. Seasonal variation in denitrification rates and nitrous oxide fluxes in intertidal sediments of the western Wadden Sea. Marine Ecology Progress Series. Oldendorf 72 (1), 145–151.

Koike, I., Sørensen, J., 1988. Nitrate reduction and denitrification in marine sediments. Nitrogen Cycling in Coastal Marine Environments 251–273.

Krauss, K.W., et al., 2008. Environmental drivers in mangrove establishment and early development: a review. Aquatic Botany 89 (2), 105–127.

Krishnan, K.P., Loka Bharathi, P.A., 2009. Organic carbon and iron modulate nitrification rates in mangrove swamps of Goa, south west coast of India. Estuarine, Coastal and Shelf Science 84 (3), 419–426.

Krithika, K., Purvaja, R., Ramesh, R., 2008. Fluxes of methane and nitrous oxide from an Indian mangrove. Current Science 94 (2), 218–224.

Kumar, G., Ramanathan, A.L., 2013. Microbial diversity in the surface sediments and its interaction with nutrients of mangroves of Gulf of Kachchh, Gujarat, India. International Research Journal of Environment Sciences 2 (1), 25–30.

Kunu, T.M., et al., 1998. Denitrification in estuarine sediments determined by membrane inlet mass spectrometry. Limnology and Oceanography 43 (2), 334–339.

Lam, P., et al., 2009. Revising the nitrogen cycle in the Peruvian oxygen minimum zone. Proceedings of the National Academy of Sciences of the United States of America 106 (12), 4752–4757.

Law, C.S., Rees, A.P., Owens, N.J.P., 1991. Temporal variability of denitrification in estuarine sediments. Estuarine, Coastal and Shelf Science 33 (1), 37–56.

Lee, R.Y., Joye, S.B., 2006. Seasonal patterns of nitrogen fixation and denitrification in oceanic mangrove habitats. Marine Ecology Progress Series 307, 127–141.

Lohse, L., et al., 1993. Nitrogen cycling in North Sea sediments: interaction of denitrification and nitrification in offshore and coastal areas. Marine Ecology-Progress Series 101, 283–296.

Mandal, S., Ray, S., Ghosh, P.B., 2009. Modelling of the contribution of dissolved inorganic nitrogen (DIN) from litterfall of adjacent mangrove forest to Hooghly–Matla estuary, India. Ecological Modelling 220 (21), 2988–3000.

Manju, M.N., et al., 2012. Assessment of water quality parameters in mangrove ecosystems along Kerala coast: a statistical approach. International Journal of Environmental Research 6 (4), 893–902.

Manna, S., et al., 2010. Dynamics of Sundarban estuarine ecosystem: eutrophication induced threat to mangroves. Saline Systems 6 (1), 1.

Maya, M.V., et al., 2011. Variations in some environmental characteristics including C and N stable isotopic composition of suspended organic matter in the Mandovi estuary. Environmental Monitoring and Assessment 175 (1–4), 501–517.

McCarty, G.W., Bremner, J.M., 1993. Factors affecting the availability of organic carbon for denitrification of nitrate in subsoils. Biology and Fertility of Soils 15 (2), 132–136.

Meyer, Louise, R., Risgaard-Petersen, N., Allen, D.E., 2005. Correlation between anammox activity and microscale distribution of nitrite in a subtropical mangrove sediment. Applied and Environmental Microbiology 71 (10), 6142–6149.

Middelburg, J.J., et al., 1995. Nitrous oxide emissions from estuarine intertidal sediments. Hydrobiologia 311 (1–3), 43–55.

Middelburg, J.J., Soetaert, K., Herman, P.M.J., 1996. Evaluation of the nitrogen isotope-pairing method for measuring benthic denitrification: a simulation analysis. Limnology and Oceanography 41, 1839–1844.

Muñoz-Hincapié, M., Morell, J.M., Corredor, J.E., 2002. Increase of nitrous oxide flux to the atmosphere upon nitrogen addition to red mangroves sediments. Marine Pollution Bulletin 992–996.

Muzuka, A.N.N., Shunula, J.P., 2006. Stable isotope compositions of organic carbon and nitrogen of two mangrove stands along the Tanzanian coastal zone. Estuarine, Coastal and Shelf Science 66 (3), 447–458.

Nakajima, J., et al., 2008. Enrichment of anammox bacteria from marine environment for the construction of a bioremediation reactor. Applied Microbiology and Biotechnology 77 (5), 1159–1166.

Naqvi, S.W.A., et al., 2000. Increased marine production of $N_2O$ due to intensifying anoxia on the Indian continental shelf. Nature 408 (6810), 346–349.

Nedwell, D.B., et al., 2002. Variations of the nutrients loads to the mainland UK estuaries: correlation with catchment areas, urbanization and coastal eutrophication. Estuarine, Coastal and Shelf Science 54 (6), 951–970.

Nielsen, L.P., 1992. Denitrification in sediment determined from nitrogen isotope pairing. FEMS Microbiology Letters 86 (4), 357–362.

Nobi, E.P., et al., 2010. Geochemical and geo-statistical assessment of heavy metal concentration in the sediments of different coastal ecosystems of Andaman Islands, India. Estuarine, Coastal and Shelf Science 87 (2), 253–264.

Owens, N.J.P., 1988. Natural variations in $^{15}N$ in the marine environment. Advances in Marine Biology 24, 389–451.

Park, Young, K., et al., 2000. Emission and control of nitrous oxide from a biological wastewater treatment system with intermittent aeration. Journal of Bioscience and Bioengineering 90 (3), 247–252.

Pawar, P.R., 2013. Monitoring of impact of anthropogenic inputs on water quality of mangrove ecosystem of Uran, Navi Mumbai, west coast of India. Marine Pollution Bulletin 75 (1), 291–300.

Pfenning, K.S., McMahon, P.B., 1997. Effect of nitrate, organic carbon, and temperature on potential denitrification rates in nitrate-rich riverbed sediments. Journal of Hydrology 187 (3), 283–295.

Potts, M., 1984. Nitrogen fixation in mangrove forests. Hydrobiology of the mangal. The ecosystem of the mangrove forest. Junk, The Hague, pp. 155–162.

Prasad, M.B.K., Ramanathan, A.L., 2009. Organic matter characterization in a tropical estuarine-mangrove ecosystem of India: preliminary assessment by using stable isotopes and lignin phenols. Estuarine, Coastal and Shelf Science 84 (4), 617–624.

Prasad, M.B.K., 2012. Nutrient stoichiometry and eutrophication in Indian mangroves. Environmental Earth Sciences 67 (1), 293–299.

Purvaja, R., et al., 2008. Nitrogen cycling: a review of the processes, transformations and fluxes in coastal ecosystems. Current Science-Bangalore 94 (11), 1419.

Rajkumar, M., et al., 2009. Phytoplankton diversity in Pichavaram mangrove waters from south-east coast of India. Journal of Environmental Biology 489–498.

Ramos e Silva, C.A., et al., 2007. Dynamics of phosphorus and nitrogen through litter fall and decomposition in a tropical mangrove forest. Marine Environmental Research 64 (4), 524–534.

Ramanathan, A.L., et al., 1999. Environmental geochemistry of the Pichavaram mangrove ecosystem (tropical), southeast coast of India. Environmental Geology 37 (3), 223–233.

Ranjan, R.K., et al., 2011. Elemental and stable isotope records of organic matter input and its fate in the Pichavaram mangrove–estuarine sediments (Tamil Nadu, India). Marine Chemistry 126 (1), 163–172.

Ranjan, Kumar, R., et al., 2013. Trace metal fractionation in the Pichavaram mangrove–estuarine sediments in southeast India after the tsunami of 2004. Environmental Monitoring and Assessment 185 (10), 8197–8213.

Ray, A.K., et al., 2006. Assessment of Godavari estuarine mangrove ecosystem through trace metal studies. Environment International 32 (2), 219–223.

Ray, R., et al., 2014. Biogeochemical cycle of nitrogen in a tropical mangrove ecosystem, east coast of India. Marine Chemistry 167, 33–43.

Routray, T.K., Satapathy, G.C., Mishra, A.K., 1996. Seasonal fluctuation of soil nitrogen transforming microorganisms in Bhitarkanika mangrove forest. Journal of Environmental Biology 17 (4), 325–330.

Rust, C.M., Marjorie Aelion, C., Flora, J.R.V., 2000. Control of pH during denitrification in subsurface sediment microcosms using encapsulated phosphate buffer. Water Research 34 (5), 1447–1454.

Sarangi, R.K., Kathiresan, K., Subramanian, A.N., 2002. Metal concentrations in five mangrove species of the Bhitarkanika, Orissa, east coast of India. Indian Journal of Marine Sciences 31 (3), 251–253.

Sardessai, S., Sundar, D., 2007. Variability of nitrate and phosphate. In: The Mandovi and Zuari Estuaries, pp. 59–66.

Sarma, V.V.S.S., et al., 2009. Influence of river discharge on plankton metabolic rates in the tropical monsoon driven Godavari estuary, India. Estuarine, Coastal and Shelf Science 85 (4), 515–524.

Satoh, T., Horn, S.S.M., Shanmugam, K.T., 1981. Production of nitrous oxide as a product of nitrite metabolism by enteric bacteria. In: Genetic Engineering Symbiotic Nitrogen Fixation and Conservation of Fixed Nitrogen. Springer U. S., pp. 473–480

Seiler, W., Conrad, R., 1987. Contribution of tropical ecosystems to the global budgets of trace gases, especially $CH_4$, $H_2$, C0, and $N_20$. In: Geophysiology of Amazonia. John Wiley & Sons.

Seitzinger, S.P., 1988. Denitrification in freshwater and coastal marine ecosystems: ecological and geochemical significance. Limnology and Oceanography 33, 702–724.

Seitzinger, S.P., 1990. Denitrification in aquatic sediments. In: Denitrification in Soil and Sediment. Springer U. S., pp. 301–322

Seitzinger, S.P., Kroeze, C., 1998. Global distribution of nitrous oxide production and N inputs in freshwater and coastal marine ecosystems. Global Biogeochemical Cycles 12 (1), 93–113.

Selvam, V., 2003. Environmental classification of mangrove wetlands of India. Current Science 84 (6), 757–765.

Senthilkumar, B., 2014. Biogeochemical and biophysical aspects of pichavaram mangrove ecosystem south India.

Senthilkumar, B., Purvaja, R., Ramesh, R., 2008. Seasonal and tidal dynamics of nutrients and chlorophyll a in a tropical mangrove estuary, southeast coast of India. Indian Journal of Marine Sciences 37 (2), 132.

Shoun, H., et al., 1992. Denitrification by fungi. FEMS Microbiology Letters 94 (3), 277–281.

Shynu, R., et al., 2015. Sources and fate of organic matter in suspended and bottom sediments of the Mandovi and Zuari estuaries, western India. Current Science 226–238.

Šimek, M., Jišová, L., Hopkins, D.W., 2002. What is the so-called optimum pH for denitrification in soil? Soil Biology and Biochemistry 34 (9), 1227–1234.

Singh, G., 2009. Nutrient dynamics in the buffer zone of the Sundarban mangrove ecosystem. India.

Singh, A.K., et al., 2012. Status, biodiversity and distribution of mangroves in India: an overview. In: Marine Biodiversity One Ocean, Many Worlds Life. Uttar Pradesh State Biodiversity Board, pp. 59–67.

Smith, M.S., 1982. Dissimilatory reduction of $NO_2^-$ to $NH_4^+$ and $N_2O$ by a soil *Citrobacter* sp. Applied and Environmental Microbiology 43 (4), 854–860.

Stockenberg, A., Johnstone, R.W., 1997. Benthic denitrification in the Gulf of Bothnia. Estuarine, Coastal and Shelf Science 45 (6), 835–843.

Tam, N.F.Y., et al., 1995. Nutrients and heavy metal contamination of plants and sediments in Futian mangrove forest. Hydrobiologia 295 (1–3), 149–158.

Thamdrup, B., Dalsgaard, T., 2002. Production of $N_2$ through anaerobic ammonium oxidation coupled to nitrate reduction in marine sediments. Applied and Environmental Microbiology 68 (3), 1312–1318.

Tilman, D., 1977. Resource competition between plankton algae: an experimental and theoretical approach. Ecology 58 (2), 338–348.

Trimmer, M., Nicholls, J.C., Deflandre, B., 2003. Anaerobic ammonium oxidation measured in sediments along the Thames estuary, United Kingdom. Applied and Environmental Microbiology 69 (11), 6447–6454.

Tuominen, L., et al., 1998. Spatial and temporal variability of denitrification in the sediments of the northern Baltic Proper. Marine Ecology Progress Series 172, 13–24.

Usui, T., Koike, I., Ogura, N., 2001. $N_2O$ production, nitrification and denitrification in an estuarine sediment. Estuarine, Coastal and Shelf Science 52 (6), 769–781.

Van Raaphorst, W., et al., 1992. Nitrogen cycling in two types of sediments of the southern North Sea (Frisian Front, Broad Fourteens): field data and mesocosm results. Netherlands Journal of Sea Research 28 (4), 293–316.

Weathers, P.J., 1984. $N_2O$ evolution by green algae. Applied and Environmental Microbiology 48 (6), 1251–1253.

Williams, P.H., Jarvis, S.C., Dixon, E., 1998. Emission of nitric oxide and nitrous oxide from soil under field and laboratory conditions. Soil Biology and Biochemistry 30 (14), 1885–1893.

Wooller, M., et al., 2003. A taphonomic study of $\delta^{13}C$ and $\delta^{15}N$ values in *Rhizophora mangle* leaves for a multi-proxy approach to mangrove palaeoecology. Organic Geochemistry 34 (9), 1259–1275.

Wrage, N., et al., 2001. Role of nitrifier denitrification in the production of nitrous oxide. Soil Biology and Biochemistry 33 (12), 1723–1732.

Wu, J., et al., 2009. Impact of COD/N ratio on nitrous oxide emission from microcosm wetlands and their performance in removing nitrogen from wastewater. Bioresource Technology 100 (12), 2910–2917.

Yoshinari, T., 1990. Emissions of $N_2O$ from various environments—The use of stable isotope composition of $N_2O$ as tracer for the studies of $N_2O$ biogeochemical cycling. In: Denitrification in Soil and sediment. Springer U. S., pp. 129–150

Zhang, Z., et al., 2012. Soil C, N and P stoichiometry of *Deyeuxia angustifolia* and *Carex lasiocarpa* wetlands in Sanjiang plain, northeast China. Journal of Soils and Sediments 12 (9), 1309–1315.

Zheng, H., Hanaki, K., Matsuo, T., 1994. Production of nitrous oxide gas during nitrification of wastewater. Water Science and Technology 30 (6), 133–141.

Zhou, S., Hou, H., Hosomi, M., 2008. Nitrogen removal, $N_2O$ emission, and $NH_3$ volatilization under different water levels in a vertical flow treatment system. Water, Air, and Soil Pollution 191 (1–4), 171–182.

# 23

# Nitrogen Assessment in Indian Coastal Systems

R. Ramesh, A. Paneer Selvam, R.S. Robin, D. Ganguly, G. Singh, R. Purvaja

*NATIONAL CENTRE FOR SUSTAINABLE COASTAL MANAGEMENT, MINISTRY OF ENVIRONMENT, FORESTS AND CLIMATE CHANGE, ANNA UNIVERSITY CAMPUS, CHENNAI, INDIA*

**CHAPTER OUTLINE**



## Introduction

Productive coastal zones have been a backbone of human development since ages. Coastal zones occupy ~8% of the world's surface area, but provide 25% of global

**The Indian Nitrogen Assessment. http://dx.doi.org/10.1016/B978-0-12-811836-8.00023-9**

productivity, 90–95% of the world's marine fish catch, 80% of global carbonate production, 50% of global denitrification, and 90% of global sedimentary mineralization (UNEP, 1992). Human interventions have significantly altered global N budget, which was in balance during the preindustrial era. Concurrently, input rates from synthetic N fertilizer production, legume crops, and combustion of fossil fuels have superseded the losses. Consequently, the inputs of newly fixed $N_2$ in the terrestrial systems (formation of bioavailable N from atmospheric $N_2$) have increased by 100% due to human interventions (Galloway et al., 2004). A considerable fraction of this newly fixed/chemical nitrogen is transported to the coastal waters through rivers where it modifies the coastal ecosystem function (Rabalais, 2002; Gilbert et al., 2005). Changes in phytoplankton species composition, alteration and loss of sea grass habitats, and increases in carbon fixation, extent and duration of anoxic and hypoxic waters, harmful algal blooms, and coral reef degradation are some of the effects associated with the excess N inputs to coastal systems (Sutton et al., 2011). The coastal and marine nitrogen cycle occupies a complex, central role within the biogeochemical cycles (Purvaja et al., 2008a).

The main sources of nitrogen to coastal ecosystem is the recycling of organic matter, fluvial input, atmospheric deposition, onshore transport of nutrients from the open sea, coastal upwelling and fixation of atmospheric $N_2$. Despite the fact that main input is in the form of inorganic N-species (DIN: nitrate, nitrite and ammonium), the input of DON has been the focus of many researchers suggesting that it can comprise up to 90% of the total nitrogen input (Sutton et al., 2011). Various biogeochemical processes such as ammonification, nitrification, denitrification, and nitrogen fixation contribute toward the biogeochemical cycling of nitrogen in the coastal region. Ammonium is consumed by phytoplankton and microbes, by removing approximately half of total $NH_4^+$ in surface water and nearly three-fourth of the subsurface waters (Bradley et al., 2010).

There is a strong coupling in the marine and terrestrial nitrogen cycles (Gruber, 2004) with consistent increase in human-induced input of nitrogen to coastal regions. Furthermore, the rate of transport from land to ocean is accelerated resulting in excess nutrient supply not only to the coastal zones (Rabalais, 2002) but also to the open ocean (Duce et al., 2008). Consequently, many of the coastal ecosystems are not limited with respect to nitrogen, whereas they are mostly limited with respect to phosphorus. However, Duce et al. (2008) suggested that overall coastal zone acts as a sink and not as a source of nitrogen for the open ocean, though the natural fluxes are often strongly perturbed resulting in release of GHGs (nitrous oxide, $N_2O$). A conceptual representation of the present day N budget (Voss et al., 2005) involving various addition and removal processes for the entire coastal and oceanic areas of the globe is given in Fig. 23.1.

In order to formulate a mass balance budget, quantification of total N export from watersheds to coastal systems may be enough, though in order to understand the dynamics of the transport and factors controlling the flux, the speciation in various forms is essential. As the bioavailability potential of each and every form of nitrogen varies according to the ambient conditions, the factors controlling the transport and mobility are not common for all forms. For example, DIN is considered readily bioavailable, whereas only a fraction of DON is immediately

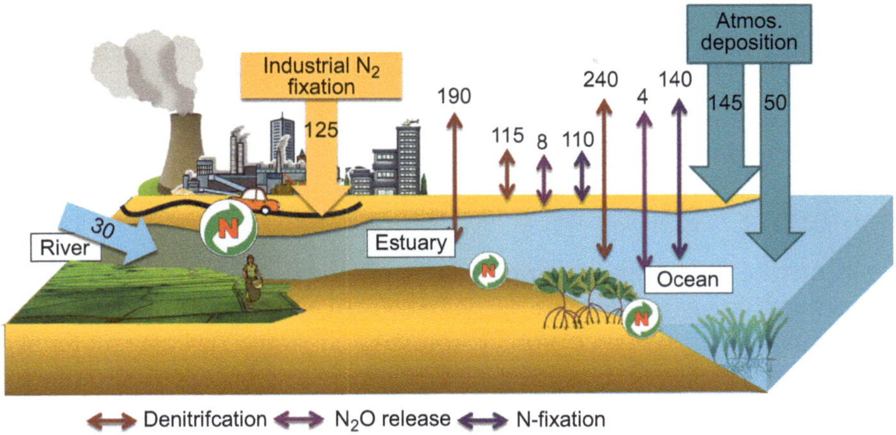


**FIGURE 23.1** Nitrogen budget in coastal ecosystem. Values are given in Tg N $year^{-1}$.

available for the biological uptake. The availability and solubility of dissolved and particulate nitrogen is different. This calls for a thorough approach while explaining nitrogen dynamics in the coastal regions.

## Nitrogen Transformation Processes in the Coastal Systems

Global nitrogen cycle has evolved more than three billion years ago through various biogeochemical and microbial processes coupled with natural inputs that keep the nitrogen cycle of the oceans in near balance (Canfield et al., 2010). The largest pool of nitrogen on earth is ammonium bound in rocks and sediments ($1.8 \times 10^{22}$ g), while atmospheric $N_2$ accounts for $0.39 \times 10^{22}$ g. Fixed nitrogen in the oceans amounts to less than 0.01% of the atmospheric pool, which is split between inorganic forms ($5.7 \times 10^{17}$ g), living biomass ($0.005 \times 10^{17}$ g), and nitrogen in organic detritus ($5.3 \times 10^{17}$ g) (Soderlund and Rosswall, 1982). However, recent findings strongly suggest a significant imbalance of the oceans' nitrogen budget towards the larger sinks than the sources. (Gruber and Galloway, 2008).

Denitrification by prokaryotes was thought to be the only removal pathway of reactive nitrogen hitherto. However, various new pathways are identified, such as the anaerobic ammonium oxidation (anammox) (van de Graaf et al., 1995) and dissimilatory nitrate reduction to ammonium (DNRA), which subsequently modified the N Cycle. Similarly, the photoautotrophic bacteria/blue-green algae of genus *Trichodesmium* sp., a species found globally in tropical and subtropical waters, are the only marine source of nitrogen fixation. Nevertheless, in recent years, various other nitrogen-fixing bacteria have been spotted which may escalate the estimate of the global nitrogen input to the oceans substantially. In proportion with the enhanced anthropogenic activities, nitrogen inputs to these marginal water bodies drastically alter and disrupt the N balance in the coastal and marine systems. Anthropogenic nutrient enrichment fuels eutrophication and degradation of estuaries and coastal water. The assemblage of various nitrogenous nutrients/components in aquatic systems and the processes by which these nutrients are transformed are principally regulated by the redox condition and play an important role in the growth and physiology of the primary producers. The various biogeochemical processes such as ammonification, nitrification, denitrification, and nitrogen fixation are discussed below with reference to their importance in coastal/marine systems.

## Ammonification

Coastal and marine sediments are important sites where biological processes pair with nutrient regeneration of organic matter. The process of ammonification is the result of the breakdown of organic matter such as dead animals and plants, or waste materials such as excrement. This process typically occurs in the soil in an aerobic environment which gives the bacteria and other microorganisms the required energy to work with.$NH_4^+$ is the major end product of organic N mineralization (amino acids), and the regenerated $NH_4^+$ is transported to the water column by diffusion and upwelling. In surface sediments, $NH_4^+$ is also assimilated as a source of N by benthic plants, or oxidized to $NO_3^-$ and $NO_2^-$ by nitrification under suitable conditions. Ammonification may be an important nitrogen transformation in mangrove ecosystems due to their generally low $NO_3^-$ sediment concentrations (pore waters = 0–21 μM; Alongi et al., 1992; Boto, 1992) and high C: N ratios (>20). Ammonification in both tropical and temperate sea grass beds can supply $NH_4^+$ in excess of that required for plant growth (Short et al., 1993; Purvaja et al., 2008b).

## Nitrification

Nitrification is a two-step process, converting reduced forms of nitrogen—ammonia to nitrite and then to nitrate, which is a significant species in the global nitrogen pathway. Most nitrification occurs aerobically and is carried out exclusively by prokaryotes (e.g., autotrophic ammonia oxidizers, autotrophic nitrite oxidizers, and heterotrophic nitrifiers).

Two distinct steps of nitrification carried out by specific types of microorganisms are (1) oxidation of ammonia to nitrite, which is carried out by ammonia oxidizers, and (2) genesis of nitrate from nitrite. Furthermore, aerobic ammonia oxidizers are also the autotrophs, competent of fixing carbon dioxide to produce organic carbon, much as photosynthetic organisms, but uses ammonia as the energy source instead of light. Nitrifying activities in marine systems occur in sediments where dissolved oxygen concentrations are low. Mostly nitrification rate is controlled by variables such as temperature, under favorable redox conditions indicated by the in situ dissolved $O_2$ and ammonium concentration (Kaplan, 1983). Nitrification occurs in the water column as well as in oxygenated sediment layers, and the fate of the generated nitrate is either assimilated by phytoplankton and algae or reduced to $N_2$ and $N_2O$ during bacterial denitrification. Therefore, in mangrove forests nitrification occurs predominantly close to the sediment surface in microoxic zones (Purvaja et al., 2008a).

The pivotal aspect of nitrification in the surface ocean N cycling is the metamorphosis between ammonium and nitrate. The benthic strata have very high nitrate concentrations, while the ocean surface is usually drained out in nitrate species, and this observation led to the long-held belief that nitrification occurred only in the deep sea. Coastal upwelling controlled by physical processes causes an intermixing of the deep water with the surface and enhances nitrate availability. Concentration of $NO_3^-$ is very significant in determining the new production, which represents the effective C sequestration potential of the coastal waters.

## Denitrification

Denitrification is the process that converts nitrate to nitrogen gas, thus removing bioavailable nitrogen and returning it to the atmosphere. Under hypoxic conditions ($O_2 < 10\ \mu mol\ L^{-1}$), the heterotrophic denitrifiers utilize nitrate as a final electron acceptor during respiration and produce $N_2$ gas as the final end product and $N_2O$ as an important byproduct. Nitrous oxide ($N_2O$) is one of the key greenhouse gases with high atmospheric life span. Unlike nitrification, denitrification is an anaerobic process, occurring mostly in soils and sediments and anoxic zones in lakes, estuaries, and oceans. Emission rate of $N_2O$ in Indian mangroves varied between 0.12 and 1.3 $\mu mol\ m^{-2}\ h^{-1}$ (Barnes et al., 2006; Krithika et al., 2008) and between 1.07–1.65 $\mu mol\ m^{-2}\ day^{-1}$ in estuaries (Rao and Sarma, 2013). Nevertheless, contribution of Indian estuaries toward annual $N_2O$ emissions (0.6 Gg $year^{-1}$) during denitrification is insignificant compared to agriculture sector (~131 Gg $year^{-1}$) (Garg et al., 2012; Rao and Sarma, 2013).

Denitrification rates reported for the mangroves are generally much lower than those reported for estuarine sediments, despite the suitability of mangrove sediments (i.e., alternating anoxic and oxic conditions with tidal inundation) to the coupling of mineralization–nitrification–denitrification processes (Rivera-Monroy and Twilley, 1996). In sea grass ecosystems, denitrification rates were found to be highly dependent on the concentration of nitrate in the water column, but are not significant systems for denitrification to occur. Both coastal and oceanic low-oxygen regions are sensitive to the environmental change brought about by changes in nutrient loading (anthropogenic) and stratification (physical). Another process that removes N from the system is the anaerobic ammonium oxidation (anammox) that couples nitrite reduction and ammonium oxidation.

## Role of Coastal Flora and Fauna in Nitrogen Cycling

Nitrogen is one of the essential micronutrients, governing biological productivity and growth in coastal and marine environments. However, enrichment of this nutrient as a result of anthropogenic inputs often leads to eutrophication of the coastal waters. Rapid proliferation of phytoplankton and pelagic macroalgae mask light available to the associate biota causing environmental stress. Several studies have reported that coastal ecosystems such as mangroves and sea grass play a vital role in regulating the nitrogen cycle in the surrounding water column. Both N and P can be effectively stored within sea grass tissues, for example, nitrate is generally stored in vacuoles within the leaf cytosol (Ferdie and Forquerean, 2004).

Nutrient enrichment of the coastal waters supports the growth of sea grass ecosystems. However, light becomes the primary limiting factor with continuous increase in nutrient inputs. Consequently, fast multiplying organisms, such as macroalgae (in shallow regime) and phytoplankton (in deeper part), compete for dominance, while sea grass declines (Burkholder et al., 2007). Reducing conditions in mangrove sediments are known to favor alternate respiratory pathways such as denitrification, sulfate reduction. Fernandes et al. (2012) showed that "denitrification and anammox operate in tandem, resulting in N loss in tropical mangrove sediments." However, denitrification is the most important process in the reduced mangrove sediments (Purvaja et al., 2008a,b).

Benthic macrofauna and microbes also play a major role in nutrient dynamics, eventually affecting the trophic status of aquatic ecosystems. Macrofaunal organisms in many coastal systems stimulate benthic nitrification and denitrification, leading to nitrogen removal. However, the release of ammonium from sediments is preferred over the sedimentary uptake of nitrate (Ji, 2008). Further, nitrogen transformation processes within shallow aquatic ecosystems are significantly influenced by macrofaunal activities (Stief, 2013) through coupled nitrification–denitrification. This can thus offset the concurrent release of (1) ammonium that stimulates aquatic primary production, and (2) nitrous oxide that contributes to global warming (Stief, 2013).

### Anthropogenic Interferences in N Transformation

Anthropogenic nitrogen loads are the sources of many coastal water quality problems globally (UNEP, 2007). During the past century, anthropogenic activities have evidently enhanced the rate of nitrogen fixation on the land and in the coastal waters. Major sources of this enhanced supply include industrial processes producing nitrogen fertilizers, combustion of fossil fuels, and power plants, which have increased atmospheric concentration of nitrous oxide ($N_2O$). Anthropogenic interference primarily modifies the redox state of the coastal waters by supplying a variable amount of inorganic and organic matter.

Enrichment of nitrogen in coastal waters has been linked to both ecological and human health issues, including impaired drinking water quality, harmful algal blooms, acidification, eutrophication, hypoxia, reduced fishery production, global climate change, and loss of biodiversity (Townsend and Ellis, 2010). The addition of excessive amounts of nitrogen from anthropogenic sources reduces water quality resulting in habitat loss in near coastal waters. In India, nitrogen loading is one of the most significant long-term threats that must be managed (Fig. 23.2).

## Sources of Nitrogen to the Coastal Systems

### Atmospheric Sources

Nitrogen is released to the atmosphere mainly as ammonia, nitrogen oxides (NO, $NO_2$), and in meager organic forms. The main sources of organic forms to atmosphere are industrial processes, agricultural activities, biogenic emissions, ocean surface, and

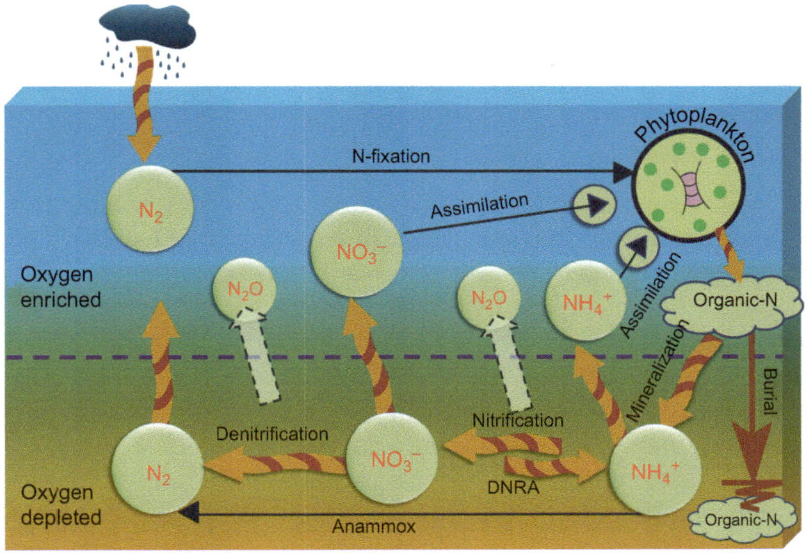


**FIGURE 23.2** Various biogeochemical transformation processes in the coastal N cycle.

grasslands. Galloway et al. (2004) showed anthropogenic activities as a major provider of both oxidized (79%) and reduced (80%) forms of N to atmospheric emissions. Thus, the transport and deposition of anthropogenic atmospheric nitrogen is of major concern, as it can affect the primary productivity and biogeochemistry of nitrogen species in coastal systems. In coastal waters, atmospheric inputs contribute significantly to the overall eutrophication pressure, but evidence that they trigger algal blooms is rather limited (Purvaja et al., 2008b). Management of atmospheric inputs to coastal waters to mitigate eutrophication pressures requires that emissions be managed over a wide area, reflecting the efficient long-range transport of atmospheric nitrogen (Jickells, 2006).

Nitrogen deposition fluxes to the coastal systems are more localized phenomena as they are profoundly and closely linked with the emission patterns and transport pathways in that region. Earlier studies of atmospheric N flux to the Arabian Sea and Bay of Bengal clearly provide the spatial and temporal variations (Sarin et al., 1999a,b). Inputs of atmospheric deposition to coastal systems are lower in comparison with other inputs from rivers, estuaries, and internal cycling in these systems. However, in the Indian scenario, there is a large gap in the atmospheric N input data to the coastal systems.

Contribution of atmospheric nitrogen to coastal waters (uptake/deposition) has been a topic of interest to several researchers. Inorganic nitrogen sources, including $NO_2^-$, $HNO_3$, $NH_4^+$, and $NO_3^-$, also contribute to the total nitrogen deposition. Global inorganic nitrogen was reported to be 34.26 Tg N year$^{-1}$ (Jia et al., 2016). Though the deposition of inorganic nitrogen has been studied in wider details, the research related to organic nitrogen is still in nascent stage. Various sources such as amino acids, particulate nitrogen, dust, pollen, organic nitrates, and bacteria contribute to reduced or oxidized form of atmospheric organic nitrogen (AON). The total load of organic nitrogen varies from site to site as well as the mode of measurement. It is suggested that global atmospheric organic nitrogen flux may vary between 10 and 50 Tg $year^{-1}$ (Neff et al., 2002).

## Geogenic Sources

Geogenic nitrogen is a potentially large pool of nitrogen, which is often not quantified. While modern nitrogen cycle considers atmospheric N input as major source compared to geologic nitrogen, nitrogen-bearing rocks are also well thought-out as potential sources as a result of weathering. Nitrogen in these rocks is present as recalcitrant organic matter, nitride minerals, salts of ammonium or nitrate, or fixed in silicate minerals (Holloway and Dahlgren, 2002). The linkage between mineral weathering and nitrogen cycle involves association and conversion of nitrogen forms in sedimentary organic matter and ammonium/nitrate minerals. In the weathered rock, soil microbial pathways such as nitrification and denitrification play critical roles in release of nitrogen. N concentrations and its isotopic signatures in the Precambrian rocks reflect the presence of nitrogen cycling mechanisms (nitrification and nitrogen fixation by chemosynthetic bacteria) in late Archean and Proterozoic (Pinti and Hashizume, 2001) times. Globally, the release of nitrogen through weathering occurs at ecologically significant rates ranging from 10 to 20 mol N $cm^{-2}$ $s^{-1}$ corresponding to nitrogen fluxes of 4–37 kg N $ha^{-1}$ $year^{-1}$ (Holloway and Dahlgren, 2002). However, these rates are often substantially altered by regional and global phenomenon such as climate change, change in precipitation, and discharges.

## Nitrogen From Rivers, Estuaries, Groundwater, and Coastal Upwelling

Riverine nitrogen loads are the most prominent nitrogen inputs to coastal systems, and 80% of the annual global riverine N fluxes are in the form of dissolved inorganic N (40%); particulate N (40%); and the remaining as dissolved organic N. This riverine nitrogen supply is largely influenced by the agricultural effluents, domestic and industrial waste from urban areas, and is important in sustaining the high rates of primary production in coastal systems. Swaney et al. (2015) indicates the dominance of nitrogen fertilizer consumption in the Indian agricultural sector, resulting in high nitrogen inputs to major Indian watersheds.

Transfer rate/diffusion of N from the agricultural soils to riverine systems is controlled by leaching and runoff factors of that particular region. There is substantial spatial variability between both the east and west coasts of India in the magnitude of N loading through rivers. Gradient in the nitrogen fertilizer consumption of the major river catchments (Ganges, Brahmaputra, Godavari, Krishna, Cauvery, Periyar, Narmada, Tapi, etc.) is attributed to the varying nitrogen loads. This is further supported by the disparity in N fertilizer consumption in catchments of major rivers in east (73%) and west coast (27%) (Swaney et al., 2015). Singh and Ramesh (2011) estimated that ~9 Tg N $year^{-1}$ is transported by the east coast rivers, of which, Ganges alone contributes >50%

(5 Tg N $year^{-1}$); whereas west coast river transports 4.3 Tg N $year^{-1}$. However, N export fluxes (of DIN) normalized by the catchment area showed that rivers along the west coast (607 ± 700.7 kg N $km^{-2}$ $year^{-1}$) exports ~4 times more N than the east coast rivers (160.9 ± 368.9 kg N $km^{-2}$ $year^{-1}$) during the southwest monsoon (Krishna et al., 2016).

Variations in the seasonal distribution of N species between the Indian rivers flowing toward the east and west coasts were studied by Rao and Sarma (2013). Due to the variations in N balance, estuaries on the east coast connected to the Bay of Bengal showed 5.10 ± 6.22 higher $NH_4$–N concentration during the wet season than the dry season; on the contrary, estuaries along the west coast showed marginally higher concentration during the dry season than the wet. Although larger rivers with intense hinterland activities are present along the east coast, studies on their N flux are rather limited and hence unaccounted for.

Studies by Charette et al. (2001) showed that nutrient flow through groundwater to coastal ocean has significant impact on the geochemical cycling. Coastal upwelling is mainly prominent in Arabian Sea (June–September) that introduces nutrient rich subsurface waters to the surface. Although some studies on the nitrogen inputs from groundwater discharge and coastal upwelling were done in the past decades, it is necessary to achieve better quantification of nitrogen fluxes through these mechanisms to reach a strong conclusion.

## N Fixation

Nitrogen fixation also contributes to N inputs to coastal systems. N addition, through biological N fixation, is a reductive process that generates ammonium, which is incorporated into organic matter. Under the N-limiting conditions and in presence of ample P, certain planktonic autotrophs (blue-green algae) outcompete other groups and survive in coastal waters by fixing atmospheric nitrogen. Addition of N through biological fixation significantly counterbalances N removal through denitrification in the global N cycle. Parab et al. (2006) in their report on the measurement of $N_2$ fixation by *Trichodesmium* sp. (130 μmol N $m^{-2}$ $day^{-1}$) in Arabian Sea suggest an important source of new nitrogen to the phytoplankton.

## Effects of Land Use on $N_2O$ and NO Emissions

Land use changes such as deforestation, forest degradation, and increase in croplands are vital concerns for both $N_2O$ and NO emissions. Growing population and its corresponding food demand is the primary driver of deforestation. The availability of fixed nitrogen in the biosphere is doubled by human activities, particularly through the N fertilizer application (Robertson and Vitousek, 2009). In the tropics, high $N_2O$ and NO emissions are observed with land use changes (van Lent et al., 2015). Dam constructions and overconsumption of nitrogen fertilizers in Indian agriculture fields have significantly influenced $N_2O$ emissions from the Indian estuaries (Rao and Sarma, 2013). Significant variation in the $N_2O$ concentration and its fluxes were noticed between the east and west coast ecosystems linked to human interventions and land use changes (Barnes et al., 2006; Krithika et al., 2008; Rao and Sarma, 2013), emphasized by different land use patterns.

# Impacts of Changes in N Balance

N balance in the coastal waters is directly associated with the land-based activities. Dissolved and particulate nutrients from various anthropogenic sources (both organic and inorganic) reach the estuaries and coastal waters primarily through rivers. Anthropogenic activities, such as extensive use of fertilizers in agricultural fields and untreated domestic sewage, significantly influence the riverine export of N and P into rivers, and subsequently to the coastal systems (Lima et al., 2010). For instance, Green et al. (2004) reported an over sixfold increase in the riverine export of dissolved inorganic N to the global ocean from preindustrial (2.1 Tg $year^{-1}$; 1 Tg = 1012 g) to the contemporary period (14.5 Tg $year^{-1}$). In the Mahanadi estuary, approximately sevenfold increase in the annual riverine N loading to the system (Dey et al., 2013) was observed between 1983 and 2008.

Compared to the dry seasons, Indian estuaries carry a substantially higher amount of nutrients to the open ocean during south-west monsoon. The rapid increase in the bioavailable nutrient load to the estuaries and coastal waters causes several ecological problems, including eutrophication (Nixon, 1995; Lima et al., 2010), changes in the ecosystem function and community structure (Bowmann et al., 1995), hypoxia, etc. Enhanced human interference in the coastal systems, in association with the variations in the environmental conditions (e.g., temperature, precipitation pattern) due to climate change, may cause dramatic changes in the riverine DIN loading and associated in situ processes. Further, changes in hydraulic residence time and surface water temperature of the rivers and estuaries may have a significant impact on the trace gas production (including $N_2O$) and their solubility in these aquatic systems. Two most distinct impacts of excessive nutrients, and the most common occurrence in Indian coastal waters are (1) eutrophication and (2) hypoxia, which are discussed in the following sections.

## Eutrophication

To date, the processes driving eutrophication (i.e., nutrient inputs) and acidification (i.e., increased atmospheric $CO_2$ concentrations) have been viewed independently, whereas new research suggests a link between the two and that eutrophication may be exacerbating the effects of increased atmospheric $CO_2$ concentrations on ocean acidification (Cai et al. 2011; Ramesh et al., 2014). Natural eutrophication is a slow process occurring over a period of centuries by the natural accumulation of growth-limiting macronutrients in closed or partially closed aquatic ecosystems. However, anthropogenic activities may stimulate the rate and extent of this event through both point–source discharges and nonpoint loadings of limiting nutrients into aquatic ecosystems, and are termed as cultural eutrophication. Several researchers identified eutrophication as the major cause of the spring bloom and the algal production during the summer months in the estuarine and coastal waters (Andersen et al., 2006; Purvaja et al., 2008b). Additionally, harmful algal blooms (HABs) caused by only 7% of the total marine phytoplankton species (~5000) have detrimental impacts to other organisms due to the production of natural toxins, as reported from over 30 countries including India.

One of the primary effects of eutrophication in coastal marine areas is a community shift from slow-growing, perennial macrophytes, and sea grass, etc. to the fast-growing ephemeral macroalgae, microphytobenthos, and phytoplankton. Slow-growing plants are the superior competitors for nutrients under low-nutrient condition and have a higher capacity to store carbon and nutrients compared to the fast-growing plants. A considerable part of Indian coastal waters are covered by sea grass and other macrophyte community well known for their high efficiency in C sequestration and storage capacity (Singh et al., 2015). Future intensification of external nutrient loading in the coastal waters therefore creates a potential threat to these ecologically important ecosystems (Purvaja et al., 2008a,b). A total of 80 algal bloom events were recorded during the period 1998–2010 from the exclusive economic zone of India, with a clear-cut increase in the occurrence during the last few decades (Padmakumar et al., 2012). The majority of the blooms reported along the west coast of India are caused by dinoflagellates, whereas diatom blooms prevail along the east coast. There have been 39 causative species reported from Indian coastal waters, responsible for the bloom formation, of which *Noctilucascintillans* and *Trichodesmium erythraeum* are the most common (D'Silva et al., 2012).

In Indian coastal waters, most of the algal blooms occurred during withdrawal of the south-west monsoon and premonsoon period. This process is mainly influenced by seasonal upwelling and monsoonal forcing that causes the high riverine discharge resulting in nutrient-enriched waters that provide a competitive edge for the algal bloom. Approximately 400% increase in the agricultural consumption rate of N fertilizers from Indian mainland was recorded between the time period of 1981–82 and 2012–14 (Fig. 23.3). A considerable fraction of these fertilizers comes from the catchment basin of various rivers and significantly contribute bioavailable N to the estuary.

Bloom conditions more often prevail in the northern Arabian Sea caused by convective mixing of deep and surface seawater during the winter monsoon that often cause nutrient enrichment of the coastal as well as estuarine waters. Internal recycling of dissolved and particulate matter in shallow coastal waters is an additional source of bioavailable nutrients. Additionally, construction of dams for irrigation, particularly along the east coast of India, along the river runoff, drastically alters the sediment export to the estuaries and coastal waters. Further, a significant fraction of dissolved nutrients can be partitioned into the suspended particulate matter (SPM). Alteration of SPM load further aggravates the eutrophic condition and may cause frequent bloom events.

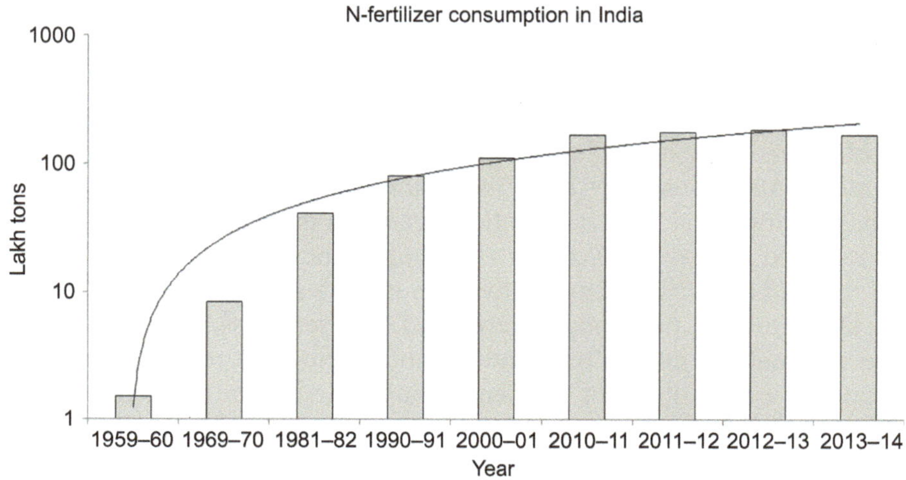


**FIGURE 23.3** N fertilizer consumption in Indian agricultural field from 1959–60 to 2013–14. (http://www.faidelhi.org/statistical-database.htm).

The concern is magnified by findings from recent studies that suggest eutrophication can enhance ocean acidification (Provoost et al., 2010; Howarth et al., 2012; Chou et al., 2012; Ramesh et al., 2014). Excess organic matter production (e.g., algal blooms) in coastal and oceanic environments eventually undergoes microbial degradation, whereby $O_2$ is consumed and $CO_2$ is produced in the water column through microbial respiration resulting in a decline in pH.

## Hypoxia

Eutrophication is the process of nutrient overenrichment of waters that can lead to hypoxia (or oxygen depletion) and harmful algal blooms, which among others can destroy aquatic life in affected areas (Selman et al., 2008; Mee, 2006; Ramesh et al., 2014). Hypoxia can be defined as a condition of aquatic environment where the dissolved oxygen concentration reaches below $2\,mg\,L^{-1}$ that induces avoidance, or altered behavior in growth, reproduction, or survivorship of many marine organisms (Table 23.1). Hypoxia often results in biodiversity loss, mass mortality of fishes, change in community structure, and other environmental and economic losses. Additionally, hypoxia and other associated $O_2$-limiting conditions often induce significant emission of some important trace gases such as $CO_2$, $H_2S$, $N_2O$, $CH_4$, etc. to the emission. This is especially true for regions experiencing coastal upwelling with shallow oxygen minimum zones that are characterized by rapid denitrification and dramatic increase in $N_2O$ production.

Hypoxia in tropical coastal waters generally exhibits large seasonality due to changes in freshwater inputs, suspended sediment load, solar insolation, surface water

**Table 23.1** Definitions of Various Stages of $O_2$ Deficiency

| Stage | Condition | Comment |
|---|---|---|
| Normoxic | $O_2 > 1.4$ mL $L^{-1}$ | Oxygenated water. Marine organisms without $O_2$ stress. |
| Hypoxia | $0.1 < O_2$ (winkler) $\leq 1.4$ mL $L^{-1a}$ $NO_2^- \approx 0$ μM, $NO_3^- > 0$ μM | The upper limit is based on physiological considerations. Oxygen concentrations <2mg $L^{-1}$ or <30% of oxygen saturation induce abnormal responses in marine organisms |
| Suboxia | $O2 \leq 0.1$ mL $L^{-1}$, $NO_2^-$, $NO_3^- > 0$ μM; $H_2S = 0$ μM | The actual $O_2$ levels are probably in the nM range. This allows reduction of elements such as N, I, Mn, and Fe (but not S) with denitrification being the dominant respiratory process. |
| Anoxia | $O_2 = 0$ mL $L^{-1}$, $NO_2^-$, $NO_3^- = 0$ μM; $H_2S > 0$ μM | The environment is completely anaerobic with sulfate reduction being the dominant respiratory process. |

[a]1 mL $L^{-1}$ = 1.43 mg $L^{-1}$ = 44.64 μM.

From Naqvi, S.W.A.,Bange, H.W., Farıas, L., Monteiro, P.M.S., Scranton, M.I., Zhang, J., 2009. Coastal hypoxia/anoxia as a source of $CH_4$ and $N_2O$, Biogeosciences Discuss 6, 9455–9523.

temperature, etc. In India, seasonal formation of coastal hypoxia has been reported from many estuaries and along open coasts as a result of natural upwelling or from anthropogenic eutrophication induced by riverine nutrient inputs. Naturally occurring permanent hypoxia is found in some isolated seas and marine basins as well as in open slope oxygen minimum zones. An area of $677.3 \times 10^3$ $km^2$ is reported as oxygen minimum zone (59% of the total global hypoxic zone) in the shelf and the bathyal sea floor of the northern part (mainly Arabian Sea and Bay of Bengal) of the Indian Ocean (Helly and Levin, 2004). It has primarily caused persistent occurrence of hypoxia, resulting in increased emission of $N_2O$.

Intense heterotrophic microbial respiration often causes severe oxygen depletion in estuarine waters. Such conditions were reported from several Indian systems, e.g., Tapi estuary (Ram et al., 2014), Thane Creek (Quadros et al, 2004), Zuary estuary (Mochemadkar et al., 2013). Ram et al. (2014) reported mass fish kill events in Tapi estuary due to hypoxic condition. However, studies on the correlation of intense hypoxia with major fish kills are meager. Further hypoxic waters, characterized with high $pCO_2$ and elevated nutrient concentration may cause severe aragonite undersaturation posing major threat to marine calcareous organisms. Increase in sea surface temperature associated with the future climate change may further reduce the solubility of $O_2$ which in turn may result in a shallower OMZ and enhanced extent of denitrification.

## Physical and Biogeochemical Processes in Estuaries

Estuaries, seaward parts of the river, significantly modify the riverine nutrient fluxes to the sea through various physical and biogeochemical processes. Estuarine processes in turn affect the sustainability of nearshore ecosystems, and perhaps over long periods of time, the open ocean ecosystems as well (Nixon et al., 1986). Physical processes such as precipitation, evaporation, fresh water and groundwater discharges, tidal mixing, wind-driven re-suspension, light availability have significant direct and indirect influence at a varying degree on the estuarine biogeochemistry and net ecosystem metabolism. Spatiotemporal alteration in the physical forcing causes significant modification in the hydraulic residence time of the estuarine system. Changes in water-residence time drastically influence the time for the biogeochemical processing of organic and inorganic material before being released to the open ocean. In relation to estuaries, residence time is defined as the time required to replace the existing freshwater received by the estuary from the river discharge.

Estuaries with shorter residence times are more dynamic and capable of processing the effluents due to higher mixing of fresh and saltwater and efficient dilution of polluting substances. For example, the risk of large nutrient accumulation (eutrophication) and subsequent depletion of dissolved $O_2$ (hypoxic/suboxic) in the estuarine waters may be substantially reduced when effective removal effluents are rapid (short residence time). Swaney et al. (2011) represented the net ecosystem metabolism of various estuaries as a function of the residence time and showed a higher dominance of respiration over the autotrophic production, with the increase of the residence time in an ecosystem. A recent study by Patra et al. (2016) showed that SPM in the shallow coastal waters may act as a source of nutrients for the primary producers during the non-monsoon season.

An advanced understanding of N dynamics in the estuarine systems is necessary to predict the ecological response of altered nutrient inputs from various sources and the subsequent changes in the ecosystem dynamics. Primary and secondary productions in the estuaries often depend on the internal N cycling regulated by the redox gradient in the water column or sediments. Ganguly et al. (2015) showed the possible dependence of the primary producers on the regenerated dissolve organic nitrogen as an additional source of nutrients in the shallow Indian coastal waters.

## Summary and Way Forward

Influence of in situ biogeochemical processes, such as ammonification, nitrification, N fixation, DNRA, denitrification, and the extent of sediment–water coupling could often be comparable to the external loadings of nutrients in determining the ecological status of the shallow estuaries and coastal waters. Further research to quantify the relative importance of these various in situ processes and incorporation into the integrated modeling framework will be useful to determine primary productivity and to sustain coastal fisheries.

Under scenarios of climate change, production of reactive N (Nr) for agricultural use may increase up to 90 to 190 Tg N year$^{-1}$ by 2100, a range that spans from slightly less to almost twice as much current fertilizer consumption rates (IPCC, 2013). Four driving parameters (i.e., population growth, consumption of animal protein, agricultural efficiency improvement, and additional biofuel productions) are used to project this future nitrogen demand. Being commonly known as the marginal filters, estuaries, and coastal waters would face a significant increase in the anthropogenic N load in the near future.

It is clear from the above discussion that the combination of eutrophication and hypoxia is to have greater adverse effects on coastal ecosystems in the future, if conditions exceed thresholds. Hence, this needs to be urgently addressed. It is beyond doubt that nitrogen is the primary cause of eutrophication and, hence, its management together with phosphorus is crucial for primary coastal productivity. Finally, the effects of nitrogen on greenhouse gas emission (production of $N_2O$) are far more deterrent to atmosphere, as it has already exceeded the planetary boundaries. Hence there is a need to manage nitrogen as an integral part of research and decision making by aiding the effective management of land-based pollution to the coasts and the oceans. An integrated approach, involving measures such as implementation of best agricultural practice (BAP), measures for waste water treatment, as well as management of coastal and estuarine vegetation as a sink of nitrogen, is needed to effectively achieve the sustainable nitrogen in the coastal environment.

# References

Alongi, D.M., Boto, K.G., Al, R., 1992. Nitrogen and phosphorous cycles. In: Al, R., Alongi, D.M. (Eds.), Tropical Mangrove Ecosystems, Coastal and Estuarine Studies, 41. American Geophysical Union, Washington, DC.

Andersen, J.H., Schlüter, Louise, Ærtebjerg, Gunni, 2006. Coastal eutrophication: recent developments in definitions and implications for monitoring strategies. Journal of Plankton Research 28 (7), 621–628.

Barnes, J., Ramesh, R., Purvaja, R., Rajkumar, A.N., Senthil Kumar, B., Krithika, K., Ravichandran, K., Uher, G., Upstill-Goddard, R., 2006. Tidal dynamics and rainfall control $N_2O$ and $CH_4$ emissions from a pristine mangrove creek. Geophysical Research Letters 33, L15405. http://dx.doi.org/10.1029/2006GL026829.

Boto, K.B., 1992. Nutrients and mangroves. In: Connell, D.W., Hawker, D.H. (Eds.), Pollution in Tropical Aquatic Systems. CRC Press Inc, Boca Raton, pp. 129–145.

Bowmann, W.D., Theodose, T.A., Fisk, M.C., 1995. Physiological and production responses of plant growth forms to increases in limiting resources in Alpine tundra: implications for differential community response to environmental change. Oecologia 101, 217–227.

Bradley, P.B., Sanderson, M.P., Frischer, M.E., Brofft, J., Booth, M.G., Kerkhof, L.J., Bronk, D.A., 2010. Inorganic and organic nitrogen uptake by phytoplankton and heterotrophic bacteria in the stratified Mid-Atlantic Bight. Estuarine, Coastal and Shelf Science 88, 429–441.

Burkholder, J.M, Tomasko, D.A., Touchette, B.W., 2007. Seagrasses and eutrophication. Journal of Experimental Marine Biology and Ecology 350, 46–72.

Cai, W.J., Hu, X., Huang, W.J., et al., 2011. Acidification of subsurface coastal waters enhanced by eutrophication. Nature Geosciences 4, 766–770.

Canfield, D.E., Glazer, A.N., Falkowski, P.G., 2010. The evolution and future of earth's nitrogen cycle. Science 330, 192–196.

Charette, M.A., Buesseler, K.O., Andrews, J.E., 2001. Utility of radium isotopes for evaluating the input and transport of groundwater-derived nitrogen to a Cape Cod estuary. Limnology and Oceanography 46, 465–470.

Chou, W.C., Gong, G., Cai, W.J., 2012. Seasonality of $CO_2$ in coastal oceans altered by increasing anthropogenic nutrient delivery from large rivers: evidence from the Changjiang-East China Sea system. Biogeoscience Discuss 9 (12), 18993–19017.

D'Silva, M.S., Chandrashekar, A.A., Naik, R.K., D'Costa, P.M., 2012. Algal blooms: a perspective from the coasts of India. Nature Hazards 63, 1225–1253.

Dey, M., Chowdhury, C., Pattnaik, A.A., Ganguly, D., Mukhopadhyay, S.K., De, T.K., Jana, T.K., 2013. Comparison of Monsoonal change of water quality parameters between 1983 and 2008 in a tropical estuary in Northeastern India: role of phytoplankton and community metabolism. Marine Ecology 34 (1), 14–29.

Duce, R.A., et al., 2008. Impacts of atmospheric anthropogenic nitrogen on the open ocean. Sciences 320, 893–897.

Ferdie, M., Forquerean, J.W., 2004. Responses of seagrass communities to fertilization along a gradient of relative availability of nitrogen and phosphorus in a carbonate environment. Limnology and Oceanography 49, 2082–2094.

Fernandes, S.O., Bonin, P.C., Michotey, V.D., Garcia, N., Loka Bharathi, P.A., 2012. Nitrogen-limited mangrove ecosystems conserve N through dissimilatory nitrate reduction to ammonium. Scientific Reports 2, 419.

Galloway, J.N., Dentener, F.J., Capone, D.G., Boyer, E.W., Howarth, R.W., Seitzinger, S.P., Asner, G.P., Cleveland, C.C., Green, P.A., Holland, E.A., Karl, D.M., Michaels, A.F., Porter, J.H., et al., 2004. Nitrogen cycles: past, present, future. Biogeochemistry 70, 153–226.

Ganguly, D., Patra, S., Muduli, P.R., Kanuri, V.V., Abhilash, K.R., Robin, R.S., Subramanian, B.R., 2015. Influence of nutrient input on the trophic state of a tropical brackish water lagoon. Journal of Earth System Sci 124 (5), 1005–1017.

Garg, A., Shukla, P.R., Upadhyay, J., 2012. $N_2O$ emissions of India: an assessment of temporal, regional and sector trends. Climatic Change 110 (3), 755–782.

Gilbert, D., Sundby, B., Gobeil, C., Mucci, A., Tremblay, G.-H., 2005. A seventy-two-year record of diminishing deep-water oxygen in the St. Lawrence estuary: the Northwest Atlantic connection. Limnology and Oceanography 50, 1654–1666.

Green, P.A., Vorosmarty, C.J., Meybeck, M., Galloway, J.N., Peterson, B.J., Boyer, E.W., 2004. Pre-industrial and contemporary fluxes of nitrogen through rivers: a global assessment based on topology. Biogeochemistry 68, 71–105.

Gruber, N., 2004. The dynamics of the marine nitrogen cycle and atmospheric $CO_2$. In: Oguz, T., Follows, M. (Eds.), Carbon Climate Interactions. Kluwer, Dordrecht, pp. 97–148.

Gruber, N., Galloway, J.N., 2008. An earth-system perspective of the global nitrogen cycle. Nature 451, 293–296.

Van de Graaf, A., Mulder, A., De Brujin, P., Jetten, M.S.M., Robertson, L.A., Kuenen, J.G., 1995. Anaerobic oxidation of ammonium is a biologically mediated process. Applied and Environmental Microbiology 61, 1246–1251.

Helly, J.J., Levin, L.A., 2004. Global distribution of naturally occurring marine hypoxia on continental margins. Deep-Sea Research Part I-Oceanographic Research Papers 51, 1159–1168.

Howarth, R.W., Swaney, D.P., Billen, G., Garnier, J., Hong, B., Humborg, C., Johnes, P., Mörth, C.-M., Marino, R., 2012. Nitrogen fluxes from the landscape are controlled by net anthropogenic nitrogen inputs and by climate. Frontiers in Ecology and the Environment 10 (1), 37–43.

Holloway, J.M., Dahlgren, R.A., 2002. Nitrogen in rock: Occurrences and biogeochemical implications. Global Biogeochemical Cycles 16 (4), 1118.

IPCC, 2013. Climate change 2013: the physical science basis. In: Stocker, T.F., Qin, D., Plattner, G.-K., Tignor, M., Allen, S.K., Boschung, J., Nauels, A., Xia, Y., Bex, V., Midgley, P.M. (Eds.), Contribution of

Working Group I to the Fifth Assessment Report of the Intergovernmental Panel on Climate Change. Cambridge University Press, Cambridge, United Kingdom and New York, NY, USA, pp. 465–570. http://dx.doi.org/10.1017/CBO9781107415324. Chapter 6.

Jia, Y., Yu, G., Gao, Y., He, N., Wang, Q., Jiao, C., Zuo, Y., 2016. Global inorganic nitrogen dry deposition inferred from ground- and space-based measurements. Scientific Reports 6, 19810. http://dx.doi.org/10.1038/srep19810.

Jickells, T., 2006. The role of air-sea exchange in the marine nitrogen cycle. Biogeosciences 3, 271–280. http://dx.doi.org/10.5194/bg-3-271-2006.

Ji, Z.-G., 2008. Hydrodynamics and Water Quality: Modeling Rivers, Lakes, and Estuaries. Wiley, p. 704.

Kaplan, W.A., 1983. Nitrification. In: Carpenter, J.E., Capone, D.G. (Eds.), Nitrogen in the Marine Environment. AcademicPress, New York, pp. 139–190.

Krishna, M.S., Prasad, M.H.K., Rao, D.B., Viswanadham, R., Sarma, V.V.S.S., Reddy, N.P.C., 2016. Export of dissolved inorganic nutrients to the northern Indian Ocean from the Indian monsoonal rivers during discharge period. Geochimica et Cosmochimica Acta 172, 430–443.

Krithika, K., Purvaja, R., Ramesh, R., 2008. Fluxes of methane and nitrous oxide from an Indian mangrove. Current Science 94, 218–224.

Lima, M.C., Souza, M.F.L., Eca, G.F., Silva, M.A.M., 2010. Export and retention of dissolved inorganic nutrients in the Cachoeira River, Ilhéus, Bahia, Brazil. Journal of Limnology 69, 138–145.

van Lent, J., Hergoualch, K., Verchot, L.V., 2015. Reviews and syntheses: soil $N_2O$ and NO emissions from land use and land-use change in the tropics and subtropics: a meta-analysis. Biogeosciences 12, 7299–7313.

Mee, L., November 2006. Reviving dead zones. Scientific American.

Mochemadkar, S., Gauns, M., Pratihary, A.K., Thorat, B.R., Roy, R., Pai, Naqvi, S.W.A., 2013. Response of phytoplankton to nutrient enrichment with high growth rates in a tropical monsoonal estuary—Zuari Estuary, India. Indian Journal of Geo-Marine Sciences 42 (3), 314–325.

Naqvi, S.W.A., Bange, H.W., Farıas, L., Monteiro, P.M.S., Scranton, M.I., Zhang, J., 2009. Coastal hypoxia/anoxia as a source of CH4 and N2O. Biogeosciences Discuss 6, 9455–9523.

Neff, J.C., Holland, E.A., Dentener, F.J., McDowell, W.H., Russell, K., 2002. The origin, composition and rates of organic nitrogen deposition: a missing piece of the nitrogen cycle? Biogeochemistry 57, 99–136.

Nixon, S.W., 1995. Coastal marine eutrophication: a definition, social causes and future concerns. Ophelia 41, 199–219.

Nixon, S.W., Oviatt, C.A., Frithsen, J., Sullivan, B., 1986. Nutrients and the productivity of estuarine and coastal marine ecosystems. Journal Limnological Society of Southern Africa 12, 43–71.

Padmakumar, K.B., Menon, N.R., Sanjeevan, V.N., 2012. Is occurrence of harmful algal blooms in the exclusive economic zone of India on the rise? International Journal of Oceanography. http://dx.doi.org/10.1155/2012/263946.

Parab, S.G., PrabhuMatondkar, S.G., Gomes, H.do R., Goes, J.I., 2006. Monsoon driven changes in phytoplankton population in the Eastern Arabian Sea as revealed by microscopy and HPLC pigment analysis. Continental Shelf Research 26, 2538–2558.

Patra, S., Raman, A.V., Ganguly, D., Robin, R.S., Muduli, P.R., Vishnu, V.K., Abhilash, K.R., CharanKumar, B., Subramanian, B.R., 2016. Influence of suspended particulate matter on nutrient biogeochemistry of a tropical shallow lagoon, Chilika, India. Limnology. http://dx.doi.org/10.1007/s10201-015-0475-2.

Pinti, D.L., Hashizume, K., 2001. 15N-depleted nitrogen in early Archean kerogens: Clues on ancient marine chemosynthetic-based ecosystems? Precambrian Research 105, 85–88.

Provoost, P., van Heuven, S., Soetaert, K., Laane, R., Middelburg, J.J., 2010. Long-term record of pH in the Dutch coastal zone: a major role for eutrophication-induced changes. Biogeoscience Discuss 7 (3), 4127–4152.

Purvaja, R., Ramesh, R., Ray, A.K., Rixen, T., 2008a. Nitrogen cycling: a review of the processes, transformations and fluxes in coastal ecosystems. Current Science 94 (11), 1419.

Purvaja, R., Ramesh, R., Shalini, A., Rixen, T., 2008b. Biogeochemistry of nitrogen in seagrass and oceanic systems. Memoir Geological Society of India 73, 435–460.

Quadros, G., Mishra, V., Borkar, M.U., Athalye, R.P., 2004. Impact of construction and reclamation activities on the water quality of the Thane Creek, Central-west coast of India. Journal of Coastal Development 7 (2), 71–78.

Rabalais, N.N., 2002. Nitrogen in aquatic environments. Ambio 31 (2), 102–112.

Ram, A., Jaiswar, R.M.J., Rokade, M.A., Bharti, S., Vishwasrao, C., Majithiya, D., 2014. Nutrients, hypoxia and mass Fishkill events in Tapi Estuary, India. Estuarine Coast Shelf Science 148, 48–58.

Ramesh, R., Lakshmi, A., Purvaja, R., Costanzo, S.D., Kelsey, R.H., Hawkey, J., Datta, A., Dennison, W.C., 2014. Policy Brief—Eutrophication and Ocean Acidification. A joint publication of GPNM, NCSCM, and IAN/UMCES.

Rao, G.D., Sarma, V.V.S.S., 2013. Contribution of $N_2O$ emissions to the atmosphere from Indian monsoonal estuaries. Tellus 65, 19660.

Rivera-Monroy, V.H., Twilley, R.R., 1996. The relative role of denitrification and immobilization in the fate of inorganic nitrogen in mangrove sediments (Terminos Lagoon, Mexico). Limnology and Oceanography 41 (2), 284–296.

Robertson, G.P., Vitousek, P.M., 2009. Nitrogen in agriculture: balancing the cost of an essential resource. Annual Reviews Environment and Resources 34, 97–125.

Sarin, M.M., Rengarajan, R., Krishnaswami, S., 1999a. Aerosol $NO_3^-$and $^{210}Pb$ distribution over the central-eastern Arabian Sea and their air-sea deposition fluxes. Tellus 51B, 749–758.

Sarin, M.M., Rengarajan, R., Krishnaswami, S., 1999b. Nitrate and $^{210}Pb$ in marine air over the central-eastern Arabian Sea and their air-sea deposition fluxes. Tellus51B 749–758.

Selman, M., et al., 2008. WRI Policy Note—Water Quality: Eutrophication and Hypoxia. World Resources Institute, Washington DC.

Short, F.T., Montgomery, J., Zimmerman, C.F., Short, C.A., 1993. Production and Nutrient Dynamics of a SyringodiumfihformeKütz. Seagrass Bed in Indian River Lagoon, Florida. Estuaries, vol. 16, pp. 323–334.

Singh, A., Ramesh, R., 2011. Contribution of Riverine dissolved inorganic nitrogen flux to new production in the coastal Northern Indian Ocean: an assessment. International Journal of Oceanography 2011. Article ID 983561, 7 pages.

Singh, G., Ganguly, D., Selvam, A.P., Kakolee, B., Purvaja, R., Ramesh, R., 2015. Seagrass ecosystem and climate change: an Indian perspective. Journal of Climate Change 1, 67–74. No. 1-2(2015).

Soderlund, R., Rosswall, T., 1982. The nitrogen cycles. In: Huntziger, O. (Ed.), The Natural Environment and the Biogeochemical Cycles. Springer-Verlag, p. 70.

Stief, P., 2013. Stimulation of microbial nitrogen cycling in aquatic ecosystems by benthic macrofauna: mechanisms and environmental implications. Biogeosciences 10, 7829–7846.

Sutton, M.A., Howard, C.,M., Erisman, J.W., Billen, G., Bleeker, A., Grennfelt, P., van Grinsven, H., Grizzetti, B., 2011. The European Nitrogen Assessment. Cambridge University Press, pp. 147–176. Chapter 8.

Swaney, D.P., Hong, B., Paneer Selvam, A., Howarth, R.W., Ramesh, R., Purvaja, R., 2015. Net anthropogenic nitrogen inputs and nitrogen fluxes from Indian watersheds: an initial assessment. Journal Marine Systems Vol 141, 45–58.

Swaney, D.P., Smith, S.V., Wulff, F., 2011. The LOICZ biogeochemical modeling protocol and its application to estuarine ecosystems. Treatise on Estuarine and Coastal Science 9, 135–159.

Townsend, D.W., Ellis, W.G., 2010. Primary production and nutrient cycling on the Northwest Atlantic continental shelf. In: Liu, K.-K., Atkinson, L., Quiñones, R., Talaue-McManus, L. (Eds.), Carbon and Nutrient Fluxes in Continental Margins: A Global Synthesis, IGBP Book Series. Springer, Berlin, pp. 234–248, 744.

UNEP (United Nations Environment Programme), 1992. The World Environment 1972–1992: Two Decades of Challenge. Chapman & Hall, New York, NY (USA), p. 884.

UNEP, 2007. Reactive Nitrogen in the Environment: Too Much or Too Little of a Good Thing. United Nations Environment Programme, Paris, p. 51.

Voss, M., Emeis, K.C., Hille, S., Neumann, T., Dippner, J.W., 2005. Nitrogen cycle of the Baltic Sea from an isotopic perspective. Global Biogeochemical Cycles 19, 1–16.

N fertilizer consumption in Indian agricultural field, 2015. The Fertilizer association of India, New Delhi. http://www.faidelhi.org/statistical-database.htm.

SECTION D

# Environmental and Climate Impacts

# 24

# Reactive Nitrogen and Its Impacts on Climate Change: an Indian Synthesis

H. Pathak, A. Bhatia

*ICAR-INDIAN AGRICULTURAL RESEARCH INSTITUTE (IARI), NEW DELHI, INDIA*

**CHAPTER OUTLINE**



## Introduction

Nitrogen (N), existing in its inert state as $N_2$, is the most abundant element in the earth's atmosphere. However, in this form, it is almost wholly unusable by the vast majority of living organisms. Thus, this element, essential for life, must be transformed into a variety of other forms, together called as "reactive nitrogen (Nr)," in order to be biologically functional. The emissions of Nr from agricultural systems in the form of $NH_3$, $NO_x$, nitrous oxide ($N_2O$), nitrates ($NO_3$), and many organic N forms are altering the nitrogen cycle. The creation of Nr occurs both via natural processes and human interventions.

**The Indian Nitrogen Assessment. http://dx.doi.org/10.1016/B978-0-12-811836-8.00024-0**

Under natural conditions, breaking the triple bond that binds the two nitrogen atoms together as $N_2$ requires high energy— either lightning (1000 kV) or the activity of microbes (at least 18 ATPs) is able to carry out the process known as biological nitrogen fixation (BNF). These microbes are generally bacteria and actinomycetes often residing in a symbiotic relationship with certain legumes, including soybeans, and many native plants in the leguminous and other cryptogamous families. However, by the end of the 1800s, with a growing human population, it became apparent that these microbial sources of Nr could not adequately support the needed expansion of agriculture.

Nitrogen is crucial for food production and nutritional security of the expanding population. The earth population has already surpassed 7 billion inhabitants, and it has been possible to feed these many people only by using N fertilizer instead of relying on the food produced from natural resources. Half of the inorganic N available today globally was applied in the last 15 years (Howarth et al., 2002). In India consumption of N fertilizer in agriculture has a close relationship ($R^2 = 0.95$) with the food production of the country (Fig. 24.1). The readily available N contributes substantially to global food security, but this is not without costs.

The use of Nr is inefficient and leakages lead to human health impacts, diminished ecosystem services, biodiversity loss, and climate change. Worldwide, use efficiency of external N supply, i.e., ratio of N harvested to N applied in crop, is as low as 30–35% (Galloway and Cowling, 2002; Ladha et al., 2005). However, N-use efficiency in terms of the amount of N actually consumed to N produced by the Haber-Bosch process in the factory is much lower. For a vegetarian the efficiency is 14%, whereas for a nonvegetarian it is merely 4% (Galloway and Cowling, 2002). The unaccounted for 86% and 96% of N, depending upon the food we consume, represent the loss of N fertilizer into the environment and cause a serious concern. Soil N dynamics is an integral component of the global N cycle, since introduction of N anywhere in the system can have cascading effects to various parts of the environment, after it has been converted to Nr forms, such as ammonia ($NH_3$), N oxides ($NO_x$), nitrous oxide ($N_2O$), nitrate ($NO_3^-$), urea,

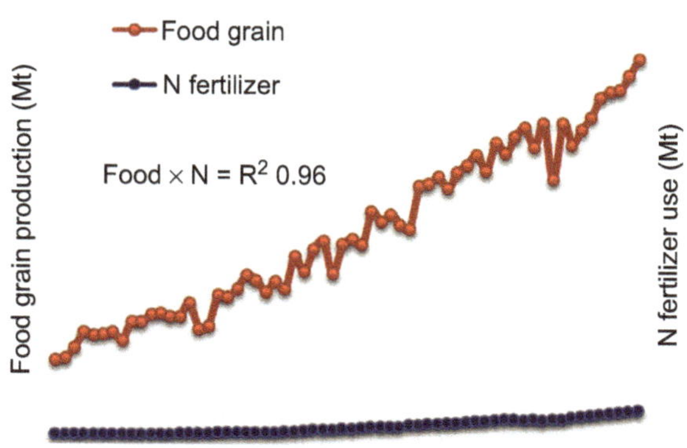


**FIGURE 24.1** Relationship of N-fertilizer use and food production in India.

amines, proteins, and nucleic acids (Galloway et al., 2008; Erisman et al., 2011; Pinder et al., 2012).

According to some estimates, humanity has already transgressed the planetary boundary for the rate of human interference with the global N cycle (Rockström et al., 2009; Steffen et al., 2015). The uncontrolled accumulation of Nr in our soil, water, and air not only causes pollution, ill health, and adversely impacts biodiversity, but also exacerbates climate change and associated challenges to our national planning and international negotiations. Gross inequalities in the development led to unequal anthropogenic contribution of Nr within and between geographical boundaries, or within and between sectors. Our biggest challenge is to apportion their contribution accurately for appropriate remedial action.

While we know N affects estuaries leading to hypoxia and fish kills and $NO_x$ as such and as ozone precursor leads to air pollution, the relation between N and climate might come as a surprise. $NO_x$ influences the radiation budget of the atmosphere indirectly through $O_3$, representing about 10–15% of the total anthropogenic greenhouse effect (Wild et al., 2001). $NO_x$ also influences the oxidation capacity of the atmosphere through OH and nitrate. Before deposition, ammonium aerosols contribute to fine particulate matter and regional haze concentrations in the atmosphere. Nitrous oxide (a reactive species of N), as a greenhouse gas, is responsible for global warming and climate change (Fig. 24.3) (IPCC, 2014). However, apart from causing global warming, Nr also has several cooling effects on the climate.

The objectives of this chapter are to assess the impacts of climate change on Indian agriculture, analyze the trend of N use in India, assess the roles of N on global warming and cooling, and identify various options for climate change mitigation and adaptation.

## Nitrogen Use in Indian Agriculture

There are a variety of sources of N in agricultural systems that are anthropogenic. These include (1) synthetic fertilizers, (2) animal manures (urine and feces), (3) N derived from enhanced biological N-fixation through $N_2$-fixing crops, (4) crop residue returned to the field after harvest, and (5) human sewage sludge application. Although some part of the N from animal manures, crop residue, and sewage may come from previous application of synthetic fertilizer, the reentry of this N into soil system again renders it susceptible to microbial processes of transformation.

Agriculture in India, until the middle of the 20th century, relied mostly on organic manures. After the introduction of urea in 1956 and di-ammonium phosphate (DAP) in 1967, fertilizer use has steadily increased along with the expansion of modern high-yielding varieties accompanied by development of irrigation facilities and availability of N fertilizer at low cost. Nitrogen is the most widely used fertilizer nutrient in this country and its consumption has grown dramatically (Fig. 24.2A). In 1950–51, consumption of fertilizer-N in the country was only 0.055 Mt (55,000 t), which increased to

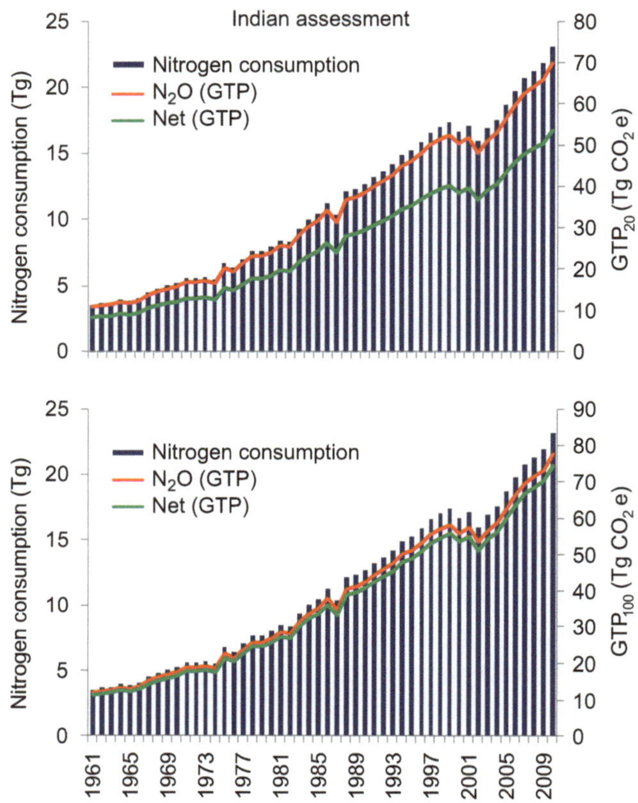


**FIGURE 24.2** Total nitrogen consumption, warming due to $N_2O$ emission and net warming, 20-year basis (upper) and 100-year basis (lower).

16.75 Mt in 2013–14, an increase of about 314-fold in the last 64 years, and the per hectare fertilizer-N consumption increased from 0.44 kg $ha^{-1}$in 1951–52 to 86.2 kg $ha^{-1}$ in 2013–14 (Fertiliser Statistics, 2015). Fertilizer price policy of the Government, provision of subsidy and most importantly the compulsion to produce more food encouraged farmers to apply more and more of fertilizer-N. At present, India ranks third in the world after China and USA in consumption of fertilizer-N.

Total nitrogen consumption in agriculture has increased from 3.58 (1961) to 23.17 Tg (2010) in India. In 1961 total 3.58 Tg N was added to Indian agriculture by different sources (Table 24.1). The highest amount of 1.77 Tg (49.44%) was supplied through crop residue, while atmospheric deposition contributed the lowest 0.17 Tg (4.75%). Animal manure and fertilizer contributed 1.39 (38.83%) and 0.25 (6.89%) Tg N, respectively. In 2010, the highest amount of 16.56 Tg (71.43%) was supplied through nitrogenous fertilizers, followed by crop residue, animal manure, and atmospheric deposition 3.9 (16.83%), 2.54 (10.96%), and 0.18 (0.78%) Tg N, respectively (Table 24.1). Percentage share of fertilizer-N application in India was increased tremendously from 6.89% in

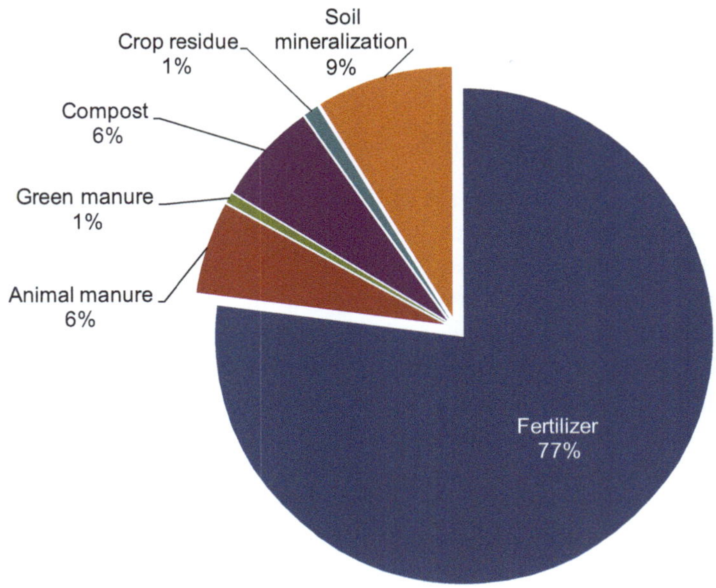


**FIGURE 24.3** Contribution of different sources to nitrous oxide emission from Indian agricultural soils in 2010. *Adapted from Pathak, H., Bhatia, A., Jain, N., 2014. Greenhouse Gas Emission from Indian Agriculture: Trends, Mitigation and Policy Needs. Indian Agricultural Research Institute, New Delhi—110012, p. 39. TB-ICN: 133/2014.*

1961–71.43% in 2010 (Table 24.1). On the other hand percentage shares of animal manure and crop residue have decreased (Fig. 24.3).

Inefficient use of fertilizers up to 300 kg $ha^{-1}$ in a few intensive cropping (irrigated) areas of India has led to the accumulation of excessive Nr in the soil, water, and air (Pathak et al., 2001). Considering that the average N-use efficiency for cereal production is as low as 33%, leakages of Nr from agricultural systems are a cause of serious concern for both economic and environmental reasons. On the other hand, chronic N-deficiency continues to be a key problem in a majority of the Indian soils. This has led to a peculiar

**Table 24.1** Contribution of Different Sources to Nitrogen Used in Global and Indian Agriculture in 1961 and 2010

| | Nitrogen in Tg (%) | | | |
|---|---|---|---|---|
| | Global | | India | |
| **Particulars** | **1961** | **2010** | **1961** | **2010** |
| Fertilizer-N | 11.59 (21.34) | 113.40 (61.59) | 0.25 (6.89) | 16.55 (71.43) |
| Animal manure | 24.20 (44.57) | 34.02 (18.48) | 1.39 (38.83) | 2.54 (10.96) |
| Crop residue | 14.05 (25.78) | 31.79 (17.27) | 1.77 (49.44) | 3.9 (16.83) |
| Atmospheric deposition | 4.46 (8.21) | 4.91 (2.67) | 0.17 (4.75) | 0.18 (0.78) |
| Total N | 54.29 (100 | 184.12 (100) | 3.58 (100) | 23.17 (100) |

situation in which genuine demands for expansion of fertilizer-N use in most (rainfed) areas coexist with concerns over the excessive and inefficient N use in other areas.

## Climate Change and Indian Agriculture

Climate change caused by an increase in the concentration of atmospheric greenhouse gases (GHGs), namely carbon dioxide ($CO_2$), methane ($CH_4$), and nitrous oxide ($N_2O$) is likely to affect the global food security and agricultural sustainability aversely. Change in the global climate scenario results in warming of atmosphere and ocean, melting and/or diminishing glaciers, rise in sea level, and increase in the concentrations of greenhouse gases. These GHGs trap the infrared radiation emitted from the earth's surface raising the atmospheric temperature. Recent observations show increase in temperatures, hot days, hot nights, and heat waves; increasing frequency of heavy precipitation events; and increased snow melt and rise in sea level. The Intergovernmental Panel on Climate Change (IPCC), in its fifth assessment report (AR5) reiterated that the warming of the climate system is unequivocal. Anthropogenic influence on the climate system is evident from the increasing greenhouse gas concentrations in the atmosphere and positive radiative forcing. Global climate change has considerable impacts on crops, soils, livestock, and pests.

Indian agriculture is highly prone to risks due to climate change; especially to drought, because 52% of the agricultural land in India is rainfed and even the irrigation system is dependent on monsoon. Flood is also a major problem in many parts of the country, especially in eastern part, where frequent flood events take place. In addition, frost in north-west, heat waves in central and northern parts (Pai et al., 2011), and cyclone in eastern coast also cause havoc. Revadekar et al. (2012) have reported an increase in the intensity and frequency of hot events and a decrease in the frequency of cold events. Heavily populated areas such as coastal regions are exposed to climatic extremes and a large fall in sown area has been observed in the arid and semiarid zones, of which nearly two-thirds are drought-prone (TERI, 2007). Droughts are expected to be more frequent in some areas, especially in north-western India, Jharkhand, Orissa, and Chhattisgarh. About 40 million hectares of land is flood-prone, including most of the river basins in the north and the north-eastern belt. Gosain et al. (2011) projected a decline in rainfall in 14 out of the 17 river basins by 2030s. O' Brien et al. (2004) observed north-western and central parts of India to be more vulnerable and southern part of India to be relatively less vulnerable. Unusual and unprecedented spells of hot weather are expected to occur far more frequently and cover much larger areas. In recent years, the frequency of these climatic extremes are getting more due to increased atmospheric temperature, resulting in increased risks with substantial loss of agricultural production. Increase in temperature reduces crop duration, increases crop respiration rates, alters photosynthesis process, affects the survival, and distributions of pest populations, and thus develops new equilibrium between crops and pests, hastens nutrient mineralization in soils, decreases fertilizer-use efficiencies, and increases evapotranspiration.

## Role of Nitrogen in Climate Change

Nitrogen has direct and indirect as well as short- and long-term effects toward global warming and cooling and thereby climate change (Fig. 24.3). The warming effects of N include (1) $N_2O$ emission, which is a greenhouse gas with long atmospheric lifetime (IPCC, 2007), and (2) $NO_x$ emission, which contributes to formation of tropospheric $O_3$, a short-lived GHG lasting several weeks and has a detrimental effect on plant C sequestration. The cooling effects include (1) C sequestration due to application of N, which increases plant $CO_2$ fixation (De Vries et al., 2011); (2) losses of N to water bodies, where freshwater and marine eutrophication can increase $CO_2$ removal from the atmosphere (Boyd et al., 2010); (3) increasing oxidation potential of the atmosphere by $O_3$, which decreases the atmospheric lifetime of $CH_4$ and increases rates of aerosol formation; and (4) $NO_x$ and $NH_3$ emissions, which contribute to formation of ammonium and nitrate aerosols. In addition, tropospheric $O_3$ and $NH_3$ both accelerate the oxidation of sulfur dioxide ($SO_2$) to sulfate aerosols. The Nr supply also affects $CH_4$ production and consumption in soils (De Vries et al., 2011) and albedo of the land surface by affecting vegetative cover and increasing chlorophyll content of vegetation (Hollinger et al., 2010). The quantitative importance of these effects is, however, uncertain. The net effect of these multiple Nr interactions suggests net global cooling of $-16$ ($-47$ to $+15$) $mW\,m^{-2}$ for European emissions and $-240$ ($-500$ to $+200$) $mW\,m^{-2}$ for the global release of Nr (Erisman et al., 2011). This global cooling effect is about 15% of the overall net radiation effect of different anthropogenic contributions of $+1600$ $mW\,m^{-2}$ (IPCC, 2007). There are strong feedback effects of climate change on Nr availability through the effects of temperature and rainfall patterns on the cycling of Nr (Erisman et al., 2012). The recent European Nitrogen Assessment (ENA) estimates that Nr emissions may be having a net cooling effect on climate, as aerosol Nr effects and forest fertilization from atmospheric Nr deposition tend to outweigh the warming effects of $N_2O$ emissions and the Nr contribution to $O_3$ formation (Butterbach-Bahl et al., 2011). However, these cooling effects are largely short-lived having more societal costs than their climate-related benefits as Nr deposition threatens biodiversity and aerosols affect human health. Under current conditions, the warming and cooling effects partially offset each other (Pinder et al., 2012).

## Emission of Gaseous-N

A major consequence of this human-driven change in global N cycle is the increased emission of N-based trace gases, such as nitrous oxide ($N_2O$), $NO_x$ ($NO + NO_2$), and ammonia, that impacted regional and global atmospheric chemistry (Mosier et al., 2004).

## Nitrous Oxide Emission

The $N_2O$ is emitted into the atmosphere from both natural (water bodies and soils) as well as from anthropogenic activities such as agriculture, transport, industries, and waste management practices. Nitrous oxide is a gaseous intermediate, produced naturally in soils during the reaction sequence of denitrification and as a by-product of nitrification process. Nitrification is the microbial oxidation of ammonium to nitrate under aerobic condition while denitrification is the microbial reduction of nitrate to nitrogen gas ($N_2$).

The soil N for these process can come from all possible sources; native soil N, N from recent atmospheric deposition, previous fertilization, crop residues, manure, $N_2$ fixation, $N_2O$ from subsurface aquifers, and current N fertilization. However, fertilizer application is the dominant source from agricultural fields (Mosier et al., 1998; Verge et al., 2007).The emissions of $N_2O$ that result from anthropogenic N inputs or N mineralization occur through both a direct pathway (i.e., directly from the soils to which the N is added/released), and through two indirect pathways following (1) volatilization of $NH_3$ and $NO_x$ from managed soils and from fossil fuel combustion and biomass burning, and the subsequent redeposition of these gases and their products $NH_4^+$ and $NO_3^-$ to soil and water; and (2) after leaching and runoff of N, mainly as $NO_3^-$ from managed soils.

Bhatia et al. (2013) estimated the direct $N_2O$ emission from Indian agricultural soils to be 0.22 million tons for the base year 2010. Total amount of $N_2O$ emissions from livestock were estimated to be 214 Gg year$^{-1}$ for the base year 2003 (Aneja et al., 2012).

The emission of $N_2O$ from Indian agriculture has increased considerably during 1970–2013 because of the application of more fertilizers and manure in soil (Fig. 24.4).

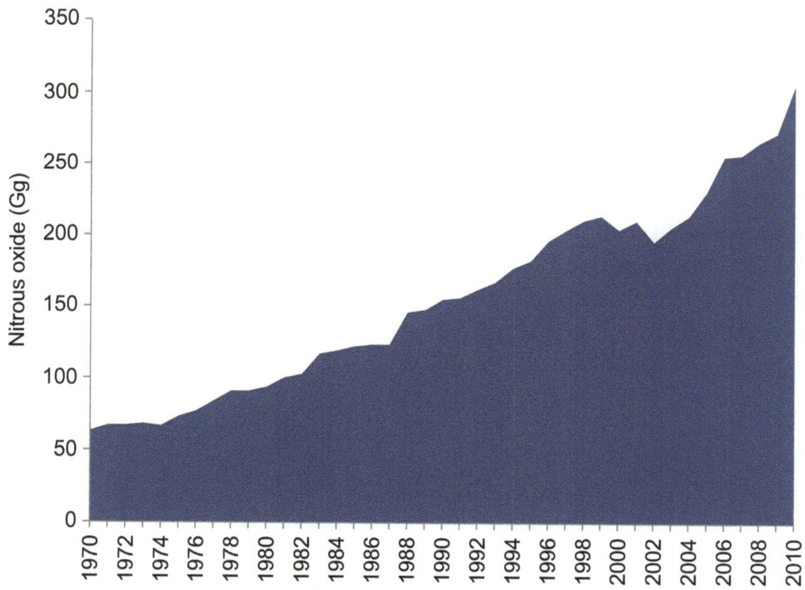


**FIGURE 24.4** Trends of nitrous oxide emissions from Indian agriculture.

Total $N_2O$–N emission from Indian agriculture was estimated to be 0.14 million tons in 2010. Application of N fertilizer is the major source of $N_2O$–N emission contributing 70%, followed by crop residue and mineralization of organic N in soil (Fig. 24.3). Agricultural $N_2O$ emission is projected to increase globally by 35–60% up to 2030 due to increased N fertilizer use and increased animal manure production. Similarly, it is estimated that $N_2O$ emissions will increase by about 50% by 2020 (relative to 1990). If demands for food increase, and diets shift as projected, then annual emissions of GHGs from agriculture may escalate further. But improved management practices and emerging technologies may permit a reduction in emissions per unit of food (or protein) produced, and perhaps also a reduction in emissions per capita food consumption (IPCC, 2007).

## Nitrogen Oxide ($NO_x$) Emission

The release of nitrogen oxides ($NO_x = NO + NO_2$) has accelerated during the last few decades primarily through increase in fossil fuel combustion. The $NO_x$ emission is predominantly contributed by sources such as vehicular exhaust, coal combustion, and biomass burning. The higher temperatures in the flame during combustion are generally responsible for the formation of $NO_x$ as it helps in breaking down the molecular nitrogen and oxygen of the air, which recombine to form $NO_x$ which includes both nitric oxide (NO) and nitrogen dioxide ($NO_2$). In addition, the nitrogen content of the fuel also gets oxidized during the combustion and releases $NO_x$. Out of all the above mentioned three sources, the vehicular exhaust is the largest contributor of $NO_x$ emission in India.

In the thermal power plants, NO formation largely depends upon the excess air used during combustion and the gas temperature in the boiler. NO is formed by the oxidation of atmospheric nitrogen during combustion of the fuel. The contribution of nitrogen that is chemically bound within the fuel is very small.

The $NO_x$ emission from the soil is primarily a result of NO production by the microbial oxidation of ammonium, the process known as nitrification. The NO production in the soil also occurs through the microbial reduction of nitrate (denitrification). This reaction only occurs in water saturated soils where little NO is released from the soil to the atmosphere.

Average concentrations of $NO_2$ at urban vs rural site in the Indo-Gangetic plains were recorded as $40.4 \pm 16.8$ and $51.57 \pm 22.8$ $gm^{-3}$, respectively (Singh and Kulshresta, 2014). The emission of $NO_x$ in India in 1995 was estimated to be 3.46 Tg and grew at about 5.6% per year (Garg et al., 2001). Sahu et al. (2015) estimated $NO_x$ emission to be around 255.4 Gg $year^{-1}$ over the national capital region of Delhi for the base year 2010. The relative contributions from power, vehicular, and industrial sources were estimated to be 6.9 Gg $year^{-1}$, 79.84 Gg $year^{-1}$, 162.3 Gg $year^{-1}$, respectively. The diesel-powered vehicles have been estimated to be the dominant contributors and account for about 84% of the total $NO_x$ emission for the road transport sector (Sharma et al., 2008). Despite increases in traffic volume, the implementation of technical measures to reduce $NO_x$

emission from vehicles contributes an approximately 40% reduction in oxidized nitrogen ($NO_y$) deposition.

## Ammonia ($NH_3$) Emission

Ammonia plays an important role in many atmospheric processes. It is the main alkaline component in the atmosphere and is highly reactive either in forming aerosols or by depositing rapidly to most of the surfaces including sensitive ecosystems (Sutton et al., 2007; Aneja et al., 2008). $NH_3$ is also an important source of atmospheric aerosols facilitating as-to-particle conversion (Baek and Aneja, 2004; Sharma et al., 2007; Behera and Sharma, 2010). Agriculture is the major sector responsible for $NH_3$ emission. In agriculture sector, application of fertilizers and livestock are mainly responsible for $NH_3$ emission. The emission estimates of $NH_3$ are highly uncertain in India as no country-specific emission factor for $NH_3$ is available as yet. The loss varied from 0.98 to 25.8 kg $ha^{-1}$ depending upon soil type, crop, and fertilizer material. Some other studies have shown that the leaching loss of N from soils in the Indo-Gangetic plain (IGP) is 10–15 kg N $ha^{-1}$ while the ammonia volatilization loss is 20–30 kg N $ha^{-1}$with an application of 120 kg N $ha^{-1}$ in rice and wheat (Aulakh and Bijay-Singh, 1997). In a study on the loss of N as $NH_3^-$from the rice–wheat system of north India, volatilization losses ranged from 38.6 kg N $ha^{-1}$ in unfertilized soil to 69.0 kg N $ha^{-1}$ with urea plus DCD treatments (Banerjee et al., 2002). Daily emission of $NH_3$-N was as high as 0.45 kg N $ha^{-1}$ $day^{-1}$ in rice and 0.5 kg N $ha^{-1}$ $day^{-1}$ in wheat. Substitution of 50% urea-N by FYM reduced $NH_3$–N emission by 10% in rice and wheat as compared to urea-N alone.

Parashar et al. (1998) estimated, using emission factors proposed by Asman (1992) that the fertilizers applications are responsible for about 1.17 Tg $NH_3$ emissions in India for 1993–94. The total amounts of $NH_3$ from livestock were estimated at 1705 Gg $year^{-1}$, for the base year 2003 (Aneja et al., 2012). The average concentrations of $NH_3$ in urban vs rural sites in the IGP has been recorded as $24.4 \pm 13.5$ and $18.8 \pm 12.6$ $gm^{-3}$, respectively (Singh and Kulshresta, 2014). Out of all the fertilizers, the urea application is responsible for the contribution of more than 90% of these emissions. The other types of fertilizers such as ammonium sulfate, calcium ammonium sulfate, DAP, ammonium phosphate, nitrophosphate, and NPK have been estimated to contribute only a small amount of $NH_3$ emission. The reason for the highest contribution from urea is not only because of its highest usages in Indian agriculture but also the largest emission factor for $NH_3$ (as percent loss of N). $NH_3$ emissions were 2697 Gg $year^{-1}$ from the fertilizer applications and the state of Uttar Pradesh emitted the highest $NH_3$ (522 Gg $year^{-1}$) followed by the state of Maharashtra (425 Gg $year^{-1}$) both from animal and crop farming (Aneja et al., 2012).

However, none of these emission estimates have taken into account the organic manure applications in the agriculture fields in India so far. Pathak et al. (2006) simulated N loss from rice–wheat cropping systems of the IGP through ammonia volatilization. Most of the $NH_3$ emissions from agricultural production are strongly influenced by climatic interactions. Volatilization of ammonia is a physical process that is highly temperature-dependent. In principle, according to solubility and dissociation thermodynamics, $NH_3$ volatilization potential nearly doubles every 5°C, equivalent to a $Q_{10}$ (the relative increase over a range of 10°C) of 3–4. At the same time, $NH_3$ emission is controlled by water availability, which allows $NH_x$ to dissolve, be taken up by organisms, and be released through decomposition. Considering these interactions, $NH_3$ emission and deposition are expected to be extremely climate-sensitive. Due to this temperature effect of local meteorology, hot years are likely to give higher ammonia emissions compared to cold years. Similarly, hot days are likely to cause higher emissions than cold days. Based on available measurements, a first empirical estimate suggests that a 5°C warming would increase emissions by 42% (28–67%). Together with increased anthropogenic activity, global $NH_3$ emissions may increase from 65 (45–85) Tg N in 2008 to reach 132 (89–179) Tg by 2100. The minimal adoption of technical measures to reduce $NH_3$ emission from agriculture is being offset by increased meat and dairy consumption, requiring more livestock and fertilizers, and increasing $NH_x$ deposition in some areas by greater than 40%.

The wet deposition of N as $NO_3^-$ and $NH_4^+$ onto agricultural soils of India was estimated to be ~0.81 Tg N by Velmurugan et al. (2008), whereas $NO_3$–N leaching and runoff was estimated to be ~0.06 Tg N annually. BNF, another source of Nr, added 1.14–1.18 Tg N to the system. The reactive forms of N leaked to the environment accumulate in groundwater and lakes, or volatilize into air, or are transported through rivers to oceans, with serious environmental consequences to earth's ecology, health, and climate change. Anoxic and hypoxic conditions in Indian coastal waters are being reported due to enhanced N loading from land and associated eutrophication. The contribution of gaseous emissions of Nr from fossil-fuel-consuming industries (especially the power sector) and automobile exhausts is also becoming increasingly significant factor in the rapidly growing economy.

The ever-growing human population and its increased generation of sewage and industrial wastes have also increased reactive nitrogen inputs to the environment. In India, the majority of wastewater is released untreated directly into waterways. Where sewage treatment facilities do exist, they often provide only primary treatment, which does little to remove nitrogen. Because populations are expanding rapidly, many communities are ill-equipped to implement effective sewage treatment processes. In some localities, wastewater is the largest source of release of reactive nitrogen into the environment. Reactive nitrogen is also introduced to the environment through multiple steps of food production prior to human consumption and subsequently as wastage.

## Global Temperature Potential of $N_2O$ Emission

The $N_2O$ itself is a greenhouse gas, with a global temperature potential of 290 on a 100-year basis ($GTP_{100}$) (Shine et al., 2005). Nitrogen (N) fertilizers contribute significantly to anthropogenic nitrous oxide ($N_2O$) emissions (Davidson, 2009). Atmospheric $N_2O$ concentration increased, particularly after 1960, because of the increasing use of N fertilizers (Davidson, 2009). Mostly, $N_2O$ is directly emitted from the agricultural soil during microbial processes, namely nitrification and denitrification (Forster et al., 2007; Davidson, 2009). It is found that $N_2O$ emission from global agriculture shows an increasing trend, and it increased from 0.60 Tg in 1961 to 2.03 Tg in 2010. Converting to $CO_2e$ on a 20-year basis $GTP_{20}$ of $N_2O$ from global agriculture was 164.23 and 556.95 Tg $CO_2e$ for the years 1961 and 2010, respectively. Using IPCC default coefficient as 1% (IPCC, 2006) $N_2O$ emission from global agriculture was estimated to be 0.85 and 2.89 Tg in 1961 and 2010, respectively. Converting to $CO_2e$ the $GTP_{20}$was estimated to be 234.64 and 795.65 Tg $CO_2e$ for the years 1961 and 2010, respectively.

Similarly, $N_2O$ is showing an increased trend of emission from Indian agriculture. The $N_2O$ emissions from Indian agriculture been estimated using the emission coefficient of 0.7% (Bhatia et al., 2004). They increased from 0.04 Tg $N_2O$ in 1961 to 0.25 Tg $N_2O$ in 2010–11). Converting to $CO_2e$ on a 20-year basis, $GTP_{20}$ of $N_2O$ from Indian agriculture are 10.86 and 70.07 Tg $CO_2e$ in 1961 and 2010 respectively. As per the IPCC default coefficient of 1%, the $N_2O$ emission from Indian agriculture was estimated to be 0.06 and 0.36 Tg in 1961 and 2010, respectively. Converting to $CO_2e$ it had a $GTP_{20}$ of 15.51 and 100.10 Tg $CO_2e$ in the years 1961 and 2010, respectively and $GTP_{100}$ of 17.20 and 111.02 in the years 1961 and 2010, respectively. Indian agriculture contributes 6.61% and 12.58% in 1961 and 2010, respectively over 20- and 100-year basis to world total GTP as a result of $N_2O$ emission.

## Net Climate Change Impact of Reactive Nitrogen Use in Indian Agriculture

As we know that apart from $N_2O$ emission from agriculture soil, nitrogen application also affects the fluxes of GHGs ($CH_4$ and $CO_2$) and emission of $NO_x$ and $NH_3$. Therefore, the climate change effects of Nr used in global and Indian agriculture from 1961 to 2010, in Tg $CO_2e$, for each of the six effects are listed in Table 24.1 and summarized in Fig. 24.2B on a 20-year and a 100-year basis.

Overall effects of reactive N used in Indian agriculture were 8.28 and 53.43 Tg $CO_2e$ on a $GTP_{20}$ in 1961 and 2010, respectively (Fig. 24.2B). On $GTP_{100}$ basis the results were 11.52 and 74.37 Tg $CO_2e$ in 1961 and 2010, respectively (Fig. 24.2D). Overall warming on $GTP_{20}$ basis was lowered by 2.58 Tg $CO_2e$ in 1961 and 16.64 Tg $CO_2e$ 2010 than the warming caused by $N_2O$ alone (Fig. 24.2B). On $GTP_{100}$-year basis it was lowered by 0.52 and 3.34 Tg $CO_2e$ in 1961 and 2010, respectively (Fig. 24.2D). On $GTP_{100}$ basis the difference between the warming caused by $N_2O$ alone and overall warming was less compared to $GTP_{20}$-year basis because of the cooling effect of short-lived aerosols. Over a 100-year scale, aerosols effects are negligible. Indian agriculture contributed 6.61% and 12.58% in 1961 and 2010, respectively, over 20- and 100-year basis to world total net GTP as a result of emission of all parameters and their subsequent atmospheric chemistry.

## Nitrogen Management for Mitigation

Intensive agricultural systems, envisaging high productivity, high input use, high rainfall or irrigated systems are widely recognized as major sources of reactive N. It is argued that there is substantial scope of mitigating Nr emission from such systems. Appropriate crop management practices, which lead to increased N-use efficiency and higher yield by optimizing the crop's natural ability to compete with the N-loss processes, hold the key to mitigating Nr emission.

Currently farmers' decision on input use is based on "field average" concept combined with visual observations. Rather than such blanket applications, there is an urgent need to fine-tune N-use based on season- and field-specific monitoring of crop and soil nutrient status. Linking soil N status and crop N demand with N supply is the key to enhance fertilizer-N-use efficiency beyond the level achieved through using blanket recommendations for large tracts or regions. Some tools and techniques to achieve this are already available, while more are being developed by researchers. For example, many approaches have been developed for increasing N-use efficiency such as:

- Optimal time, rate, and methods of application for matching N supply with crop demand
- Split fertilizers application, customized fertilizers at different sites
- Slow-release fertilizers including those with urease and nitrification inhibitors
- Integrated use of fertilizers, manures, and crop residues
- Demand-driven N supply using leaf color chart (LCC) and GreenSeeker
- Water-use-efficient technologies such as fertigation, sprinkler, drip
- Other specially formulated fertilizers or methods of use
- Adoption of N-use efficient crop cultivars
- Provide continuous plant cover for increasing soil fertility
- Site-specific nitrogen management for reducing losses
- Balanced fertilization for increasing N-use efficiency
- Breeding input-efficient crop varieties and cultivars
- Integrated crop management
- Optimizing irrigation, drainage, and tillage

Studies at the Indian Agricultural Research Institute have shown that urea, which is widely used as a nitrogenous fertilizer in the country, contributes maximum to the nitrous oxide emission. It is followed by ammonium sulfate, ammonium chloride, and potassium nitrate from alluvial soil at submerged and field capacity moisture regimes (Majumdar et al., 2000). Applications of nitrate ($NO_3$–N) fertilizers, such as calcium ammonium nitrate in crops with aerobic conditions, and of ammonium ($NH_4$–N) fertilizers, such as ammonium sulfate, urea, in wetland crops, also help in reducing the emission. Plant uptake of fertilizer-N can be improved by various methods of application such as deep placement and placing fertilizer in bands, thereby minimizing the loss of N and mitigating nitrous oxide emission (Pathak et al., 2006). The most efficient management practices to reduce nitrous oxide emission are site-specific nutrient management and use of nitrification inhibitors such as nitrapyrin and dicyandiamide. There are some plant-derived organics such as neem oil, neem cake, and karanja-seed extract which can also act as nitrification inhibitors (Table 24.2). Demand-driven N use with the help of an LCC reduced nitrous oxide emission and GWP by about 10% (Bhatia et al., 2010). Use of urea super granules (USG) and nitrification inhibitors based on neem products improves N-use efficiency but these still have not gained popularity with farmers. Advantage of deep-placed USG in transplanted rice over split application of prilled urea has not gained popularity due to nonavailability of suitable machinery for placing granules into the soil. N-use efficiency could be increased by retarding the hydrolysis of urea to ammonia by the use of urease inhibitors or by regulating nitrate accumulation by nitrification inhibitors. This results in the reduction of direct nitrous oxide emission from nitrifiers, and reduces the availability of $NO_3^-$, necessary for denitrification. Nitrapyrin, acetylene, dicyandiamide (DCD), 2-amino-4-chloro–6-methyl-pyrimidine (AM), and 2-sulphanilamide-thiazole are some of the promising nitrification inhibitors in agriculture and have also been found effective in reducing nitrous oxide emission from soils (Pathak and Nedwell, 2001). Pathak and Nedwell (2001) found that in the presence of nitrapyrin, AM, and DCD emission of nitrous oxide was reduced by 12%, 24% and 63%, respectively, whereas sodium thiosulphate, sulfur, thiourea, and acetylene had no effect on emission of nitrous oxide. Under submerged conditions, no inhibitor remains effective in reducing nitrous oxide emission. There are some plant-derived organics, such as neem oil and neem cake, which can act as nitrification inhibitors (Malla et al., 2005). These are being experimented in the fields to find their efficacy in reducing the emission of nitrous oxide and increasing the fertilizer-use efficiency. Some biocidal inhibitors, such as karanja-seed extract, have been found to retard

**Table 24.2** Efficiency of Nitrification Inhibitors in Mitigating Nitrous Oxide Emission in Rice–Wheat System

| Nitrification inhibitor | Mitigation (%) |
|---|---|
| Dicyndiamide (DCD) | 13–42 |
| Neem cake | 10–21 |
| Neem oil | 15–21 |
| Nimin | 25–35 |
| Coated Ca-carbide | 12–29 |
| Thiosulphate | 15–20 |

Pathak, H., Bhatia, A., Jain, N., Aggarwal, P.K., 2010. Greenhouse gas emission and mitigation in Indian agriculture—a review, In ING Bulletins on Regional Assessment of Reactive Nitrogen, Bulletin No. 19, Eds. B., Singh, SCON-ING, New Delhi, pp. i-iv, 1–34.

nitrification by 60–70% (Majumdar et al., 2001). Similarly, wax-coated calcium carbide and ATC (4-amino 1,2,4-triazole) reduce the rate of nitrification and hence, the emission of nitrous oxide. Most of these nitrification inhibitors, however, have limitations. For example, the efficacy of the most commonly used nitrification inhibitor nitrapyrin gets reduced due to sorption on soil colloids, hydrolysis and loss through volatilization.

There is a need to combine the use of inorganic fertilizers and organic sources of nutrients, namely manures, green manures, crop residues, and biofertilizers in a synergistic manner. Incorporation of manure within 2 days of application is necessary to significantly reduce N volatilization. In cases where incorporation is not possible or practical, manure should be applied immediately before a nonrunoff-producing rain event or incorporated by conservation tillage methods, or proper irrigation. Incorporation of crop residues with high C:N ratio is found to stimulate immobilization of N, and decrease in the availability of nitrogen substrates for nitrification and denitrification (Singh et al., 2010), thus can also reduce $N_2O$ emission. Mineralization of plant residues and $N_2O$ emission has been reported to negatively correlated with C:N ratio of residue (Huang et al., 2004). Bhattacharyya et al. (2012) reported that application of rice straw + urea and rice straw + green manure significantly reduced $N_2O$ emission by 16% and 28% over urea treatment. Pathak et al. (2002) observed that application of manure also plays an important role in reducing $N_2O$ emission. Substituting 50% of inorganic N with organic sources such as FYM also has potential to reduce emission by 16%. Singh and Ahlawat (2012) concluded that substitution of 25% recommended dose of N (RDN) through FYM recorded the greatest agronomic-use efficiency (ANUE) and apparent nitrogen recovery, improved N uptake of crop because of the increased humus content of soil, slowed down the release of ammoniacal N and its conversion to nitrates, thereby reducing the leaching loss of N and $N_2O$ emission.

## Nitrogen Management for Climate Change Adaptation

The most common strategies suggested for climate change adaptation includes alternate land-use management, intensification of agriculture, crop diversification, and conservation agriculture. All of these require a suitable N management strategy, which might be different from the conventional way of N management. Management of N can also help in improving crop quality, which is likely to be deteriorated under climate change scenarios. As suggested, because of higher level of $CO_2$ concentration in the atmosphere and lower N availability in soil under climate change scenarios, the crops would have higher carbohydrate content, lower protein content, and higher C:N. Additional and efficient N application under elevated $CO_2$ can compensate the deteriorating quality of crop (SD Singh et al., unpublished). The adverse impact of climate change and variability on crop yield could also be compensated with more and efficient use of N. For example, yield reduction because of late sowing of rice as a result of delayed onset of monsoon or terminal heat impact on wheat can be compensated with demand-driven N application (H Pathak et al., unpublished).

# Conclusion

$NH_4$-N Nitrogen is one of the most critical input that defines crop productivity and yields under field conditions, and must be supplemented to meet the food production demands of an ever-increasing population. Efficient use of N is absolutely essential for attaining the goal of sustainability in agriculture. Better management of N in agriculture by enhanced implementation of all the available tools and techniques could immensely help the farmers, consumers, and government. It could well be the driving force for climate change mitigation. There is an urgent need to assess the possibilities of N management for climate abatement and at the same time increase food security, while minimizing environmental and human health impacts. The combination of weak international commitments to mitigate N emissions and increasing per capita consumption represents one of the greatest challenges for future management of the nitrogen cycle. The quantification of all the significant sources of reactive N and their relative contributions to the various sinks are most important for informed decisions on domestic policy on sustainable N management for food security, energy, industry, health, and environment, as well as strengthen our negotiating positions on international platforms. Although there have been some recent efforts to summarize the available knowledge in this area, a lot more needs to be done to integrate the formal, peer-reviewed literature with the institutional publications, local reports, and other documents. At present, there is no comprehensive source for reliable nationwide data on various aspects of reactive N, although localized surveys and estimates are available from many regions that enable rudimentary reconstruction of the Indian N cycle. We need a thorough understanding of the use, misuse, and release of Nr with reference to various point and nonpoint sources and the biogeochemical N cycle.

# References

Aneja, V.P., Schlesinger, W.H., Erisman, J.W., Behera, S.N., Sharma, M., Battye, W., 2012. Reactive nitrogen emissions from crop and livestock farming in India. Atmospheric Environment 47, 92–103.

Aneja, V.P., Arya, S.P., Rumsey, I.C., Kim, D.-S., Bajwa, K.S., Williams, C.M., 2008. Characterizing ammonia emissions from swine farms in eastern North Carolina: reduction of emissions fromwater-holding structures at two candidate superior technologies for waste treatment. Atmospheric Environment 42, 3291–3300.

Asman, W.A.H., 1992. Ammonia Emission in Europe: Updated Emission and Emission Variations. Report 228471008. National Institute of Public Health and Environmental Protection, Bilthoven, The Netherlands.

Aulakh, M.S., Bijay-Singh, 1997. Nitrogen losses and fertilizer N use efficiency in irrigated porous soils. Nutrient Cycling in Agroecosystems 7, 1–16.

Baek, B.H., Aneja, V.P., 2004. Measurement and analysis of the relationship betweenammonia, acid gases, and fine particles in Eastern North Carolina. Journal of Air and Waste Management Association 54, 623–633.

Banerjee, B., Pathak, H., Aggarwal, P.K., 2002. Effects of dicyandiamide, farmyard manure and irrigation on ammonia volatilization from an alluvial soil in rice (*Oryza sativa* L.)-wheat (*Triticum aestivum* L.) cropping system. Biology and Fertility of Soils 36, 207–214.

Behera, S.N., Sharma, M., 2010. Investigating the potential role of ammonia in ion chemistry of fine particulate matter formation for an urban environment. Science of the Total Environment 408, 3569–3575.

Bhatia, A., Pathak, H., Aggarwal, P.K., 2004. Inventory of methane and nitrous oxide emissions from agricultural soils of India and their global warming potential. Current Science 87, 317–324.

Bhatia, A., Sasmal, S., Jain, N., Pathak, H., Kumar, R., Singh, A., 2010. Mitigating nitrous oxide emission from soil under conventional and no-tillage in wheat using nitrification inhibitors. Agriculture, Ecosystems & Environment. http://dx.doi.org/10.1016/j.agee. 2010.01.004.

Bhatia, A., Jain, N., Pathak, H., 2013. Methane and nitrous oxide emissions from Indian rice paddies, agricultural soils and crop residue burning. Greenhouse Gas Science Technology 1–16.

Bhattacharyya, P., Roy, K.S., Neogi, S., Adhya, T.K., Rao, K.S., Manna, M.C., 2012. Effects of rice straw and nitrogen fertilization on greenhouse gas emissions and carbon storage in tropical flooded soil planted with rice. Soil and Tillage Research 124, 119–130.

Boyd, P.W., Strzepek, R., Fu, F., Hutchins, D., 2010. Environmental control of open-ocean phytoplankton groups: now and in the future. Limnology and Oceanography 55, 1353–1376.

Butterbach-Bahl, K., Nemitz, E., Zaehle, S., Billen, G., Boeckx, P., Erisman, J., Garnier, J., Goddard, R., Kreuzer, M., Oenema, O., Reis, S., Schaap, M., Simpson, D., Vries, de, Winiwater, W., Sutton, M., 2011. Nitrogen as a threat to the European greenhouse gas balance In The European nitrogen assessment: sources, effects and policy perspectives, Cambridge University Press 443–462.

Davidson, E.A., 2009. The contribution of manure and fertilizer nitrogen to atmospheric nitrous oxide since 1860. Nature Geoscience 2, 659–662.

De Vries, W., Kros, J., Reinds, G.J., Butterbach-Bahl, K., 2011. Quantifying impacts of nitrogen use in European agriculture on global warming potential. Current Opinion in Environmental Sustainability 3, 291–302.

Erisman, J.W., Galloway, J.N., Seitzinger, S., Bleeker, A., Butterbach-Bahl, K., 2011. Reactive nitrogen in the environment and its effect on climate change. Curr.ent Opinion in Environmental Sustainability. ISSN: 18773435 5 (3), 281–290. http://dx.doi.org/10.1016/j.cosust.2011.08.012.

Erisman, J.W., Butterbach-Bahl, K., Bleeker, A., Davidson, E., Dentener, F., Elsiddig, E., Galloway, J.N., Goodale, C.L., van Grinsven, H., Holland, E.A., Howard, C.M., Morton, J., Nemitz, E., Oenema, O., Pathak, H., Prins, W., Reis, S., Roy, J., Seitzinger, S.P., Vries, W.de, Winiwarter, W., Zaehle, S., Sutton, M.A., 2012. Nitrogen management to mitigate climate change and deliver future global food security. Nature Climate Change (Communicated).

Fertiliser Statistics, 2015. Fertilizer Association of India, New Delhi.

Forster, P., et al., 2007. Changes in Atmospheric Constituents and in Radiative Forcing in Climate Change 2007: The Physical Science Basis. Cambridge Univ Press, Cambridge, UK.

Galloway, J., Raghuram, N., Abrol, Y.P., 2008. A perspective on reactive nitrogen ina global, Asian and Indian context. Current Science 94, 1375e1381.

Galloway, J.N., Cowling, E.B., 2002. Reactive nitrogen and the world: 200 years of change. Ambio: A Journal of the Human Environment 31 (2), 64–71.

Garg, A., Bhattacharya, S., Shukla, P.R., Dadhwal, V.K., 2001. Regional and sectoral assessment of greenhouse gas emission in India. Atmospheric Environment 35, 2679–2695.

Gosain, A.K., Rao, S., Arora, A., 2011. Climate change impact assessment of water resources of India. Current Science 101 (3).

Hollinger, D.Y., Ollinger, S.V., Richardson, A.D., Meyer, T.P., Dails, D.P., Martin, M.E., Scott, N.A., Arkebauer, T.K., Baldocchi, D.D., Clark, K.L., Curtis, P.P., Davis, K.J., Desai, A.R., Dragoni, D., Goulden, M.L., Gu, L., Katul, G.G., Pallardy, S.G., Pawu, K.T., Schmid, H.P., Stoy, P.C., Suyker, A.E., Verma, S.B., 2010. Albedo estimates for land surface models and support fora new paradigm based on foliage nitrogen concentration. Global Change Biology 16, 696–710. http://dx.doi.org/10.1111/j.1365-2486.2009.02028.x.

Howarth, R.W., Sharply, A., walker, D., 2002. Sources of nutrient pollution to coastal waters in the United States: Implications for achieving coastal water quality goals. Estuaries 25 (4b), 656–676.

Huang, Y., Zou, J., Zheng, X., Wang, Y., Xu, X., 2004. Nitrous oxide emissions as influenced by amendment of plant residues with different C: N ratios. Soil Biology and Biochemistry 36, 973–981.

IPCC, 2007. Climate change: impacts, adaptation and vulnerability. In: Parry, M.L., Canziani, O.F., Palutikof, J.P., van der Linden, P.J., Hanson, C.E. (Eds.), Contribution of Working Group II to the Fourth Assessment Report of the Intergovernmental Panel on Climate Change. Cambridge University Press, Cambridge, UK, p. 976.

IPCC (Intergovernmental Panel on Climate Change), 2006. Guidelines for National Greenhouse Gas Inventories. http://www.ipcc-nggip.iges.or.jp/public/2006gl/index.html.

IPCC, 2014. Climate change 2014: impacts, adaptation, and vulnerability. In: Field, C.B., Barros, V.R., Dokken, D.J., Mach, K.J., Mastrandrea, M.D., Bilir, T.E., Chatterjee, M., Ebi, K.L., Estrada, Y.O., Genova, R.C., Girma, B., Kissel, E.S., Levy, A.N., MacCracken, S., Mastrandrea, P.R., White, L.L. (Eds.), Part A: Global and Sectoral Aspects, Contribution of Working Group II to the Fifth Assessment Report of the Intergovernmental Panel on Climate Change. Cambridge University Press, Cambridge, United Kingdom and New York, NY, USA,.

Ladha, J.K., Pathak, H., Krupnik, T.J., Six, J., van Kessel, C., 2005. Efficiency of fertilizer nitrogen in cereal production: retrospects and prospects. Advances in Agronomy 87, 85–176.

Majumdar, D., Dutta, A., Kumar, S., Pathak, H., Jain, M.C., 2001. Mitigation of $N_2O$ emission from an alluvial soil by application of karanjin. Biology and Fertility of Soils 33, 438–442.

Majumdar, D., Kumar, S., Pathak, H., Jain, M.C., Kumar, U., 2000. Reducing nitrous oxide emission from rice field with nitrification inhibitors. Agriculture, Ecosystems & Environment 81, 163–169.

Malla, G., Bhatia, A., Pathak, H., Prasad, S., Jain, N., Singh, J., Kumar, V., 2005. Mitigating nitrous oxide and methane emissions from soil under rice-wheat system with nitrification inhibitors. Chemosphere 58, 141–147.

Mosier, A., Wassmann, R., Verchot, L., Palm, C., 2004. Greenhouse gas emissions from tropical agriculture: sources, sinks and mechanisms. Environment, Development and Sustainability 6, 11–49.

Mosier, A.R., Duxbury, J.M., Freney, J.R., Heinemeyer, O., Minami, K., 1998. Assessing and mitigating N2O emissions from agricultural soils. Climatic Change 40, 7–38.

O'Brien, et al., 2004. Mapping Vulnerability to multiple stressors: climate change and globalisation in India. Global Environmental Change 14.

Pai, D.S., Nair, S.A., Ramanathan, A.N., 2011. Long term climatology and trends of heat waves over India during the recent 50 years (1961–2010). Mausam 64, 585–604.

Parashar, D.C., Kulshrestha, U.C., Sharma, C., 1998. Anthropogenic emission of $NO_x$, $NH_3$ and $N_2O$ in India. Nutrient Cycling in Agroecosystems 52, 255–259.

Pathak, H., Bhatia, A., Jain, N., 2014. Greenhouse Gas Emission from Indian Agriculture: Trends, Mitigation and Policy Needs, 110012. Indian Agricultural Research Institute, New Delhi, p. 39. TB-ICN: 133/2014.

Pathak, H., Nedwell, D.B., 2001. Strategies to reduce nitrous oxide emission from soil with fertiliser selection and nitrification inhibitor. Water Air & Soil Pollution 129 (4), 217–228.

Pathak, H., Bhatia, A., Jain, N., Aggarwal, P.K., 2010. Greenhouse gas emission and mitigation in Indian agriculture—a review. In: Singh, B. (Ed.), ING Bulletins on Regional Assessment of Reactive Nitrogen, Bulletin No. 19. SCON-ING, New Delhi, pp. i–iv, 1–34.

Pathak, H., Bhatia, A., Prasad, S., Jain, M.C., Kumar, S., Singh, S., Kumar, U., 2002. Emission of nitrous oxide from soil in rice-wheat systems of Indo-Gangetic plains of India. Environmental Monitoring and Assessment 77 (2), 163–178.

Pathak, H., Gupta, P.K., Bhatia, A., Sharma, C., Kalra, N., Mitra, A.P., 2004. Nitrous oxide emission from soil-plant system. In: Mitra, A.P., Sharma, S., Bhattacharya, S., Garg, A., Devotta, S., Sen, K. (Eds.), Climate Change and India, Uncertainty Reduction in Greenhouse Gas Inventory Estimates. Universities Press (India) Pvt. Ltd, Hyderabad, pp. 244–257.

Pathak, H., Singh, R., Bhatia, A., Jain, N., 2006. Recycling of rice straw to improve wheat yield and soil fertility and reduce atmospheric pollution. Paddy and Water Environment 4, 111–117.

Pindera, R.W., Davidsonb, E.A., Goodalec, C.L., Greavera, T.L., Herricka, J.D., Liud, L., 2012. Climate change impacts of US reactive nitrogen. Proceedings of the National Academy of Sciences of the United States of America 109 (20), 7671–7675.

Revadekar, J.V., Kothawale, D.R., Patwardhan, S.K., Pant, G.B., Kumar, K.R., 2012. About the observed and future changes in temperature extremes over India. Natural Hazards 60 (3), 1133–1155.

Rockström, J., Steffen, W.L., Noone, K., Persson, Å., Chapin III, F.S., Lambin, E.F., Lenton, T.M., Scheffer, M., et al., 2009. "Planetary boundaries: exploring the safe operating space for humanity" (PDF). Ecology and Society 14 (2), 32.

Sahu, S.K., Beig, G., Parkhi, P., 2015. High Resolution emission inventory of NO*x* and CO for mega city Delhi. India Aerosol and Air Quality Research 15, 1137–1144.

Sharma, C., Tiwari, M.K., Pathak, H., 2008. Estimates of emission and deposition of reactive nitrogenous species for India. Current Science 11 (4), 1439–1446.

Sharma, M., Kishore, S., Tripathi, S.N., Behera, S.N., 2007. Role of atmospheric ammonia in the formation of inorganic secondary particulate matter: a study at Kanpur, India. Journal of Atmospheric Chemistry 58, 1–17.

Shine, K.P., Berntsen, T.K., Fuglestvedt, J.S., Sausen, R., 2005. Scientific issues in the design of metrics for inclusion of oxides of nitrogen in global climate agreements. Proceedings of the National Academy of Sciences of the United States of America 102, 15768–15773.

Singh, B.P., Hatton, B.J., Singh, B., et al., 2010. Influence of biochars on nitrous oxide emission and nitrogen leaching from two contrasting soils. Journal of Environmental Quality 39, 1224–1235.

Singh, R.J., Ahlawat, I.P.S., 2012. Dry matter, nitrogen, phosphorous, and potassium partitioning, accumulation and use efficiency in transgenic cotton based cropping systems. Communications in Soil Science and Plant Analysis 43 (20), 2633–2650.

Singh, S., Kulshrestha, U.C., 2014. Rural versus urban gaseous inorganic reactive nitrogen in the Indo-Gangetic plains (IGP) of India. Environmental Research Letters 9, 1–9.

Steffen, W., Richardson, K., Rockstrom, J., Cornell, S.E., Fetzer, I., Bennett, E.M., Biggs, R., Carpenter, S.R., De Vries, W., De Wit, C.A., Folke, C., Gerten, D., Heinke, J., Mace, G.M., Persson, L.M., Ramanathan, V., Reyers, B., Sorlin, S., 2015. Planetary boundaries: guiding human development on a changing planet. Science 347 (6223), 1259855. http://dx.doi.org/10.1126/science.1259855.

Sutton, M.A., Nemitz, E., Erisman, J.W., Beier, C., Bahl, K.B., Cellier, P., de Vries, W., Cotrufo, F., Skiba, U., Di Marco, C., Jones, S., Laville, P., Soussana, J.F., Loubet, B., Twigg, M., Famulari, D., Whitehead, J., Gallagher, M.W., Neftel, A., Flechard, C.R., Herrmann, B., Calanca, P.L., Schjoerring, J.K., Daemmgen, U., Horvath, L., Tang, Y.S., Emmett, B.A., Tietema, A., Penuelas, J., Kesik, M., Brueggemann, N., Pilegaard, K., Vesala, T., Campbell, C.L., Olesen, J.E., Dragosits, U., Theobald, M., 2007. Challenges in quantifying biosphere-atmosphere exchange of nitrogen species. Environ Pollut. 150:125-39.

TERI (The Energy and Resources Institute), 2007. South Asian Regional Study on Climate Change Impacts and Adaptation: Implications for Human Development. http://www.cseindia.org/challenge_balance2012/pdf/Adap_South%20Asia%20HDR%202007%20READ.pdf.

Velmurugan, A., Dadhwal, V.K., Abrol, Y.P., Galloway, J.N., 2008. Regional nitrogen cycle: an Indian perspective. Current Science 94 (11), 1455–1468.

Verge, X.P.C., DE Kimpe, C., Desjardins, R.L., 2007. Agricultural production, greenhouse gas emissions and mitigation potential. Agricultural and Forest Meteorology 142, 255–269.

Wild, M., Ohmura, A., Gilgen, H., Morcrette, J., Slingo, A., 2001. Evaluation of downward longwave radiation in general circulation models, Journal of Climate 14, 3227–3239.

# 25

# Reactive Nitrogen and Air Quality in India

G. Beig[1], S. Maji[1], A.S. Panicker[1], S.K. Sahu[2]

[1]INDIAN INSTITUTE OF TROPICAL METEOROLOGY, PUNE, INDIA; [2]UTKAL UNIVERSITY, BHUBANESWAR, INDIA

CHAPTER OUTLINE



## Introduction

Nitrogen oxides ($NO_x$) are considered as precursors of ozone which is one of the GHGs in the troposphere. Out of the five oxides of nitrogen, only three, namely nitrogen dioxide ($NO_2$), nitric oxide (NO), and nitrous oxide ($N_2O$), are present in the atmosphere in significant amounts (other two oxides of nitrogen are nitrogen trioxide ($NO_3$) and nitrogen pentaoxide ($N_2O_5$). However, only NO and $NO_2$ are considered as the air pollutants. Nitrogen oxides ($NO_x = NO + NO_2$) play a key role in tropospheric chemistry. $NO_x$ also controls the level of acid deposition and acts as a precursor to ozone formation in troposphere. Thus, indirect GHGs, e.g., $NO_x$, can play a significant role in contributing to climate change as it is one of the major sources of ground-level ozone which

The Indian Nitrogen Assessment. http://dx.doi.org/10.1016/B978-0-12-811836-8.00025-2

contributes to global warming. Most of the fuels emit $NO_x$ when burned (van Aardenne et al., 1999). Another important nitrogen compound is ammonia ($NH_3$).

The increasing demand of energy will result in ejection of large amounts of pollutants such as $NO_x$ in developing countries like India. About 81% of global energy is derived from fossil fuels with coal being the prime source (IEA, 2016). In brief, $NO_x$ can act as source of surface ozone as well as sink if its concentration crosses a particular limit which forces titration. This tropospheric ozone so produced can affect human health and crop production (Beig et al., 2008). Asia is reported to be characterized by its rapid increase of per capita emissions and population in recent decades. India occupies only 2.5% of the total global land area even then it supports 16% of the world population. Large coal reserve in India is driving the shift toward the construction of electricity generation that burn coal. Latest statistics of coal consumption (GESY, 2016) shows that India (990 Mt) is the second largest coal consumer in the world after China (3732 Mt), followed by USA (730 Mt). India's energy sector mainly depends on coal, and $NO_x$ emission may continue to indicate a growing trend which is found to be the largest from the fossil fuel sector. In this scenario, in this chapter we discuss the air pollution caused by nitrogen compounds in India and the necessary measurement steps carried out by the scientific community in the country.

## Emission Inventory of Reactive Nitrogen

### Sources of Oxides of Nitrogen ($NO_x$) in India

Emission inventory is an important tool for identifying the source of pollutants and quantitative expression of pollution load in a defined area at a particular time. In order to achieve a better understanding of atmospheric chemistry, air quality, and its impact on the regional climate change, more accurate data of spatial distribution and temporal change of emission of ozone precursors, such as $NO_x$, over India needs to be developed. A few global emission estimates of $NO_x$ are available (EDGAR, GEIA, etc.). However, these emission inventories are based on gross national values and seldom account for microlevel activity data of Indian region. Some work has been carried out to develop $NO_x$ inventory at national level in India (Kato and Akimoto, 1992; Akimoto and Narita, 1994; Garg et al., 2001, 2006; Gadi et al., 2003). However, these emission inventories are also prepared on broader scale (state-level) by using the gross values without involving district-level activity data, except some studies where the district-level data is used. Garg et al. (2001) have reported the district-level emission inventory of $NO_x$ using earlier census data for the years 1990 and 1995 available at that time. They have incorporated around 466 district-level activity data. Garg et al. (2006) also reported the $NO_x$ emissions over India, but these only included the gross values for the years 1995, 2000, and 2005 at national level.

We have used district-level activity data for the base year 1991 and 2001, and used GIS software to further distribute it spatially at $1° \times 1$. Emission factors for different sources are taken from Garg et al., 2001. Emission inventories developed in this work differ from the previous inventories in several aspects. In this work, for the first time, the inventories are developed by incorporating the latest microlevel activity data. The EFs suited to Indian conditions and latest district boundaries are also accounted for properly by us. Emission distribution for year 1991 is shown in Fig. 25.1. Emission from Indo-Gangetic plains (IGP) region accounts for 26% (around 732 Gg $year^{-1}$) of total emissions where vehicular source contributed 270 Gg $year^{-1}$ (32.6%), followed by coal at 227 Gg $year^{-1}$ (19%), and biofuel at 183 Gg $year^{-1}$ (30.4%). The $NO_x$ emissions from coal, petrol/diesel, LPG/kerosene, and biofuel are estimated to be around 1192 Gg $year^{-1}$ (42%), 827 Gg $year^{-1}$ (29%), 231 Gg $year^{-1}$ (8%), and 601 Gg $year^{-1}$ (21%), respectively, in 1991 (Fig. 25.1B).

Similar to the case of 1991, high $NO_x$ emission of 30–150 Gg/yr is found over the upper and lower IGP, and some parts of the western, southern, and eastern-central India for the year 2001 (Fig. 25.2). Among the high $NO_x$ emitting regions, IGP region contributes more than 27% (around 1172 Gg $year^{-1}$) to the total (4335 Gg $year^{-1}$) in which vehicular source contributed 526 Gg $year^{-1}$ (35%), followed by coal at 387 Gg $year^{-1}$ (20.5%), and biofuel at 184 Gg $year^{-1}$ (29.3%) due to low population and industrial activity.

In addition to the IGP region, large numbers of industrial activities over the states, such as Maharashtra, Gujarat, Andhra Pradesh and Tamil Nadu, are responsible for $NO_x$ emission of the order of 40–150 Gg $year^{-1}$. Most of the regions in central India and northeast show low $NO_x$. The relative contribution of each sector in 2001 indicated that those from coal and transport sectors are increasing at a faster rate as compared to other sources. However, coal is still found to dominate Indian $NO_x$ emission followed by transport and biofuel.

Fig. 25.3 shows the gridded decadal growth of $NO_x$ during 1991–2001). A decadal increase of 1487 Gg decade (52%) was found during this period. The maximum growth was from transport sector (83%) followed by coal (58%). Although, the vehicle numbers have gone up by more than twofold during 1990s, the fuel efficiency and technology have reduced the $NO_x$ emission up to some extent, which was clearly indicated by the emission growth of 83% during same decade. All the major cities showed a high $NO_x$ emission of the order of 20–70 Gg decade where the vehicular sources contributed a significant amount. During 1990s, the vehicle numbers increased more than twofold. The population density map of India for 2001 (as shown in Fig. 25.4) clearly indicates that the IGP region is one of the most densely populated regions. This could be directly or indirectly linked with the high vehicular density, high biofuel practices, and load of domestic coal consumption along with the presence of industrial plants, e.g., thermal, steel, and cement plants. A trend of high decadal growth of $NO_x$ emissions of the order of 15–50 Gg decade was found over the IGP region where the maximum contribution came from the vehicular source.

In order to find the dominating source over selected hotspot regions in India during 2001, we have identified five boxes marked by 1–5 in Fig. 25.5A. This figure indicates the

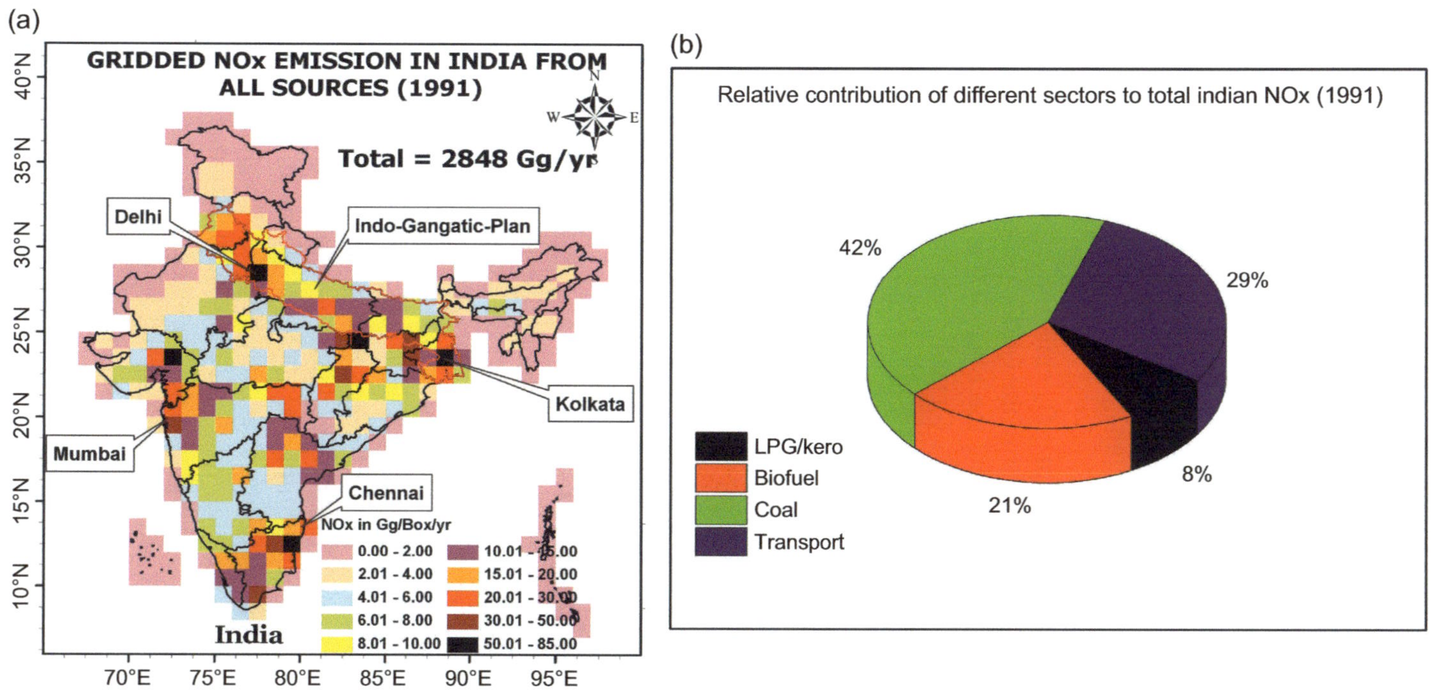


**FIGURE 25.1** (a) Gridded $NO_x$ emission for India from all sources and (b) the relative contribution from different sectors during 1991.

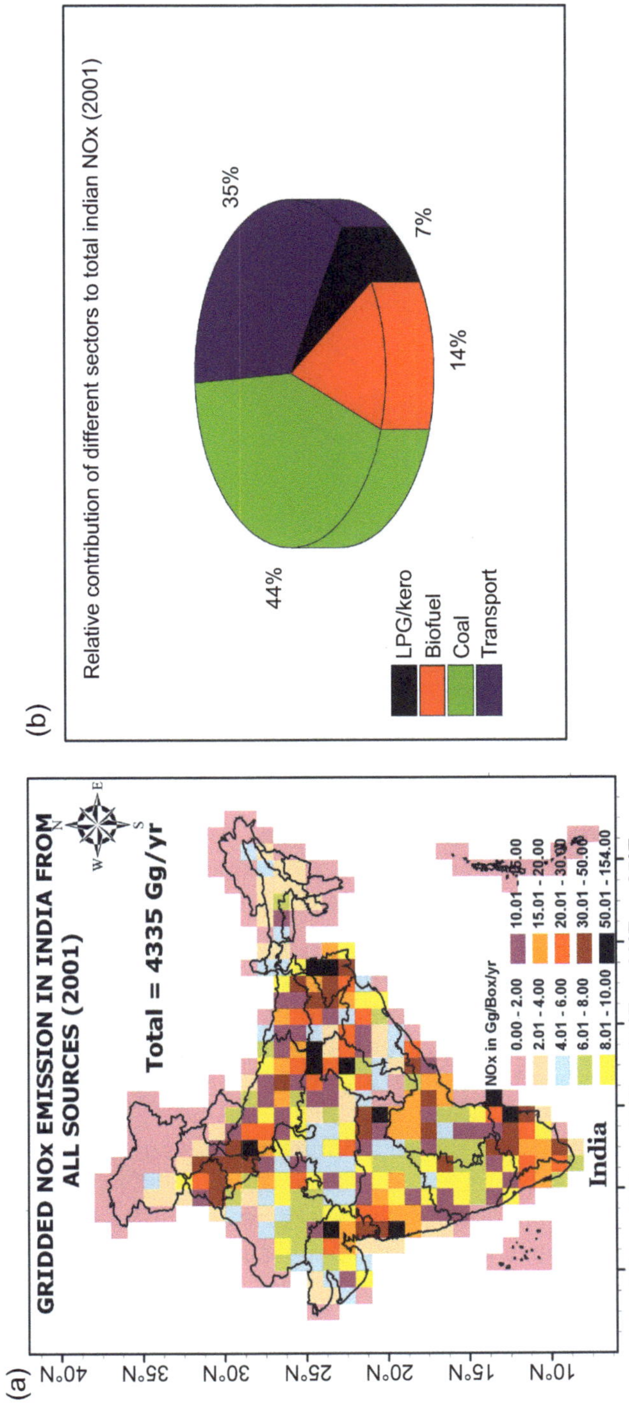


**FIGURE 25.2** (a) Gridded $NO_x$ emission for India from all sources and (b) the relative contribution from different sectors during 2001.

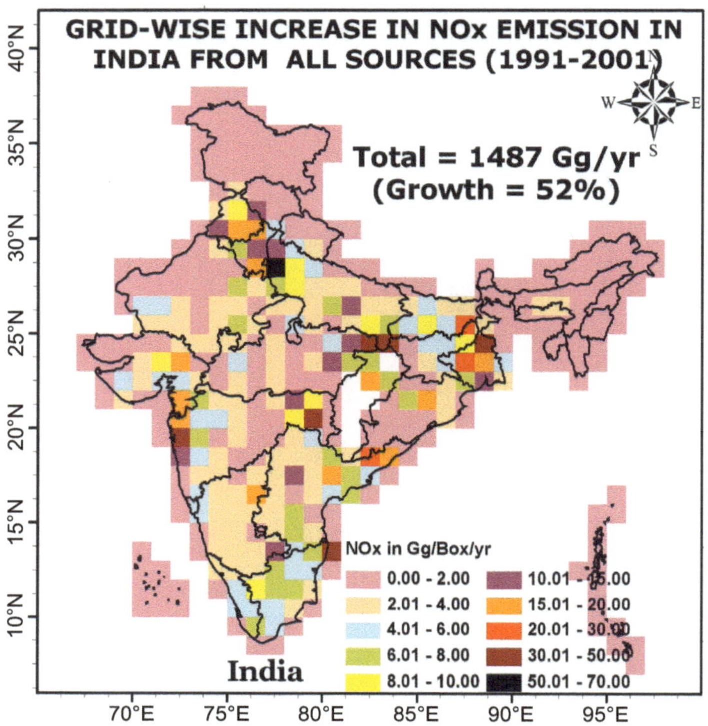


**FIGURE 25.3** Geographical distribution in gridded form of the decadal growth in total $NO_x$ emissions during 1990s.

$NO_x$ emissions over India as obtained in this work. As the coal combustion in thermal power stations and diesel vehicles are the two major sources of $NO_x$ emission, the geographical distribution of their locations along with capacities is plotted in Figs. 25.5B and C, respectively. The observed distribution of $NO_2$ column density ($10^{13}$ molecules $cm^{-2}$) as obtained by SCIAMACHY satellite data is shown in Fig. 25.5D. The identified hotspots from satellite image (Fig. 25.5D) were reasonably well reproduced by the emission inventory of $NO_x$ developed in this work (shown in Fig. 25.5A). This clearly indicates that the developed emission inventory was able to identify major hotspot regions.

However, when we try to find the dominating sources contributing to total emission in each box, it is found that boxes 2, 3, and 4 were dominated by thermal power–related $NO_x$ emission as the correlation of these regions were found to be perfect with Fig. 25.5D. This is due to the large amount of coal combustion in thermal power stations located in these areas. The diesel hotspots were very limited over these regions as depicted in Fig. 25.5C, whereas, box number 1 and 5 were dominated by diesel vehicular source where thermal power stations were not much densely located (Fig. 25.5C). The low emissions were found over the northeast, southwestern, and northwestern regions

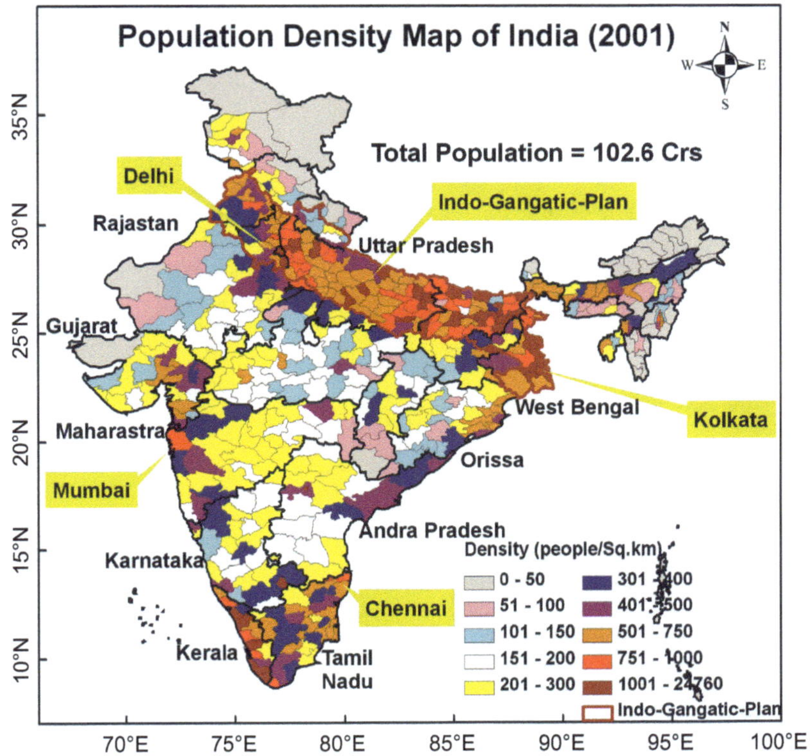


**FIGURE 25.4** Population density map of India for 2001.

and some parts of the Central India in both satellite-derived $NO_2$ column densities as well as in the present $NO_x$ emissions.

The gridded distribution of $NO_x$ emissions from all sources over Indian geographical region for the projected year 2011 is depicted in Fig. 25.6A. The total emission of $NO_x$ for India was estimated to be 7307 Gg year$^{-1}$in 2011, where emissions from coal, vehicular, LPG/kerosene, and biofuel source were estimated to be 3829 Gg year$^{-1}$, 2575 Gg year$^{-1}$, 284 Gg year$^{-1}$, and 628 Gg year$^{-1}$, respectively. A trend of high emission ranging in between 30 and 230 Gg year$^{-1}$ was also found over the IGP region which contributed to nearly 26% (1913 Gg year$^{-1}$) of total Indian $NO_x$ emission in 2011.

The spatial pattern of the decadal growth of $NO_x$ during 2000s over India is given in Fig. 25.7. During 2000s the decadal growth of $NO_x$ was found to be around 69% (2972 Gg decade). The large decadal growth in 2000s can be attributed to the huge demand of coal in the thermal sectors and other industries. The relative contribution of different sectors to Indian $NO_x$ for base years 2011 along with 1991 and 2001 is shown by the bar diagram given in Fig. 25.8.

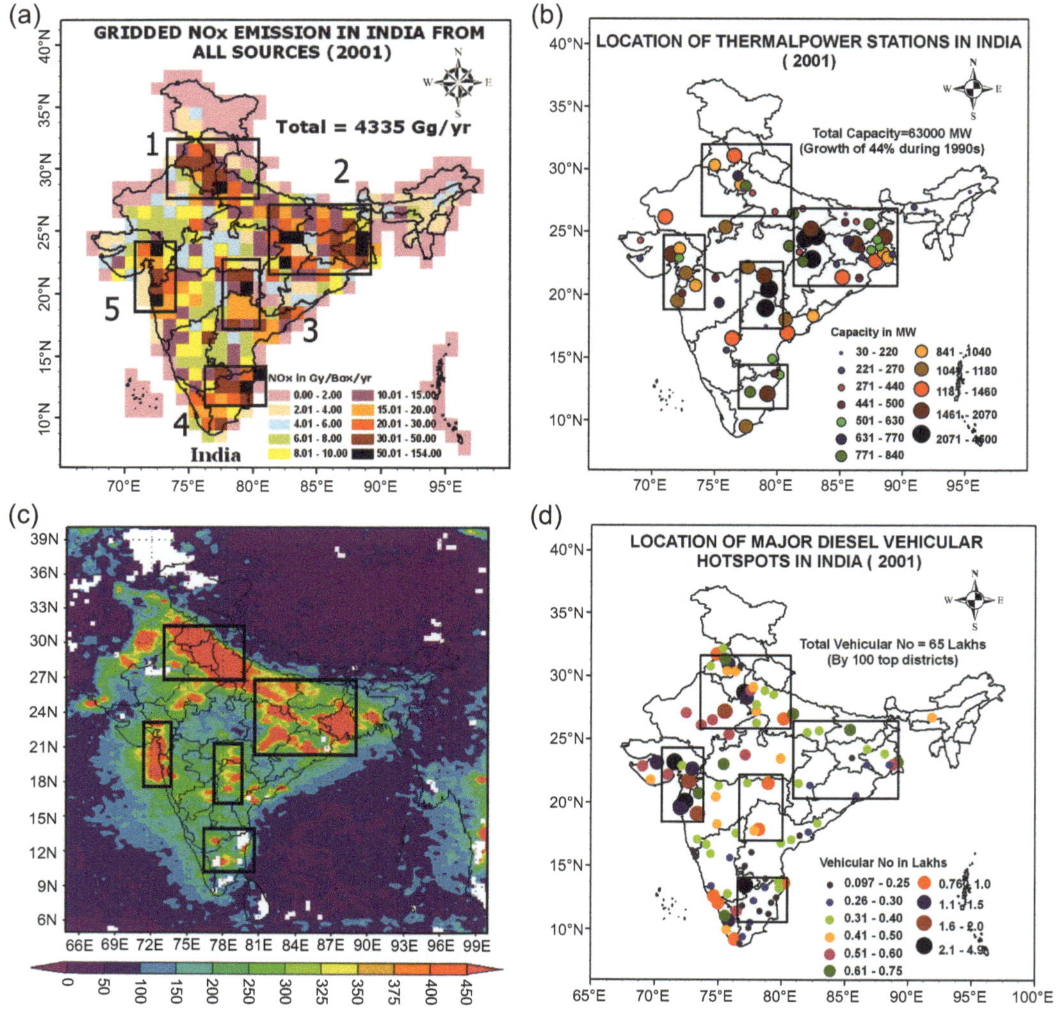


**FIGURE 25.5** The distribution of various parameters with respect to $NO_x$ in the five identified regions in India (a) gridded (1°×1°) distribution of $NO_x$ emissions (in Gg $box^{-1}year^{-1}$) as obtained in this work; (b) location of thermal power plants along with their capacity; (c) identification of hot spots of major diesel based on vehicular population and (d) the tropospheric column density of $NO_2$ in ($10^{13}$ molecules $cm^{-2}$) as observed by SCIAMACHY satellite.

## Decadal Trends in Emission of $NO_x$

Very few studies are carried out over Indian region to characterize the decadal trends in $NO_x$. Ghude et al. (2008) calculated the trend of tropospheric $NO_2$ from GOME (March 1996 to December 2002) and SCIMACHY (January 2003 to December 2006) satellite data. They observed surface hotspots of tropospheric $NO_2$ over five regions in India as explained in Fig. 25.5A. The observed trends for regions 1, 2, 3, 4, and 5, were $2.4 \pm 1.2\%\ a^{-1}$,

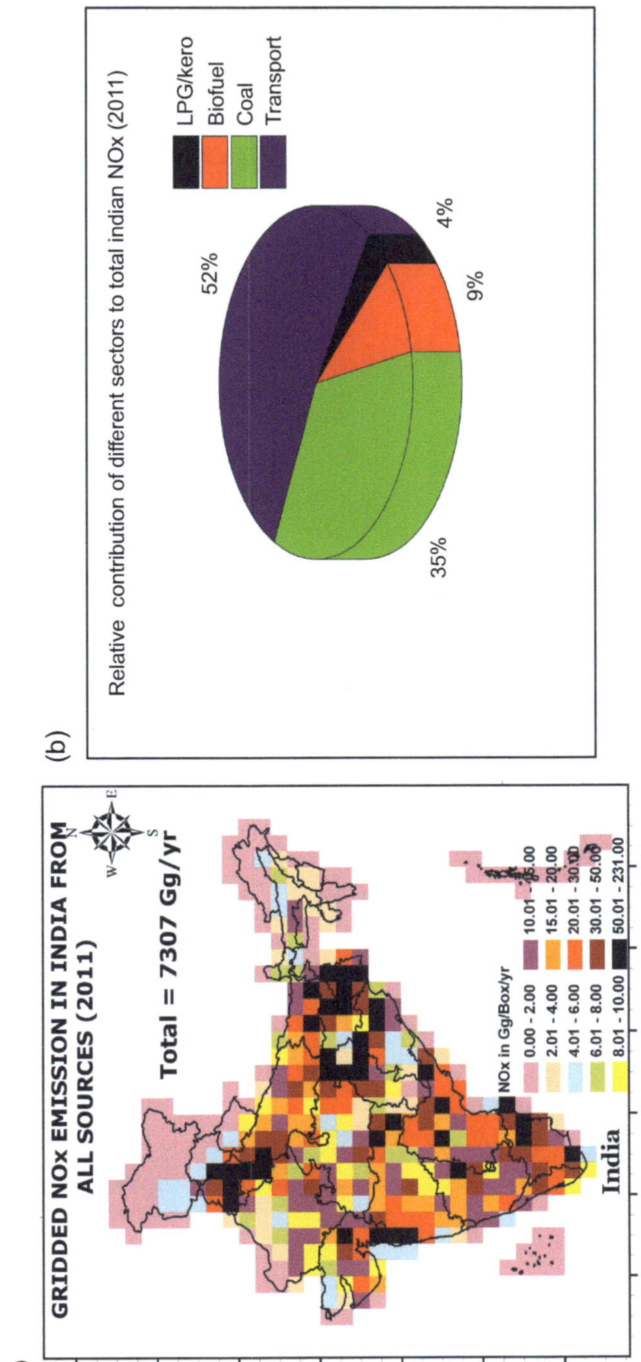


**FIGURE 25.6** (a) Gridded $NO_x$ emission for India from all sources and (b) the relative contribution from different sectors during 2011.

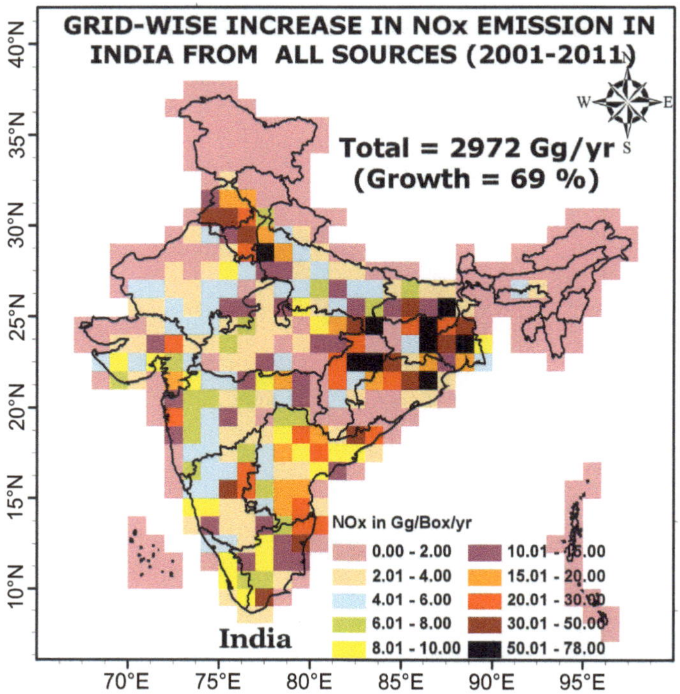


**FIGURE 25.7** Geographical distribution in gridded form of the decadal growth in total $NO_x$ emissions during 2000s.

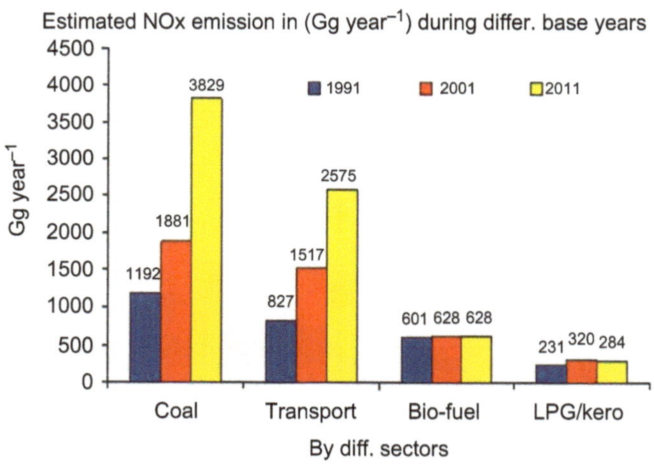


**FIGURE 25.8** Relative contribution of different sectors to Indian $NO_x$ during 1991, 2001, and 2011.

$1.57 \pm 0.87\%$ $a^{-1}$, $1.24 \pm 1.1\%$ $a^{-1}$, $1.38 \pm 1.2\%$ $a^{-1}$, and $2.1 \pm 1.1\%$ $a^{-1}$, respectively. Regions 1, 2, and 5 showed higher annual trends in $NO_x$ attributed to increase in population density, increase in transport related to population in city areas, and increase in number of thermal plants. Regions 3 and 4 show lesser trends because the growth of population is less and so the corresponding demand for transport and energy.

Beig and Brasseur (2006) reported 22% increase in $NO_x$ emissions from India during 1991–2001, where emissions from coal and transport were the highest. This study also pointed out the overall low contribution from Biofuel and biomass burning. The trend of $NO_x$ emission from coal and transport sector was around 70% in 1991–2001. Haq et al. (2015) reported a decadal increase of 14% of tropospheric $NO_2$ from OMI satellite data during the past decade (2004–2015) over South Asia region. Sahu et al. (2012) reported a decadal growth of 52% during 1990s and 69% during 2000s. Emissions from coal burning sectors, transport, biofuel, and LPG/Kerosene showed an increasing trend of 58.8%, 83.4%, 1.1%, and 38.5%, respectively, in 1990s, and 103.5%, 69.7%, 0%, and −12.6%, respectively, during 2000s. Increase in emission due to coal is supposed to increase in future due to increase in industry as already seen in 2000s. Rapid population growth and urbanization will also increase the emission from hotspot regions shown in Fig. 25.5A.

Shifting of more population to suburban areas will affect the emission from biofuel sources. Increase in slum areas will increase $NO_x$ emission from biofuel. Similar trends for $NO_x$ from coal sector, increased trend from transport, decreased or stagnant trend from biofuel, and further increased trend from LPG/Kerosene is expected in future. Hence, continuous measurements of $NO_x$ are necessary to monitor the changes. In this perspective, a network of stations (12 stations) is strategically installed in different parts of India considering the emission scenarios from different sectors as a part of Modeling Air quality and Networking (MAPAN) project. The observed station averaged annual trends (2013–2016 periods) are depicted in Fig. 25.9.

Annual trend in $NO_x$ concentration averaged over all MAPAN stations showed an overall increasing trend of 4.09% per year. Ghude et al. (2008) have previously mentioned an increasing trend of tropospheric $NO_2$ of (1.65 ± 0.52)% per year. The growing number of industries, roads, and vehicles as a part of economic development is the primary reason for this increase in $NO_x$ concentrations.

## Sources of Emission of $NH_3$

Globally, livestock industry and fertilizer application are the main sources of ammonia emission (Bouwman et al., 1997). Recent studies showed these two as major $NH_3$ sources

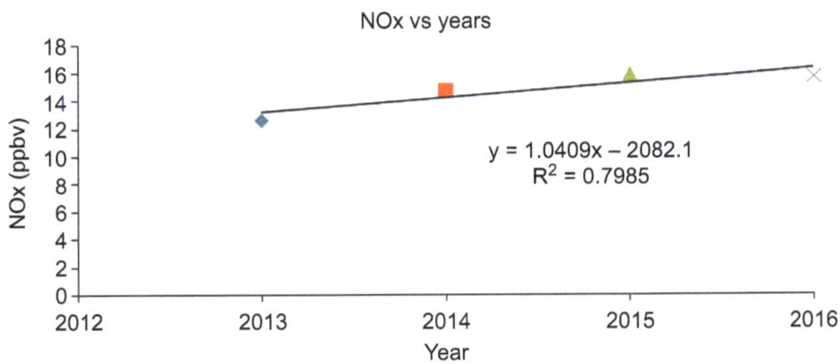


**FIGURE 25.9** Average $NO_x$ concentration over MAPAN stations (12 station average) in 2013-16 period.

in China (Huang et al., 2012; Kang et al., 2016). Ammonia emission inventory studies in India (Prashar et al., 1998; Aneja et al., 2012) showed the similar figure. While emission from livestock emission is the major contributor to $NH_3$ emission globally, emission from agricultural fertilizer application is the major contributor in India (Aneja et al., 2012). Reported Emission from different major livestock sources (in Aneja et al., 2012) are as: 956.9 Gg year$^{-1}$ (56.1%) from cattle, 403 Gg year$^{-1}$ (23.6%) from buffalo, 171.3 Gg year$^{-1}$ (10%) from goat, 23.4 (1.4%) from pig, and 29.4 (1.7%) from poultry. Among the fertilizer sources, urea has the highest contribution (2481.5 Gg year$^{-1}$, 92%) because of its high emission factor and amount of usage, followed by di-ammonium Phosphate (124.8 Gg year$^{-1}$, 4.6%), NPK (89.7 Gg year$^{-1}$, 3.3%), and ammonium sulfate (0.6, 0.02%) (Aneja et al., 2012). In this study, three states with the highest $NH_3$ emission from livestock are Uttar Pradesh (215.9 Gg year$^{-1}$), Madhya Pradesh (148.4 Gg year$^{-1}$), and West Bengal (142.5 Gg year$^{-1}$); and similarly, states with highest $NH_3$ emission from fertilizer application are Uttar Pradesh (306 Gg year$^{-1}$), Punjab (294 Gg year$^{-1}$), and Maharashtra (283 Gg year$^{-1}$). Study by Aneja et al. (2012) included the effect of temperature in calculating $NH_3$ emission. Due to short lifetime of $NH_3$ (few hours to few days) local meteorology and seasonal variability plays a great role in its emission and fate (deposition or chemical reaction), and control of $NH_3$ emission is important at regional scale. The calculated values by Aneja et al. (2012) may be an overestimation during winters or underestimation in summers when the temperature is high as seen in Chinese emission inventories (Kang et al., 2016). Future emission studies should include temporal variation in the formulation of ammonia emission inventory.

Very few measurements of $NH_3$ were carried out in Indian region in 1990s. However, past 2000, few measurements on seasonal variation of $NH_3$ were done. Such major measurements are summarized in Table 25.1. Most of the studies were carried out in campaign mode or by passive sampling. Sharma et al. (2014) carried out a campaign mode study (NH3 was measured along with $NO_x$ over 10 locations) from January 2014 to June 2014 in areas influenced by traffic and agriculture in Delhi. The $NH_3$ concentration averaged over 10 sites was found to be 20.9 ppbv. The highest mean concentration during this study was observed near sites adjacent to industrial areas (28.8 $\pm$ 3.0 ppb), followed by agricultural areas (27.5 $\pm$ 2.1 ppb). It is also seen that in the traffic areas, diurnal variation of $NH_3$ is correlated with $NO_x$ (Sharma et al., 2014). Seasonal variation of $NH_3$ showed higher concentration in winter followed by autumn seasons. However, the values were lower in summer season (Sharma et al., 2010). A measurement of $NH_3$ gas and $NH_4^+$ particulate collected concurrently was done for Delhi in 2010 (Singh and Kulsheshtra, 2012). Both $NH_3$ and $NH_4^+$ were measured using UV-Vis spectroscopy (indophenol blue method). Seasonal averaged values indicated highest concentration in summer period (June 2010–August 2010, 33.2 μg m$^{-3}$), followed by post-monsoon period (October 2010–November 2010, 32.5 μg m$^{-3}$), and pre-monsoon (March 2010 to June 2010, 26.4 μg m$^{-3}$) for $NH_3$; and summer (22.9 μg m$^{-3}$) followed by pre-monsoon (11.6 μg m$^{-3}$) and post-monsoon (8.5 μg m$^{-3}$) for $NH_4^+$. This summertime high is due to

**Table 25.1** Reported Measurement of Ambient $NH_3$ (in ppbv) in Different Places in India

| Site | Average ± Standard Deviation (ppbv) | % Deviation From NAAQS | Maximum (ppbv) | Minimum (ppbv) | Reference | Measurement Method |
|---|---|---|---|---|---|---|
| NCR-Delhi (10 locations) | 20.9 ± 1.6 | −85.4 | 44.00 | 1.84 | Sharma et al. (2014) | Chemiluminescence |
| Delhi (JNU campus)* | 48.02 (summer) 47.01 (post-monsoon) 38.19 (pre-monsoon) | −69.2 | | | Singh and Kulshrestha (2012) | Indophenol Blue |
| Delhi (winter) | 20.23 ± 0.42 | −85.9 | 46.17 | 1.02 | Sharma et al. (2010) | Chemiluminescence |
| Delhi (summer) | 6.38 ± 4.33 | −95.5 | 23.98 | 0.32 | Sharma et al. (2010) | Chemiluminescence |
| Delhi (autumn) | 17.47 ± 7.17 | −87.86 | 28.32 | 1.40 | Sharma et al. (2010) | Chemiluminescence |
| Agra | 16.4 ± 3.5 | −88.6 | 15.6 | 10.3 | Parmar et al. (2001) | Indophenol blue |
| Dibrugarh | 13.5 ± 2.3 | −90 | 26.5 | 1.3 | Sharma et al. (2011) | Chemiluminescence |
| Trivandrum | 12.7 ± 1.5 | −91.1 | 22.7 | 1.1 | Sharma et al. (2011) | Chemiluminescence |
| Cochin | 19 | −86.8 | – | – | Carmichael et al. (2003) | IVL Passive sampling |
| Bhubhaneswar | 7 | −95.1 | – | – | Carmichael et al. (2003) | IVL Passive sampling |
| Rampur | 9.7 ± 6.1 | −93.3 | – | – | Gupta et al. (2003) | Colorimetry |
| Mumbai | 24.02 ± 18.02 | −83.05 | 107.8 | 2.0 | SAFAR Mumbai | Chemiluminescence |
| Kanpur | 26.5 (summer) 27 (winter) | −81.4 | – | – | Sharma et al. (2007) | Colorimetry |

the emission from degradation of waste, plant, and biogenic material for $NH_3$ and because of increased partitioning between solid and gas phases for $NH_4^+$.

In southern India (Cochin and Trivandrum), the average $NH_3$ concentrations showed variation of 12–19 ppbv in different months (Carmichael et al., 2003). Bhubaneswar, in eastern part of India, also showed lower $NH_3$ values (7 ppbv); however, higher values were reported over sites in IGP (Kanpur, Agra, and Delhi) in northern India (Carmichael et al., 2003). A detailed account of $NH_3$ concentration reported across different Indian regions is summarized in Table 25.1.

The studies by Sharma et al. (2010) reported that the major source of $NH_3$ emission is from agricultural fields of IARI in all seasons over Delhi. The influence of boundary layer height change (stabilization of boundary layer due to temperature inversions) was reported to be responsible for wintertime high values of $NH_3$. However, latest study by Sharma et al. (2014) reported agricultural and road traffic as major sources of $NH_3$ over Delhi region. Parmer et al. (2001) observed higher concentration in summer (16.4 ± 3.5 ppbv) than winter (11.8 ± 4.4 ppbv) in rural areas of Agra and attributed it to the increased emission from natural vegetation and volatilization of $NH_3$ from cattle sheds located around observational site. Carmichael et al. (2003) concluded emission from agricultural fertilizer use, livestock, and use of biofuel as sources of $NH_3$ with a major contribution from agriculture as main $NH_3$ sources in Bhubaneswar. Sharma et al. (2007) indicated intense agricultural activity as the reason for high $NH_3$ concentrations over Kanpur. Gupta et al. (2003) observed higher $NH_3$ concentrations in winter over Rampur, attributed mainly to excess $NH_3$ emission from vegetation in winter due to dry conditions.

Most of these reported values of $NH_3$ from different sites are within the NAAQS limit for $NH_3$ (CPCB, 2012) of 400 $\mu g\ m^{-3}$ (or 576 ppbv at 1 atm and 298 K) for 24 h average and 100 $\mu g\ m^{-3}$ for yearly average.

## Distribution of Reactive Nitrogen Species

### $NO_x$ Measurement over India: Modeling and Measurements

Several model-based studies have been carried out to outline the distribution of $NO_x$ over Indian region. A particular study by Beig and Ali (2006) reported model-simulated surface $NO_x$ distribution over Indian region for pre-monsoon (May), monsoon (July), and post-monsoon (October) period for the year 2001 using MOZART-2 global chemical transport model. Meteorological data from ECMWF were used to drive meteorological fields. They have used national inventory available for India for CO and NOx devoloped by Dalvi et al. (2006). Consistent enhancement in $NO_x$ was observed over IGP region than the rest of India in all seasons (1.5–6 ppbv in May; 2.5–6 ppbv in July; 0.4–2.5 ppbv in October) (Fig. 25.10).

The major concentration of $NO_x$ in IGP region during May is attributed to the burning of seeds and residues of agriculture during April and May. High lightning flash count reported over the region (Kandalgaonkar et al., 2005, 2006) was found to be another

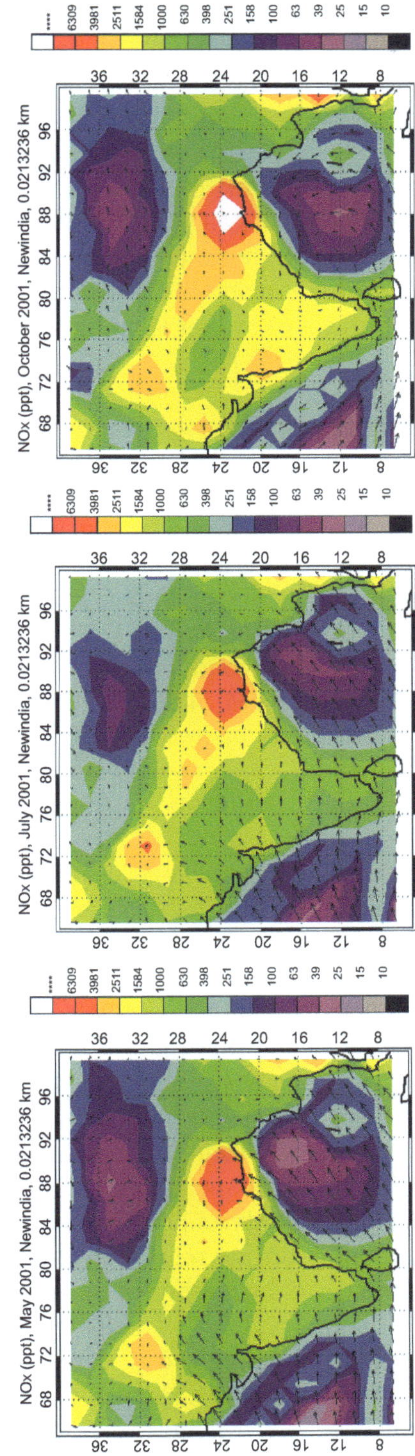


**FIGURE 25.10** Distribution of $NO_x$ as simulated by MOZART-2 over India in May, July, and October 2001, from left to right.

potential reason for high $NO_x$ concentration over IGP region. However, the south westerly wind bringing clean marine air was found to decrease $NO_x$ mixing ratios in monsoon. The $NO_x$ concentration in October was attributed to the emissions from local sources trapped in low levels due to inversions. Lifetime of $NO_x$ was also reported to increase during cold season, which led to higher concentrations. Very high level of $NO_x$ (~7.5 ppbv) was observed in lower portion of IGP, whereas minimum is observed around 25N, 75E. The spatial variability of $NO_x$ over IGP region is similar for all the seasons reported—quantitative difference between seasons is due to meteorology, solar radiation, and loss due to ozone formation and titration with ozone. Spatial variability is closely related with the presence of industry and local biomass burning sources.

*Pan India—MAPAN Monitoring*

As explained in previous sections, several model-based studies have reported $NO_x$ variations over Indian region (e.g., Beig and Ali, 2006; Beig and Brassuer, 2006; Ghude et al., 2008). However, measurements are sparse in this aspect over the subcontinent. Modeling air quality and networking (MAPAN) is a project intended to synchronize both observations and modeling. Here we exclusively explain main observational results from MAPAN, as several model-based studies are available. $NO_x$ measurements have been carried out over a network of 12 stations as a part of MAPAN project. Measurements of nitrogen oxides ($NO_x = NO + NO_2$) were conducted using chemiluminescence technique (Environmental S.A. model no. AC32 M). The detection limit of $NO_x$ instrument is reported to be 0.4 ppb. Tests and calibrations for zero and span have been conducted twice a month during the study period.

The average monthly mean concentration of $NO_x$ (average of 2013–2016 periods) over different MAPAN sites is shown in Fig. 25.11. $NO_x$ concentration showed lower values over Trivandrum in southern India and Aizwal and Teapur in northeast India, where concentrations hardly crossed 15 ppbv, as these regions are free from considerable industrial activities and huge traffic issues. Highly urbanized cities such as Bangalore and Hyderabad showed consistent high trends in $NO_x$ concentrations (above 30 ppbv) throughout the year attributed mainly to vehicular and industrial activities (Also explained in Section 2.1). However, other medium urbanized regions (Guwahati, Srinagar, Udaipur, Jabalpur, Vizag) showed moderate $NO_x$ concentrations (below 30 ppbv) throughout the year. The $NO_x$ concentration showed a clear seasonal variation with viable high values in winter and lower values in monsoon months. The high values in winter (December–February) could be associated with boundary layer meteorology, where inversions bring down the boundary layer height, leading to trapping of pollutants in lower levels. However, the rainouts during rainfall in monsoon along with entrainment of fresh air to the subcontinent are the reasons for lower values in monsoon season (June–September).

$NO_x$ concentrations in places like Bangalore and Hyderabad crossed the NAAQS limit of $NO_2$ (40 $\mu g\ m^{-3}$ (21.3 ppbv at 1 atm and 298 K) annually, and 80 $\mu g\ m^{-3}$ (42.6 ppbv at 1 atm and 298K) daily. Previously reported values from different places in India are

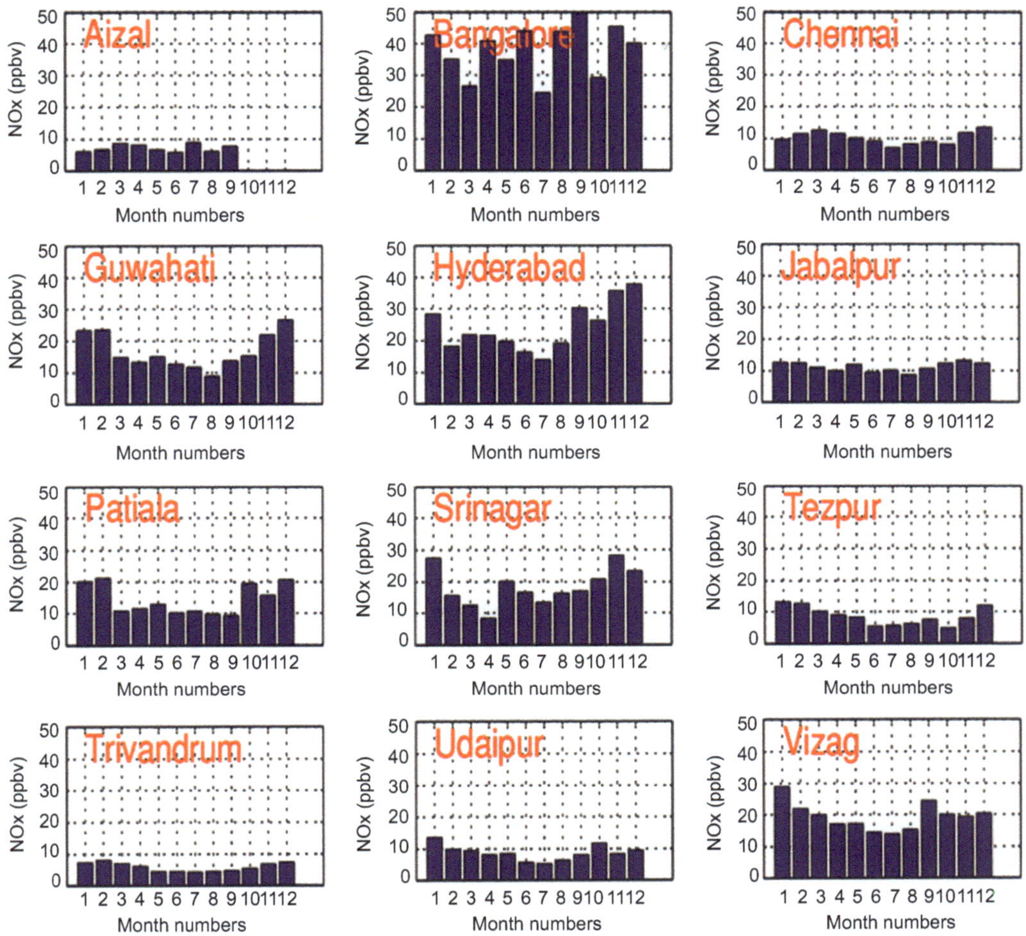


**FIGURE 25.11** Monthly variation of $NO_x$ volume mixing ratio (in ppbv) in different MAPAN stations across India.

shown in Table 25.2. Remote places like Gadanki and Anantapur showed very small $NO_x$ values as in case of MAPAN stations in Aizal,Tezpur. Places adjacent to sea (e.g., Kannur) showed less $NO_x$ values. City areas such as Kanpur, Pune, and Ahmedabad showed high values because of the local industry and biomass burning activities.

### *$NO_x$ Measurements in Microenvironment over Mega Cities: SAFAR Project*

System of air quality forecasting and research (SAFAR) is a project oriented to measure the air pollutants at microlevel in metropolitan cities. SAFAR has been operational in Pune, an urban city in western India, since 2013. The $NO_x$ measurements in SAFAR stations were also conducted using the same methodology mentioned in Section 3.1.

The monthly variation in $NO_x$ over a semiurban site as a part of SAFAR Pune (Pashan) is shown in Fig. 25.12. Significant seasonal variation in $NO_x$ was observed over Pune with

**Table 25.2** Reported Measurement of Ambient $NO_x$ (in ppbv) in Different Places in India

| Location | Period | $NO_x$ (ppbv) (Mean ± Std. Dev.) | % Deviation From NAAQS | References |
|---|---|---|---|---|
| Kanpur | 06/2009–05/2013 | 5.7 ± 4.5 | −73.2 | Gaur et al. (2014) |
| Pune | 01/2001–12/2005 | 9.6 | −54.9 | Debaje and Kakade (2009) |
| Pune | 06/2003–05/2004 | 9.6 ± 5.9 | −54.9 | Beig et al. (2007) |
| Ahmedabad | 06/2003–05/2004 | 9.5 ± 2.7 | −55.4 | Beig et al. (2007) |
| Anantapur | 01/2010-12/2010 | 5.1 ± 0.7 | −74.6 | Reddy et al. (2013) |
| Gadanki | 11/1993–12/1996 | 2.1 ± 1.8 | −90.14 | Naja and Lal (2002) |
| Kannur | 11/2009–10/2010 | 2.5 | −88.26 | Nishanth et al. (2012) |
| Mt Abu | 01/1993–12/2000 | 1.5 ± 1.4 | −93 | Naja et al. (2003) |

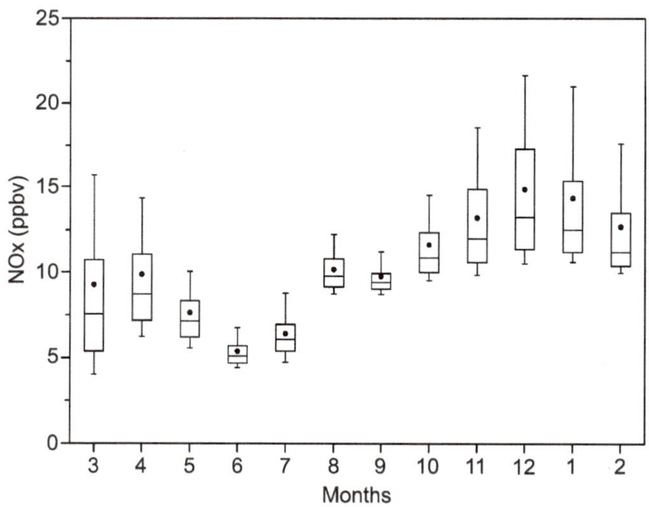


**FIGURE 25.12** Monthly variation in $NO_x$ at Pune during March 2014 to February 2015. Here the whiskers represent 10–90 percentile, box represents 25–75 percentiles, circle represents average, and the dash inside the box represents median.

high concentration during winter and low concentration during monsoon months. High influx of clean marine air during monsoon months and washout could be the major reasons for low $NO_x$ observed during summertime. Reduced boundary layer and enhanced local traffic, industrial emission from sugar industry (especially in winter), and increased biomass burning over the site were the reasons for high $NO_x$ values in winter. Beig et al. (2007) have also reported similar seasonal variation of $NO_x$ over the region. Comparison between MOZART-2 model study and observation of $NO_x$ values at Pune indicates that the effect of local emissions has high role in variation of $NO_x$ due to its short lifetime. During monsoon long-range transport plays a major role whereas in wintertime local effect was more important than long-range (Beig et al., 2007).

## Monitoring of $NH_3$ in India

Because of the spatial heterogeneity of $NH_3$ sources (agricultural, livestock, industry, and traffic) and importance of $NH_3$ in acid deposition, aerosol formation, and eutrophication, it is very necessary to have a full spatial picture of $NH_3$. In this regard, continuous systematic measurements of $NH_3$ have been started very recently as a part of SAFAR project in Mumbai. Mumbai is the economic capital of India and one of the most urbanized cities in India. $NH_3$ is measured in the same instrument mentioned in the previous section (Environmental S.A. model no. AC32 M) in which $NO_x$ is measured with addition of a module $CNH_3S_2$. $NO_y$ (=$NO + NO_2 + NH_3$) is first converted to NO, and then measured with chemiluminescence technique. $NO_x$ is subtracted from $NO_y$ to get $NH_3$. Detectable limit is 1 ppbv. Calibration is same as for $NO_x$.

The different sites of Ammonium measurement around Mumbai and their observed variations are depicted in Table 25.3. These sites are carefully selected to map different sources of $NH_3$ emission. The measurements were continuously carried out during October 2015 to June 2016. Mixing ratio values of $NH_3$ showed low values of 13.37 ppbv at less polluted sites at Kolaba to high values (above 30 ppbv) near industrial areas like Mazgaon, BKC, and Navi Mumbai. Previous reported highest $NH_3$ value was 35 $\mu g\ m^{-3}$ in Trombay, Mumbai (Zutshi et al., 1970).

Monthly variation of $NH_3$ in all the sites is shown as box and whisker plot in Fig. 25.13. All sites showed higher $NH_3$ concentration in winter months. Monthly mean values exceeding 50 ppbv were observed over few sites (Andheri, BKC, Borivali, and Malad). The higher $NH_3$ values in these sites are attributed to their proximity to $NH_3$ point sources, such as traffic, industry, and livestock. Other plausible reason for higher $NH_3$ is that its closeness to fertilizer industry (Chembur site) leads to volatilization of $NH_3$ from fertilizer. Other reasons for higher $NH_3$ content could be high population density along with poor sanitization.

$NH_3$ is also emitted from animal waste in animal husbandry (livestock) in the measurement sites (Except Kolaba). Altogether it may be noted that there is microlevel heterogeneity in $NH_3$concentration associated with local sources.

**Table 25.3** Mixing Ratios of $NH_3$ (in ppbv) Measured During October 2015 to June 2016 at Different Sites of SAFAR Mumbai

| Stations | Mean | Std Dev | Median | Max | Min |
|---|---|---|---|---|---|
| Andheri | 17.01 | 15.17 | 10.20 | 87.91 | 0.52 |
| Bhandup | 12.81 | 5.26 | 12.02 | 33.31 | 1.61 |
| BKC | 35.73 | 25.22 | 28.63 | 125.54 | 5.15 |
| Boriwali | 18.80 | 11.98 | 17.15 | 81.16 | 2.34 |
| Chembur | 19.99 | 21.80 | 9.62 | 159.14 | 1.94 |
| Colaba | 13.37 | 9.23 | 12.46 | 52.36 | 2.16 |
| Malad | 21.05 | 21.60 | 12.64 | 192.36 | 1.33 |
| Mazgaon | 39.37 | 26.38 | 34.58 | 122.26 | 1.34 |
| Navi Mumbai | 38.07 | 26.68 | 32.64 | 116.79 | 1.69 |

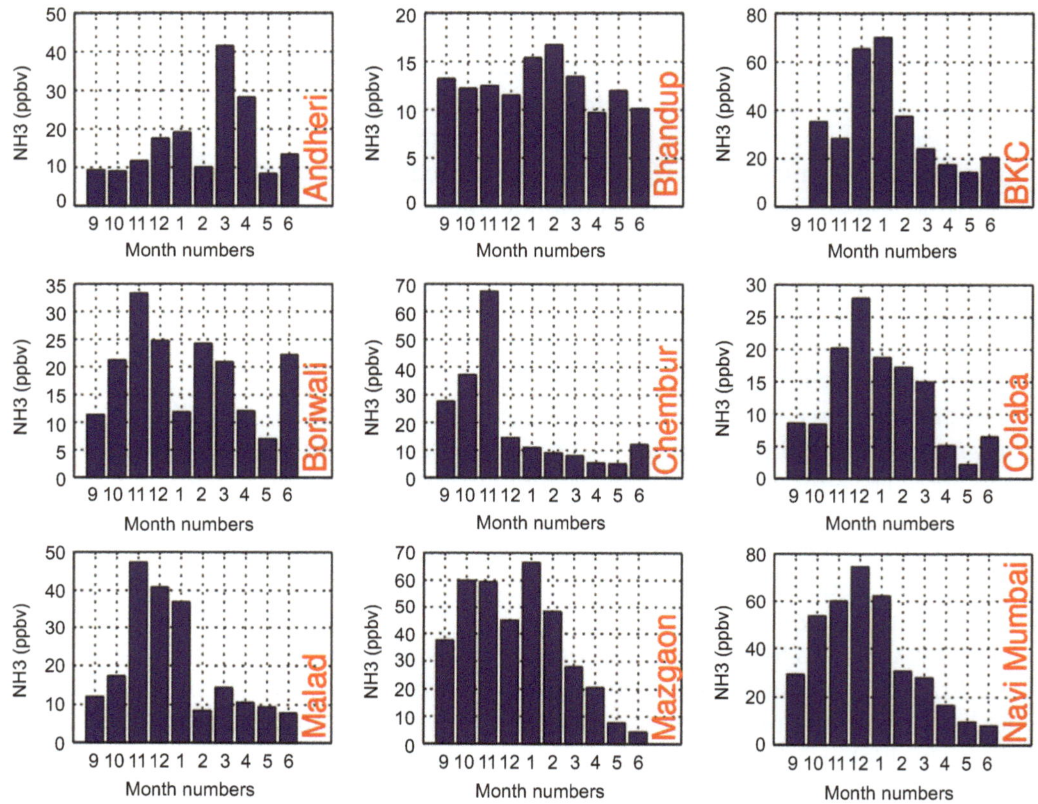


**FIGURE 25.13** Monthly averaged mixing ratios of $NH_3$ in ppbv at different sites in Mumbai during September 2015 to June 2016.

# Future Nitrogen Challenges in India

Continuous measurements of $NO_x$ were established in several important cities as a part of MAPAN project and the data was acquired. Satellite-based studies showed an increasing trend of $NO_x$ at the rate of 1.65% per year (Ghude et al., 2008). However, our measurements in past 3 years showed a further increasing trend of $NO_x$ (4.09% per year) over measurement sites (12 sites across India). Increase in Number of vehicles, roads, and industries with economic development of India are leading to increased emission of $NO_x$. This alarming trend in increase in $NO_x$ needs to be addressed and more rigorous measures have to be taken to control $NO_x$ emission values.

In addition to the measurement of $NO_x$, importance of $NH_3$ as an agent of reactive nitrogen to the earth system also needs to be examined. As the population is increasing rapidly, agriculture and food production is also going to increase in coming decades. Hence, application of fertilizer input and livestock industry both are going to increase. These two factors are most important sources of $NH_3$ on global scale (Bouwman et al., 1997). $NH_3$ has a very short lifetime and local sources, dry deposition, reemission from soil are the important factors controlling atmospheric concentration of $NH_3$ (Sutton et al., 1998). Thus, it is very necessary to have measurements of $NH_3$ over different regions in India. $NH_3$ leads to acidification of soil (Jougebreur and Voorburg, 1992), eutrophication in water bodies, and destruction of leaf (at high concentration). $NH_3$ forms aerosols such as ammonium sulfate, ammonium nitrate, and ammonium chloride, which have negative radiative forcing (cooling effect on climate). Thus it is necessary to create an emission inventory over Indian region using bottom-up approach to be used in regional chemical transport and climate models. Also federated network to measure ammonium concentration across the country also need to be established to monitor its potential role in climate change.

## Summary

This chapter discusses measurements and review of the previous studies on the reactive nitrogen compounds in India. The discussions are mainly concentrated on $NO_x$ and ammonia. The summary of the chapter is as follows.

1. Most of the studies reported on $NO_x$ over Indian region were majorly based on satellite observations or with model simulations, and there were very few observations available for validation.
2. However, considerable work has been carried out to measure $NO_x$ across India in the past 5 years with the implementation of air quality stations across major sites in India.
3. The study based on data sets from these networks has shown that there is enhancement in increasing trend in $NO_x$ concentration over India in the past 3 years (4.09% per year), compared to what was reported in the previous decade (1.65% per year).
4. The increase in industries, roads, and vehicles as a part of economic development are the primary reasons for this increase in $NO_x$ concentrations.
5. An emission inventory to describe plausible $NO_x$ sources has also been developed with the modern available techniques.
6. However, the monitoring of $NH_3$ still remains as a challenge in the country. The available $NH_3$ measurements in the past decade (mostly in campaign mode) are also discussed in the chapter.
7. Measurements of $NH_3$ at microlevel were started as a part of SAFAR project in Mumbai in recent years. The chapter concludes with the recommendation of expanding the $NH_3$ monitoring network across the India subcontinent.

# References


Akimoto, H., Narita, H., 1994. Distribution of $SO_2$, $NO_x$ and $CO_2$ emissions from fuel combustion and industrial activities in Asia with 1°×°1 resolution. Atmospheric Environment 28, 213–225.

Aneja, V.P., Schlesinger, W.H., Erisman, J.W., Behera, S.N., Sharma, M., Battye, W., 2012. Reactive nitrogen emissions from crop and livestock farming in India. Atmospheric Environment 47, 92–103, 2012.

Beig, G., Brasseur, G.P., 2006. Influence of anthropogenic emissions on tropospheric ozone and its precursors over the Indian tropical region during a monsoon. Geophysical Research Letters 33, L07808. http://dx.doi.org/10.1029/2005GL024949.

Beig, G., Ali, K., 2006. Behavior of boundary layer ozone and its precursors over a great alluvial plain of the world: Indo-Gangetic Plains. Geophysical Research Letters 33, L24813. http://dx.doi.org/10.1029/2006GL028352.

Beig, G., Ghude, S.D., Polade, S.D., Tyagi, B., 2008. Threshold exceedances and cumulative ozone exposure indices at tropical suburban site. Geophysical Research Letters 35, 1944–8007.

Beig, G., Gunthe, S., Jadhav, D.B., 2007. Simultaneous measurements of ozone and its precursors on a diurnal scale at a semi urban site in India. Journal of Atmospheric Chemistry 57, 239–253. http://dx.doi.org/10.1007/s10874-007-9068-8.

Bouwman, A.F., Lee, D.S., Asman, W.A.H., Dentener, F.J., van der Hoek, K.W., Olivier, J.G.J., 1997. A global high-resolution emission inventory for $NH_3$. Global Biogeochemical Cycles 11, 561–587.

Carmichael, G.R., Ferm, M., Thongboonchoo, N., Woo, J.H., Chan, L.Y., Murano, K., 2003. Measurements of sulfur dioxide, ozone and $NH_3$ concentration in Asia, Africa and South America using passive samplers. Atmospheric Environment 37, 1293–1308.

CPCB, 2012. National Ambient Air Quality Status & Trends in India-2010 (Report No. NAAQMS/35/2011–2012). http://www.cpcb.nic.in/.

Dalvi, M., Beig, G., Patil, U., Kaginalkar, A., Sharma, C., Mitra, A.P., 2006. A GIS based methodology for gridding large scale emission inventories: application to carbon-monoxide emissions over Indian region. Atmospheric Environment 40, 2995–3007. http://dx.doi.org/10.1016/j.atmosenv.2006.01.013.

Debaje, S.B., Kakade, A.D., 2009. Surface ozone variability over western Maharashtra, India. Journal of Hazardous Materials 161 (2–3), 686–700. http://dx.doi.org/10.1016/j.jhazmat.2008.04.010.

EDGARv4.2. Emission Database for Global Atmospheric Research Report. http://edgar.jrc.ec.europa.eu/overview.php?v=42.

Gadi, R., Kulshrestha, U.C., Sarkar, A.K., Garg, S.C., Parashar, D.C., 2003. Emissions of $SO_2$ and $NO_x$ from biofuels in India. Tellus Series B Chemical and Physical Meteorology 55, 787–795.

Garg, A., Shukla, P.R., Bhattacharya, S., Dadhwal, V.K., 2001. Sub-region (district) and sector level $SO_2$ and $NO_x$ emissions for India: assessment of inventories and mitigation flexibility. Atmospheric Environment 35, 703–713.

Garg, A., Shukla, P.R., Kapshe, M., 2006. The sectoral trends of multigas emissions inventory of India. Atmospheric Environment 40, 4608–4620.

Gaur, A., Tripathi, S.N., Kanawade, V.P., Tare, V., Shukla, S.P., 2014. Four-year measurements of trace gases ($SO_2$, $NO_x$, CO, and $O_3$) at an urban location, Kanpur, in Northern India. Journal of Atmospheric Chemistry 283–301.

GESY, 2016. Global Energy Statistical Yearbook. https://yearbook.enerdata.net/coal-and-lignite-world-consumption.html.

Ghude, S.D., Fadnavis, S., Beig, G., Polade, S.D., van der A., R.J., 2008. Detection of surface emission hot spots, trends, and seasonal cycle from satellite-retrieved $NO_2$ over India. Journal of Geophysical Research 113, D20305. http://dx.doi.org/10.1029/2007JD009615.

Gupta, A., Kumar, R., Maharaj, K.K., Srivastava, S.S., 2003. Measurement of $NO_2$, $HNO_3$, $NH_3$ and $SO_2$ and related particulate matter at a rural site in Rampur, India. Atmospheric Environment 37, 4837–4846.

Haq, Z., Tariq, S., Ali, M., 2015. Tropospheric $NO_2$ trends over south Asia during the last decade (2004–2014) using OMI data. Advances in Meteorology 959284. http://dx.doi.org/10.1155/2015/959284.

Huang, X., Song, Y., Li, M., Li, J., Huo, Q., Cai, X., Zhu, T., Hu, M., Zhang, H., 2012. A high-resolution ammonia emission inventory in China. Global Biogeochemical Cycles, 26, GB1030. http://dx.doi.org/10.1029/2011GB004161.

International Energy Agency, 2016. Energy and Air Pollution 2016-World Energy Outlook Special Report. http://www.worldenergyoutlook.org. www.iea.org.

Jougebreur, A.A., Voorburg, J.H., 1992. The role of $NH_3$ in acidification. Perspectives for the prevention and reductions of emissions from livestock operations. Studies in Environmental Science 50, 55–64.

Kandalgaonkar, S.S., Tinmaker, M.I.R., Nath, A., Kulkarni, M.K., Trimbake, H.K., 2005. Study of thunderstorm and rainfall activity over Indian region. Atmósfera 18, 91–101.

Kandalgaonkar, S.S., Tinmaker, M.I.R., Kulkarni, M.K., Nath, A., 2006. Lightning and rainfall activity over Gangetic West Bengal. Journal of Atmospheric Electricity 26, 37–50.

Kang, Y., Liu, M., Song, Y., Huang, X., Yao, H., Cai, X., Zhang, H., Kang, L., Liu, X., Yan, X., He, H., Zhang, Q. , Shao, M., Zhu, T., 2016. High-resolution ammonia emissions inventories in China from 1980 to 2012. Atmospheric Chemistry and Physics 16, 2043–2058. http://dx.doi.org/10.5194/acp-16-2043-2016.

Kato, N., Akimoto, H., 1992. Anthropogenic emissions of $SO_2$ and $NO_x$ in Asia emission inventories. Atmospheric Environment Part A. General Topics 26, 2997–3017.

Naja, M., Lal, S., 2002. Surface ozone and precursor gases at Gadanki (13.5 N, 79.2 E), a tropical rural site in India. Journal of Geophysical Research: Atmospheres 107 (D14).

Naja, M., Lal, S., Chand, D., 2003. Diurnal and seasonal variabilities in surface ozone at a high altitude site Mt Abu(24.6°N, 72.7°E, 1680 m asl) in India. Atmospheric Environment 37 (30), 4205–4215. http://dx.doi.org/10.1016/S1352-2310(03)00565-X.

Nishanth, T., Satheesh Kumar, M.K., Valsaraj, K.T., 2012. Variations in surface ozone and $NO_x$ at Kannur: a tropical, coastal site in India. Journal of Atmospheric Electricity 69 (2), 101–126. http://dx.doi.org/10.1007/s10874-012-9234-5.

Parashar, D.C., Kulshrestha, U.C., Sharma, C., 1998. Anthropogenic emissions of $NO_x$, $NH_3$ and $N_2O$ in India. Nutrient Cycling in Agroecosystem 52, 255–259.

Parmar, R.S., Satsangi, G.S., Lakhani, A., Srivastava, S.S., Prakash, S., 2001. Simultaneous measurements of $NH_3$ and nitric acid in ambient air at Agra (27°10N and 78°05E) (India). Atmospheric Environment 35, 5979–5988.

Reddy, B.S., Kumar, K.R., Balakrishnaiah, G., Gopal, K.R., Reddy, R.R., Sivakumar, V., Lingaswamy, A.P., Arafath, S.M., Umadevi, K., Kumari, S.P., Ahammed, Y.N., Lal, S., 2013. Analysis of diurnal and seasonal behavior of surface ozone and its precursors ($NO_x$) at a semi-arid rural site in southern India. Aerosol and Air Quality Research 12, 1081–1094. http://dx.doi.org/10.4209/aaqr.2012.03.0055.

Sharma, M., Kishore, S., Tripathi, S.N., Behra, S.N., 2007. Role of atmospheric $NH_3$ in the formation of inorganic secondary particulate matter: a study at Kanpur, India. Journal of Atmospheric Chemistry 58, 1–17.

Sharma, S.K., Datta, A., Saud, T., Mandal, T.K., Ahammed, Y.N., Arya, B.C., 2010. Study on concentration of ambient $NH_3$ and interactions with some other ambient trace gases. Environmental Monitoring and Assessment 162, 225–235.

Sharma, S.K., Kumar, M., Rohtash, Gupta, N.C., Saraswati, Saxena, M., Mandal, T.K., 2014. Characteristics of ambient $NH_3$ over Delhi, India. Meteorology Atmospheric Physics 124, 67–82.

Sharma, S.K., Saxena, M., Mandal, T.K., Ahammed, Y.N., Pathak, H., Datta, A., Saud, T., Arya, B.C., 2011. Variations in mixing ratios of ambient $NH_3$, nitric oxide and nitrogen dioxide in different environments of India. Journal of Earth Science & Climatic Change 1, 1–5.

Sahu, S.K., Beig, G., Parkhi, N.S., 2012. Emerging pattern of anthropogenic NOx emission over Indian subcontinent during 1990s and 2000s. Atmospheric Pollution Research 3 (3), 262–269. http://dx.doi.org/10.5094/APR.2012.021.

Singh, S., Kulshrestha, U.C., 2012. Abundance and distribution of gaseous ammonia and particulate ammonium at Delhi, India. Biogeosciences 9, 5023–5029.

Sutton, M.A., Lee, D.S., Dollard, G.A., Fowler, D., 1998. Introduction atmospheric $NH_3$: emission, deposition and environmental impacts. Atmospheric Environment 32, 269–271.

van Aardenne, J.A., Carmichael, G.R., Levy II, H., Streets, D., Hordijk, L., 1999. Anthropogenic $NO_x$ emissions in Asia in the period 1990–2020. Atmospheric Environment 33, 633–646.

Zutshi, P.K., Sequaria, R., Mahadevan, T.N., Banerjee, T., 1970. Environmental concentrations of some of the major pollutants at the BARC site, Trombay. Indian Journal of Meteorology & Geophysics 21, 473–478.

# 26

# Assessment of Atmospheric Emissions and Depositions of Major Nr Species in Indian Region

U. Kulshrestha

*JAWAHARLAL NEHRU UNIVERSITY, NEW DELHI, INDIA*

## CHAPTER OUTLINE



**The Indian Nitrogen Assessment. http://dx.doi.org/10.1016/B978-0-12-811836-8.00026-4**

## Introduction

Recently, it has been realized that with the growing demand of food supply, nitrogenous fertilizer consumption is increased globally. This has resulted in significant alterations in the global N cycle as the Haber-Bosch process of fertilizer production converts inert atmospheric $N_2$ into more reactive forms such as $NH_3$ and urea (Velmurugan et al., 2008; Gu et al., 2015; Galloway and Cowling, 2002; Murugan and Dadhwal, 2007). A comprehensive review by Behera et al. (2013) provides detailed information about global $NH_3$ emission sources. According to various estimates, the global N fertilizer consumption rate is increasing every year. The upsurge is estimated to be 3.7% in 2014, 5.4% in 2015, 6.9% in 2016, 8.8% in 2017%, and 9.5% in 2018 (FAOSTAT, 2015). Similarly, there has been a high demand of fertilizers in India too, which has resulted in Indian consumption of nitrogenous fertilizer around 15 million tons in 2010–11 as compared to the earlier consumption of 1 million ton in 1960s. According to FAOSTAT (2015), the consumption figure has reached to 16.8 Tg for the year 2012 in India contributing about 14.4% of total N fertilizer through Haber-Bosch N fixation (HBNF) process. Such a high consumption of N fertilizers ranks India as second country on the global list of N fertilizer consumers leaving behind 11.7 Tg N fertilizer consumption of USA. However, Indian consumption is almost half of the Chinese consumption of fertilizers (35 Tg). The rising demand of nitrogenous fertilizer is a direct indicator of increased emissions of $NH_3$ and other related species in the atmosphere. Significant amount of gaseous $NH_3$ is contributed by several other sources such as agriculture, humans, sanitation, disposal of untreated sewage and municipal wastes, biomass burning (Bouwman et al., 1997; Behera et al., 2013; Sharma et al., 2008). According to the estimates, total ammonium emissions from anthropogenic sources such as fertilizers and livestock amounted to 2039 Gg $NH_3$–N as compared to global emission of 44,000 Gg representing approximately 4% of the global budget (Parashar et al., 1998).

In addition, increasing consumption of fossil fuel in the thermal power plants, industries, automobiles, etc., is also a source of huge $NO_x$ further enhancing Nr accumulation in the atmosphere (Fig. 26.1). In 2011, total global coal production was 1905 Mt accounting for 32% hike in its total production of 2005 (IEA, 2012). In the year 2000, global anthropogenic NO emission budget was estimated as 38 Tg N per year which is expected to reach at 64.7 Tg by 2030 under A2 scenario of development (Dentener et al., 2006). Anthropogenic emission trends of $N_2O$, which is another major reactive N species, have revealed a significant increase in $N_2O$ budget during 1995–2008 (Saikawa et al., 2014) and between 1981 and 2010 (Tian et al., 2015). Global land ecosystem emit around 12 Tg $N_2O$–N per year. A remarkable increase in $N_2O$ emissions has been estimated from Asian agriculture sector during recent years primarily due to increased application of nitrogenous fertilizers (Saikawa et al., 2014).

$NO_x$, $N_2O$, and $NH_3$, being the major Nr contributors have been attempted more vigorously worldwide as compared to other Nr species(Dentener et al., 2014; Hicks et al., 2014). Their global budgets are more reliable too. However, such information from the

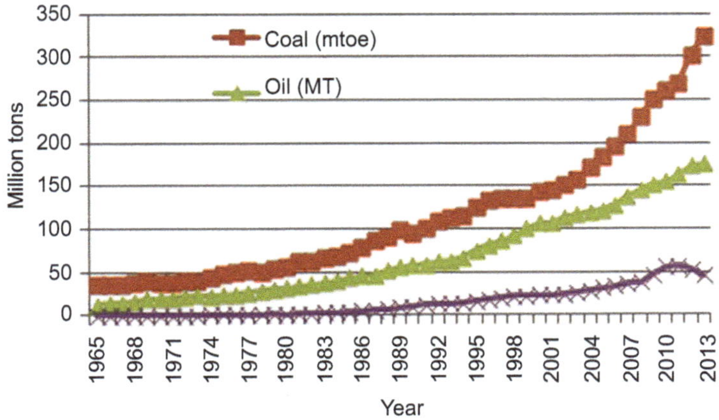


**FIGURE 26.1** Trends of fossil-fuel consumption in India during 1965–2013. *BP Statistical Review of World Energy, 2015.*

developing world is still very limited. Apart from the above major Nr species, nitrous acid (HONO), peroxyacetyl nitrate (PAN), and nitric acid ($HNO_3$), etc., also play significant roles in atmospheric transformation and deposition of different reactive nitrogen precursors and products. Reduced and oxidized inorganic forms of these species along with organic forms are given in Table 26.1.

# Trends of Reactive Nitrogen

Fig. 26.2 shows that during past few decades, anthropogenic contributions of Nr species have surpassed the natural rate of conversion from inert $N_2$ to Nr via natural processes. The most common natural $N_2$ conversion to reactive nitrogen is considered as nitrogen fixation through *Rhizobium* bacteria. According to estimates, around 100–300 Tg of nitrogen is fixed by natural and nonagricultural processes by these types of bacteria per year on the earth surface (Galloway et al., 2004). Natural N fixation takes place almost at

**Table 26.1** Different Chemical Species of Reactive Nitrogen

| | | |
|---|---|---|
| Inorganic | Inorganic chemically reduced forms of N | Ammonia ($NH_3$) |
| | | Ammonium ion ($NH_4$) |
| | Inorganic chemically oxidized forms of N | Nitrogen oxides ($NO_x$) |
| | | Nitric acid ($HNO_3$) |
| | | Nitrous oxide ($N_2O$) |
| | | Nitrogen pentaoxide ($N_2O_5$) |
| | | Nitrous acid (HONO) |
| | | Peroxyacetyl nitrate (PAN) ($CH_3COOONO_2$) |
| | | Nitrate ion ($NO_3$) |
| Organic | Organic compounds | Urea |
| | | Amines |
| | | Amino acids |
| | | Proteins |

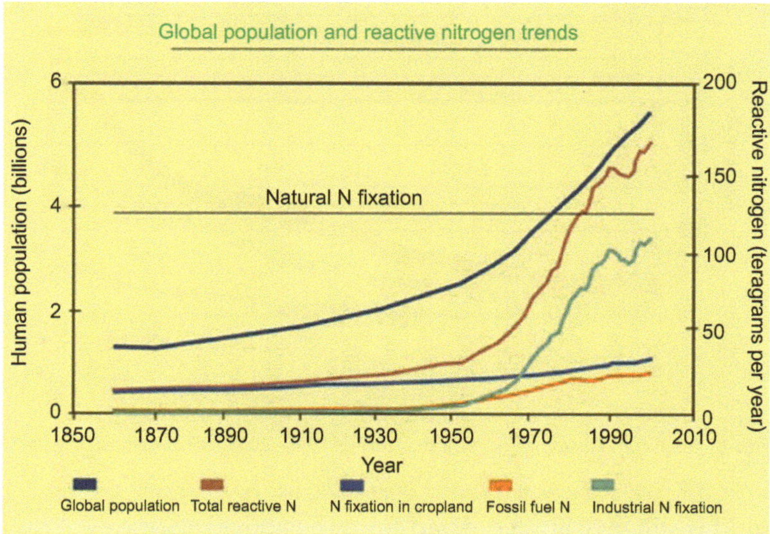


**FIGURE 26.2** Change in population and reactive nitrogen after industrialization. *Galloway and Cowling, 2002 with permission of The Royal Swedish Academy of Sciences.*

a constant rate without much fluctuation. But in case of anthropogenic fixation of nitrogen, scenario is changed drastically after industrialization especially during post-1950s when the use of fossil fuel was at peak. In addition, the increasing demand of food for the growing population has also forced to increase the HBNF application worldwide. As shown in Fig. 26.2, the rate of increase of reactive nitrogen is very high during past few decades crossing over natural nitrogen fixation showing an imbalance in N cycle. Similarly, consumption of fossil fuel, which is a significant source of $NO_x$, is also increased during past few decades (Fig. 26.1). According to available estimates, Indian $NO_x$ emission was 7.583 Tg in 2011 as compared to 2.9526 Tg in 1991 (Sahu et al., 2012). Total $N_2O$ emissions of India were reported to be 267 Gg in 2005, showing an annual increase of 3.1% as compared to 144 Gg in 1985 (Garg et al., 2012). However, according to INCCA report, total Indian $N_2O$ emissions were 239 Gg for the year 2007 (INCCA, 2010). Similarly, $NH_3$ emissions are increasing in India but its reliable trends are not available due to lack of concern of scientists and policy makers.

## Emission of Reactive Nitrogen

In Indian region, the contributions of reactive nitrogen from various sectors have been discussed as follows.

### $N_2O$ Emissions

Table 26.2 gives $N_2O$ emissions from different source categories in India. Most of $N_2O$ is contributed by the soils due to nitrogenous fertilizer application. Paddy cultivation is one

**Table 26.2** $N_2O$ Emissions From Different Source Categories in India

| Sector | Source | $N_2O$ emission (Gg) |
|---|---|---|
| Agriculture | Soils | 140 |
| | Crop residue burning | 6 |
| | Livestock manure management | 0.07 |
| | Total | **146.07** |
| Energy | Electricity generation | 10.66 |
| | Other energy industries | 0.07 |
| | Transport | 8.67 |
| | Road transport | 6 |
| | Railways | 2.35 |
| | Aviation | 0.28 |
| | Navigation | 0.04 |
| | Residential | 36.29 |
| | Commercial | 0.04 |
| | Agriculture/fisheries | 1.15 |
| | Total | **56.88** |
| Industry | Mineral | 0.46 |
| | Chemicals | 17.33 |
| | Metals | 1.11 |
| | Other industries | 2.37 |
| | Total | **20.56** |
| Domestic | Domestic waste water | 15.8 |
| Bunkers | Bunkers | 0.1 |
| Grand total | | 239.31 |

INCCA, 2010.

among major contributors of $N_2O$ emissions (Adhya et al., 2007; Sharma et al., 2008). As shown in Fig. 26.3, agriculture sector contributes maximum (61%) to the total $N_2O$ emissions in India which is followed by energy sector (24%). It is to be mentioned that livestock manure management is a significant source of $N_2O$ which is not indicated by INCCA estimates. There are huge discrepancies among various reports about $N_2O$ livestock manure emissions. According to a recent report, $N_2O$ emissions from livestock manure management have been 8.67 Gg in 1961 and 15.3 Gg in 2010, having an average annual growth rate (AGR) of 1.17% only (Patra, 2014) (Table 26.3). With this rate of emission, $N_2O$ emissions from livestock manure management will be around 17.5 and 21.4 Gg in 2025 and 2050, respectively. On the other hand, another study (Aneja et al., 2012) reports very high $N_2O$ emissions and AGR from different livestock categories indicating huge discrepancy in the budget in comparison to other estimates (Table 26.4). This suggests an immediate need of reliable data through in situ measurements from different livestock categories. However, the AGR values given by Patra (2014) seem more realistic as per weight and health of Indian animals together with the apparent changes seen in their demographic distribution in India.

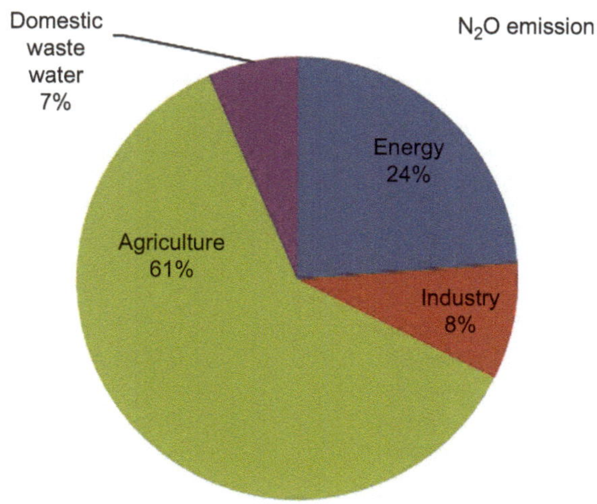


**FIGURE 26.3** Major sectors contributing $N_2O$ in India (INCCA, 2010).

**Table 26.3** Estimated and Projected Nitrous Oxide Emissions (Gg) From Livestock in India

| Livestock Category | 1961 | 2010 | 2025 | 2050 |
|---|---|---|---|---|
| Cattle | 4.2 | 4.1 | 4.1 | 3.9 |
| Buffalo | 2.2 | 4.8 | 5.4 | 6.8 |
| Sheep | 0.6 | 1.2 | 1.3 | 1.5 |
| Goat | 1 | 2.4 | 2.7 | 3.4 |
| Swine | 0.29 | 0.54 | 0.94 | 1.2 |
| Poultry | 0.3 | 2.3 | 3 | 4.6 |
| Camel | 0.014 | 0.007 | 0.006 | 0.005 |
| Ass | 0.017 | 0.005 | 0.004 | 0.003 |
| Mule | 0.001 | 0.002 | 0.002 | 0.002 |
| Horse | 0.004 | 0.002 | 0.002 | 0.001 |
| Total | 8.67 | 15.3 | 17.5 | 21.4 |

Patra, A.K., 2014. Trends and projected estimates of GHG emissions from Indian livestock in comparisons with GHG emissions from world and developing countries. Asian-Australasian Journal of Animal Sciences 27(4), 592–599, http://dx.doi.org/10.5713/ajas.2013.13342.

## $NO_x$ Emissions

Mostly, $NO_x$ is contributed by combustion of fossil fuels in thermal power plants, automobiles, and industries, etc. According to Garg et al. (2001) most of the $NO_x$ is contributed by transport sector (32%), followed by power sector (27.9%), industrial sector (19.2%), biomass burning (18.7%), and others (2.2%). The relative contribution of $NO_x$ from transport sector is increased in urban areas due to higher density of automobiles consuming greater amount of petroleum. According to Sahu et al. (2015), Delhi emits 255 Gg of $NO_x$ every year out of which 63% is contributed by transport sector, 31%

**Table 26.4** Level of Uncertainty in Two Different $N_2O$ Budgets and Annual Growth Rates (Patra, 2014; Aneja et al., 2012)

| | Patra (2014) | % AGR | Aneja et al. (2012) | %AGR |
|---|---|---|---|---|
| Cattle | 4.1 | −0.06 | 90.2 | 42.3 |
| Buffalo | 4.8 | 1.6 | 59.9 | 28.1 |
| Sheep | 1.2 | 1.25 | 20.3 | 9.5 |
| Goat | 2.4 | 1.25 | 33.1 | 15.5 |
| Pig/swine | 0.54 | 1.28 | 3.8 | 1.8 |
| Poultry | 2.3 | 4.28 | 2.7 | 1.3 |
| Camel | 0.007 | −1.43 | 1.1 | 0.5 |
| Mule and ass | 0.007 | −0.45 | 1.1 | 0.5 |
| Horse | 0.002 | −1.88 | 1.1 | 0.5 |

by industrial sector, and 3% each by power and residential sector. Table 26.5 gives $NO_x$ emission estimates as reported by different workers from fossil fuel and biofuels in India. Estimates given by Garg et al. (2006) and Sahu et al. (2012) are very close and seem reliable for further utilization of data for different purposes.

## $NH_3$ Emissions

Most of the atmospheric ammonia is contributed by nitrogenous fertilizers and livestock. Other sources of $NH_3$ include biomass burning, crop land, human excreta, etc. According to Aneja et al. (2012), fertilizer contribution of $NH_3$ was 2696.6 Gg, while livestock contribution was estimated as 1704.8 Gg for India. As given in Table 26.6 urea contributes the highest $NH_3$ in the atmosphere. Global $NH_3$ emissions from fertilizers and livestock were reported to be 9000 Gg and 21,700 Gg, respectively (Bouwman et al., 1997). Contrary to global highs from livestock as compared to fertilizer, Indian emissions of $NH_3$ are largely contributed by fertilizer application as compared to livestock. However, Parashar et al. (1998) have reported higher $NH_3$ emissions from livestock than from fertilizer which are on the lines of other global estimates (Schlesinger and Hartley, 1992; Dentener and Crutzen, 1994; Bouwman et al., 1997). Another issue is pointed out through the estimates given by Aneja et al. (2012) that the rate of increase of urea in agricultural fields is estimated as 92% per year. Such increase is surprisingly very high which sets an alarm for N

**Table 26.5** $NO_x$ Emission From Fossil Fuel and Biofuel in India

| Reference | $NO_x$ emission (Gg $year^{-1}$) | | |
|---|---|---|---|
| Garg et al. (2006) | 2640 (1990) | 4310 (2000) | 5020 (2005) |
| Sahu et al. (2012) | 2952 (1991) | 4487 (2001) | 7583 (2011) |
| Ohara et al. (2007) | 4377 (1995) | 4730 (2000) | |

**Table 26.6** $NH_3$ Emission From Fertilizers and Livestock (Aneja et al., 2012)

| $NH_3$ emission | | | | | |
|---|---|---|---|---|---|
| **Fertilizer application** | | | **Livestock** | | |
| | **Gg year$^{-1}$** | **%AGR** | | **Gg year$^{-1}$** | **%AGR** |
| Urea | 2481.5 | 92 | Cattle | 956.9 | 56.1 |
| Ammonium sulfate | 0.6 | 0.02 | Buffalo | 403 | 23.6 |
| NPK | 89.7 | 3.3 | Sheep | 101.8 | 6 |
| Di-ammonium phosphate (DAP) | 124.8 | 4.6 | Goat | 171.3 | 10 |
| | | | Pigs/swine | 23.4 | 1.4 |
| | | Poultry | 29.4 | 1.7 | |
| | | Camel | 5.7 | 0.3 | |
| | | Mules and asses | 7.1 | 0.4 | |
| | | Horses | 6.2 | 0.4 | |
| Total | 2696.6 | | Total | 1704.8 | |

cycle imbalance in the region very soon. Further application of fertilizer needs the policy and programs favoring its use with maximum efficiency having least impact on the environment in terms of N accumulation (Pathak and Ladha, 2007; Singh et al., 2007).

## Atmospheric Deposition of Reactive Nitrogen

Once emitted, gaseous Nr species are transported and transformed in the atmosphere before their deposition to any surface. These are removed through wet and dry deposition processes. The relative importance of these two removal mechanisms varies in time and space. For example wet deposition is an effective mean of scavenging of pollutants round the year in temperate regions due to precipitation, while in the regions where dry weather conditions persist having no or very limited rain, dry deposition process is more effective mean of pollutant removal. Latter conditions are qualified by Indian region where wet deposition is effective during monsoon season as almost 90% precipitation occurs during June–September. Rest of the months weather remains almost dry. Hence, during dry weather conditions, most of the pollutants including reactive nitrogen species are scavenged through dry deposition process. However, there are always very high water vapors and solar radiations available in the atmosphere in tropical regions which actively favor various transformation reactions changing gaseous Nr to particulate Nr specie, e.g., NO forming $NO_3^-$. However, the fate of Nr species also depends on the land-use dynamics, photochemistry, atmospheric acidity, and particulate loadings in the atmosphere (Aas et al., 2014). The suspended soil-dust and mineral-dust are reported as the typical scavenger for acidic species of reactive nitrogen through heterogeneous chemistry pathways (Singh and Kulshrestha, 2012; Kulshrestha et al., 2009). Relative importance of their removal through wet/dry deposition processes also changes from one region to other (Budhavant et al., 2012; Kulshrestha, 2013; Vet et al., 2014). Hence, such causes and assumptions obstruct in making any regional comparisons. Due to this reason, the studies reported from temperate regions cannot be replicated with the same level of confidence over the tropical region.

## Wet Deposition of Nr

Previous measurements related to atmospheric deposition of reactive nitrogen were a part of precipitation chemistry and acid deposition research worldwide. Very recently, it has been realized that N cycle alteration studies need dedicated efforts. Deposition of $NH_4$ and $NO_3$ aerosols can cause acidification and eutrophication of the ecosystem. Similarly, dry deposition of $NH_3$ and $NO_2$ gases can have adverse impacts on the ecosystems. The Nr deposition provides extra load to soil, vegetation surfaces, etc. The cause of concern is that with the global change, Nr depositions are increasing with time in India too due to increasing emissions of the precursors (Rao et al., 2006; Kulshrestha, 2013). In this regard, Dentener et al. (2006) have reported a comprehensive analysis of Nr deposition under different emission scenarios. However, the role of base cations is important in controlling the overall acidity (Rodhe et al., 2002). Factors affecting net and potential acidity and role of soil dust in India have been reported in couple of studies (Granat et al., 2001; Kulshrestha, 2013). In fact, in order to assess ecological impacts of deposition of acidic N species ($NO_3$, $NO_x$) a soil map of India with different soil types needs to be overlaid with the total deposition taking place over it. Generally, in India, atmospheric deposition measurements are carried out in urban areas with fewer data reported from rural areas. The data from the latter category are also important for assessing the overall impact of precipitation on soils and nutrient levels.

### *Wet Deposition Fluxes at Sites of Different Characteristics*

Wet deposition fluxes of $NH_4$ and $NO_3$ at four categories (rural, suburban, urban, and industrial) of sites have been given by Kulshrestha et al. (2005), which are based on numerous reports from India. Out of four categories of sites, the highest $NH_4-N$ is deposited at the industrial sites, while the highest $NO_3-N$ is deposited at urban sites clearly indicating the influence of surrounding sources according to the site characteristics (Fig. 26.4). The annual depositions of 2.38 kg $ha^{-1}$, 2.10, 2.38, and 5.04 kg $ha^{-1}$ $NH_4-N$ occur at rural, suburban, urban, and industrial sites, respectively. Similarly, annual depositions of 4.06, 2.10, 4.48, and 3.92 kg $ha^{-1}$ $NO_3-N$ occur at rural, suburban, urban, and industrial sites respectively. Total average $NH_4-N$ and $NO_3-N$ wet fluxes are estimated as 6.01 kg $ha^{-1}$ N deposition which is equivalent to 1.97 Tg N wet deposition per year for India. This deposition of N is much less than the quantity of N fertilizer applied in the agricultural fields (Singh and Kulshrestha, 2014).

### *Increase of Nr in Urban Areas During Past Decades*

It is to be mentioned that the above estimates need extrapolation of figures as during recent decades, the process of shifting toward cities has induced extremely higher

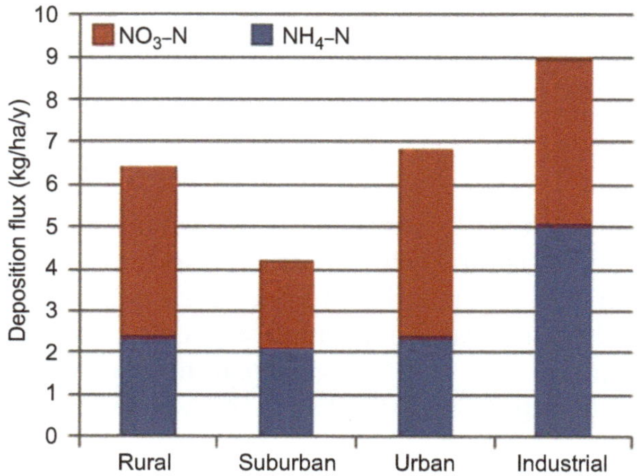


**FIGURE 26.4** Wet deposition fluxes of $NH_4$–N and $NO_3$–N at different categories of sites in India. *Kulshrestha, U.C., Granat, L., Engardt, M., Rodhe, H., 2005. Review of precipitation monitoring studies in India—a search for regional patterns. Atmospheric Environment, 39, 7403–7419.*

emissions at urban sites. For example, there is a report on remarkable increase in $NH_4$ and $NO_3$ fluxes in Delhi (Singh et al., 2014). Fig. 26.5 shows that during 1994 $NH_4$ and $NO_3$ wet deposition fluxes were 3.5 kg ha$^{-1}$ and 2.3 kg ha$^{-1}$, respectively which are increased to 13.44 kg ha$^{-1}$ and 13.49 kg ha$^{-1}$ in 2011, respectively. $NH_4$–N and $NO_3$–N together deposit about 13 kg ha$^{-1}$ N through wet removal process. As 90% of rain occurs during monsoon season, these N deposition fluxes become important for June–September period. This increase of $NH_4$ and $NO_3$ follows a pattern of increase in population, fossil-fuel consumption, and number of automobiles as shown in Fig. 26.5, correlating the deposition patterns with these activities and sources. Data show that the human

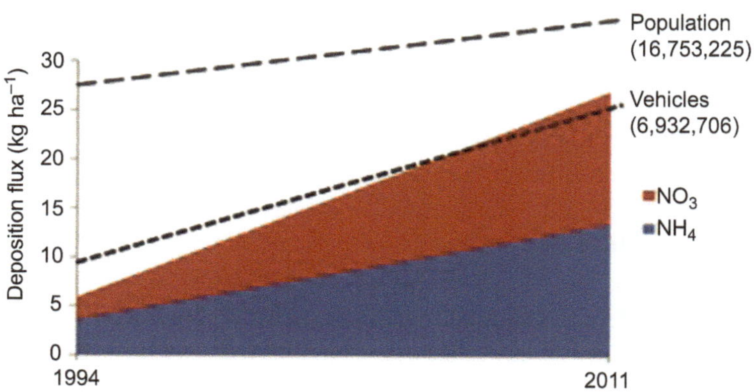


**FIGURE 26.5** Increase of $NH_4$ and $NO_3$ wet deposition fluxes at Delhi between 1994 and 2011. *Singh, S., Kumar, B., Gupta, G.P., Kulshrestha, U.C., 2014. Signatures of increasing energy demand of past two decades as captured in rain water composition and airmass trajectory analysis at Delhi (India). Journal of Energy, Environment and Carbon Credit, 4(3), 43–61.*

population of Delhi is increased from 10,748,000 in 1994 to 16,753,225 in 2011. The number of automobiles in Delhi was 2,432,295 in 1994, which has reached to 6,932,706 in 2010 (DSHB, 2011), which is the major source of $NO_x$, finally responsible for increased deposition of $NO_3{-}N$.

### *Nr Deposition in Rural Areas*

In general, Indo-Gangetic region is considered as "high deposition" area for $NO_3$ and $NH_4$, and even for $SO_4$, due to their large emissions from various sources (Kulshrestha et al., 2005). The region also has the highest density of human population. Recent measurements by Singh et al. (2016) have also demonstrated the effects of strong sources showing very high deposition of $NH_4{-}N$ and $NO_3{-}N$ in Delhi and Mai (a rural site near Lucknow) in the Indo-Gangetic region. Depositions of $NH_4{-}N$ as 10.42 and 3.89 kg $ha^{-1}$, and $NO_3{-}N$ as 3.94 and 1.98 kg $ha^{-1}$ occur in Delhi and Mai, respectively. These estimates also verify that their depositions are significantly increasing at urban sites in recent years due to rapid urbanization. Such patterns of reactive nitrogen deposition have been reported in south India too (Kulshrestha and Kulshrestha, 2007; Kulshrestha et al., 2014).

As expected, rural sites in Indo-Gangetic region show higher concentration of ambient ammonia as compared to urban sites having mean $NH_3$ concentration as 40.4 $\mu g\ m^{-3}$ and 51.6 $\mu g\ m^{-3}$ at urban and rural sites, respectively (Singh and Kulshrestha, 2014). This is primarily due to the fact that rural site has more vigorous activities, such as fertilizer applications, agricultural activities, livestock, which emit $NH_3$. However, the location, seasonal meteorology, and relative position of the site from the source region can also result in significant spatial variations in ambient levels of pollutants within the same city or area. This effect is very well noticed for $NH_3$ values in the ambient air of Delhi (Singh and Kulshrestha, 2014; Sharma et al., 2012).

### *$NH_3/NH_4$ Ratios*

In order to understand the fraction of gaseous $NH_3$ converting into aerosol $NH_4$ in Delhi, percent fraction was calculated as follows (Singh and Kulshrestha, 2012)-

$$\%\mathrm{NH_4 fraction} = \frac{\mathrm{NH_{4(aerosol)}} \times 100}{\left(\mathrm{NH_{4(aerosol)}} + \mathrm{NH_{4(ammonia)}}\right)}$$

Results showed that aerosol to gaseous $NH_3$ ratios are lower in Delhi as compared to their relative abundance in temperate regions. This is probably due to the stronger source strength, elevated temperature, and low atmospheric acidity which favor $NH_3$ to remain in gas phase (Singh and Kulshrestha, 2012). It is to be noted that the pH of rain water is significantly higher in Indian region due to abundance of atmospheric dust rich in calcium carbonate (Kulshrestha et al., 2003; Kulshrestha, 2013). Unlike $NH_3$, $SO_2$ being an acidic gas is preferably scavenged by its adsorption onto the dust particle surface. Due to this reason, $SO_2$ levels are recorded very low in India especially in north India where soil-derived atmospheric dust is found abundant (Sharma and Kulshrestha, 2014;

Kulshrestha et al., 2009). Dry deposition and dust-fall deposition of different Nr species along with the influence of soil-dust are described in the next section.

## Dry Deposition of Nr

### *Nr Deposition Through Dust Fall at Delhi*

As mentioned earlier, besides wet deposition, the dry deposition of particles and gases is also equally important in India. Abundance of atmospheric dust provides a sink to the gaseous $NO_2$ and $NH_3$ depending upon the atmospheric conditions and chemistry of the particles. Hence, in addition to dry deposition of $NH_3$ and $NO_2$, dust-fall fluxes of $NH_4$ and $NO_3$ provide information about gas to coarse particle interactions. According to Tiwari and Kulshrestha (2016), out of six sites in Delhi, the $NH_4$ dust-fall fluxes varied from zero at Connaught Place site to 16.3 mg $m^{-2}$ day at SMA industrial site, while $NO_3$ dust-fall fluxes varied from 1.2 mg $m^{-2}$ day at Connaught Place site to 16.4 mg $m^{-2}$ day at SMA industrial site indicating that dust fall is a significant contributor of Nr species.

### *Dry Deposition of $NH_4$ and $NO_3$ Aerosols*

The values of dry deposition velocities of $NH_4$ and $NO_3$ are noticed to be very high, indicating that the dry deposition of N-particles is a very effective mechanism of removal in Indian region. Kumar et al. (2003) recorded 1.60 cm $s^{-1}$ and 1.09 cm $s^{-}$ as the dry deposition velocities of $NH_4$ and $NO_3$, respectively, on the natural surface (Cassia leaf). Table 26.7 gives the dry deposition fluxes of Nr species in the form of aerosols, dust fall and gases. Dry deposition of N as $NH_4$ is significantly higher that $NO_3$ mainly due to its larger contribution from dry deposition of gaseous ammonia. However, these estimates have very large uncertainties because of several reasons: (1) very less number of sites and reports about gaseous ammonia in India, (2) QA/QC concerns about the reported measurements, (3) varying methodology, and (4) nonseriousness of the authorities about Nr issue. This again emphasizes the need for extensive $NH_3$ measurements in the country. Total dry deposition budget of N as combined ($NH_4$–N + $NO_3$–N) is about

**Table 26.7** Average total deposition of $NH_4$–N and $NO_3$–N for India

| | | kg $ha^{-1}$ | | India (Tg $year^{-1}$) | | |
|---|---|---|---|---|---|---|
| **Deposition** | **Type** | **$NH_4$–N** | **$NO_3$–N** | **$NH_4$–N** | **$NO_3$–N** | **$NH_4$–N + $NO_3$–N** |
| Dry deposition | Dust fall | 0.37 | 0.84 | 0.12 | 0.27 | 0.40 |
| | Aerosols | 0.28 | 1.65 | 0.09 | 0.54 | 0.63 |
| | Gases | 4.72 | 0.28 | 1.55 | 0.09 | 1.64 |
| | Total dry deposition | **4.72** | **0.28** | **1.55** | **0.09** | **1.64** |
| Wet deposition | Total wet deposition | 2.6824 | 3.33 | 0.88 | 1.09 | 1.97 |
| Total deposition (Tg) | | | | | | **3.61** |

**Table 26.8** Total Emissions Versus Total Depositions of Nr in India

| Major Nr species | Emissions (Gg) | N Emissions (Gg) | Total Emissions (Tg) | Total Deposition (Tg) |
|---|---|---|---|---|
| $NH_3$ | 4401 | 3624 | 6.24 N | 3.61 N |
| $NO_2$ | 5020 | 2196 | | |
| $N_2O$ | 239 | 418 | | |
| | Total N emissions | 6238 | | |

1.64 Tg for India against the total wet deposition budget of 1.97 Tg per year for India, making a total deposition (wet + dry) budget of 3.61 Tg for India.

Table 26.8 gives total emissions of Nr gases and their N contribution per year for India. Emissions of gaseous $NH_3$, $NO_2$, and $N_2O$ in (Gg) are given in one column, while their emissions in terms of N are given separately in another column. Accordingly, total Indian contribution of reactive N is around 6 Tg per year. These estimates seem on higher side due to large uncertainties in input data used in the inventories by different researchers. As pointed out earlier, there is a huge difference between different estimates, e.g., $N_2O$ emissions from livestock. However, total Indian Nr emissions are about 17.9% of the total Nr emission of the United States wherein agriculture, industry, fossil fuel along with natural sources contributed 34.9 Tg N in 2002 (Rudek et al., 2016).

### *Total Emission Versus Total Deposition*

The total deposition of N (3.61 Tg) in India is almost 57% of the total emissions of Nr as N (6.24 Tg). This indicates that remaining 43% of N is distributed among various intermediate stages between emission and deposition. This needs to be attempted in order to calculate the fractions of N distributed among different processes in the N cycle such as cultivation biological nitrogen fixation, HBNF, natural biological nitrogen fixation, Nr losses as gases, runoff, leaching, and accumulation. This will help in preparing complete budget of reactive N in India. In this direction, probably, the design suggested by Rudek et al. (2016) is an appropriate approach for integrated nitrogen assessment in India.

## Gap Areas

The following gap areas need to be filled to understand the role of Nr in the Indian region.

- Good quality data availability is limited on N depositions in Indian region. The major reason is the delay in chemical analysis due to which both $NH_4$ and $NO_3$ are underestimated.
- A systematic countrywide network of reactive nitrogen monitoring is needed covering sites of characteristics. Such network should be supported for long-term research. However, some efforts have been made by the Indian Space Research Organization (ISRO) and Ministry of Earth Sciences (MoES), India in this direction.

System for Air Quality and Weather Forecasting and Research (SAFAR) which is an initiative of MoES has recently started ambient $NH_3$ measurements.

- Trajectory analysis coupled with deposition studies is missing. in order to understand local vs long-range transport of pollution.
- Studies reporting estimates of dry deposition of $NO_x$ and $NH_3$ gases are very few.
- Impact studies of high concentrations of ambient ammonia are not reported in the region.
- Very limited number of studies have been reported on gas–aerosol interactions in the region.

## Conclusion

Total Nr depositions for India are around 3.61 Tg against the total global Nr emissions of 6.24 Tg. The wet deposition (1.97 Tg) of Nr species is almost equal to the dry deposition (1.67 Tg) imparting almost equal importance to both the processes. Among $N-NH_3$, $N-NH_4$ and $N-NO_3$, gaseous ammonia contributes highest nitrogen in the atmosphere in India. Indo-Gangetic region has the highest wet deposition of $N-NH_3$.

## Recommendations

- There is a need of proper design for an integrated assessment of Nr in this region. We need to put our serious efforts in $NH_3$ measurements. Due to its major emissions from fertilizers and other agricultural activities, we need to develop a rural network for ambient $NH_3$. Since, agricultural scientists/organizations (IARI/ICAR) have very effective rural network in India for the measurements of parameters related to fertilizers, crops, and weather effects etc., their network sites can be used for the proposed integrated N assessment in India. The measurements from these rural sites of network can be complimented by the input of SAFAR which is mainly focusing upon urban areas.
- Appropriate emphasis needs to be given on dry deposition studies of Nr in order to reduce uncertainties in Nr budget of this region.
- Researchers need to be encouraged to report QA/QC in the datasets. Interlaboratory comparison exercises can be held to achieve good quality data.
- Potential scientist/groups should be identified for modeling and measurements of atmospheric Nr for which flow of government funding is necessary in order to run the countrywide network and complete the objectives in time.

## Acknowledgment

The author acknowledges the benefit of discussions with Prof. Y. P. Abrol, Prof. T. K. Adhya, and Prof. M. Sutton.

# References


Aas, W., Carou, S., Alebic-Juretic, A., Aneja, V.P., Balasubramanian, R., Berge, H., Cape, J.N., Delon, C., Denmead, O.T., Dennis, R.L., Dentener, F., Dore, A.J., Du, E., Forti, M.C., Galy-Lacaux, C., Geupel, M., Haeuber, R., Iacoban, C., Komarov, A.S., Kubin, E., Kulshrestha, U.C., Lamb, B., Liu, X., Patra, D.D., Pienaar, J.J., Pinho, P., Rao, P.S.P., Shen, J., Sutton, M.A., Theobald, M.R., Vadrevu, K.P., Vet, R., 2014. Progress in nitrogen deposition monitoring and modelling. In: Sutton, M., et al. (Eds.), Nitrogen Deposition, Critical Loads and Biodiversity. Springer, pp. 455–463.

Adhya, T.K., Pathak, H., Chhabra, A., 2007. N fertilizers and gaseous N emissions from rice based cropping system. In: Abrol, Y.P., Raghuram, N., Sachdev, M.S. (Eds.), Agricultural Nitrogen Use and Its Environmental Implications. I K International Pvt. Ltd, New Delhi, pp. 459–476.

Aneja, V.P., Schlesinger, W.H., Erisman, J.W., Behera, S.N., Sharma, M., Battye, W., 2012. Reactive nitrogen emissions from crop and livestock farming in India. Atmospheric Environment 47, 92–103.

Behera, S.N., Mukesh, S., Aneja, V.P., Balasubramanian, R., 2013. Ammonia in the atmosphere: a review on emission sources, atmospheric chemistry and deposition on terrestrial bodies. Environmental Science and Pollution Research International 20, 8092–8131.

Bouwman, A.F., Lee, D.S., Asman, W.A.H., Dentener, F.J., van der Hoek, K.W., Olivier, J.G.J., 1997. A global high-resolution emission inventory for ammonia. Global Biogeochemical Cycles 11, 561–578.

Budhavant, K.B., Rao, P.S.P., Safai, P.D., Gawhane, R.D., Raju, M.P., Mahajan, C.M., Satsangi, P.G., 2012. Atmospheric wet and dry depositions of Ions over an urban location in south-west India. Aerosol and Air Quality Research 12, 561–570.

BP Statistical Review of World Energy, 2015. http://www.bp.com/en/global/corporate/energy-economics/statistical-review-of-world-energy/2015-in-review.html.

Dentener, F., Crutzen, P., 1994. A three dimensional model of global ammonia cycle. Journal of Atmospheric Chemistry 19, 331–369.

Dentener, F., Vet, R., Dennis, R.L., Du, E., Kulshrestha, U.C., Galy-Lacaux, C., 2014. Progress in monitoring and Modelling estimates of nitrogen deposition at local, regional and global scales. In: Sutton, M., et al. (Eds.), Nitrogen Deposition, Critical Loads and Biodiversity. Springer, pp. 7–22.

Dentener, F., Drevet, J., Lamarque, J.F., Bey, I., Eickhout, B., Fiore, A.M., Hauglustaine, D., Horowitz, L. W., Krol, M., Kulshrestha, U.C., Lawrence, M., Galy-Lacaux, C., Rast, S., Shindell, D., Stevenson, D., Van Noije, T., Atherton, C., Bell, N., Bergman, D., Butler, T., Cofala, J., Collins, B., Doherty, R., Ellingsen, K., Galloway, J., Gauss, M., Montanaro, V., Müller, J.F., Pitari, G., Rodriguez, J., Sanderson, M., Solmon, F., Strahan, S., Schultz, M., Sudo, K., Szopa, S., Wild, O., 2006. Nitrogen and sulfur deposition on regional and global scales: a multimodel evaluation. Global Biogeochemical Cycles 20. http://dx.doi.org/10.1029/2005GB002672. No. 4, GB4003.

DSHB, 2011. Delhi Statistical Handbook 2009-11. Directorate of Economics and Statistics Government of National Capital Territory of Delhi.

FAOSATAT: Food and Agriculture Organization o the United Nations, 2015. Food and Agriculture Database. Food, Agriculture Organ, UN, Rome.

Galloway, J.N., Dentener, F.J., Capone, D.G., Boyer, E.W., HowarthR, W., SeitzingerS, P., AsnerG, P., Cleveland, C.C., Green, P.A., Holland, E.A., Kari, D.M., Michaels, A.F., Porter, J.H., Townsend, A.R., Vöosmarty, C.J., 2004. Nitrogen cycles: past, present, and future. Biogeochemistry 70 (2), ,153–226.

Galloway, J.N., Cowling, E.B., 2002. Reactive nitrogen and the world: 200 years of change. Ambio 31, 64–71.

Garg, A., Shukla, P.R., Kapshe, M., 2006. The sectoral trends of multigas emissions inventory of India. Atmospheric Environment 40, 4608–4620.

Garg, A., Shukla, P.R., Bhattacharya, S., Dadhwal, V.K., 2001. Sub-region (district) and sector level $SO_2$ and $NO_x$ emissions for India: assessment of inventories and mitigation flexibility. Atmospheric Environment 35 (4), 703–713.

Garg, A., Shukla, P.R., Upadhyay, J., 2012. $N_2O$ emissions of India: an assessment of temporal, regional and sector trends. Climatic change 110, 755–782.

Granat, L., Das, S.N., Thakur, R.S., Rodhe, H., 2001. Atmospheric deposition in rural areas of India-Net and Potential acidity. Water, Air and Soil Pollution 130, 469–474.

Gu, B., Ju, X., Chang, J., Ge, Y., Vitousek, M., 2015. Integrated reactive nitrogen budgets and future trends in China. PNAS Early Edition. http://dx.doi.org/10.1073/pnas/1510211112.

Hicks, K., Haeuber, R., Sutton, M., 2014. Nitrogen deposition, critical loads and biodiversity: Introduction. In: Sutton, M., et al. (Eds.), Nitrogen Deposition, Critical Loads and Biodiversity. Springer, pp. 1–4.

IEA, 2012. Key Statistics 2012, 2010 and 2006 IEA.

INCCA, 2010. Indian Network for Climate Change Assessment Report Climate Change and India: A 4x4 Assessment. Ministry of Environment and Forests. Govt of India, New Delhi.

Kulshrestha, U.C., Kulshrestha, M.J., Satyanarayana, J., Reddy, L.A.K., 2014. Atmospheric deposition of reactive nitrogen in India. In: Sutton, M., et al. (Eds.), Nitrogen Deposition, Critical Loads and Biodiversity. Springer, pp. 75–82.

Kulshrestha, U., 2013. Acid Rain. In: Jorgensen, S.E. (Ed.), Encyclopedia of Environmental Management, vol. 1. Taylor & Francis, New York, pp. 8–22.

Kulshrestha, U.C., Reddy, L.A.K., Satyanarayana, J., Kulshrestha, M.J., 2009. Real-time wet scavenging of major chemical constituents of aerosols and role of rain intensity in Indian region. Atmospheric Environment 43, 5123–5127.

Kulshrestha, U.C., Kulshrestha, M.J., 2007. Reactive N species in air, precipitation and soil and their role in nutrient cycling in south-central India. In: Proceedings, Nitrogen 2007 Conference, Bahia, Brazil, October 1–5, 2007.

Kulshrestha, M.J., Kulshrestha, U.C., Parashar, D.C., Vairamani, M., 2003. Estimation of $SO_4$ contribution by dry deposition of $SO_2$ onto the dust particles in India. Atmospheric Environment 37, 3057–3063.

Kulshrestha, U.C., Granat, L., Engardt, M., Rodhe, H., 2005. Review of precipitation monitoring studies in India—a search for regional patterns. Atmospheric Environment 39, 7403–7419. http://doi.org/10.1016/j.atmosenv.2005.08.035.

Kumar, R., Rani, A., Maharaj, A.K., Kumari, K.M., Srivastava, S.S., 2003. Direct measurement of atmospheric dry deposition to natural surfaces in a semiarid region of north central India. Journal of Geophysical Research. http://dx.doi.org/10.1029/2002JD003194.

Murugan, A.V., Dadhwal, V.K., 2007. Indian agriculture and nitrogen cycle. In: Abrol, Y.P., Raghuram, N., Sachdev, M.S. (Eds.), Agricultural Nitrogen Use and Its Environmental Implications. I K International Pvt. Ltd, New Delhi, pp. 9–28.

Ohara, T., Akimoto, H., Kurokawa, J., Horii, N., Yamaji, K., Yan, X., Hayasaka, T., 2007. An asian emission inventory of anthropogenic emission sources for the period 1980-2020. Atmospheric Chemistry and Physics 7, 4419–4444.

Parashar, D.C., Kulshrestha, U.C., Sharma, C., 1998. Anthropogenic emissions of $NO_x$, $NH_3$ and $N_2O$ in India. Nutrient Cycling in Agroecosystem 52, 255–259.

Pathak, H., Ladha, J.K., 2007. Agricultural Nitrogen Use and Its Environmental Implications. In: Abrol, Y. P., Nandula Raghuram (Eds.). IK Int. Pub. House, New Delhi, pp. 279–302.

Patra, A.K., 2014. Trends and projected estimates of GHG emissions from Indian livestock in comparisons with GHG emissions from world and developing countries. Asian-Australasian Journal of Animal Sciences 27 (4), 592–599. http://dx.doi.org/10.5713/ajas.2013.13342.

Rao, P.S.P., Safai, P.D., Momin, G.A., Ali, K., Chate, D.M., Tiwari, S., 2006. Precipitation chemistry at different locations in India. In: Proceedings of Composition of Asian Atmospheric Deposition (CAD) Workshop Held at Indian Institute of Chemical Technology Hyderabad during November 30-December 02, 2006.

Rodhe, H., Dentener, F., Schulz, M., 2002. The global distribution of acidifying wet deposition. Environmental Science & Technology 36 (20), 4382–4388. http://dx.doi.org/10.1021/es020057g.

Rudek, J., Aneja, V.P., Abrol, Y.P., 2016. Concept for Considerations in the Design of an Indian Integrated Nitrogen Assessment. A chapter in this book.

Sahu, S.K., Beig, G., Parkhi, N., 2012. Emerging pattern of anthropogenic $NO_x$ emission over Indian subcontinent during 1990s and 2000s. Atmospheric Pollution Research 3, 262–269.

Sahu, S.K., Beig, G., Parkhi, N., 2015. High Resolution emission inventory of $NO_x$ and CO for mega city Delhi, India. Aerosol and Air Quality Research 15, 1137–1144.

Saikawa, E., Prinnl, R.G., Dlugokencky, E., Ishijima, K., Dutton, G.S., Hall, B.D., Langenfelds6, R., Tohjima, Y., Machida, T., Manizza, M., Rigby, M., O'Doherty, S., Patra, P.K., Harth, C.M., Weiss, R.F., Krummel, P.B., van der Schoot, M., Frase, P.J., Steele, L.P., Aoki, S., Nakazawa, T., Elkins, J.W., 2014. Global and regional emissions estimates for $N_2O$. Atmosheric Chemistry and Physics 14, 4617–4641. http://dx.doi.org/10.5194/acp-14-4617-2014. www.atmos-chem-phys.net/14/4617/2014/.

Schlesinger, W.H., Hartley, A.E., 1992. A global budget for atmospheric $NH_3$. Biogeochemsitry 15, 191–211.

Sharma, C., Tiwari, M.K., Pathak, H., 2008. Estimates of emission and deposition of reactive nitrogenous species for India. Current Science 94, 1439–1446.

Sharma, D., Kulshrestha, U., 2014. Spatial and temporal patterns of air pollutants in rural and urban areas of India. Environmental Pollution. http://dx.doi.org/10.1016/j.envpol.2014.08.026.

Sharma, S.K., Saxena, M., Saud, T., Korpole, S., Mandal, T.K., 2012. Measurement of $NH_3$, NO, $NO_2$ and related particulates at urban sites of Indo Gangetic Plain (IGP) of India. J Scientific & Industrial Research 71, 360–362.

Singh, R., Kumar, V., Vyas, B.N., 2007. Nitrogen use efficiency improvement through USG and initiative of the Indian industry. In: Abrol, Y.P., Raghuram, N., Sachdev, M.S. (Eds.), Agricultural Nitrogen Use and Its Environmental Implications. I K International Pvt. Ltd, New Delhi, pp. 249–262.

Singh, S., Sharma, A., Kumar, B., Kulshrestha, U.C., 2016. Wet Deposition Fluxes of Atmospheric Inorganic Reactive Nitrogen at Urban and Rural Site in the Indo-Gangetic Plain (Communicated).

Singh, S., Kulshrestha, U.C., 2012. Abundance and distribution of gaseous ammonia and particulate ammonium at Delhi, India. Biogeosciences 9, 5023–5029. http://doi.org/10.5194/bg-9-5023-2012.

Singh, S., Kulshrestha, U.C., 2014. Rural versus urban gaseous inorganic reactive nitrogen in the Indo-Gangetic plains (IGP) of India. Environmental Research Letters 9(12) (125004). http://doi.org/10.1088/1748-9326/9/12/125004.

Singh, S., Kumar, B., Gupta, G.P., Kulshrestha, U.C., 2014. Signatures of increasing energy demand of past two decades as captured in rain water composition and airmass trajectory analysis at Delhi (India). Journal of Energy, Environment and Carbon Credit 4 (3), 43–61.

Tian, H., Chen, Guangsheng, Lu, Chaoqun, Xu, Xiaofeng, Ren, Wei, Zhang, Bowen, Banger, Kamaljit, Tao, Bo, Pan, Shufen, Liu, Mingliang, Zhang, Chi, Bruhwiler, Lori, Wofsy, Steven, 2015. Global methane and nitrous oxide emissions from terrestrial ecosystems due to multiple environmental changes. Ecosystem Health and Sustainability 1 (1), 4. http://dx.doi.org/10.1890/EHS14-0015.1.

Tiwari, R., Kulshrestha, U.C., 2016. Reactive nitrogen species in the atmospheric dust over the megacity of Delhi (India). In: Proceedings of the National Conference on "Environmental Pollutants: Impact Assessment and Remediation" (NCEPIAR-2016), Held at Jawaharlal Nehru University, New Delhi on 18–19 March 2016.

Velmurugan, A., Dadhwal, V.K., Abrol, Y.P., 2008. Regional nitrogen cycle: an Indian perspective. Current Science 94, 1455–1468.

Vet, R., Artz, R.S., Carou, S., Shaw, M., Ro, C.U., Aas, W., et al., 2014. A global assessment of precipitation chemistry and deposition of sulfur, nitrogen, sea salt, base cations, organic acids, acidity and pH, and phosphorus. Atmospheric Environment 93, 3–100. http://dx.doi.org/10.1016/j.atmosenv.2013.10.060.

SECTION E

# Managing Nitrogen in Relation to Key Societal Effects

# 27

# Dietary Patterns and Implications for Reactive N Flows in India

T.K. Adhya[1,2], E.J.M. Joy[3,4], S. Agrawal[5], M. Tak[4,6]

[1]SOUTH ASIA NITROGEN CENTRE, DPS MARG, NEW DELHI, INDIA; [2]KIIT UNIVERSITY, BHUBANESWAR, ODISHA, INDIA; [3]LONDON SCHOOL OF HYGIENE & TROPICAL MEDICINE, LONDON, UNITED KINGDOM; [4]LEVERHULME CENTRE FOR INTEGRATIVE RESEARCH ON AGRICULTURE AND HEALTH (LCIRAH), LONDON, UNITED KINGDOM; [5]PUBLIC HEALTH FOUNDATION OF INDIA (NEW DELHI NCR), GURGAON, INDIA; [6]SCHOOL OF AFRICAN AND ORIENTAL STUDIES, UNIVERSITY OF LONDON, LONDON, UNITED KINGDOM

**CHAPTER OUTLINE**



## Introduction

The population of India is ~1.27 billion (17% of the global population) and the country is the second largest producer of wheat and rice globally (FAOSTAT, 2016). India's food system thus has massive, globally significant implications for human health and environment. Managing the food system to achieve improved human health and reduced impacts on the environment presents a huge and complex challenge to policy makers, and this is likely to become even more challenging with a growing population and changing food consumption patterns.

There is a growing body of research seeking to quantify the impacts of diets and food systems on the environment, particularly emissions of greenhouse gases (GHGs), land use, and use of fresh water (Jones et al., 2016). The food system also influences flows of reactive nitrogen (N), i.e., biologically active forms of N including nitrogen oxides ($NO_x$), nitrous oxide ($N_2O$), nitrate ($NO_3^-$), nitrite ($NO_2^-$), ammonia ($NH_3$), and ammonium

The Indian Nitrogen Assessment. http://dx.doi.org/10.1016/B978-0-12-811836-8.00027-6

($NH_4^+$); through the processes of industrial and biological fixation of $N_2$ to $NH_3$, and application of $NO_3^-$ or $NH_4^+$ to crops via organic or mineral fertilizers. These processes contribute to major environmental issues including eutrophication via N-leaching and anthropogenic global warming via $NH_3$ volatilization. However, N inputs are central to current food production systems and food security. In addition, N is an essential component of amino and nucleic acids and is required for tissue protein synthesis and production of several compounds involved in a variety of functions, including hormones, immune mediators, neurotransmitters, antioxidant defenses, etc. (Tessari, 2006). Thus, consumption of adequate N is important for normal body functions. This chapter presents an overview of the changing Indian food system and discusses implications for reactive N flows, with consideration for health and the environment.

## An Overview of the Indian Food System

A food system comprises all the actors and actions involved in the production, processing, transport, retail, and consumption of the food. In terms of the environmental issues related to reactive N flows, the most significant stage is the agricultural production. It is during crop and livestock production that significant quantities of N are leached or volatilized, contributing to eutrophication and global warming. However, assessing reactive N flows through a consumption perspective (i.e., considering *diets* rather than *agricultural production*) is useful for two main reasons: first, diets are a driver of agricultural production and as diets change this is likely to influence crop and livestock areas and modes of production; second, dietary patterns determine nutritional N status. The limitations of this approach are that: (1) at a national scale, trends in dietary consumption data are only a good proxy of trends in production, so long as imports/exports are relatively insignificant; (2) dietary data gives no information on production of nonfood crops, e.g., cotton.

Agriculture is the largest sector of the rural economy in India and contributes to economic growth and employment. Though its contribution to the Gross Domestic Product has declined over the last five decades from 42% to 18%, ~58% of rural households practice agriculture, of whom ~68% see it as their main source of income (NSSO, 2014). There are a number of distinct food production systems in India. For example, areas receiving tropical monsoon rains typically grow rice in flooded paddy systems while semiarid areas grow wheat. The spatial variation in production systems overlaps with food availability and dietary choices. An important distinct production system is found in the largely flat and fertile Indo-Gangetic plains in north-west India (e.g., States of Punjab, Haryana, and western U.P.). Partly due to extensive canal and groundwater irrigation infrastructure, many farmers practice a rice–wheat double cropping system here. Approximately 63 Mha representing 36% of cropping land are now irrigated in India, with 97% of crop land irrigated in Punjab (Department of Agriculture, 2013). Punjab and Haryana are termed as the "bread basket" of India and these net-producing areas are interlinked with the dietary patterns of a wider geographical area including Delhi-NCR. Millet and sorghum are widely produced and consumed in arid regions, including large areas of Central and Western India (e.g., states of Maharashtra and Rajasthan).

The population of India has distinct and diverse dietary patterns that are strongly associated with region, socioeconomic status, religion, and other cultural factors (Green et al., 2016). Populations residing in northern states of India tend to consume wheat-based diets whereas individuals from southern states tend to consume rice-based diets. This is consistent with the broad areas of crop production in India. A significant proportion of the population consumes rice-based diets characterized by low energy consumption and low dietary diversity, particularly rural dwellers with little or no formal education. Meat consumption is low across the population, although is greater among non-Hindus. Meat consumption in India is closely associated with religion and culture, and may not rise with increasing incomes and urbanization to the extent that it has in China.

Socioeconomic inequalities and cultural factors are reflected in diets including dietary diversity, consumption of animal-source foods such as dairy and eggs, and consumption of refined cooking oils and sugar (Fig. 27.1). Meat consumption is typically low in India, partly due to the preference for (lacto-ovo-)vegetarian diets along religious lines (Fig. 27.2). Thus, the dietary patterns of meat-eating communities tend to be distinctive. Meat consumption is greatest in the north-east of India (Seven Sister states and West Bengal), the far north (Jammu and Kashmir) and the south (Kerala, Tamil Nadu, and Andhra Pradesh) where fish consumption is also widespread (Agrawal et al., 2014a,b). Milk and other dairy products, derived from cows and buffalo, are widely consumed across India and India is the largest producer of milk globally (FAOSTAT, 2016). Dairy consumption per capita is greatest in the Punjab region and is associated with socioeconomic status with greater consumption among wealthier and urban communities. There are stark differences in food consumption across various income levels in India. Typically, households consume a greater quantity of food as incomes rise. This is particularly the case for food items with a high price elasticity of demand, including animal-source foods, fruits, and vegetables (Pingali and Khwaja, 2004). However, it is also true for cereal consumption among the poorer income groups, although consumption may decline among the wealthiest groups as diets diversify. This observation may be expected, as poor households first increase calorie consumption before diversifying away from cereals to more expensive calories, such as milk and pulses.

India has experienced rapid changes in its agricultural production in recent decades, tied strongly to the changes in consumption. The Green Revolution was an incredibly important period. A primary political objective of the Green Revolution was to achieve domestic self-sufficiency in the production of wheat and rice. This was achieved through adoption of high-yielding hybrid varieties in combination with increased use of mineral fertilizers and development of irrigation infrastructure. More recently, the Government of

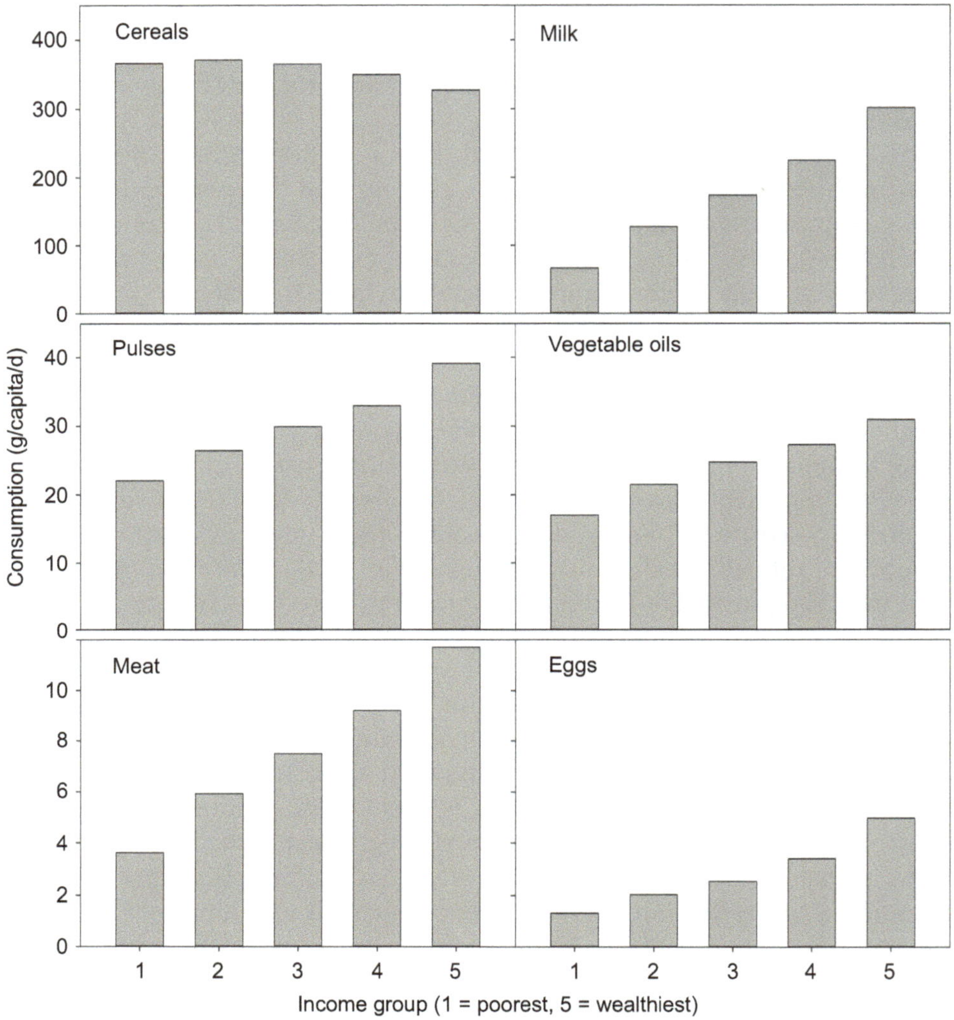


**FIGURE 27.1** Mean consumption of selected food groups by income group. *Data are derived from The Household Consumer Expenditure Survey, National Sample Survey Office of India, 60th Round (July 2011–12). (Source: author own).*

India spear-headed the "Operation Flood" (http://www.nddb.org/about/genesis/flood) project with the objective of achieving self-sufficiency in milk production.

## Nitrogen in Food Production in India

In recent times, concerted efforts by planners, agricultural scientists, and farmers have led to food self-sufficiency for the country and this represents a great achievement. Secure food production relies on the application of plant macronutrients including N, phosphorus (P), and potassium (K) and India relies heavily on the use of mineral

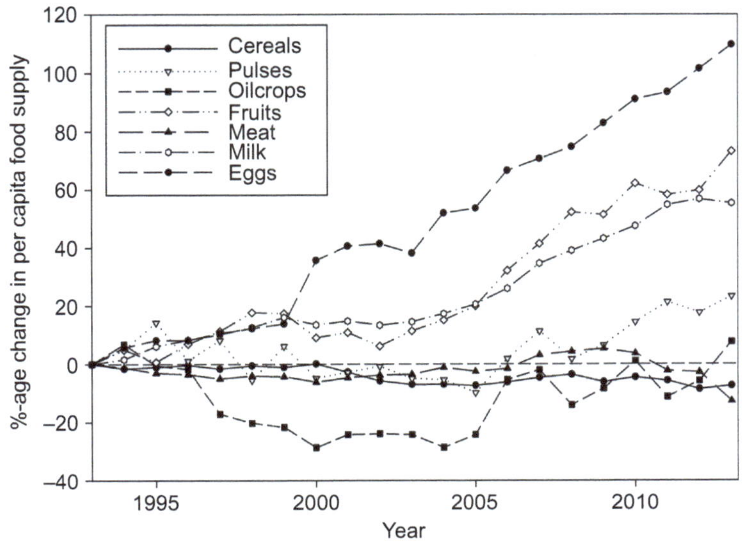


**FIGURE 27.2** Change in food supply *per capita* in India between 1993 and 2012, relative to 1993 as a baseline. Cereal and meat supply declined marginally over the period. The most striking change is the rapid increase in supply of eggs, fruit and milk. *Data are derived from FAO Food Balance Sheets FAOSTAT, 2016. (Source: author own).*

fertilizers. Thus, annual fertilizer N consumption in India was just 0.2 Mt during 1960–61, which increased to 17.4 Mt by 2015–16 and is expected to rise up to 23.5 Mt by 2030 (Tewatia and Chanda, 2016). The high absolute N-use combined with a cropping system that achieves an average N-use efficiency of <40% means that a large amount of reactive N is leached or volatilized into the environment (Sutton et al., 2011).

Overall N-use efficiency is quite low in India with an average of >80% of N lost to the environment. This compares to ~36% in Europe (Erisman, 2011). The N flow is mostly linked to the production and use of mineral fertilizers, although leaching of N from manure may also be significant in some systems (Antikainen, 2007; Kahiluoto et al., 2014). The proportion of N that is leached depends largely on agronomy, soil conditions, and factors affecting crop yield. However, the environmental emissions of N associated with the production of different food commodities are highly variable (Shibata et al., 2016; Vetter et al., 2017), and the human diet is a crucial driver of the N cycle. In particular, feeding crops to livestock provides further opportunity for N losses, and a vegetarian diet is considered to retain 14% of N originally applied to the crop whereas for a nonvegetarian diet it is merely 4% (Galloway and Cowling, 2002).

The N-footprint provides a metric of the quantity of reactive N released to the environment as a result of the production and consumption of a food item (Leach et al., 2012), whether produced domestically or imported (Oita et al., 2016). A food N-footprint has two parts: the food consumption N-footprint (i.e., the N actually embedded in the consumed food product, which is ultimately returned to the environment as human waste), and the food production N-footprint (i.e., the N released to the environment

**Table 27.1** Per Capita N Footprints of Various Countries (kg N $capita^{-1}$ $year^{-1}$)

| Category | Japan (Shibata et al., 2014) | Germany (Leach et al., 2012) | USA (Leach et al., 2012) |
|---|---|---|---|
| Food | 26 | 19 | 28 |
| Housing | 0.8 | 1.6 | 3 |
| Transport | 0.7 | 1.8 | 6 |
| Goods and services | 1.0 | 0.7 | 2.5 |
| Total | 28 | 24 | 39 |

Shibata, H., Galloway, J.N., Leach, A.M., Cattaneo, L.R., Noll, L.C., Erisman, J.W., Gu, B., Liang, X., Hayashi, K., Ma, L., Dalgaard, T., Graversgaard, M., Chen, D., Nansai, K., Shindo, J., Matsubae, K., Oita, A., Su, M.-C., Mishima, S.I., Bleeker, A., 2016. Nitrogen footprints: regional realities and options to reduce nitrogen loss to the environment. Ambio http://dx.doi.org/10.1007/s13280-016-0815-4.

**Table 27.2** Protein Conversion Efficiency and Virtual Nitrogen Factors (VNF) in Various Regions

| Food Category | Protein Conversion Efficiency (%) (Smil, 2002) | USA (Shibata et al., 2016) | Europe (Shibata et al., 2016) | Japan (Shibata et al., 2016) |
|---|---|---|---|---|
| Meat (pork) | 13 | 4.4 | 4.4 | 6.7 |
| Meat (beef) | 5 | 7.9 | 7.9 | 12.4 |
| Meat (mutton) | | 5.2 | | |
| Meat (chicken) | 25 | 3.2 | 3.2 | 6.0 |
| Egg | 30 | | | |
| Fish | 30 | 4.1 | 2.9 | 2.9 |
| Milk | 40 | 4.3 | 3.9 | 2.7 |
| Vegetables | | 9.6 | 8.2 | 5.5 |
| Starchy roots | | 1.5 | 1.1 | 4.9 |
| Legumes | | 0.5 | 0.5 | 1.3 |
| Fruits | | 9.6 | 8.2 | 5.5 |
| Cereals | | 1.4 | 1.3 | 1.5 |

during food production, such as from fertilizer leaching and processing wastes). At a regional scale, N-footprints have been reported for Asia–Pacific, North America, and Europe (Table 27.1), and these regions exhibit the dominance of the food N-footprint for the overall *per capita* N-footprint of the region (Leach et al., 2012; Galloway et al., 2014; Shibata et al., 2014) (Table 27.2). In all countries, virtual N-footprints for crop products are lower than for animal products (Leach et al., 2012; Shibata et al., 2014) due to the additional recycling into animal protein and $N_2O$ emissions from manure. In India, reported emissions of $N_2O$ from manure were 1.4 and 0.9 kg *per* kg for poultry and mutton meat and ~0.1 kg $L^{-1}$ for cattle and buffalo milk (Vetter et al., 2017). In comparison, emissions of $N_2O$ from soil tillage, fertilizer, and crop residue management combined were 0.07 and 0.16 kg *per* kg of paddy rice and wheat, respectively. Feed conversion ratios for poultry meat, mutton, and cattle milk are typically ~2, 7, and 1.5, respectively. Thus, if emissions associated with crop production for feed are also attributed to livestock, plus leaching of N from manure, then N-footprints of livestock products are likely to be ~2 to 10-fold greater for livestock products than cereal crops. The reported N-footprint in the United States of America (USA) is 5.2 kg N $capita^{-1}$ $year^{-1}$ (Leach et al., 2012), and is likely to be much less for India where average dietary protein supply of animal origin during 2011–13 was ~12 g $capita^{-1}$ $day^{-1}$ compared to ~70 g $capita^{-1}$ $day^{-1}$ in the USA (FAOSTAT, 2016).

## The Changing Diets and Food System of India and Implications for Reactive Nitrogen Flows

The food system in India is rapidly changing with the effects of urbanization, globalization, population growth, technological and economic development, and the greater integration of markets and cultures. The process has been termed the nutrition transition. But what does India's nutrition transition entail? A major feature was the increased diversity of diets with reduced consumption of traditional staples such as coarse grains, and increased consumption of animal products, fats and oils, and sugar (Kumar et al., 2007). This shift in diets has been linked to the growing burden of obesity and noncommunicable diseases (Pingali and Khwaja, 2004). However, a significant proportion of the population remains undernourished and India faces a double burden of diseases related to under- and overnutrition (Agrawal et al., 2014a,b).

Indian diets rely heavily on cereals, milk, and pulses, while meat and eggs constitute only a small portion. Indeed, traditional Indian diets combine cereals and pulses, e.g., *idli–sambhar* and *rajma–chawal*. However, diets are changing. Dietary supply of cereals has declined by ~10% from 440 to 400 g $capita^{-1}$ $day^{-1}$ between 1993 and 2012. During the same time period, supply of milk increased from 149 to 234 g $capita^{-1}$ $day^{-1}$ (FAOSTAT, 2016) and supply of eggs has more than doubled. Supply of pulses also marginally increased during this time period, while supply of cereals declined (Figs. 27.2 and 27.3). An increasing reliance on animal-source foods is likely to increase the N-footprint of Indian diets, unless improved N-use efficiency is achieved through production-side changes. In addition to increasing consumption of animal-source foods, dietary supply of sugar has increased by ~40% since 1993 (FAOSTAT, 2013). Sugar cane received 125 kg N $ha^{-1}$ in 2003/04, a greater rate of N application than any other major crop. (e.g., 82 and 100 kg N $ha^{-1}$ for paddy rice and wheat, respectively) (FAO, 2005). Thus, the increase in contribution of sugar to Indian diets is likely to further increase the N-footprints with little or no nutritional benefit.

Dietary protein deficiency is common, particularly among vegetarians. A recent survey conducted by the Indian Market Research Bureau (IMRB) across seven major cities of India reported that 91% of the vegetarians' diets were deficient in protein compared to 85% of nonvegetarian diets. It was also found that in northern India, 98% of the respondents were not aware of the recommended protein intakes for an average adult, although other parts of India fared better (Mahajan, 2015). In contrast,

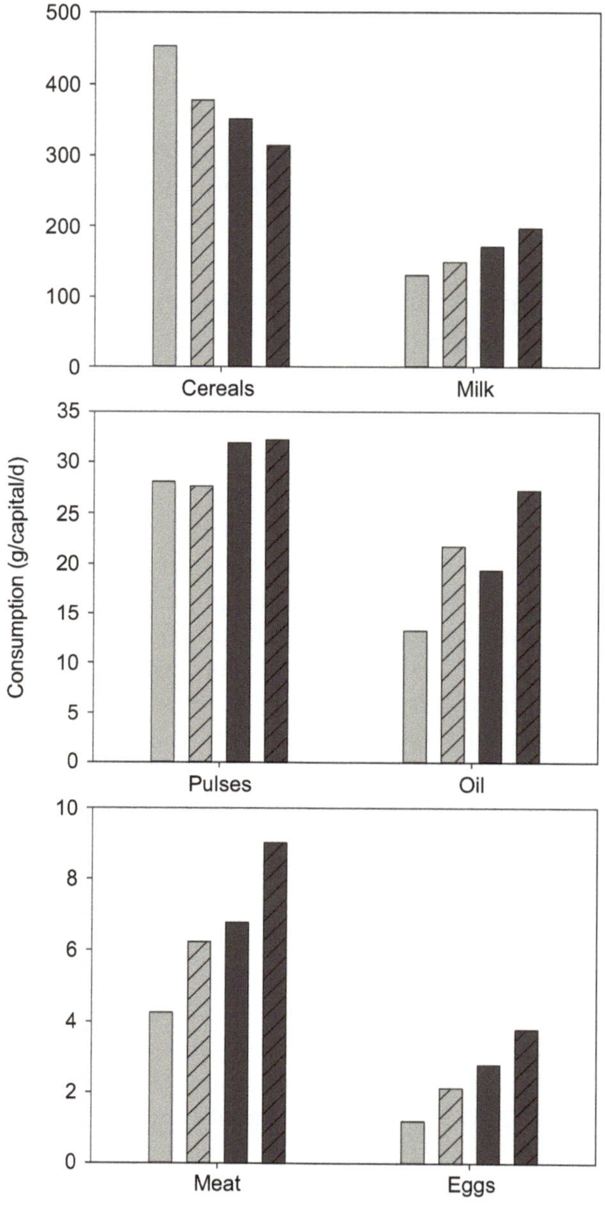


**FIGURE 27.3** Food consumption in rural (*light gray*) and urban (*dark gray*) India in 1993–94 (open) and 2011–12 (hatched). *Data are derived from The Household Consumer Expenditure Survey, National Sample Survey Office of India, 50th Round (July 1993–June 1994) and 60th Round (July 2011–12). (Source: author own).*

consumption of protein is typically greater than requirements in many regions globally (Ranganathan et al., 2016). Global average protein consumption was ~68 g $capita^{-1}$ $day^{-1}$ compared to the average daily adult male requirement of 56 g $capita^{-1}$ $day^{-1}$. The *per capita* food consumption of animal-based foods may be peaking in most developed countries but is also projected to rise in developing countries, especially in emerging economies and in urban areas. Health benefits may be realized in India through reduced protein deficiency and improved amino acid profile of diets, and greater consumption of milk, curd, or eggs at breakfast may be culturally appropriate options to increase dietary protein supplies. Other dietary diversification strategies include greater consumption of millets, pulses, whole grains, green leafy vegetables, cheese, and *paneer*. Greater protein consumption will increase the embedded N in diets.

As we look ahead, income growth, urbanization, and other long-term social and economic transformations are likely to influence the composition of the food basket (Mohanty, 2013). This provides an ongoing production challenge and future supplies may increasingly rely on imports. Certainly, with greater market integration there is evidence to suggest that the Indian agricultural system is becoming more export-oriented, e.g., Basmati rice and sugar cane production are promoted as export-oriented commodity crops. India is the leading exporter of Basmati rice to the global market and has exported 4.0 Mt of Basmati Rice to the world during 2015–16 (http://www.apeda.gov.in/apedawebsite/SubHead_Products/Basmati_Rice.htm). According to US Department of Agriculture (USDA) figures, India also exported 2.9 Mt of sugar in 2015–16, accounting for 5.3% of global exports next only to Brazil (http://www.fas.usda.gov/data/search?search_api_views_fulltext=India). Beef is also an important commodity to consider; although not widely consumed in India, beef production is effectively a by-product of dairy production and is likely to rise alongside the dairy sector. Globally, international trade of agricultural products including food and feed has increased by 10-fold during the past six decades, and it is expected to continue to grow in the future (Schmitz et al., 2012). The implications for N global fluxes are more complex than merely the N flows associated with import or export of food and feed. Impacts on the environment or human health of N leached into water bodies are located where the goods are produced and emissions of $NH_3$ and $N_2O$ contribute to the carbon budgets of producer countries. Thus, as international trade increases, the consumption location is increasingly far removed from the environmental implications of production (Billen et al., 2010; Lassaletta et al., 2012, 2013; de Ruiter et al., 2016; Oita et al., 2016).

## Greenhouse Gas Emission and Food Consumption

With increasing purchasing power in recent decades in developing countries, animal protein consumption has gone up. This trend will add to the emissions of GHGs including $N_2O$ (Pathak, 2015). Pathak et al. (2010) evaluated GHG emissions during the life cycle of 24 Indian food items. The GHG emissions during the life cycle of cooked rice was 2.8-fold greater than of chapatti, a product of wheat flour. Mutton emitted 11.9-fold more GHGs than milk, 12.1-fold more than fish, 12.9-fold more than rice and 36.5-fold more than chapatti. As Indians mostly consume fresh foods produced locally, 87% of emissions came from food production followed by preparation (10%), processing (2%), and transportation (1%). For a balanced diet (vegetarian), an adult Indian man consumed 1165 g food and emitted

724 g $CO_2$ eq. day$^{-1}$. Pathak et al. (2010) evaluated that a nonvegetarian meal with mutton emitted 1.8-fold more $CO_2$ eq. than a vegetarian meal, 1.5-fold that of a nonvegetarian meal with chicken and an ovo-vegetarian meal, and 1.4-fold that of a lacto-vegetarian meal.

Changes in food-habits thus could offer a possibility for GHG mitigation. At the household level, options to reduce GHG emissions could consist of substitution from animal food products to crop products. Shifting from ruminants to poultry-based products would also be useful to mitigate GHG emissions. Options to minimize the GHG emissions of food consumption include consumption of locally produced foods; less mutton consumption; and substitution of meat and milk by other vegetable protein (Pathak, 2015). The Intergovernmental Panel on Climate Change (IPCC) in its Fifth Assessment Report also pointed out that lifestyle changes and behavior patterns can contribute to climate change mitigation across all sectors.

## Conclusion

Dietary patterns are an important consideration for nutrition and the environment. This chapter provided an overview of food systems and diets in India and discussed the potential implications of changing diets for reactive N flows. Dietary patterns are diverse in India and are typically characterized by low consumption of animal-source foods, in particular red meat. Thus, Indian diets are likely to have low N-footprints relative to global diets. However, a large proportion of diets in India are likely to be protein deficient.

Diets are changing rapidly in India, with particularly pronounced increases in egg and dairy consumption. Dietary protein supplies and amino acid profiles are likely to be improving for many, although there remains a wide disparity between income groups and urban/rural households. Unfortunately, the improvements in nutritional value derived from animal-source foods delivers a trade-off for many environmental indicators, including the N-footprint of diets, and this may have implications for eutrophication, groundwater quality, and global warming potential. In terms of mitigation strategies, production-side improvements in N-use efficiency are paramount, while policies to discourage the overconsumption of animal-source foods may deliver combined health and environmental benefits for particular population groups (e.g., wealthy, urban consumers). Another opportunity is through reversing the rapid increase in sugar consumption. Sugar cane is an N-hungry crop and excessive sugar consumption contributes to obesity and diabetes.

## References

Agrawal, S., Millett, C.J., Dhillon, P.K., Subramanian, S.V., Ebrahim, S., 2014a. Type of vegetarian diet, obesity and diabetes in adult Indian population. Nutrition Journal 13, 89. http://dx.doi.org/10.1186/1475-2891-13-89.

Agrawal, S., Millett, C.J., Subramanian, S.V., Ebrahim, S., 2014b. Frequency of fish intake and diabetes among adult Indians. Journal of the American College of Nutrition 33, 215–230. http://dx.doi.org/10.1080/07315724.2013.867420.

Antikainen, R., 2007. Substance Flow Analysis in Finland – Four Case Studies on N and P Flows. Finnish Environment Institute, Helsinki, Finland.

Billen, G., Beusen, A., Bouwman, L., Garnier, J., 2010. Anthropogenic nitrogen autotrophy and heterotrophy of the world's watersheds: past, present, and future trends. Global Biogeochemical Cycle 24, GB0A11.

de Ruiter, H., Macdiarmid, J.I., Matthews, R.B., Kastner, T., Smith, P., 2016. Global cropland and greenhouse gas impacts of UK food supply are increasingly located overseas. Journal of the Royal Society Interface 13 (114), 20151001.

Department of Agriculture, 2013. The Ninth Agricultural Census of India. Agricultural Census Division, Department of Agriculture, Government of India. Available online: http://agcensus.nic.in/.

Erisman, J.W., 2011. The European nitrogen problem in a global perspective. In: Sutton, M.A., et al. (Eds.), The European Nitrogen Assessment: Sources, Effects and Policy Perspectives. Cambridge University Press, New York, pp. 9–31.

FAO [Food and Agriculture Organization of the United Nations], 2005. Fertilizer Use by Crop in India. FAO, Rome. Available online: http://www.fao.org/docrep/009/a0257e/A0257E00.htm#TOC.

FAOSTAT [Food and Agriculture Organization of the United Nations], 2016. FAO Statistics. Available online: http://faostat3.fao.org/home/E.

FAOSTAT [Food and Agriculture Organization of the United Nations], 2013. Time-Series and Cross Sectional Data Relating to Food and Agriculture. http://faostat.fao.org/site/291/default.aspx.

Galloway, J.N., Winiwarter, W., Leip, A., Leach, A.M., Bleeker, A., Erisman, J.W., 2014. Nitrogen footprints: past, present and future. Environmental Research Letters 9, 115003. http://dx.doi.org/10.1088/1748-9326/9/11/115003.

Galloway, J.N., Cowling, E.B., 2002. Reactive nitrogen and the world: 200 years of change. Ambio 31 (2), 64–71.

Green, R., Milner, J., Joy, E.J.M., Agrawal, S., Dangour, A.D., 2016. Dietary patterns in India: A systematic review. British Journal of Nutrition 116, 142–148.

Jones, A.D., Hoey, L., Blesh, J., Miller, L., Green, A., Shapiro, L.F., 2016. A systematic review of the measurement of sustainable diets. Advances in Nutrition 7, 641–664.

Kahiluoto, H., Kuisma, M., Kuokkanen, A., Mikkilä, M., Linnanen, L., 2014. Taking planetary nutrient boundaries seriously: can we feed the people? Global Food Security 3, 16–21.

Kumar, P., Mruthyunjaya, Dey, M.M., 2007. Long-term changes in Indian food basket and nutrition. Economic and Political Weekly 42, 3567–3572.

Lassaletta, L., Romero, E., Billen, G., Garnier, J., Garcia-Gomez, H., Rovira, J.V., 2012. Spatialized N budgets in a large agricultural Mediterranean watershed: high loading and low transfer. Biogeosciences 9, 57–70.

Lassaletta, L., Billen, G., Romero, E., Garnier, J., Aguilera, E., 2013. How changes in diet and trade patterns have shaped the N cycle at the national scale: Spain (1961–2009). Regional Environmental Change 14, 785. http://dx.doi.org/10.1007/s10113-013-0536-1.

Leach, A.M., Galloway, J.N., Bleeker, A., Erisman, J.W., Kohn, R., Kitzes, J., 2012. A nitrogen footprint model to help consumers understand their role in nitrogen losses to the environment. Environmental Development 1, 40–66.

Mahajan, M., 2015. Protein consumption in diet of adult Indians: a general consumer survey (PRODIGY). Indian Medical Gazette,. April 2015, 149–150.

Mohanty, S., 2013. Trends in global rice consumption. Rice Today 12, 44–45.

NSSO [National Sample Survey Office], 2014. Key Indicators of Situation of Agricultural Households in India. NSS 70th Round (January–December 2013). NSSO, Government of India. Available online: http://mospi.nic.in/mospi_new/upload/KI_70_33_19dec14.pdf.

Oita, A., Malik, A., Kanemoto, K., Geschke, A., Nishijima, S., Lenzen, M., 2016. Substantial nitrogen pollution embedded in international trade. Nature Geoscience 9, 111–115.

Pathak, H., 2015. Greenhouse gas emission from Indian agriculture: trends, drivers and mitigation strategies. Proceedings of the Indian National Science Academy 81, 1133–1149.

Pathak, H., Jain, N., Bhatia, A., Patel, J., Aggarwal, P.K., 2010. Carbon footprints of Indian food items. Agriculture Ecosystems and Environment 139, 66–73.

Pingali, P., Khwaja, Y., 2004. Globalisation of Indian Diets and the Transformation of Food Supply Systems. ESA Working Paper No. 04–05. Available online: http://www.eldis.org/vfile/upload/1/document/0708/doc15357.pdf.

Ranganathan, J., Vennard, D., Waite, R., Dumas, P., Lipinski, B., Serachinger, T., GlobAgri_WRR Model Authors, 2016. Shifting Diets for a Sustainable Food Future. Working Paper, Installment 11 of Creating a Sustainable Food Future. World Resources Institute, Washington, DC. Accessible at: http://www.worldresourcesreport.org.

Schmitz, C., Biewald, A., Lotze-Campen, H., Popp, A., Dietrich, J.P., Bodirsky, B., Krause, M., Weindl, I., 2012. Trading more food: implications for land use, greenhouse gas emissions, and the food system. Global Environmental Change 22, 189–209.

Shibata, H., Cattaneo, L.R., Leach, A.M., Galloway, J.N., 2014. First approach to the Japanese nitrogen footprint model to predict the loss of nitrogen to the environment. Environmental Research Letters 9, 115013. http://dx.doi.org/10.1088/1748-9326/9/11/115013.

Shibata, H., Galloway, J.N., Leach, A.M., Cattaneo, L.R., Noll, L.C., Erisman, J.W., Gu, B., Liang, X., Hayashi, K., Ma, L., Dalgaard, T., Graversgaard, M., Chen, D., Nansai, K., Shindo, J., Matsubae, K., Oita, A., Su, M.-C., Mishima, S.I., Bleeker, A., 2016. Nitrogen footprints: regional realities and options to reduce nitrogen loss to the environment. Ambio. http://dx.doi.org/10.1007/s13280-016-0815-4.

Smil, V., 2002. Nitrogen and food production: proteins for human diets. Ambio 31, 126–131.

Sutton, M.A., Oenema, O., Erisman, J.E., Leipt, A., van Grinsven, H., Winiwarter, W., 2011. Too much of a good thing. Nature 472, 159–161. http://dx.doi.org/10.1038/472159a.

Tessari, P., 2006. Nitrogen balance and protein requirements: definition and measurements. In: Mantovani, G., Anker, S.D., Inui, A., Morley, J.E., Fanelli, F.R., Scevola, D., Schuster, M.W., Yeh, S.-S. (Eds.), Cachexia and Wasting: A Modern Approach. Springer, Milan, pp. 73–79 (Chapter 3.2).

Tewatia, R.K., Chanda, T.K., 2016. Trends in Fertilizer Nitrogen Production and Consumption in India. In (Chapter 3), this volume, pp.

Vetter, S.H., Sapkota, T.B., Hillier, J., Stirling, C., Macdiarmid, J.I., Aleksandrowicz, L., Green, R., Joy, E.J.M., Dangour, A.D., Smith, P., 2017. Greenhouse gas emissions from food production to supply Indian diets: Implications for climate change mitigation. Agriculture, Ecosystems and Environment 237, 234–241.

28

# Pathophysiology of Nitrate Toxicity in Humans in View of the Changing Trends of the Global Nitrogen Cycle With Special Reference to India

S.K. Gupta[1], A.B. Gupta[2], R. Gupta[3]

[1]*CONSULTANT PEDIATRICIAN & NEONATOLOGIST AND SCIENTIST OF ENVIRONMENTAL MEDICINE, JAIPUR, RAJASTHAN, INDIA;* [2]*MALAVIYA NATIONAL INSTITUTE OF TECHNOLOGY, JAIPUR, RAJASTHAN, INDIA;* [3]*GOVERNMENT POST GRADUATE COLLEGE, DAUSA, INDIA*

**CHAPTER OUTLINE**



## Introduction

Nitrate levels in surface and groundwater are generally of few milligrams per liter; however, higher concentrations of nitrates have been observed due to sewage

**The Indian Nitrogen Assessment. http://dx.doi.org/10.1016/B978-0-12-811836-8.00028-8**

percolation or use of excessive chemical fertilizers. In many groundwater samples higher levels are found due to water percolating through nitrate-rich rocks and also due to an excessive use of chemical fertilizers. World Health Organization's (WHO) *Guidelines for Drinking-water Quality* (GDWQ) report, 2011 have laid down permissible limit for nitrate-nitrogen to 10 mg $L^{-1}$ or for nitrate to 50 mg $L^{-1}$ (WHO, 2011) for human consumption whereas IS-10500 prescribes maximum permissible limits of nitrate in drinking water as 45 mg of $NO_3$ per liter (IS 10500, 1995). The guideline (WHO, 2011) for nitrite is 3 mg $L^{-1}$ as nitrite and ranges from 0.4 or more.

The nitrate contamination of water is endemic, internationally as well as nationally, but systematic compilation of data at national level is almost non existent. For India, the data available either pertain to small areas or to scattered zones, though some information is available in case of Rajasthan state (Kumar et al., 2002; Malik, 2000; PHED Habitat Survey, 1991–93).

## Epidemiology

Nitrate is a widespread contaminant of ground and surface waters worldwide (Prakasa Rao and Puttanna, 2006). Water quality studies carried out by NEERI (National Environmental Engineering Research Institute, 1991) identified the states with high nitrate concentration in groundwater in India. Data available on concentrations of nitrate in groundwater samples from Punjab, Haryana, UP, Delhi, Orissa, Tamil Nadu, Bihar, AP, MP, Maharashtra, and Karnataka indicate that the nitrate content in groundwater is much more than permissible limits. In many parts of Punjab and Haryana nitrate level in groundwater is continuously increasing beyond the critical standard. Nitrate levels in groundwater over vast agricultural areas can be correlated with intensive irrigated agriculture, use of nitrogenous fertilizers, and groundwater exploitation. Nitrate pollution of groundwater by fertilizer use in India has also been reviewed (Singh et al., 2007; Gupta et al., 2008, 2011; 2013).

## Sources of Reactive-N

Nitrate is ingested by human beings both from exogenous as well as endogenous sources. Nitrates found generally in groundwater and nitrate-containing compounds in the soil are generally soluble and readily percolate through groundwater. Nitrogen is a major essential element for plant growth (Galloway et al., 2014) and farmers usually apply large doses of fertilizers. Nitrogen fertilizers can cause an increase in the concentration of nitrates in fruits, vegetables, and food. About 70% of ingested nitrate comes from vegetables (cauliflower, spinach, collard greens, broccoli, and root vegetables), 21% from water, and 6% from meat and meat products. Nitrate content in vegetables reaching the markets of India has been reported to be as high as 4451 mg $kg^{-1}$ fresh weight in chenopodium and 4293 mg $kg^{-1}$ fresh weight in spinach (Anjana et al., 2007). The nitrate ion ($NO_3^-$) is the most stable form of nitrogen in oxygenated environments, thus all nitrogen-containing molecules can act as sources of nitrates.

Industrial effluents and polluted waters disposed of to surface water sources (rivers, lakes, canals, etc.), stored in lagoons or ponds for percolation, or spread or sprayed onto the ground surface percolates either directly from the surface water bodies or from agricultural fields (when such water is used for irrigation), and pollutes the ground water. Under certain circumstances forests also contribute to nitrate pollution.

Endogenous nitrates are also produced in the body. A major pathway for endogenous nitrate production is conversion of arginine by macrophages to nitric oxide and citrulline. This is followed by oxidation of the nitric oxide to nitrous anhydride, and then reaction of nitrous anhydride with water to yield nitrite. Gastrointestinal infections and non specific diarrhea increases endogenous (non bacterial) nitrate synthesis, probably induced by activation of the mammalian reticuloendothelial system. The mean dietary intake ranges from 43 to 131 mg of nitrate per day and 1.2–3 mg of nitrite per day. Vegetarians usually have a two- to four fold higher intake of nitrates than nonvegetarians (WHO, 1985, 1996, 2011).

## Metabolism of Ingested Nitrate in Human Body at Cellular Level

About 20% of ingested nitrate is reduced to nitrite by nitrate-reducing microflora present in saliva at the base of the tongue (Walker, 1995). Ingested nitrate is also reduced to nitrite by nitrate-reducing micro flora in stomach (in favorable conditions viz. $pH \geq 4$) and upper part of intestine, and may be in other parts of the human gastrointestinal tract.

Ingested inorganic or organic nitrates cause increased oxidation of hemoglobin to methemoglobin and increased production of nitric oxide (Craven PA and De Rubartis FR, 1978) (Fig. 28.1) by nonenzymatic process (Smith et al., 1997; Lowestein et al., 1994). The oxidation of hemoglobin to methemoglobin results in the formation of the superoxide radical, by the transfer of single electron. The enzyme superoxide dismutase, present in the erythrocytes, catalyzes the conversion of superoxide radical ($O^{\cdot -}$) to $H_2O_2$ and $O_2$. The $H_2O_2$, so produced, is decomposed by glutathione peroxidase or catalase as both are present in erythrocytes (Winterbourn et al., 1976). Once the rate of oxidation of hemoglobin increases sufficiently in erythrocytes and overwhelms the protective and reductive capacities (e.g., cytochrome $b_5$ reductase system) of the cells, there is increased production of reactive free radicals of nitric oxide ($NO^{\cdot}$) and oxygen (Winterbourn et al., 1976).

Thus, increased consumption of nitrate will leads to (1) increased production of nitrite, (2) enhanced absorption of sodium from the intestinal lumen, (3) excess $NO^-$ (nitric oxide free radical) generation, having vasodilatory effect (Hsia, 1998; Berger et al., 1997), nitric oxide, and (4) increased production of $O^{\cdot -}$, which will react with other cell constituents, possibly causing irreversible cell damage (Berger et al., 1997; Gupta AB and Gupta SK, 1998).

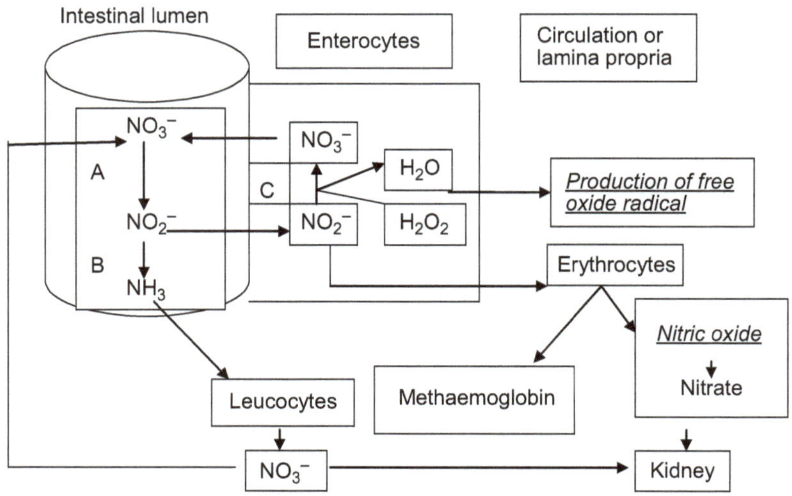


**FIGURE 28.1** Metabolism of ingested nitrate in Human body at cellular level. *A*, bacterial nitrate reductase; *B*, bacterial nitrite reductase; *C*, catalase (Gupta et al., 2001; Gupta et al., 2013).

# Excretion

About 70–75% of ingested nitrate is excreted within 24 h of ingestion irrespective of amount of intake. The maximum excretion was noticed high in first 5 h (Bartholomew and Hill, 1984). Low levels of nitrate and nitrite were detected in the feces of humans (Saul et al., 1981). A small amount of ingested nitrate converting to nitrite, changes to oxides of nitrogen (nitric oxide, nitrous oxide, nitrogen, etc.) in blood and gets excreted by exhalation through lungs (as shown in flow chart given below).

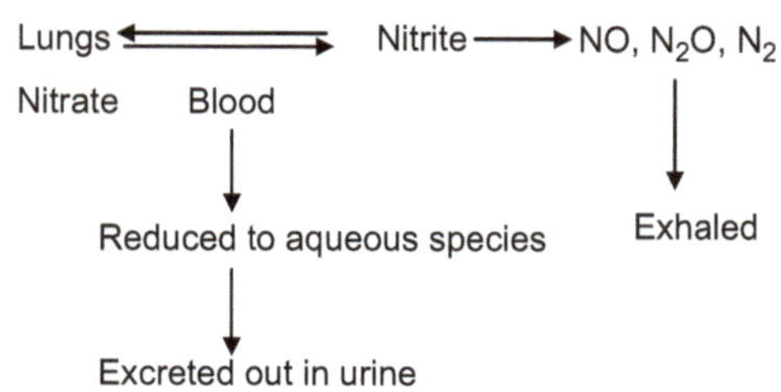


# Acute Toxic Effects

Human lethal doses of 4–50 g $NO_3^-$ (equivalent to 67–833 mg$NO_3^-$ $kg^{-1}$ body weight) have been reported. Toxic doses range from 2 to 5 g (RIVM, 1989; WHO, 2011) of $NO_3$. These values are equivalent to 33–83 and 100–150 mg $NO_3^-$ $kg^{-1}$ body weight, respectively. Severe gastroenteritis with abdominal pain, blood in the urine, and feces (RIVM,

1989; WHO, 2011) occurred rapidly. Repeated doses gave rise to dyspepsia, mental depression, headache, and weakness. Cyanosis (bluish discoloration) will be present.

In animals (mice, rats, and rabbits) $LD_{50}$ values of 1600–9000 mg of sodium nitrate per kg of body weight have been reported. $LD_{50}$ values of 85–220 mg of sodium nitrite per kg of body weight have been reported for mice and rats (RIVM, 1989; WHO, 1996). Ruminants are more sensitive to the effects of nitrate as a result of high nitrate reduction in the rumen; the $LD_{50}$ for cows was 450 mg of sodium nitrate per kg of body weight.

## Chronic Toxic Effect

Chronic toxic effects are caused due to long-term exposure of nonlethal doses of nitrate. The effects reported include methemoglobinemia (presence of nitrates in drinking water cause methemoglobinemia in infants up to 6 months of age); abortion, birth defects and infant mortality (increased infant mortality due to increasing exposure of pregnant mothers and infants to nitrate through either undetected toxic methemoglobinemia or malformations, and weaknesses in the infant caused by fetal nitrate exposure); carcinoma (nitrate acting as a "procarcinogen" reacts with amines and amides to form carcinogenic N-nitroso compound); cardio-respiratory impacts (cardiac toxicity of nitrate mediated through its metabolic effects and metabolic products, increased asthmatic attacks due to high air borne nitrate concentrations and presence of inflammatory cells and breakdown of alveoli); gastrointestinal system (recurrent diarrhea in children and recurrent stomatitis); thyroid function and morphology (nitrate-inhibiting iodine uptake by sodium–iodine symporter - Hence causing Hypothyroidism - Jahreis et al., 1987); diabetes; nephrotic syndrome; and other immunity and physiological disorders.

Inhalation of $NO_x$ causes a wide variety of health and environmental impacts because of various compounds and derivatives of the nitrogen oxides family, including nitrogen dioxide, nitric acid, nitrous oxide, nitrates, and nitric oxide. It reacts to form nitrate particles, acid aerosols, and contributes to formation of acid rain. Health effects of $NO_x$ inhalation are related to levels as well as the duration of exposure (Bernstein et al., 2004). Low levels of nitrogen oxides in the air irritate eyes, nose, throat, and lungs leading to cough and shortness of breath, tiredness, and nausea. Breathing high levels of $NO_x$ can cause rapid burning, spasms, and inflammatory swelling of tissues in the throat and upper respiratory tract. High exposures of $NO_x$ may lead to pulmonary edema and hypoxemia and even death. Industrial exposure to nitrogen dioxide may cause genetic mutations, damage a developing fetus, and decrease fertility in women. Industrial exposure to nitric oxide can cause unconsciousness, vomiting, mental confusion, and damage to the teeth. So far, there is no evidence that nitrogen oxides are potential carcinogens.

The EPA has established that the average concentration of nitrogen dioxide in ambient air in a calendar year should not exceed 0.053 parts of nitrogen dioxide per million parts of air (0.053 ppm). The Occupational Safety and Health Administration (OSHA) has set a limit of 25 ppm of nitric oxide in workplace air during an 8-h workday,

40-h work week. OSHA has also set a 15-min exposure limit of 5 ppm for nitrogen dioxide in workplace air.

## Indian Scenario on Health Effects of Reactive-N

Recently, methemoglobinemia has been reported to be prevalent among all age groups, but was more severe in infants and people of old age groups (>45 years) (Gupta et al., 2000). The methemoglobin levels in blood, when exceed 10% of total hemoglobin, manifest as clinical cyanosis and cause cellular anoxia. At less than 10% methemoglobin, no signs and symptoms have been observed; at 10–25% cyanosis is present; at 25–50% cyanosis, dyspnea, and headache have been reported; at 50–60% dyspnea even on lying, cyanosis, and disorientation are observed; and at 60% or more it is lethal (WHO, 1977). In India breastfeeding is a common practice especially up to the age of 6 months, which protects the infants from nitrate toxicity. The probability of toxicity to fetus in pregnant women cannot be ruled out as transplacental passages of nitrate metabolite (Nitrite) have been documented in animal experiment.

Recurrent diarrhea (Gupta et al., 2001) was well correlated in children up to 8 years of age with increasing nitrate concentration in drinking water. High nitrate ingestion causes hyperemia and edema in the enteric mucosa and later on possibly causes irreversible mucosal damage, and therefore provide high-risk conditions suitable for recurrent diarrhea (Gupta et al., 2008, 2010, 2013). It is further reported that WHO ORS could be of grave concern, if prepared with water containing high nitrate.

An epidemiological field study indicated high percentage (40–82%) of cases of acute respiratory tract infection with history of recurrence reported in children consuming high nitrate in water (Gupta et al., 2000). Effects such as earlier onset of hypertension and increase of blood pressure in school children have been reported due to nitrate ingestion (Gupta et al., 2010).

In comparison with healthy controls, children with MCNS had increased urinary nitrite excretion (Trachtman et al., 1996). Plasma nitrite levels were high in relapse compared with controls. Rabbit experiment indicated increasing damage to kidney with increasing nitrate concentration in drinking water of the animal (Gupta et al., 2010). Histopathological changes at different nitrate concentration indicated tubular changes, eosinophilic changes, granular changes, and increased cellularity.

## Treatment

Ascorbic acid, methionine, alpha-tocopherol supplementation, and methylene blue have been found to be effective in overcoming the problem of nitrate toxicity. These substances are effective in inhibiting nitrosation if present in gastric juices at all time in significant concentrations. However, clinical trials testing this hypothesis showed little benefits in reducing respiratory, GIT, cardiovascular, etc., events or mortality; some of the treatments even caused harm (Dotan et al., 2009).

## Prevention

1. Avoid water pollution: (1) Minimize the contamination of water supplies by nitrates originating from agricultural practice; (2) avoid the habit of open defecation and thereby contamination of water source; (3) avoid the stagnation of waste water around the source of water; and (4) avoid the sewage disposal directly to ground water table.
2. Removal of nitrate from drinking water by denitrification: Although the safest way to have potable water free of nitrate, removal of nitrate from water is costly and is difficult to implement and maintain both at domestic as well as community level.
3. At human level: Promote breastfeeding, at least up to the age of 4 months. If top feed is necessary, then preferably use cow milk (of course unadulterated) or dry milk preparations reconstituted with water containing low level of nitrate. Avoid the use of high nitrate-containing food for weaning. Avoid use of nitrate and nitrite as preservative to the minimum extent possible, especially to cured and canned meats and fish. Use drinking water with low level of nitrates especially for pregnant mothers. Avoid WHO ORS preparation in locally available high nitrate water during diarrhea. Avoid long-term use of antacid secretary agents, especially in pregnant mothers and children. In case if it is necessary to use these drugs, they should be used cautiously and preferably with antioxidants.
4. Reducing nitrate levels in vegetables: By peeling, e.g., in potatoes because nitrates are concentrated in the skin and just below, in leafy vegetables, such as lettuce, discarding the stem and midrib can decrease up to 40% of the nitrates. A study of French fries showed that peeling reduced levels by 30%; preheating and cutting reduced the nitrate content by a further 20% and blanching by 30%. After final frying, only about 5% of the original nitrate content remained. For homegrown vegetables, choose low nitrate varieties; use slower releasing nitrogen sources such as manure and compost, and avoid fertilizing just before harvest.

## Policy Issues and Recommendation

Considering the toxicities of nitrate and limited role of the protective system (adaptation of enzyme cytochrome $b_5$ reductase - Gupta SK et al., 1999b) available in human body against nitrate toxicity, probably the present WHO guidelines for short exposure may be adhered to. Looking at the ground reality that available technology options are not economically viable for public distribution, it is recommended that simple policy changes to minimize the toxicity of nitrate may be urgently introduced in the government campaign at least in high nitrate belt. It is highly desirable that more detailed health-related epidemiological studies covering a large population should be carried out to arrive at new acceptable standards. A simple and cost-effective method for nitrate removal should be developed. Health education system should be developed to make people aware about the toxic effects of nitrate ingestion and the ways of its prevention.

## Conclusion

Nitrate levels in surface and groundwater have been found to increase due to sewage percolation or use of excessive chemical fertilizers. The major sources of nitrates for humans are drinking water, food, and air. There are reports of methemoglobinemia in infants due to high nitrate ingestion. Cancer has also been documented as a part of nitrate toxicity, but no firm conclusions have been drawn so far. The other effects observed are increased infant mortality, abortions, birth defects, recurrent diarrhea, recurrent stomatitis, early onset of hypertension, ischemia, CVD, histopathological changes in cardiac muscles, alveoli of lungs and adrenal glands, recurrent respiratory tract infection in children, nephrotic syndrome in children, hypothyroidism, and diabetes. Recent studies have indicated that high nitrate ingestion adversely affects the immune system of the body.

Since it is difficult and costly to remove nitrates from nitrate-rich water because it is chemically nonreactive in dilute aqueous solutions, it is recommended that a simple change in habits and adoption of simple preventive measures to be urgently introduced through a government campaign at least in high nitrate belts to combat the problem of nitrate toxicity. The most important strategy is to promote breastfeeding up to the age of at least 6 months. Indiscriminate use of nitrogenous fertilizers in agriculture should also be avoided.

## References

Anjana, U.S., Iqbal, M., Abrol, Y.P., 2007. Are nitrate concentrations in leafy vegetables within safe limits? Current Science 92 (3), 355–359.

Bartholomew, B., Hill, M.J., 1984. The pharmacology of dietary nitrate and the origin of urinary nitrate. Food and Chemical Toxicology 22, 789–795.

Berger, H.M., Moison, R.M.W., Van Zoeren-Grobben, D., 1997. Pathophysiology of respiratory distress syndrome. In: David, T.J. (Ed.), Recent Advances in Pediatrics, vol. 14. Churchill livingstone, New Delhi, pp. 117–119.

Bernstein, J.A., Alexis, N., Barnes, C., Bernstein, I.L., Bernstein, J.A., Nel, A., Peden, D., Diaz-Sanchez, D., Tarlo, S.M., Williams, P.B., 2004. Health effects of air pollution. The Journal of Allergy Clinical Immunology 114, 1116–1123.

Craven, P.A., De Rubertis, F.R., 1978. Restoration of the responsiveness of purified guanylate cyclase to nitrosoguanidine, nitric oxide, and related activators by heme and hemeproteins. Evidence for involvement of the paramagnetic nitrosyl-heme complex in enzyme activation. The Journal of Biological Chemistry 253, 8433–8443.

Dotan, Y., Pinchuk, I., Lichtenberg, D., Leshno, M., 2009. Decision analysis supports the paradigm that indiscriminate supplementation of vitamin E does more harm than good. Arteriosclerosis, Thrombosis Vascular Biology 29, 1304–1309.

Galloway, J.N., Winiwarter, W., Leip, A., Leach, A.M., Bleeker, A., Erisman, J.W., 2014. Nitrogen footprints: past, present and future. Environmental Research Letters 9, 115003. http://dx.doi.org/10.1088/1748-9326/9/11/115003.

Gupta, A.B., Gupta, S.K., 1998. Removal of organics and nitrogen from sewage in an aerobic bio-film containing thiosphaera pantotropha. In: Proceedings Asian Conference on Water and Wastewater Management, Tehran, March, 2–4, pp. 562–580.

Gupta, A.B., Gupta, S.K., Jajoo, B.S., Kadaverugu, A., March 22–25, 2011. Synergistic effect of fluoride and aluminium in manifestation of fluorosis - a case study of (Chaksu) Jaipur, India. The 1st Environment Asia International Conference, Bangkok, Thailand.

Gupta, S.K., Gupta, R.C., Gupta, A.B., Seth, A.K., Bassin, J.K., Gupta, A., Sharma, M.L., 2001. Recurrent diarrhea in areas with high nitrate in drinking water. Archives of Environmental Health 56, 369–373.

Gupta, S.K., Gupta, R.C., Seth, A.K., Gupta, A.B., Bassin, J.K., Gupta, A., 1999b. Adaptation of cytochrome-b5 reductase activity and methaemoglobinaemia in areas with a high nitrate concentration in drinking-water. Bulletin of the World Health Organization 77, 749–753.

Gupta, S.K., Gupta, R.C., Seth, A.K., Gupta, A.B., Bassin, J.K., Gupta, A., 2000. Methaemoglobinaemia in areas with high nitrate concentration in drinking water. The National Medical Journal of India 13, 58–61.

Gupta, S.K., Gupta, R.C., Gupta, A.B., 2013. Pathophysiology of nitrate toxicity in human and its mitigation measures. In: Umar, S., Anjum, N.A., Khan, N.A. (Eds.), Nitrate in Leafy Vegetables – Toxicity and Safety Measures. I.K. International Publishing House Pvt. Ltd., New Delhi, India, pp. 52–92.

Gupta, S.K., Gupta, R.C., Chhabra, S.K., Eskiocak, S., Gupta, A.B., Gupta, R., 2008. Health issues related to N pollution in water and air. Current Science 94, 1469–1477.

Gupta, S.K., Gupta, R.C., Gupta, A.B., SevgiEskiocak, Prakasa Rao, E.V.S., Puttanna, K., Singhvi, A., 2010. Pathophysiology of nitrate toxicity in human and its mitigation measures. In: Raghuram, N., Abrol, Y.P. (Eds.), ING Bulletins on Regional Assessment of Reactive Nitrogen, Series No. 20. Published by Indian Nitrogen Group (ING) and Society for Conservation of Nature (SCON) in association with South Asia Nitrogen Centre (SANC) and International Nitrogen Initiative (INI).

Hsia, C.C., 1998. Respiratory function of hemoglobin. The New England of Journal of Medicine 338, 239–248.

IS 10500, 1995. Drinking Water Specification. Bureau of Indian Standards, New Delhi, p. 3.

Jahreis, G., Schöne, F., Lüdke, H., Hesse, V., 1987. Growth impairment caused by dietary nitrate intake regulated via hypothyroidism and decreased somatomedin. Endocrinologia Experimentalis 21, 171–180.

Kumar, S., Gupta, A.B., Gupta, S., 2002. Need for revision of nitrates standards for drinking water: a case study of Rajasthan. Indian Journal of Environmental Health 44, 168–172.

Lowestein, C.J., Dinerman, J.L., Snyder, S.H., 1994. Nitric oxide: a physiological messenger. Annals of Internal Medicine 120, 227–237.

Malik, R.P.S., 2000. Agriculture and water quality in India towards sustainable management. In: Water Quality Management and Control of Water Pollution. FAO, Rome, Italy, pp. 73–85. Water Reports 21.

NEERI [National Environmental Engineering Research Institute], 1991. Water Mission Report on Participation of NEERI. Nagpur, India.

PHED [Public Health Engineering Department], 1995. Habitat Survey (1991–93): Nitrate Concentration in Water in Villages of Rajasthan. Public Health Engineering Department, Government of Rajasthan.

Prakasa Rao, E.V.S., Puttanna, K., 2006. Strategies for combating nitrate pollution. Current Science 91, 1335–1339.

RIVM, 1989. Integrated Criteria Document Nitrate; Effects. Appendix to RIVM Report No. 758473012. Bilthoven, Rijksintituutvoor de Volksgenzondheid en Milieuhygiëne. National Institute of Public Health and Environmental Protection. RIVM Report No. A758473012.

Saul, R.L., Kabir, S.H., Cohen, Z., Bruce, W.R., Archer, M.C., 1981. Reevaluation of nitrate and nitrite levels in the human intestine. Cancer Reserach 41, 2280–2283.

Singh, B., PrakasaRao, E.V.S., Singh, Y., Puttanna, K., 2007. Nitrate pollution of ground water vis-a-vis fertilizer use in India. In: Abrol, Y.P., Raghuram, N., Sachdev, M.S. (Eds.), Agricultural Nitrogen Use

& its Environmental Implications. L.K. International Publishing House Pvt. Ltd., New Delhi, India, pp. 439–458.

Smith, A.D., Datta, S.P., Smith, G.H., 1997. Nitric oxide. In: Smith, A.D., Datta, S.P., Smith, G.H., et al. (Eds.), Oxford dictionary of Biochemistry and Molecular Biology. Oxford University Press, Oxford, New York, Tokyo, p. 451.

Trachtman, H., Gauthier, B., Frank, R., Futterweit, S., Goldstein, A., Tomczak, J., 1996. Increased urinary nitrite excretion in children with minimal change nephrotic syndrome. The Journal of Pediatrics 128, 173–176.

Walker, R., 1995. The conversion of nitrate into nitrite in several animal species and man. In: Health Aspects of Nitrate and its Metabolites (Particularly Nitrite). Proceedings of an International Workshop, Bilthoven (Netherlands), 8–10 November 1994. Strasbourg. Council of Europe Press, Strasbourg, pp. 115–123.

WHO [World Health Organization], 1977. Nitrate, nitrite and N-nitroso compounds. In: Environmental Health Criteria, 5, Geneva, pp. 15–82.

WHO [World Health Organization], 1985. Environmental health series No. 1. In: Health Hazards from Nitrate in Drinking-Water: Report on a WHO Meeting, Copenhagen, 5–9 March 1984. WHO Regional Office for Europe, Copenhagen, pp. 14–40.

WHO [World Health Organzation], 1996. Toxicological evaluation of certain food additives and contaminants. In: Prepared by the Forty-fourth Meeting of the Joint FAO/WHO Expert Committee on Food Additives (JECFA). International Programme on Chemical Safety (WHO Food Additives Series 35), Geneva.

WHO [World Health Organization], 2011. Nitrate and Nitrite in Drinking-water. Guidelines for Drinking-water Quality, fourth ed. Geneva World Health Organization, Geneva.

Winterbourn, C.C., McGrath, B.M., Carrell, R.W., 1976. Reaction involving superoxide and normal and unstable haemoglobins. The Biochemical Journal 155, 493–502.

29

# Assessment of Reactive Nitrogen Emissions From Indian Transport Sector

B.R. Gurjar[1], V. Sahu[1], A. Nagpure[2], C. Sharma[3], A. Singh[4], S. Bhattacharya[5]

[1]*INDIAN INSTITUTE OF TECHNOLOGY (I.I.T.) – ROORKEE, ROORKEE, INDIA;* [2]*UNIVERSITY OF MINNESOTA, UNITED STATES;* [3]*CSIR-NATIONAL PHYSICAL LABORATORY, NEW DELHI, INDIA;* [4]*CSIR-CENTRAL ROAD RESEARCH INSTITUTE, NEW DELHI, INDIA;* [5]*IORA ECOLOGICAL SOLUTIONS PVT. LTD., NEW DELHI, INDIA*

**CHAPTER OUTLINE**



## Introduction

Transport system of India is one of the largest in the world that serves a land mass of 3.3 million km$^2$ and a population of over one billion. The distinct modes of transportation comprise road, railways, aviation, and marine navigation/shipping. Railways and roads carry more than 95% of the total traffic generated in the country and are

**The Indian Nitrogen Assessment. http://dx.doi.org/10.1016/B978-0-12-811836-8.00029-X**

expected to dominate the transport landscape in the coming decades (Planning Commission, 2001; TERI, 2009). However, the inherent advantage of adaptability, reliability, and flexibility of operation makes road transport the most convenient choice among other modes of transportation. The road and rail transport have carried 4252 and 616 bpkm of passenger traffic and handled 659 and 442 btkm of freight traffic, respectively, during 2005–06 (MoSRTH, 2009). The transport sector contributes a share of 6.4% to India's gross domestic product, out of which road transport has emerged as a dominant mode with a share of 4.5% in comparison to 1.2% of railways, and 0.2% each by air and water transport during 2006–07 (Planning Commission, 2007).

## Energy Demand in Transport Sector

Different modes of transportation use different forms of energy with varying efficiency and intensity (Planning Commission, 2001). Among various modes of transport (road, rail, aviation, and navigation), air transport is the most energy intensive, followed by road transport. Road transport is the largest consumer of diesel and petrol fuel energy in the country. Aviation turbine fuel (ATF) is the only fuel type used in aviation, diesel and electricity are the major energy sources used for traction purposes in railways, and marine navigation (shipping) uses fuel oil (FO), diesel, and light diesel oil (LDO) in varying proportions.

The consumption of diesel is expected to increase in future due primarily to preference of passenger and freight traffic through road transport. The rapid expansion of the passenger transport industry in the past decade has led to significant increase in the demand for petrol. Further, the high level of growth of Indian economy, both in industrial and agriculture sectors, has led to substantial growth in diesel demand by the freight transport industry, resulting in greater demand for these fuels. In addition, oil gradually replaced coal as a fuel in this sector. This is evident from the declining share of coal and increase in the share of oil in railways. This is because of a greater dependence on road transport vis-à-vis rail transport, and a shift from steam traction to diesel and electric traction. Similarly, cleaner fuels such as CNG and auto LPG are also used in road transport in smaller quantities as compared to conventional liquid fuels (MoPNG, 2010; PPAC, 2009).

## Fuel Consumption

Out of the major modes of transportation in India, road transport has the largest fossil fuel energy consumption potential. Hence, the largest uncertainty in actual fuel consumption, particularly in diesel consumption, lies in this subsector. The major reason for this uncertainty is that the fuel (petrol and diesel) is sold at two distinct levels: dedicated sale and sale through network of retail outlets.

## Road Transportation

The increasing demand in freight and passenger transport over the past two decades is primarily met by road transport. This is evident by the increase in number of registered motor vehicles. Number of motor vehicles have increased from 21.3 million in 1991 to nearly 99.6 million (estimate) in 2007 at compound annual growth rate of 10.1% (Fig. 29.1). The relative proportions of commercial- and personal-use vehicles between 1991 and 2007 do not appear to have changed much. Personal-use vehicles such as two wheelers and cars constitute nearly 87% of the total vehicles at the national level (SIAM, 2008; MoSRTH, 2009). Since one-time registration practice with a 15-year validity period is followed, no documentation is available pertaining to actual numbers of on-road private vehicles. Similarly, registration authorities do not keep record of the scrapping of unused private vehicles. However, the information on actual number of on-road commercial vehicles (trucks, buses, LCVs) is available fairly accurately since their registration is renewed every year. Nonetheless, it has been reported that only a small percentage of private vehicles older than 20 years are in active use, and at any given time not more than 60% of the total registered vehicles are in effective use. The new technology vehicles have been observed to dominate the vehicle mix (CRRI, 2002). However, not much information is available for the new technology vehicles pertaining to fuel efficiency and emission controls.

## Rail Transportation

Rail transport has played a dominant role in the Indian transport sector. However, despite being more energy efficient, its share in freight and passenger traffic has declined

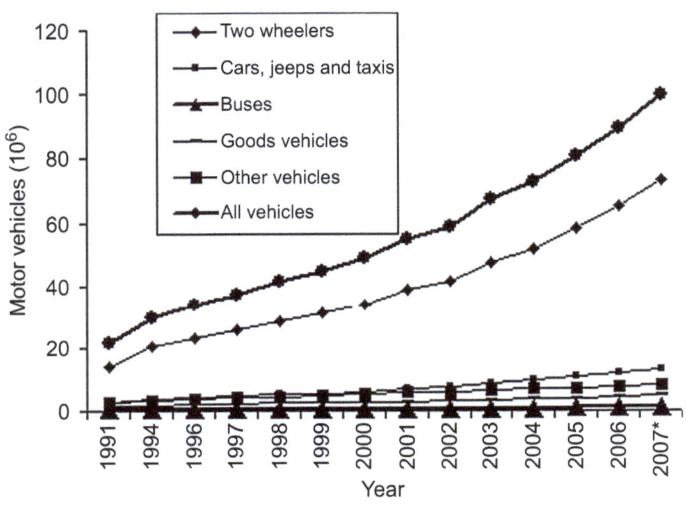


**FIGURE 29.1** Registered motor vehicles in India.

over the years. Rail transport accounted for $428 \times 10^9$ passenger-km (pkm) and $302 \times 10^9$ ton-km (tkm) of the transport demand in 2000 (Kapoor, 2002), which increased to $770 \times 10^9$ pkm and $521 \times 10^9$ tkm in 2007–08 (IRYB, 2009). The reduction in direct consumption of fossil fuel energy has been observed since early 1990s with decreasing share within the transportation sector. The resultant energy savings are because of the replacement of coal by diesel and electricity for locomotive traction, replacement of rolling stock, modernization and technological upgrade, etc. In railways, mostly diesel and coal have been used as fuel in locomotives in addition to the electricity (generated elsewhere). Other fuels LDO, FO, kerosene, and petrol have reportedly been consumed in smaller quantities.

## Aviation

Domestic aviation and aviation-related activities such as ground operations contribute to nearly 4% of the total energy consumed in the transportation sector. The airports managed nearly 1.01 million domestic and 0.24 million international aircraft movements during 2007. In addition to transporting 0.6 Mt of domestic and 1.1 Mt of international cargo, the airports handled 83 million domestic and nearly 29 million international passenger traffic in 2007 (TERI, 2009; DGCA, 2009). ATF is used in aircrafts but diesel is used for ground operations at the airports for activities related to support services. Other fuels, viz. LDO and FO, have reportedly been used earlier, but only diesel is consumed for support services (ground operations) currently.

## Marine Navigation

India has 7517 km long coastline, 12 major ports, and 200 minor/intermediate (nonmajor) ports. The cargo traffic handled at major ports has increased from 281.1 Mt in 2000–01 to nearly 469.3 Mt in 2006–07 (TERI, 2009). The cargo traffic handled by nonmajor ports was found to increase in recent past, and has escalated from 180 Mt in 2002 to 151 Mt in 2006. Out of the total energy consumed in transportation sector, the relative share of energy consumption for coastal marine navigation was observed to decrease from nearly 2% during 1991 to less than 1% in 1999 (Planning Commission, 2007). The fuel types used for marine navigation/shipping include diesel, LDO, and FO.

## Energy Consumptions in Transport Sector and Reactive Nitrogen Emissions

The total transport sector energy consumption in India was estimated at 1917.3 PJ for the year 2007. Road transport and aviation together were estimated to account for 95% of the total transport sector energy in 2007. Diesel is the major fuel type consumed, which constituted 66% of the total transport sector energy followed by petrol (23%) and ATF

(~8%). The share of alternative fuels (CNG and auto LPG) was estimated at 2.4%. Singh et al. (2008) had published a detailed national inventory of greenhouse gas emissions from Indian transport sector for the year 2000. Singh and Gangopadhyaya (Personal Communication) estimated total reactive nitrogen (N) emissions, comprising both $N_2O$-N and $NO_x$-N, from different subsectors of transport sector for the year 1980 and 2007 (Table 29.1).

## $NO_x$ Emission Assessment From Transport Sector

Transport sector plays an important role in $NO_x$ emission, contributing 32% of total $NO_x$ emissions in India, out of which around 28% is from road transport. From 1990 to 1995, all India $NO_x$ emission from transport sector has increased by 46% (Garg et al., 2001). The urban population of India is predominantly dependent on road transport. Due to increase in vehicle numbers with rising population density, all the major cities of India show higher values of $NO_x$. In all the major cities of India, $NO_x$ loading per square kilometer is found to be very high. Among the high $NO_x$ emitting regions of India, the Indo-Gangetic plains (IGP) region contributed more than 26% (around 1172 Gg year$^{-1}$) to total Indian $NO_x$ emissions during 2001, where the vehicular source contributed 526 Gg year$^{-1}$ (45%), followed by coal at 387 Gg year$^{-1}$ (33%) and biofuel at 259 Gg year$^{-1}$ (22%) to total $NO_x$ emissions over IGP (Sahu et al., 2012). According to Ramachandran and Shwetmala (2009), the total transport emissions of $NO_x$ during 2003–2004 was 2.298 Tg, out of which road transport, aviation, railways, and shipping contributed 2.213 Tg (96.3%), 0.0126 Tg (0.54%), 0.0568 Tg (2.5%) and 0.0155 Tg (0.68%), respectively (Table 29.2).

Vehicular emissions vary with its type, efficiency, and nature of fuel used. Trucks, lorries, and buses contribute higher $NO_x$ emissions compared to other vehicles. The $NO_x$ emissions calculated by Ramachandra and Shwetmala (2009) for different types of road transport vehicles are summarized in Table 29.3. Among different types of vehicles, trucks and lorries contribute 39% $NO_x$ (0.86 Tg), which constitute 25% of the total vehicular $NO_x$ emissions computed for India for 2003–2004.

**Table 29.1** Reactive Nitrogen Emissions From Transport Sector in India in 2007

| | Total Reactive N Emissions (kt) | |
|---|---|---|
| **Subsector** | **1980** | **2007** |
| Road transport | 90 | 360 |
| Railways | 39 | 34 |
| Aviation | 3 | 12 |
| Coastal marine navigation | 4 | 9 |

**Table 29.2** $NO_x$ and $N_2O$ Emissions From Indian Transport for 2003–04 (Gg) (Ramachandra and Shwetmala, 2009)

| | $NO_x$ | $N_2O$ |
|---|---|---|
| **Shiping** | | |
| High speed diesel oil | 8.5273 | 0.0064 |
| Light diesel oil | 1.7679 | 0.0013 |
| Fuel oil | 5.2408 | 0.0039 |
| **Railways** | | |
| Coal | 0.0121 | |
| Electricity | Not considered | |
| High speed diesel oil | 56.5370 | 0.0424 |
| Light diesel oil | 0.0693 | 0.0001 |
| Fuel oil | 0.2572 | 0.0002 |
| **Aviation** | | |
| High speed diesel oil | 0.9359 | 0.0007 |
| Light diesel oil | 0.0693 | 0.0001 |
| Fuel oil | 2.2828 | 0.0017 |
| Aviation turbine fuel | 8.7331 | |
| Road transport | 2213.8552 | |
| Total | 2298.2880 | 0.0568 |

**Table 29.3** $NO_x$ Emissions From Different Vehicle Types in India

| Vehicle Types | $NO_x$ (Gg) |
|---|---|
| Bus | 679.73 |
| Omni vans | 200.53 |
| Two wheelers | 62.15 |
| Light motor vehicles (passenger) | 92.93 |
| Cars and jeeps | 22.14 |
| Taxi | 5.68 |
| Trucks and lorries | 859.51 |
| Light motor vehicles (goods) | 110.94 |
| Trailers and tractors | 115.69 |
| Others | 64.54 |

Ramachandra, T.V., Shwetmala, 2009. Emissions from India's transport sector: statewise synthesis. Atmospheric Environment 43(34), 5510–5517.

The total number of registered vehicles is about 20 million in major Indian cities, having a significant share of 28% in the total vehicles of India. Delhi, Kolkata, Chennai, Bengaluru, and Mumbai have about 10.2 million total registered vehicles—around 14% of total registered vehicles of India. Total $NO_x$ emission in metropolitan cities of India is about 2287.85 Mg $km^{-2}$. Chennai with 353 Mg $km^{-2}$ of $NO_x$ emission showed the largest are averaged emissions, followed by Bangalore with 323.75 Mg $km^{-2}$, Kolkata with 273.55 Mg $km^{-2}$, Delhi with 129.99 Mg $km^{-2}$, and Hyderabad with 197.12 Mg $km^{-2}$. Table 29.4 presents $NO_x$ emissions from vehicular transport along with the number of all on-road vehicles in the metropolitan cities of India computed for 2003–04 (Ramachandran and Shwetmala, 2009).

Due to increasing ambient concentrations of nitrogen oxides, it is one of the most important gaseous pollutants of concern in metropolitan cities of India. Fig. 29.2 shows the relative contributions of industrial and vehicular sources to the $NO_x$ emissions in some of the Indian cities, which clearly reflect the significance of vehicular emissions. As can be seen, except Mumbai and Delhi, all cities have maximum contribution from vehicular sources. Mumbai and Delhi exhibit high percentage of $NO_x$ from industries owing to presence of power plants within the city boundary (CPCB, 2011). The ambient $SO_2$ values in Indian cities have continuously declined due to reduction of sulfur content in the automobile fuel. Therefore, after particulate matter (PM), $NO_x$ appears to be the next important pollutant. Unlike PM, $NO_x$ is mainly contributed by anthropogenic sources such as vehicles, industries, and other fuel combustion processes.

**Table 29.4** $NO_x$ Emissions From Vehicular Transport in Indian Metropolitan Cities

| Metropolitan City | Number of On-Road Vehicles in Metropolitan City | $NO_x$ (mg $km^{-2}$) |
|---|---|---|
| Hyderabad | 1,355,700 | 197.12 |
| Visakhapatnam | 411,691 | 54.35 |
| Patna | 335,623 | 149.03 |
| Delhi | 4,236,675 | 129.99 |
| Ahmedabad | 1,075,213 | 93.20 |
| Surat | 692,118 | 82.18 |
| Vadodara | 586,438 | 29.16 |
| Bangalore | 1,890,692 | 323.75 |
| Bhopal | 319,579 | 27.29 |
| Indore | 644,771 | 86.31 |
| Mumbai | 1,199,416 | 67.80 |
| Nagpur | 543,322 | 63.00 |
| Pune | 754,902 | 42.50 |
| Ludhiana | 770,784 | 98.33 |
| Jaipur | 823,715 | 65.61 |
| Chennai | 2,014,776 | 353.67 |
| Kanpur | 424,652 | 32.01 |
| Lucknow | 614,974 | 39.34 |
| Varanasi | 365,913 | 79.64 |
| Kolkata | 875,165 | 273.55 |
| Total | 20,007,930 | 2287.85 |

Ramachandra, T.V., Shwetmala, 2009. Emissions from India's transport sector: statewise synthesis. Atmospheric Environment 43(34), 5510–5517.

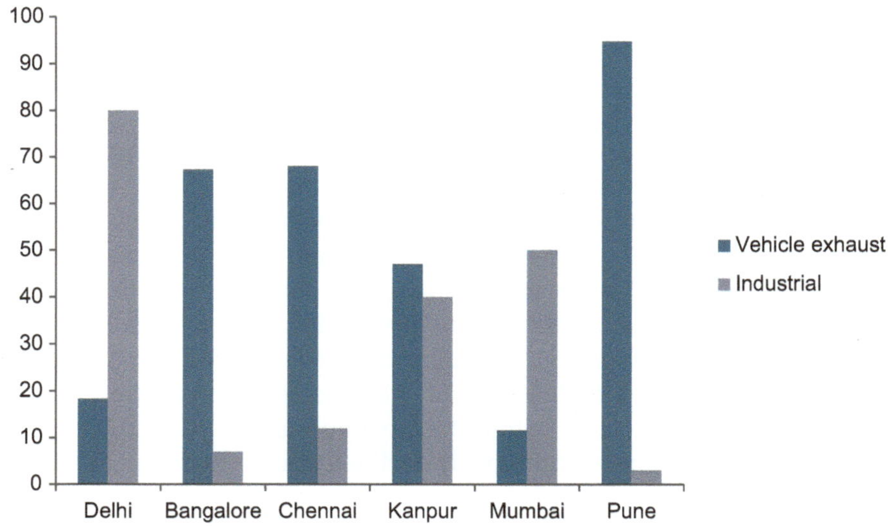


**FIGURE 29.2** Prominence of sources of $NO_x$ (CPCB, 2011).

**Table 29.5** Source Contributions in Respect of $NO_x$ Emissions

| City | $NO_x$ From Vehicle Exhaust (TPD) | Total $NO_x$ Emissions (TPD) |
|---|---|---|
| Delhi | 84.28 | 460 |
| Bangalore | 146.36 | 217.4 |
| Chennai | 8.27 | 12.16 |
| Kanpur | 10.571 | 22.5 |
| Mumbai | 25.12 | 215.6 |
| Pune | 39.19 | 41.4 |

CPCB [Central Pollution Control Board] 2011. Air Quality Monitoring, Emission Inventory and Source Apportionment Study for Indian Cities, Available at: http://cpcb.nic.in/FinalNationalSummary.pdf.

The industrial contribution to $NO_x$ emissions excluding power plants is also highest in Delhi, i.e., 65 tons per day (TPD), followed by Mumbai (23 TPD), Bangalore (17 TPD), and Kanpur (8 TPD). The least industrial contribution of 0.9 TPD is in Pune. Delhi has about 295 TPD of $NO_x$ contributions from power plant alone. Table 29.5 presents the total $NO_x$ emission from all sectors and the $NO_x$ emission from vehicular exhaust (CPCB, 2011) (Table 29.6).

## Contribution of Different Types of Vehicles to $NO_x$ Emissions

Fig. 29.3 shows contributions to $NO_x$ emissions from different vehicle types within road transport sector in six metropolitan cities. It can be observed that in transport sector,

**Table 29.6** $N_2O$ and $NO_x$ Emissions From Different Vehicles in Delhi During 2001 and 2009

| | $N_2O$ Emissions (Gg) | | $NO_x$ Emissions (Gg) | |
|---|---|---|---|---|
| **Vehicle Type** | **2001** | **2009** | **2001** | **2009** |
| 2-Wheelers | 0.009 | 0.013 | 9 | 11 |
| Cars | 0.006 | 0.009 | 6.1 | 8.6 |
| Light commercial vehicles | 0.005 | 0.006 | 7.9 | 8.2 |
| Heavy commercial vehicles | 0.013 | 0.014 | 18.5 | 19.1 |

heavy duty diesel vehicles are the major $NO_x$ emitters (43–75%) in almost all the cities. It may be pertinent to mention that these contributions depend on share in terms of number of vehicles plying in the city and, therefore, present relative contribution among various vehicle types (CPCB, 2011).

In India, an estimated total of 76,000 tons of $NO_x$ was released by in-use vehicles in 1991, which further increased to 120,500 tons in 2001, and 184,000 tons in 2012. Around 92% of the increase in $NO_x$ emissions is contributed by Heavy-Duty Vehicles (HDVs) and Light-Duty Vehicles (LDVs) (60%), three wheeler (3Ws) (17%), and two wheeler (2Ws) (14%) from 1991 to 2012. The $NO_x$ emissions from cars reached a peak of 13,200 tons in 2003, and reduced to 10,400 tons in 2012. In the case of 3 Ws, due to conversion of fuel as well as technology, from petrol and diesel with 2S engines to CNG with 4S engines, the most drastic increase in $NO_x$ emissions occurred from 3000 tons in 1998 to 15,000 tons in 2012. Because of CNG replacement with diesel-based buses, $NO_x$ emissions from CNG increased from near zero in late 1990s to more than 50% in 2007, while those from diesel vehicles reduced from 95% to less than 50% (Goel and Guttikunda, 2015).

## Reactive Nitrogen Emissions From Transport Sector in Delhi

Delhi, the national capital of India, is one of the most polluted cities in the world (WHO, 1992) having transport as major source of emissions of criteria pollutants (Gurjar et al., 2004). Automobile traffic has increased at a rapid rate in the urban areas during last few decades in India. The annual growth rate of motor vehicles in India is around 10%. The most challenging fact about it is its sharp growth rate especially in urban areas. About 32% of these vehicles are plying in metropolitan cities, which constitute 11% of the total Indian vehicle population (Singh, 2005). Mashelkar et al. (2002) stated that the emissions range of $NO_x$ from transport sector was 66%–74% in Delhi. Central Pollution Control Board (CPCB, 1995) data shows that almost 50% of the $NO_x$ emissions in Delhi are from vehicular activities, followed by domestic, industrial, and power plants. It was found that from 1990–91 to 1995–96, annual $NO_x$ emissions from gasoline consumption increased from 3.5 to 4.5 Gg, respectively, whereas it was 8 Gg in 1990–91 and 12.8 Gg in 1995–96 from diesel (Sharma et al., 2002).

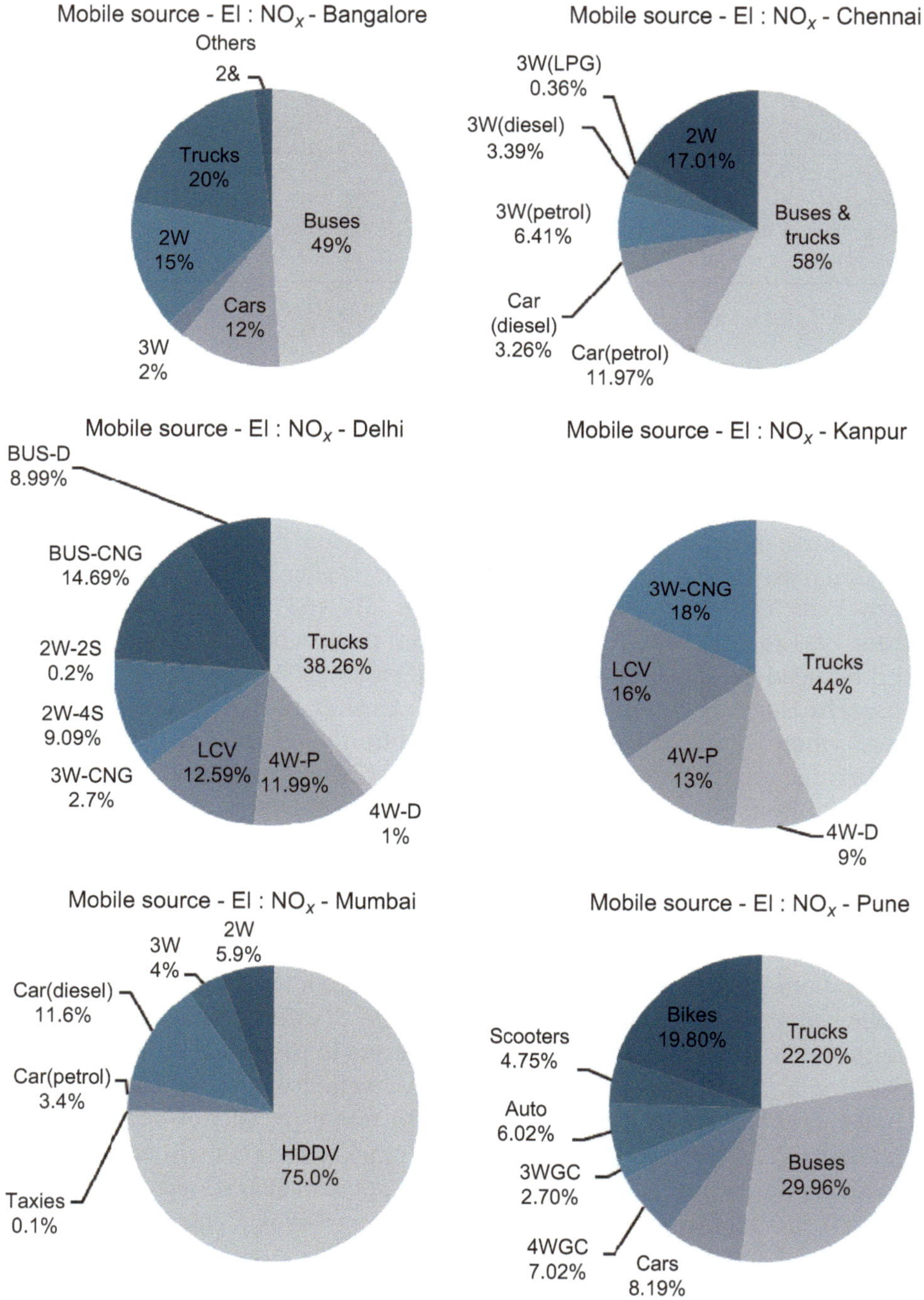


**FIGURE 29.3** Contribution of different vehicle types in $NO_x$ emissions in six cities of India (CPCB, 2011).

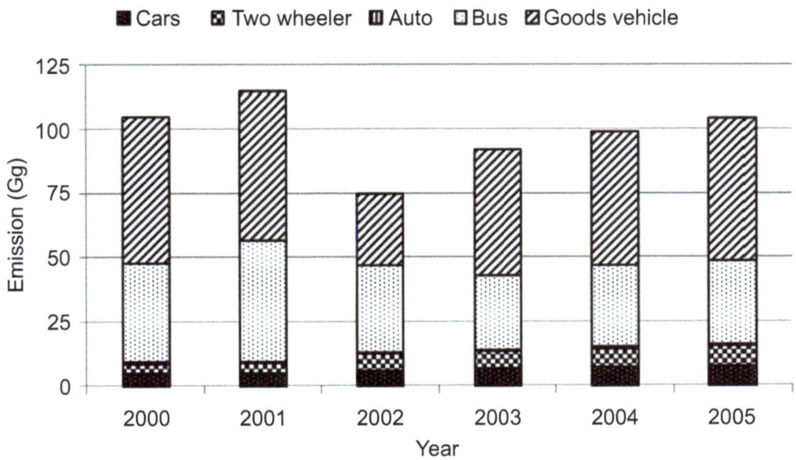


**FIGURE 29.4** Annual $NO_x$ emissions from transport sector in Delhi (2000–2005) (Nagpure et al., 2013).

$NO_x$ emission trend showed a different scenario in comparison to other pollutants (Fig. 29.4). About 0.5% of decline was observed in $NO_x$ emission (105 Gg in 2000 to 104 Gg in 2005). During 2000–02, steep decreases were observed followed by steadily rising emission trend until 2005. The highest increase (23%) was found during 2002–2003, followed by 6% annual average increase till 2005. Goods vehicles were the predominated source of $NO_x$ emission (51%), followed by buses (36%), cars (6%), two wheelers (6%), and autos (1%) during the study period (Nagpure et al., 2013). Singh and Sharma (2012) have estimated the $N_2O$ and $NO_x$ emissions from different types of vehicles including two wheelers, cars, light commercial vehicles, and heavy commercial vehicles in Delhi for the period during 2001–2009 (Table 29.5).

The number of registered cars and motorized two wheelers (2 Ws) increased from 1.6 million in 1990–91 to 7.0 million in 2011–12 (SoE-Delhi, 2012). In the emission totals, the share of $NO_x$, emissions from vehicles aged 10 years increased from 16% in 1991to 23% in 2012, respectively. The contributions in 2012 by vehicle type and pollutant are given in Fig. 29.5.

## Future Challenges

It is evident from the foregoing discussions that the information available on reactive nitrogen emissions from transport sector is quite fragmentary. This is because of the fact that these assessments have been carried out either to address issues related to climate change, $N_2O$ emissions, or to look into air quality issues, $NO_x$, in their specific domains. There is a strong need to undertake a holistic assessment of reactive nitrogen emissions from all the sources including both anthropogenic as well as natural sources to understand science and implications of reactive nitrogen on the environment.

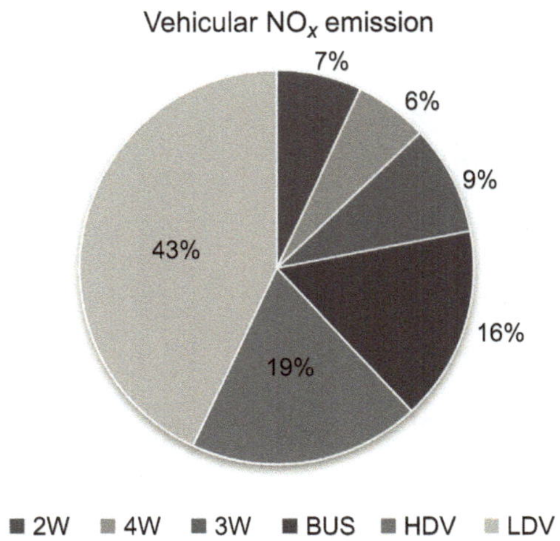


**FIGURE 29.5** Share in $NO_x$ emissions by different vehicle types for year 2012 (Goel and Guttikunda, 2015).

# References

CPCB [Central Pollution Control Board], 2011. Air Quality Monitoring, Emission Inventory and Source Apportionment Study for Indian Cities. Available at: http://cpcb.nic.in/FinalNationalSummary.pdf.

CRRI [Central Road Research Institute], 2002. Urban Road Traffic and Air Pollution in Major Cities. Report submitted to Expert Committee on Auto Fuel Policy. Central Road Research Institute, New Delhi, India.

DGCA [Director General of Civil Aviation], 2009. India Air Transport Statistics 2007-08. Directorate General of Civil Aviation, New Delhi.

Garg, A., et al., 2001. Sub-region (district) and sector level $SO_2$ and NO(x) emissions for India: assessment of inventories and mitigation flexibility. Atmospheric Environment 35 (4), 703–713.

Goel, R., Guttikunda, S.K., 2015. Evolution of on-road vehicle exhaust emissions in Delhi. Atmospheric Environment 105, 78–90.

Gurjar, B.R., et al., 2004. Emission estimates and trends (1990–2000) for megacity Delhi and implications. Atmospheric Environment 38 (33), 5663–5681.

IRYB, 2009. Indian Railways Year Book 2007–08. Railway Board, Ministry of Railways, Government of India, New Delhi, India.

Kapoor, M., 2002. Vision 2020 Transport. Prepared for Planning Commission.

Mashelkar, R.A., et al., 2002. Urban Road Traffic and Air Pollution in Major Cities, Report of the Expert Committee on Auto Fuel Policy. CRRI, New Delhi.

MoPNG [Ministry of Petroleum and Natural Gas], 2010. Indian Petroleum and Natural Gas Statistics. Economics and Statistics Division, Ministry of Petroleum and Natural Gas, Government of India, New Delhi, India.

MoSRTH [Ministry of Shipping, Road Transport and Highways], 2009. Road Transport Year Book 2006/07. Transport Research wing, Ministry of Shipping, Road Transport and Highways, Government of India.

Nagpure, A.S., Sharma, K., Gurjar, B.R., 2013. Traffic induced emission estimates and trends (2000–2005) in megacity Delhi. Urban Climate 4, 61–73.

Planning Commission, 2001. Task Force on Integrated Transport Policy. Planning Commission, Government of India, New Delhi, India.

Planning Commission, 2007. Report of the Working Group on Shipping and Inland Water Transport for the Eleventh Five-year Plan. Planning Commission. Government of India.

PPAC, 2009. Personal Communication with V. Sethi. Petroleum Planning and Analysis Cell, Ministry of Petroleum and Natural Gas, New Delhi.

Ramachandra, T.V., Shwetmala, 2009. Emissions from India's transport sector: statewise synthesis. Atmospheric Environment 43 (34), 5510–5517.

Sahu, S.K., Beig, G., Parkhi, N.S., 2012. Emerging pattern of anthropogenic $NO_x$ emission over Indian subcontinent during 1990s and 2000s. Atmospheric Pollution Research 3 (3), 262–269.

Sharma, C., Dasgupta, A., Mitra, A.P., 2002. Future scenarios of inventories of GHGs and urban pollutants from Delhi and Calcutta. In: Proceedings of IGES/APN Mega-City Project, 23–25 January 2002 (Rihga Royal Hotel Kokura, Kitakyushu, Japan).

SIAM, 2008. The Indian Autobile Industry. Statistical Profile 2007–2008. Society of Indian Automobile Manufacturers, New Delhi.

Singh, S.K., 2005. Review of urban transportation in India. Journal of Public Transportation 8 (1), 79–97.

Singh, R., Sharma, C., 2012. Assessment of emissions from transport sector in Delhi. Journal of Scientific & Industrial Research 71, 155–160.

Singh, A., Gangopadhyay, S., Nanda, P.K., Bhattacharya, S., Sharma, C., Bhan, C., 2008. Trends of greenhouse gas emissions from road transport sector in India. Science of the Total Environment 390 (1), 124–131.

SoE-Delhi, 2012. State of the Environment Report for the National Capital Region of Delhi. The Government of Delhi, New Delhi, India.

TERI, 2009. TERI Energy Data Directory and Year Book 2009. The Energy and Resources Institute, New Delhi, pp. 352–353.

WHO, 1992. Air Quality in Mega Cities of World, Urban Air pollution.Pdf. Available at: http: file:///C:/Users/user/Downloads/063118404X_eng.pdf.

# 30

# Emissions of Reactive Nitrogen From Energy and Industry Sectors in India

C. Sharma[1], Richa Singh[1], B.R. Gurjar[2], V. Sahu[2], S. Bhattacharya[3]

[1]*CSIR-NATIONAL PHYSICAL LABORATORY, NEW DELHI, INDIA;* [2]*INDIAN INSTITUTE OF TECHNOLOGY (I.I.T.) – ROORKEE, ROORKEE, INDIA;* [3]*IORA ECOLOGICAL SOLUTIONS PVT. LTD., NEW DELHI, INDIA*

**CHAPTER OUTLINE**



## Introduction

Energy is a vital component of any economy. It ensures that the energy policy cannot be separated from the national developmental strategy. India's per capita energy consumption is nearly a fourth of the global average. In the year 2010–11, India's per capita primary energy consumption was 441 kgoe compared to the world's per capita primary energy consumption of 1885.2 kgoe (EIA, 2012). Per capita energy consumption in India grew at an average annual rate of 5.30% between 1970–71 and 2010–11 (KPMG, 2012). Coal and crude oil are the important sources for the primary energy supply in the country. Coal is the favorite fuel for the electricity generation in countries like India and China, although government is now emphasizing on the sourcing of clean energy like solar and wind energy. Abundant local supply of coal and sustained high prices for imported natural gas and oil make the coal-fired generation of electricity more

The Indian Nitrogen Assessment. http://dx.doi.org/10.1016/B978-0-12-811836-8.00030-6

economically attractive. Coal is approximately 90% of the total fuel mix for electricity generation in India. Coal- and lignite-based power plants contribute approximately 54.42% of the total electric power generation capacity in India.

The Indian economy is the 10th largest in the world while industrial activities in India account for 28.8% of India's GDP. This sector provides employments to the 22% of the total Indian workforce. The industry sector consists of several activities like manufacturing, production, mining, quarrying, electricity generation, etc. During 2014–15, industry sector registered a growth of 5.9% against a growth of 5.0% recorded during 2013–14. The growth of industrial sector was estimated to be 7.3% during 2015–16 with manufacturing sector growing at 9.5% (IBEF, 2016).

Reactive nitrogen (Nr) includes inorganic species like $NH_3$, $NH_4^+$, $NO_x$, $N_2O$, $NO_2^-$, $NO_3^-$ and organic species such as urea and amines that readily interact in the environment. Anthropogenic emissions of Nr in the energy and industry sectors are primarily from the combustion of fossil fuels and production of chemicals like nitric acid and adipic acid in India. Large part of this Nr leaks into the environment and impacts the nitrogen (N) biogeochemical cycle. With the increasing demands of the society spatially and temporally, emissions of Nr from the energy and industry sectors are also increasing.

## Energy Sector

The Indian power sector is driven by state, central, and private players. As of March 2010, India had 105 thermal power plants with total installed generation capacity of 93,770 MW. This capacity included thermal power plants of less than 100 MW (CEA, 2010). Relatively lower calorific values of Indian coal, coupled with high ash content and inefficient combustion technologies aggravates emission of pollutants from India's coal- and lignite-based thermal power plants. The formation of $NO_x$ during coal combustion is a complex process involving both homogeneous and heterogeneous reactions. Most ($\geq$90%) of the $NO_x$ emitted during combustion process is in the form of NO, also known as "thermal NO," which is formed by oxidation of atmospheric nitrogen at high temperatures.

### Nitrous Oxide ($N_2O$)

India's first biennial update report (BUR) to the UNFCCC prepared by the Ministry of Environment, Forests and Climate Change, Government of India had total emissions as 12.06 Gg of $N_2O$ for the year 2010 from the Indian energy sector. However, majority of these emissions are from the electricity production which accounts for 11.68 Gg $N_2O$ while the other industries like refinery (0.07 Gg), manufacturing of solid fuel (0.30 Gg), cement (0.57 Gg), iron and steel (1.50 Gg), nonferrous metal (0.03 Gg), chemicals (0.08 Gg), pulp and paper (0.10 Gg), mining querying (0.04 Gg), textile and leather (0.03 Gg), bricks (0.02 Gg), fertilizer (0.17 Gg), engineering (0.03 Gg), and nonspecific industries (1.82 Gg) account for the remaining 0.38 Gg $N_2O$ from energy industry.

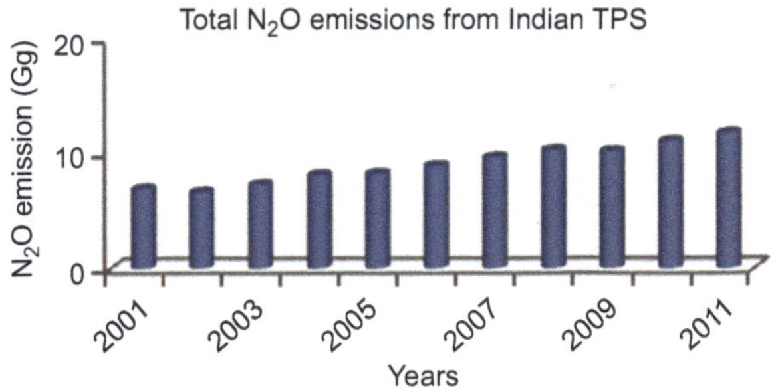


**FIGURE 30.1** Total $N_2O$ emissions from Indian Thermal Power Plants in 2010.

In another estimation, using the IPCC (2006) methodology Singh and Sharma (personal communication) estimated the $N_2O$ emissions from coal- and lignite-based power plants in India ranging from 6.3 Gg to 11.8 Gg during the period from 2001 to 2011 (Fig. 30.1). The power plants in the Indo-Gangetic plains (IGP) region in India have been estimated to contribute about 5.2 Gg $N_2O$ emissions in 2010 (Fig. 30.2).

## Oxides of Nitrogen ($NO_x$)

Formation of NO during coal combustion in power plants is influenced by the oxygen concentration (and thus depends on the excess air) in the system and the flame temperature. Using modeling approach, Mittal et al. (2014) estimated NO emissions from 86 coal-based thermal power plants in India having a total installed capacity of 77,682 MW or about 83% of the total installed thermal power capacity (CEA, 2010). The estimations were based on equilibrium reaction calculated at an average gas temperature of 1200 K. The estimates of annual NO emissions for India—national and regional—computed by Mittal et al. (2014) are given in Fig. 30.3. As per these estimates, annual NO emissions in India have been found to increase from 1 Mt in 2001–02 to 1.6 Mt in 2009–10 at an average annual increase rate of 5.1%. The eastern region has been estimated to have the lowest NO emissions while the western region showed the highest NO emissions during the 2001–02 to 2009–10 period (Mittal et al., 2014). National average annual NO

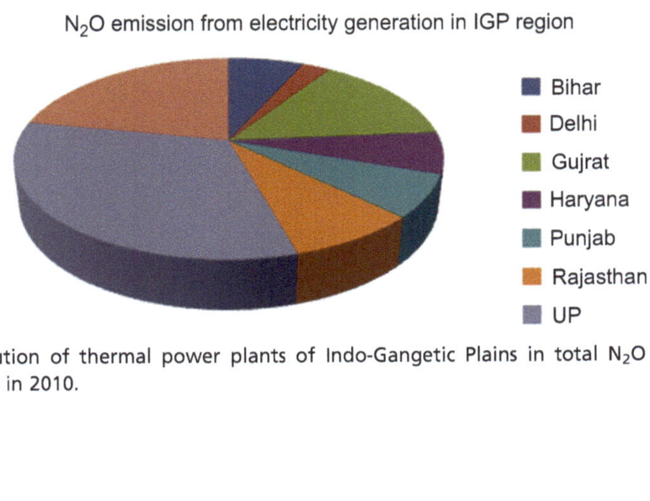


**FIGURE 30.2** Contribution of thermal power plants of Indo-Gangetic Plains in total $N_2O$ emissions from Indian Thermal Power Plants in 2010.

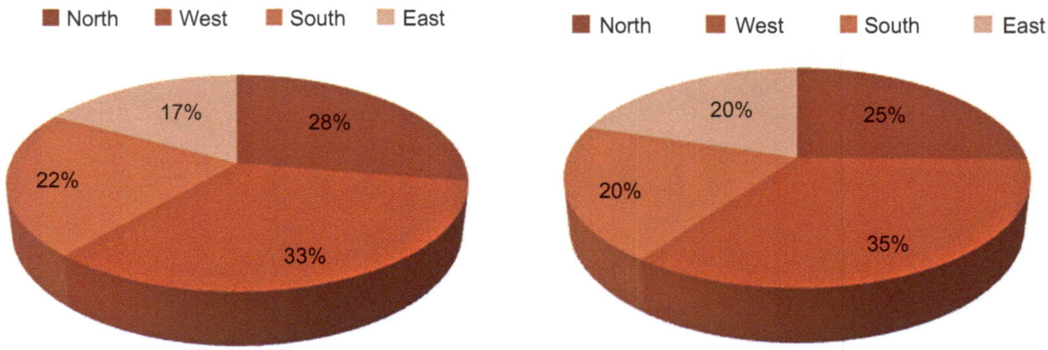


**FIGURE 30.3** Percentage regional emissions of nitric oxide (NO) from coal- and lignite-based power plants in India during the period of 2001–02 to 2009–10.

emissions per unit of generated electricity have been found to be in the range of 2.8–2.9 g $kWh^{-1}$ between 2001–02 and 2009–10. These estimates show NO emission coefficients from plants in the eastern region in the range of 2.2–3.8 g $kWh^{-1}$ with an average coefficient value of 3.0 g $kWh^{-1}$ for 2003/04. The derived NO emission coefficients in these estimates were very close to the values given by Chakraborty et al. (2008) earlier who measured NO emission coefficients in the range of 1.5–3.3 g $kWh^{-1}$ in 2003 with an average value of 2.4 g $kWh^{-1}$ for thermal power plants in India's eastern region. Plantwise NO emissions varied between 1.77 g $kWh^{-1}$ at the Neyveli plant to 4.85 g $kWh^{-1}$ at the Faridabad Plant. Two plants were reported to show the NO emission coefficients between 1.5 and 2.0 g $kWh^{-1}$; 45 plants in the range of emission coefficient between 2.0 and 3.0 g $kWh^{-1}$; and 36 plants in the range between 3.0 and 4.0 g $kWh^{-1}$, while 3 plants showed the values ranging between 4.0 and 5.0 g $kWh^{-1}$ (Mittal et al., 2014).

Sahu et al. (2012) have estimated total $NO_x$ emission from all sectors in India as 2952 Gg $year^{-1}$ for 1991. The $NO_x$ emissions from coal, petrol/diesel, LPG/kerosene, and biofuel are estimated to be around 1190, 931, 231, and 600 Gg $year^{-1}$, respectively. Contribution of coal combustion–related $NO_x$ emission is found to be around 40% during 1991 followed by transport sector (31.5%). The estimated $NO_x$ emissions from coal, petrol/diesel, LPG/kerosene, and biofuel sources were around 1871, 1669, 319 and 628 Gg $year^{-1}$, respectively, for the year 2001 with a corresponding decadal growth of 57%, 79%, 38%, and 5% during 1990s by Sahu et al. (2012) who have also stated that the five largest hotspots from coal use correspond to electric power generation sector. In the BUR (MOEF&CC, 2016), the NO emission was estimated at 2.32 Tg from electricity production. In the SNC (NATCOM, 2012), the NO emission was estimated as 7.75 Gg from electricity production.

## Industrial Sector

In this sector, chemical industries which are the main contributors of $N_2O$ emissions include nitric acid and caprolactum production.

## Nitric Acid

Nitric acid production is mainly driven by fertilizer industry. Nitric acid is used in the production of ammonium nitrate and calcium ammonium nitrate which find applications as fertilizers. Ammonium nitrate is also used in the manufacturing of urea ammonium nitrate (UAN) used in fertilizer formulations and also as explosives. Ammonium nitrate is also used as a blasting and explosive agent in mining and other applications. Calcium ammonium nitrate is used in the manufacturing of solutions and chemicals for the production of synthetic fibers, dyes, and plastics, etc. Nitric acid is also used as an intermediate in the manufacturing of toluene diisocyanate (TDI). Toluene diisocyanate is used in the manufacturing of polyurethane foams that are used in automotive, transportation, carpet, and furniture industries. Nitric acid is used as an intermediate in the manufacturing of nitrobenzene and nitrochlorobenzenes among others. Nitric acid is also used as an oxidizer in rocket propellants. However, the market is facing difficulties owing to environmental and health hazards associated with the nitric acid industry and its future growth potential will continue to be strongly dependent upon the needs of the agricultural industry. Nitric acid is produced in a number of factories in India. Out of these, NFL-NangaL RCF-Trombay, GNFC-Bharuch, and DFPCL-Taloja produce acid for fertilizer manufacture. Combined installed capacity of these nitric acid plants is 1.231 million tons of dilute nitric acid (Indian fertilizer, 2016). These plants primarily use ammonia and air for the manufacture of nitric acid.

The emission of $N_2O$ from nitric acid production has been estimated at about 11.7 Gg in 2007 (NATCOM, 2012) which has been downsized to 7 Gg in year 2010 (MoEF&CC, 2016).

## Adipic Acid

Adipic acid production is one of the applications, which uses nitric acid during its manufacturing process. Adipic acid is used in the manufacturing of nylon 6,6 for various end-user industries such as automotive, textiles, and tire among others. Thus the production trend of adipic acid is closely related with nylon consumption trends. In 1990 adipic acid production was the largest sources of industrial $N_2O$ emissions (IPCC, 2006). However, after the implementation of $N_2O$ abatement technologies from 1999, its emission has been decreased to a great extent. According to Caprolactam (adipic acid) producers of India 10,000 to 12,000 million tons of adipic acids are been annually produced.

India had earlier reported emission 1.05 Gg from adipic acid productions in the Second National Communications of India (NATCOM, 2012) that has been refreshed to about 1.11 Gg of $N_2O$ in 2010 (MoEF&CC, 2016). Therefore, according to India's first BUR (MOEF&CC, 2016) the chemical industries contributed about 8.11 Gg of $N_2O$ emissions in 2010.

## Conclusions

The energy and industry sectors are the major contributors to the national economy. While India is pushing for the development of these sectors in a big way for the economic prosperity of the country, the issues of containing the emissions of reactive nitrogen species from this sector are quite important to ensure sustainable development and better air quality. This would require a judicious mix of technological and policy options. Realizing this, a number of technological and policy instruments have already been initiated in India.

## References


CEA, Central Electricity Authority, 2010. Reviews of Performance of Thermal Power Stations 2009–10. http://www.cea.nic.in/reports/yearly/thermal_perfm_review_rep/0910/highlights.pdf.

MoEF&CC, 2016. India First Biennial Update Report to the United Nations Framework Convention on Climate Change. Ministry of Environment, Forest and Climate Change, Government of India. www.moef.gov.in.

Chakraborty, N., Mukherjee, I., Santra, A.K., Chowdhury, S., Chakraborty, S., Bhattacharya, S., Mitra, A.P., Sharma, C., 2008. Measurement of $CO_2$, CO, $SO_2$, and NO emissions from coal-based thermal power plants in India. Atmospheric Environment 42, 1073–1082.

IBEF, 2016. Economic Survey 2015–16. www.ibef.org/economy/economic-survey-2015-16.

EIA, 2012. IES Database. International Energy Statistics. https://www.eia.gov/cfapps/ipdbproject/iedindex3.cfm?tid=44&pid=45&aid=2&cid=regions&syid=2008&eyid=2012&unit=QBTU.

IPCC, 2006. IPCC Guidelines for National Greenhouse Gas Inventories. Prepared by the National Greenhouse Gas Inventories Programme. In: Eggleston, H.S., Buendia, L., Miwa, K., Ngara, T., Tanabe, K. (Eds.). IGES, Hyama, Kanagawa, Japan (2006).

Indian fertilizer, 2016. http://indianfertilizer.com/frontend/statistics/sectionView? section=statistics&page=Ferttechpart1/ferttech5.htm.

KPMG, 2012. Indian Economic Survey 2011–12 – Key Highlights. KPMG, India.

Mittal, M.L., Sharma, C., Singh, R., 2014. Decadal emission estimates of carbon dioxide, sulfur dioxide, and nitric oxide emissions from coal burning in electric power generation plants in India. Environmental Monitoring and Assessment 186, 6857. http://dx.doi.org/10.1007/s10661-014-3894-3.

NATCOM, 2012. India's Second National Communication to the United Nations Framework Convention on Climate Change, Ministry of Environment and Forests. Government of India.

Sahu, S.K., Beig, G., Parkhi, N.S., 2012. Emerging pattern of anthropogenic $NO_x$ emission over Indian subcontinent during 1990s and 2000s. Atmospheric Pollution Research 3 (3), 262–269.

SECTION

F

# Nitrogen Policies and Future Challenges

# 31

# Issues and Policies for Reactive Nitrogen Management in the Indian Region

S. Bhattacharya[1], T.K. Adhya[2,3], H. Pathak[4], N. Raghuram[5], C. Sharma[6]

[1]IORA ECOLOGICAL SOLUTIONS PVT. LTD., NEW DELHI, INDIA; [2]SOUTH ASIA NITROGEN CENTRE, DPS MARG, NEW DELHI, INDIA; [3]KIIT UNIVERSITY, BHUBANESWAR, ODISHA, INDIA; [4]ICAR-NATIONAL RICE RESEARCH INSTITUTE, CUTTACK, ODISHA, INDIA; [5]GURU GOBIND SINGH INDRAPRASTHA UNIVERSITY (IPU), NEW DELHI, INDIA; [6]CSIR-NATIONAL PHYSICAL LABORATORY, NEW DELHI, INDIA

## CHAPTER OUTLINE



**The Indian Nitrogen Assessment. http://dx.doi.org/10.1016/B978-0-12-811836-8.00031-8**

# Introduction

Nitrogen (N) is the most critical nutrient for primary productivity of agro-ecosystems. Reactive nitrogen (Nr) includes inorganic forms of N such as ammonia ($NH_3$), ammonium ($NH_4^+$), nitric oxide (NO), nitrogen dioxide ($NO_2$), nitric acid ($HNO_3$), nitrous oxide ($N_2O$), nitrate ($NO_3^+$), and organic compounds such as urea, amines, proteins, and nucleic acids. Because of the limited availability of Nr in natural systems, external application of N is required to enhance the productivity of the intensively cultivated agricultural systems. Globally, consumption of fertilizer-N has increased from 14 million tons (Mt) in 1950 to the present level of around 155 Mt (IFA, 2014) and is anticipated to increase further by 2050 to meet the demand of increased food, feed, and fiber production (Sutton et al., 2013). In India, food grain production is closely related to use of fertilizer-N. However, because of the low N-use efficiency of the crops and subsequent losses through higher levels of the food chain including animal and human wastes, a large quantity of this Nr leaks to the environment, adversely affecting the ecosystem (Aneja et al., 2010). Sources of Nr other than fertilizers and manures include sewage and solid wastes, effluents, emissions from firewood or residue burning (agricultural or municipal), fossil-fuel burning for transportation, and electricity and industrial production in India (Abrol et al., 2007, 2008, 2012).

The atmospheric Nr that returns as surface deposits corrodes buildings, bridges and other infrastructures, and also acidifies soils, fertilizes trees and grasslands creating unnatural growth rates, nutrient imbalances ultimately decreasing ecosystem health and biodiversity. Leaching out of the soil, Nr can pollute groundwater and surface water. It promotes eutrophication in coastal ecosystems (Swaney et al., 2015) reducing biodiversity due to lack of oxygen needed for the survival of many aquatic plants and animals species. Human health concerns are also associated with Nr in water and air (Braun, 2007; Gupta et al., 2010). Nitrate in drinking water can be converted to N-nitroso-compounds (NOC) in vivo, which are potent animal carcinogens. Nr in the form of $NO_x$ is an air pollutant that creates ozone and smog, which can trigger respiratory conditions ranging from asthma to even death due to chronic exposure. Increasingly smoggy conditions reported along the Indo-Gangetic plains in India, especially during conditions of tropospheric inversions in winter, impact ~465 million people living in Punjab, Haryana, Delhi, Uttar Pradesh, Bihar, and West Bengal constituting about 40% of the Indian population (Box 31.1). Considering that continued increase in concentration of Nr in the atmosphere impacts ecosystems and public health, it is now imperative to come up with national policies that enable interventions for safe levels of Nr.

This chapter reviews the sources of reactive nitrogen and estimates the baseline $NO_x$ and $N_2O$ emission trends from these sources. Further, based on a review of possible mitigation strategies, the chapter quantifies the possible reduction that can be achieved with respect to business-as-usual scenario.

**BOX 31.1 THE GREAT INDIAN SMOG OF NOVEMBER 2016**

In a recent episode of smog that continued through 7 days in the first week of November 2016, in Delhi, $NO_2$ concentrations measured in the air was in excess of 500 $\mu g/m^3$ which is more than 6 times the ambient standard prescribed by the Central Pollution Control Board (CPCB) (80 $\mu g/m^3$ for residential, industrial, and eco-sensitive areas in a 24 h period). The NO and $NO_x$ concentrations also shot up simultaneously (no ambient air quality standard exists for these gases yet in India). Additionally, dangerous levels of PM2.5 and PM10 were also recorded.

According to the Indian Meteorological Department, the smog of November 2016 was the worst in 17 years. The pollutants were a combination of sources, such as crop-residue burning in Punjab and Haryana and high density of vehicles on road, trapped by anticyclonic weather preventing dispersion of smog. This became public health concern. Schools were closed for 3 days continuously till the pollutants dispersed to avoid exposure. For the first time office goers resorted to wearing masks.

# Drivers of Reactive Nitrogen in India

India's National Communications to UNFCCC maps the emission of $N_2O$ by sources (INCCA, 2010; NATCOM, 2012; BUR1, 2015) and in these N fertilizer application in soils, waste water, commercial and residential sector, electricity generation, chemicals, road transport and crop residue burning together emerge as the sources contributing 95% of the total $N_2O$ emission from the India. The $N_2O$ emission by source is reproduced in Table 31.1 for the years 2000 and 2010. Other Nr species are not part of reporting portfolio of the National Communication. However several papers exist that have individually ascertained the quantities of reactive nitrogen emitted from various sources in India (Singh et al., 2008; Aneja et al, 2010; Adhya et al, 2016). In this section we review the major drivers of Nr ($N_2O$ and $NO_x$) and their trends in India.

## Synthetic N-Fertilizer-Consumption

In India food grain production has increased five times between 1951–52 and 2013–14 (MoA, 2015), whereas the use of synthetic fertilizer has grown by over tenfold (Fig. 31.1) and currently stands at 17.4 Mt in 2015–16, making India the second largest producer and consumer of N in the world. Five major crops namely rice, wheat, maize, sugarcane, and cotton account for about 70% of total N consumption. About 50% of total N is consumed in five states—Uttar Pradesh, Maharashtra, Punjab, Gujarat, and Madhya Pradesh.

**Table 31.1** $N_2O$ Emissions by Source in India as of 2010

| | 2000-$N_2O$ (Gg) | 2010-$N_2O$ (Gg) | CAGR (%) |
|---|---|---|---|
| **Energy industries** | **7.78** | **11.85** | **4.30** |
| Electricity | 7.75 | 11.68 | 4.19 |
| Refinery | 0.03 | 0.07 | 8.84 |
| Solid fuels | NE | 0.3 | – |
| **Manufacturing industries** | **15.2** | **12.28** | **−2.11** |
| Cement | NE | 0.57 | — |
| Iron and steel | 0.85 | 1.5 | 5.84 |
| Nonferrous metals | 0.03 | 0.03 | 0.00 |
| Chemicals | 13.14 | 8.19 | −4.62 |
| Pulp and paper | 0.08 | 0.1 | 2.26 |
| Food and beverage | 0.2 | NE | — |
| Nonmetallic minerals | 0.01 | NE | — |
| Mining and quarrying | 0.02 | 0.04 | 7.18 |
| Textile and leather | 0.08 | 0.03 | −9.34 |
| Nonspecific industries | 0.79 | 1.82 | 8.70 |
| **Transport** | **6.22** | **10.46** | **5.34** |
| Road | 4 | 7.4 | 6.35 |
| Aviation | 0.12 | 0.35 | 11.30 |
| Railways | 2.06 | 2.58 | 2.28 |
| Navigation | 0.04 | 0.13 | 12.51 |
| **Others** | **22.26** | **21.55** | **−0.32** |
| Comm/Instt | 0.03 | 1.08 | 43.10 |
| Residential | 22 | 20.42 | −0.74 |
| Agri/fisheries | 0.23 | 0.02 | −21.67 |
| Biomass energy | NE | 0.03 | |
| **Agriculture** | **192.73** | **268.7** | **3.38** |
| Manure management | 0.07 | 0.08 | 1.34 |
| Agriculture soils | 186.49 | 261.55 | 3.44 |
| Field burning of crop residue | 6.17 | 7.07 | 1.37 |
| **Forest fires** | **6.74** | **1.87** | **−12.03** |
| **Domestic wastewater** | **13.23** | **43.67** | **12.68** |
| **Grand total** | **264.16** | **370.38** | **3.44** |

*NE*, not estimated.

NATCOM II [National Communication II] 2012. India Second National Communication to the United Nations Framework Convention on Climate Change. Ministry of Environment and Forests, Government of India; India-BUR [India First Biennial Update Report to the United Nations Framework Convention on Climate Change] 2015. Ministry of Environment, Forest and Climate Change, Government of India.

Nitrogen use efficiency (NUE, ratio of N-output as biological or economic yield and N-input as external N supply) is, however, quite low and is usually in the range of 30%–35%.

In order to ensure the basic food security, countries tend to subsidize agriculture and the nature of the subsidy varies between developing and developed countries. Most of the developed countries have either moved on or are moving from price support to

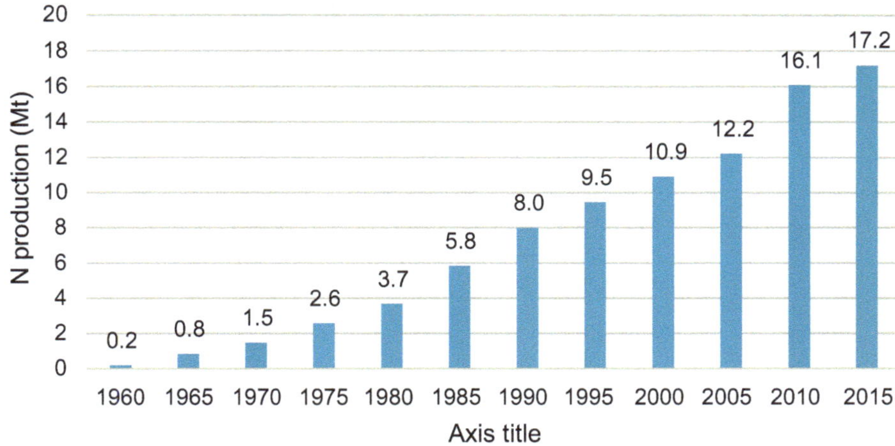


**FIGURE 31.1** Trends of Fertilizer-N consumption in India. *FAI [Fertilizer Association of India] 2015. Fertilizer Statistics 2015-15. sixtieth ed. The Fertilizer Association of India, New Delhi.*

income support. India extends support to agriculture primarily through price policy. The fertilizer sector in India is subsidized by the central government. As per the union budget for FY2016 subsidy for fertilizer is INR 730 billion (US$10.73 billion approx. when estimated at the rate of INR 68 = 1US$). As of November 2016, urea prices still remain subsidized (farmer pays 25% of cost price). On the contrary, the government subsidizes only 34% and 37%, respectively, in the case of DAP and MOP that are decontrolled and pegged to international markets making them much higher priced products as compared to urea. As a result

- There is a tendency of skewed use of NPK ratio among farmers, so much so that the demand has outgrown the domestic production capacities and India is increasingly importing Urea.
- Decreasing response of crops, particularly food grains, to fertilizer use.
- Overuse of urea releases excess of reactive $NH_3$, $N_2$, and $NO_x$ gases leading to degradation of the environment.

## Domestic Sewage and Municipal Solid Waste

Domestic sewage generation is tied to the population of the country and sewage is treated only in urban centers in India, i.e., catering to 0.37 billion out of 1.21 billion people (Census of India, 2011). As per the report of CPCB (2015), sewage generated from urban areas is estimated to be around 62,000 million L per day (MLD) (Fig. 31.2).

Domestic sewage is a major source of water pollution in India, particularly in and around large urban centers (BOBLME, 2011). Out of about 38,000 MLD of sewage generated in 2009, treatment capacity existed for only about 12,000 MLD. The existing treatment capacity was also plagued with operation and maintenance problems (CPCB, 2009). The city of Mumbai (Maharashtra) on the west coast alone generates 2400 MLD,

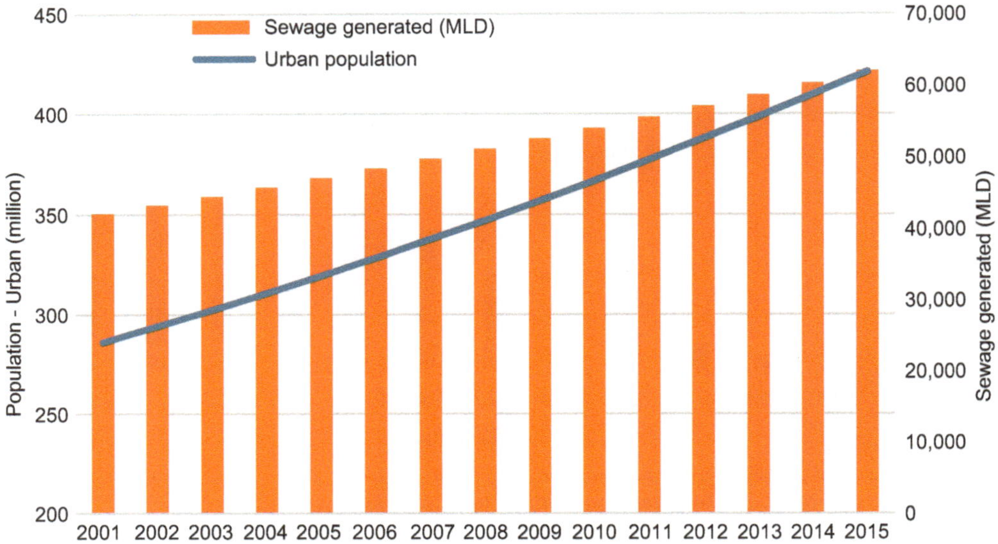


**FIGURE 31.2** Urban population and estimated sewage generation trends.

i.e., 38% of the total wastewater generated. It also has treatment capacity of 2130 MLD (88% of the sewage generated). Similarly, Chennai (Tamil Nadu) and Kolkata on the east coast generate large amounts of wastewater. It is therefore clear that if the wastewater generated by the larger cities is treated properly before disposal, a large proportion of the pollution due to municipal waste disposal could be addressed.

The sewage treatment capacity built up in India so far is 23,277 MLD (CPCB, 2015), of which only 30% is operational. This is because out of the 816 Sewage treatment plants (STPs) set up, 522 are operational, 79 nonoperational, 145 are under construction, and 70 are proposed (CPCB, 2015). Technologies such as activated sludge process (ASP), upflow anaerobic sludge blanket reactor (USAB), moving bed bioreactor (MBBR), sequential batch reactor (SBR) and waste stabilization on pond systems (WSOPS), and aerated lagoon and Karnal technology/sewage-fed agroforestry are used for sewage treatment. In the identified stretches of various rivers under the National River Conservation Plan (NRCP) and National Ganga River Basin Authority (NGRBA) program, STPs are being setup on a cost-sharing basis between the central and state governments. Various pollution abatement schemes taken up under these programs include interception and diversion of raw sewage, construction of sewerage systems, setting up of sewage treatment plants, low-cost sanitation facilities, improved electrical crematoria, and river-front development. As a result, by July 2016, an additional sewage treatment capacity of 5169 MLD was created under these programs. STPs are also being set up in various cities under Atal Mission for Rejuvenation and Urban Transformation (AMRUT) and the Namami Gange Program (Lok Sabha, 2016).

Municipal solid waste management (MSWM) is another major environmental problem of Indian cities (Sharholy et al., 2008). Of the total 62 million tons of municipal solid waste generated annually, 82% is collected and the rest is littered. Only 28% of the total waste collected is being treated and disposed. The average MSW generation in India is 450 gm/per capita/day, which ranges between 200-300 gm/capita in small towns, between 300-400 gm/capita in medium sized cities and 400-600 gm/capita in large cities (MNRE, 2016). Improper management of municipal solid waste (MSW) is hazardous for soil, water, biodiversity and human habitation in and around the dump site.

## Commercial and Residential Sector

A study by the Petroleum Planning & Analysis Cell of the Ministry of Oil and Natural Gas (PPAC, 2013) indicates that to overcome the peak demands of electricity supply gaps, the commercial and institutional sector in smaller towns, cities, and villages, use diesel and heavy fuel oil to power generators. Additionally, mobile towers in the telecom sector need diesel to operate base transceiver stations (BTS). Table 31.2 indicates the distribution of nontransport diesel use, and share of commercial use of diesel within that across North, East, West and South zones of India.

The residential sector in India is one of the largest consumers of fuel outside the energy industries. Firewood use for cooking and space heating is a key component of fuel mix for this category, especially in the rural areas, and is a source of GHGs and other pollutants. As per the 68th round of all-India level NSSO survey on fuel used for cooking and lighting (NSSO, 2015), firewood and chips were used by 67.3% of rural households, while 15.0% households used LPG. Only 9.6% and 1.1% of the rural households used dung cake and coke and coal, respectively, as a primary source. In the urban areas, 68.4% of the urban households used LPG, 14.0% of the households used firewood and chips, and 5.7% of the households used kerosene. Fig. 31.3 indicates the trend of firewood and LPG use between 1990 and 2011 among rural and urban households.

## Electricity Generation

Total power generation capacity in India has increased from about 97.875 GW in 2000 to 298.06 GW in 2016. It registered a 7.3% CAGR. Coal-based energy generation dominates but its share has only increased by 1 percentage point between 2000 and 2016 (Fig. 31.4)

**Table 31.2** Zone-Wise Share of Commercial Use of Diesel Within Total Diesel Use in the Nontransport

| Categories | North Zone (%) | East Zone (%) | West Zone (%) | South Zone (%) |
|---|---|---|---|---|
| Nontransport | 36.7 | 26.9 | 18.8 | 23.7 |
| Mobile towers | 1.82 | 2.6 | 1.45 | 1.60 |
| Gensets | 6.42 | 2.8 | 2.18 | 5.58 |
| Agriculture pump sets | 4.55 | 4.33 | 2.26 | 1.98 |

PPAC [Petroleum, Planning and Analysis Cell of the Ministry of Oil and Natural Gas] 2013. Report – All India Study on Sectoral Demand of Diesel & Petrol, Ministry of Oil and Natural Gas, Govt. of India, Nielsen, New Delhi, pp. 1–103.

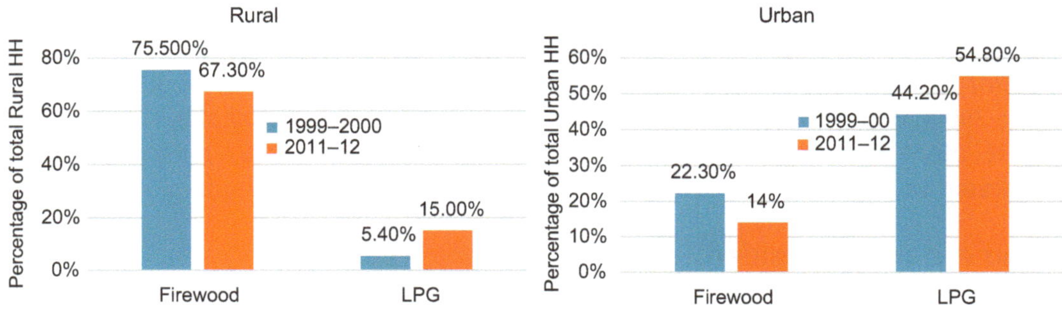


**FIGURE 31.3** Consumption trends of firewood and LPG in rural and urban households.

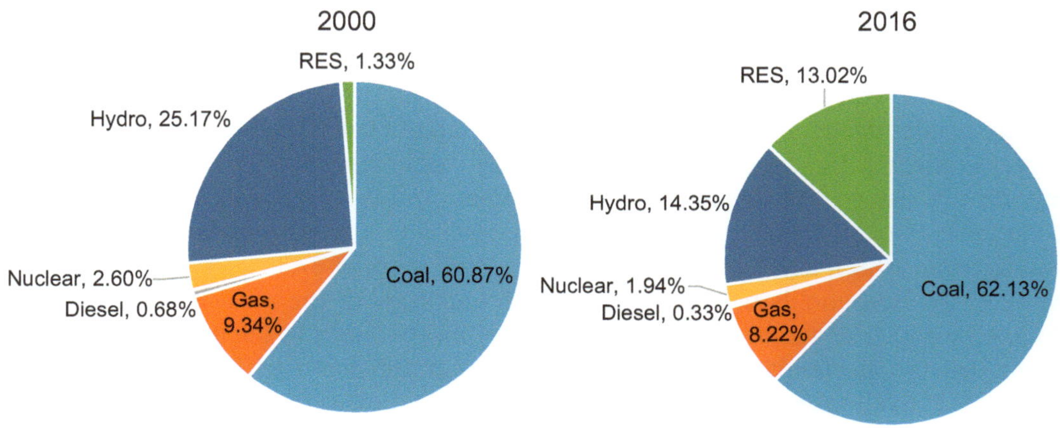


**FIGURE 31.4** Trend of energy mix for electricity generation in India.

whereas the share of renewable energy has increased from 1% to 13% between the same period. In spite of the increase in the share of renewable energy, $N_2O$ emissions continue to rise due to continued use of gas, diesel, and coal for electricity generation in the country.

## Nitric Acid Production

In india as of 2015-16 1.2 milion tons of nitric acid was produced by 5 companies (Anual reports of respective companies). The principal chemical produced from nitric acid is ammonium nitrate. Most of this is used in fertilizers, and a small amount is used to make explosives. Some nitric acid is used as an intermediate in the polymer industry for manufacturing polyamides and polyurethanes. In 2015-16 about 1.2 million tons of nitric acid was produced in India. Emission of $N_2O$ from nitric acid production has been estimated at 11.7 Gg in 2007 (NATCOM II, 2012) and has been subsequently revised at 7 Gg (India-BUR, 2015).

## Road Transport

The total number of registered vehicles in India has grown more than three times from 55 million in 2001 to around 159 million in 2012 (MoSPI, 2016). A compounded annual growth rate of 10% is registered between the period. The private vehicle feet continuous to grow at a fast rate. The high level of infusion of private vehicles on road is due to easy accessibility to loans from banks and low penetration of mass rapid transport system in cities with last mile connectivity.

At the moment all class 1 cities have introduced metro's, with Delhi having an extensive coverage, covering Delhi with the satellite cities of Gurugram and Noida within the National Capital Region. The other cities are yet to catch up, though many class II cities have MRTS projects in the pipeline. Bus Rapid Transit Systems has been successful in Ahmedabad but has been a failure in Delhi because of design faults.

The road transport sector In 2003–04 emitted 2.2 million tons of $NO_x$ (Ramachandran and Shwetmala, 2009). In the same year $N_2O$ emissions were of the order of 3 Gg (authors estimate) from the sector. The major contributors to the $NO_x$ and $N_2O$ emissions being the diesel operated heavy vehicles and cars.

## Crop Residues

Crop-residue burning is an annual event when farmers prepare the field for next cycle of wheat cropping in the winter. This is particularly debilitating for the environment as an inversion layer at the starting of the winters does not allow the pollutants to disperse resulting in corridors of smog created across the Indo-Gangetic plains (Fig. 31.5). It is

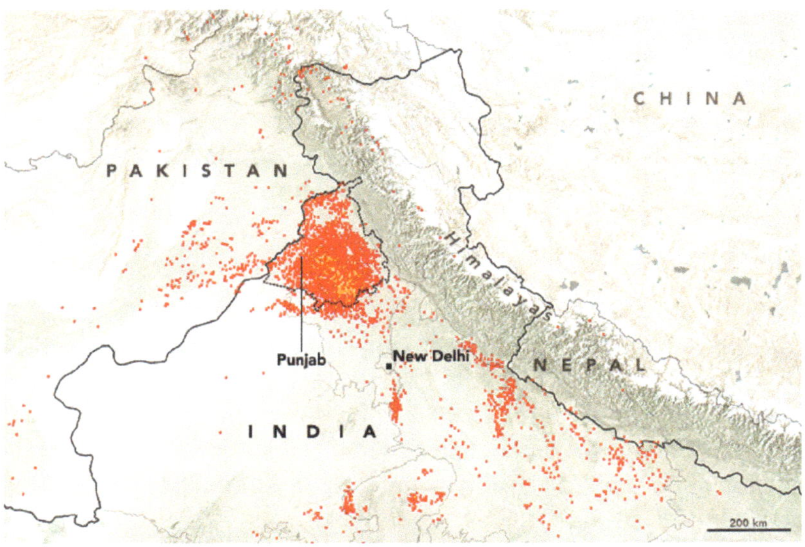


**FIGURE 31.5** Crop-residue burning visible from space during November 2016. *http://earthobservatory.nasa.gov/IOTD/view.php?id=89052*

estimated that an amount of 51.1 million tons of crop residue was generated in 2000–2004 representing 76.7% of total agricultural and forest residues. A major part of crop residues is burnt every year (Biomass Resource Atlas of India: http://lab.cgpl.iisc.ernet.in/atlas/Tables/Tables.aspx) for reasons of convenience and commonly adopted agro-practice.

The fraction of brunt crop residues varied across states and crops. Rice straw is not burnt uniformly and varies from 8% to 80% depending on the usage and knowledge of the farmer, and is estimated to release 779 Mg of $N_2O$ (Gadde et al., 2009). The percentage of wheat residue burnt in Haryana, Punjab, Uttar Pradesh, and Himachal Pradesh is 13% more than rest of the India, i.e., 10% (Jain et al., 2014). Approximately, 10% of cotton and maize residue are burnt, while sugarcane residue is as high as 25% (Jain et al., 2014). Policies to curb the burning of crop residue are not really effective.

## Policy Options for Scientific Nitrogen Management and Potential Emission Reduction

Developing policies to control unwanted Nr release is difficult because much of the release is related to food and energy production, and the Nr species can be transported to a great distance in the atmosphere and aquatic systems. There are various possibilities for limiting Nr emissions from fuel combustion for energy production, transport, and chemical industries. However, reducing the entry of new Nr and curbing the movement of this Nr in food production is more difficult. This particular problem comes from the fact that most of the N introduced into the global food production system is not converted into usable product, and rather reenters the biosphere as a surplus. It is difficult to establish a global policy on N in agriculture because all the countries need to increase food production to raise nutritional levels or to keep up with population growth, which will require enhanced use of fertilizer-N. Although N cycling occurs at regional and global scales, policies are implemented and enforced at the national or provincial/state levels. Multinational efforts to control N loss to the environment are necessary, but these efforts will require commitments from individual countries and the policy makers within those countries.

### Agriculture Soils

Agriculture in India highly relies on fertilizer-N. Fertilizer-N consumption on a per-hectare basis increased from 0.44 kg $ha^{-1}$ in 1951–52 to 86.2 kg $ha^{-1}$ in 2013–14 (FAI, 2015). This value has now increased to 98.5 kg $ha^{-1}$ and the country stands third among its peers in the South Asia with Bangladesh (144.7 kg $ha^{-1}$) as the leading country, followed by Pakistan (101.8 kg $ha^{-1}$) (Adhya et al., 2016). Annual budget of N in agricultural soils in India is showed in Fig. 31.6.

Both Govt. of India and the fertilizer industry are involved in devising a pragmatic fertilizer policy and its implementation enabling easy access of fertilizers to the farming

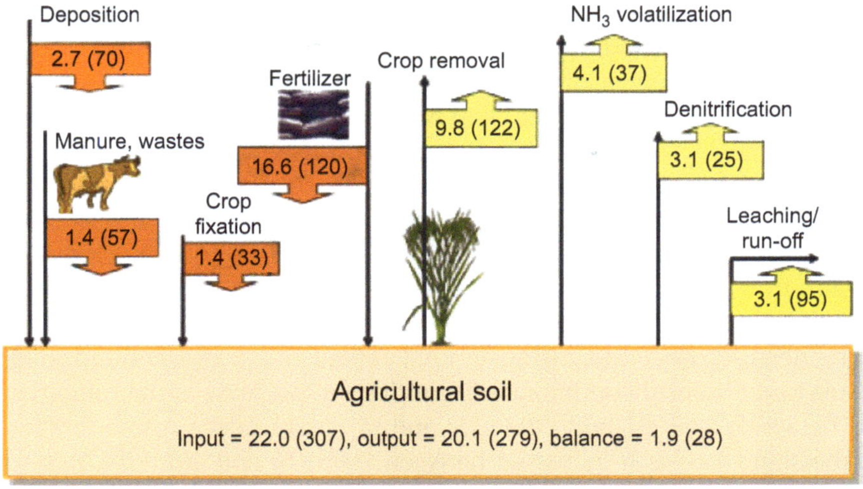


**FIGURE 31.6** Annual budget of N (Mt) in agricultural soils of India (values in parenthesis indicate global values from Sutton et al. (2013) during the year 2010. *Pathak, H., 2016. Is nitrogen fertiliser a polluter or producer? The answer lies in its use-efficiency. Indian Journal of Fertility 12, 106–117.*

community. A plethora of schemes on behalf of the government with focus on fertilizer nutrients especially fertilizer-N such as National Food Security Mission (2007), *Rastriya Kishi Vikas Yojana* (2007), National Mission for Sustainable Agriculture (2010), Nutrient-based subsidy (2010), Bringing Green Revolution in Eastern India (2010–11), Soil health card scheme (2015) are put into operation.

In its 2005–06 report, the Task Force on "Balanced Use of Fertilizers" recommended nutrient-based subsidy instead of the current product-based regime. In addition to it, the task force also recommended for (1) encouragement to the usage of organic manure, biofertilizers, and NPK-mixture fertilizers along with judicious use with chemical fertilizers; (2) application of soil-specific and crop/climate–specific nutrients; (3) the basket of the fertilizers subsidy to be extended to other fertilizers, covering secondary and micronutrients, to ensure their optimal use by the farmer for mitigating the deficiencies and sustaining balanced fertilization; (4) strengthening the soil-testing laboratories, fertilizers quality control laboratories, and prepare soil-fertility maps in the country; and (5) fortification of the major fertilizers with the appropriate grade of secondary/micronutrients, customized and value-added fertilizers. The fertilizer industry also on its own recognized 4Rs nutrient stewardship, covering (1) right source to match the fertilizer type to crop needs; (2) right rate to match the amount of fertilizer to crop needs; (3) right time making nutrient available when crops need them; and (4) right place keeping nutrients where crops can use them, and desires to improve farmers' awareness on the issue.

Indirect emissions following deposition of $NH_3$ and $NO_x$ from fertilizers are subsequently reemitted as $N_2O$. Based on the trends of N applications in agricultural fields in India (see Chapter 3), it is estimated that the amount of $N_2O$ emitted has increased from

0.001 million tons in 1960–61 to around 0.26 million tons in 2015–16. The estimate is based on the standard IPCC Tier-1 methodology (IPCC, 2006) for direct $N_2O$ emissions from soils and a default emission factor (Mosier et al., 2002) that encompasses diverse crops and soils (0.0125 kg $N_2O$-N per kg of N applied). The CAGR of $N_2O$ emissions was 13% between 1960–61 and 2000–2001 due to rapid rise in fertilizer-N application, but was around 2% till 2015–16 as N application has nearly plateaued, though at very high values. In a BAU (Business As Usual) scenario, it is assumed that N application will continue to have the same rate between 2000–01 and 2015–16. In such a scenario, the $N_2O$ emission from N application in India is likely to rise to 0.34 million tons.

In this background, we assume that (1) from 2018 onward, 100% urea manufactured in India will be neem-coated; (2) subsidy on urea as well as P and K continues; (3) by 2030 the total area under rainfed conditions will continue to decrease at the same rate as it has decreased between 2000–01 and 2014–15; and (4) area under organic fertilizer will increase at the same rate as it increased between 2009–10 and 2013–14. Based on these assumptions and considering that the measures mentioned in the schemes as listed in Table 31.3, it is estimated that around 30% or 0.08 million tons of $N_2O$ emission can be avoided as against BAU by 2030 (Fig. 31.7).

In spite of extensive use of fertilizer-N, there is a huge disparity in Indian agriculture both in terms of quantum of fertilizer-N applied vis-à-vis judicious use of nutrients. In spite of a large array of policies, harnessing actual benefit lies in their useful implementation. The basic philosophy of scientific management of fertilizer-N lies in increasing the N-use efficiency. An integrated whole-farm approach is necessary for reversing the declining N-use efficiency in agriculture. It remains to be seen whether the $N_2O$ emissions from unused synthetic fertilizers are significantly impacted by the recent policy initiatives of the government of India, such as nutrient-based subsidy and neem-coated urea (as the only form of urea allowed to be manufactured in India). However, there is a growing realization of the need for improvement in fertilizer-N-use efficiency and the Indian government has stepped up its R&D funding and international collaborations in this area.

## Wastewater

The items with the highest N fraction in this system are sewage and wastewater along with food wastes (Reis et al., 2016). Human waste contains proteins that are eventually converted to ammonia-N and, if left untreated, this ammonia flows into surrounding water bodies polluting the same. The estimates available in India's national Communication to UNFCCC indicate that wastewater (domestic and industrial together) is the second largest emitter of $N_2O$ next only to soils. Using an updated Global NEWS model, Seitzinger et al. (2010) projected that between 1970 and 2000, South Asia alone accounted for over half of the global increase in dissolved inorganic N (DIN) export to seas, with India having the highest contribution just because of high density of population. It is therefore clear that if the wastewater generated by the larger cities is

**Table 31.3** Scenario Constructs Enabling Reduction of $N_2O$ Emissions From Application of Urea

| Measure | Status | Timelines |
|---|---|---|
| Policy to encourage usage of organic manure and biofertilizers | A central government scheme, *Paramparagat Krishi Vikas Yojana* (2015), is initiated to promote organic farming in clusters of villages across India. The scheme also ensures Participatory Guarantee System for organic certification. Around 60% of the rainfed cropland including the hilly areas is the focus of this scheme. As a result, the certified cultivated area under organic farming has grown from 0.455 m ha in 2009–10 to 0.723 m ha in 2013–14, with around 0.6 million farmers practicing it. The government will run this scheme for the next 3 years. *Ref:* http://agricoop.nic.in/sites/default/files/RevisedGuidlinePKVY_10122015.pdf | It is assumed that 50% of the total area cultivated may remain unirrigated by 2030. We are assuming that no synthetic N application will take place here and whatever organic fertilizer is applied, the N from the same will be fully absorbed by the crop. |
| Fertilizer subsidy extended to micronutrients to ensure optimal use of nutrients and fertilizer-N | Given that India needs to ensure its food security, it might not be entirely possible to do away with urea subsidy. However, extending subsidy to other necessary nutrients will encourage judicious use of all types of fertilizers. This is a major policy recommendation that is yet to start. | It is proposed that the subsidy on other essential nutrient will start taking effect from 2018 to 19 onward, and by 2030 at least 50% of the area under irrigation will reduce its urea and hence N intake by 50% (from an NPK ratio of 8:3:1 to 4:2:1). |
| Fortification of the major fertilizers with the appropriate grade of secondary/micronutrients, customized and value-added fertilizers. | Again, this is a major area of policy intervention which is yet to start. | It is assumed that the subsidy policy for other nutrients will make business sense for fortification of fertilizers. Locally customized mixtures can be sold on the basis of the soil-data analysis that is going on. It is assumed that this intervention will cover the entire country by 2015 if it is allowed to start from 2018 onward. |

treated properly before disposal, a large proportion of the pollution due to municipal waste disposal could be addressed.

There are 7935 towns in India (Census of India, 2011) which are all potential wastewater generators and hence emitters of $N_2O$. The CPCB has issued directions under Section 18 1(b) of the Water (Prevention and Control of Pollution) Act, 1974 in April, 2015 to the State Pollution Control Board/Pollution Control Committees, directing them to set up sewage treatment plants and utilize treated sewage generated in their respective

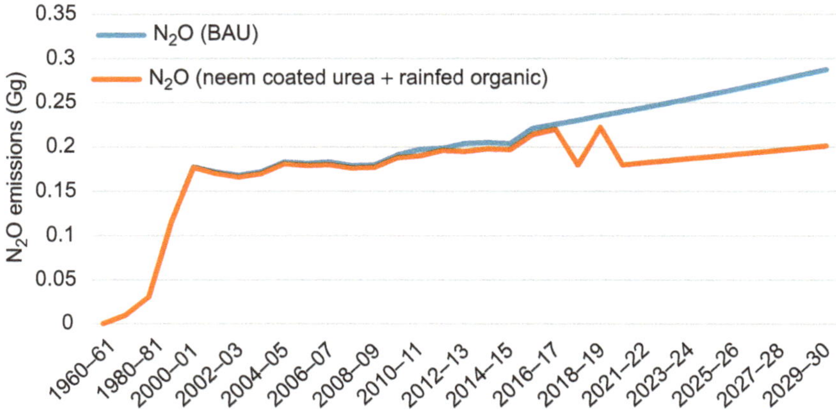


**FIGURE 31.7** $N_2O$ emission due to urea application in a BAU and mitigation scenario.

states. Additionally, CPCB has issued directions in October, 2015 to 178 most polluting towns under Section 5 of Environment (Protection) Act, 1986 for proper treatment and disposal of sewage generated in their jurisdiction. While the nutrient benefits of domestic sewage provides a viable option for farmers from semiurban and urban areas to adopt municipal wastewater-based agricultural practices, it may also give rise to serious environmental and public health hazards in addition to environmental impacts in terms of groundwater pollution and biological accumulation of various emerging pollutants.

To control discharge of industrial effluents, CPCB and respective State Pollution Control Boards/Pollution Control Committees monitor industries with respect to effluents discharge standards and take action for noncompliance under the Water (Prevention and Control of Pollution) Act, 1974 and the Environment (Protection) Act, 1986. To improve the monitoring of compliance, directions are issued to specific industries to install online 24 × 7 effluent and emission monitoring devices. Steps have also been taken by CPCB to promote low waste and no waste concepts leading to zero liquid discharge by grossly water-polluting industries, particularly those located on the river banks.

The Union Ministry of Environment, Forest and Climate Change (MoEF&CC), Govt. of India revised Solid Waste (Management and Handling) Rules 2016 replacing Municipal Solid Wastes (Management and Handling) Rules 2000 after 16 years. The jurisdiction of the rules was extended beyond municipal areas to cover outgrowths in urban agglomerations, census towns, notified industrial townships, areas under the control of Indian railways, airports, airbase, port and harbor, defense establishments, special economic zones, state and central government organizations, and places of pilgrimage, religious, and historical importance. Although the revised dispensation identifies the role of wastes as nutrients and emphasizes on its recycling, it puts focus on waste as organic C source as compost. The regulation identified the Department of Fertilizers, Ministry of Chemicals and Fertilizers, Govt. of India to provide market

**Table 31.4** Probable Scenario for Mitigating $N_2O$ From Waste Water

| Action | Level of Mitigation | References |
|---|---|---|
| Maintaining low concentrations of ($NH_4^+$)and $NO_2^-$ in the reactor through step feeding of wastewater | Can result in a 50% reduction in $N_2O$ production | Yang et al. (2009) |
| Avoiding transient pH changes under aerobic conditions by slow feeding rather than pulse feeding and an enriched ammonia oxidation bacteria culture | Significant reduction in $N_2O$ production can be achieved | Law et al. (2012). |
| Application of longer solids retention time to increase the AOB biomass concentration (greater than 5 days) and higher DO (>0.5 mg $O_2$ l−1) | Can minimize $N_2O$ production from nitrification | Zheng et al. (1994) |
| Minimize $N_2O$ emission through sequential aeration in a membrane-aerated biofilm reactor where the $N_2O$ produced by AOB within the membrane bundle is consumed by heterotrophic bacteria outside the bundle | Almost 100% mitigation can be achieved | Pellicer-Nàcher et al. (2010) |
| Minimize $N_2O$ production during denitrification by adding methanol that prevents $N_2O$ accumulation by eliminating electron competition from other denitrifying enzymes | Almost 100% mitigation can be achieved | Park et al. (2000). |

development assistance on city compost and ensure promotion and co-marketing of compost with chemical fertilizers in the ratio of 1:2 by the fertilizer companies to the extent compost is made available for marketing. The Ministry of Agriculture, Govt. of India was requested to provide flexibility in fertilizer-control order for manufacturing and sale of compost, propagating utilization of compost on farmland, and set up laboratories to test quality of compost produced by local authorities or their authorized agencies. This will make the compost plants economically viable and improve the gainful utilization of wastes.

Several mitigation strategies were experimented to minimize $N_2O$ emissions. Some of them are listed in Table 31.4. None of them are yet commercially available. However, given the imperative that India is on the path toward installing CTPS and STPs to depollute its rivers polluted due to direct discharge of wastewater in them, it is assumed that actions will be taken in the near future to abate the $N_2O$ production from sewage and create facilities that can enable safe use of N in the sludge as fertilizer for agriculture purpose, as worldwide sewage can provide 25% of the total N requirement of agriculture.

Some of the mitigation strategies that are being experimented to reduce $N_2O$ in waste water are listed in Table 31.4. We assume that between 2018 and 2030 about 50% of the urban population wastewater will be treated through sequential aeration in a membrane-aerated biofilm reactor, where the $N_2O$ produced by AOB within the membrane bundle is consumed by heterotrophic bacteria outside the bundle so as to make $N_2O$ emission zero. Consequently, 50% of emission reduction can be achieved by 2050 with respect to BAU scenario (see Fig. 31.8).

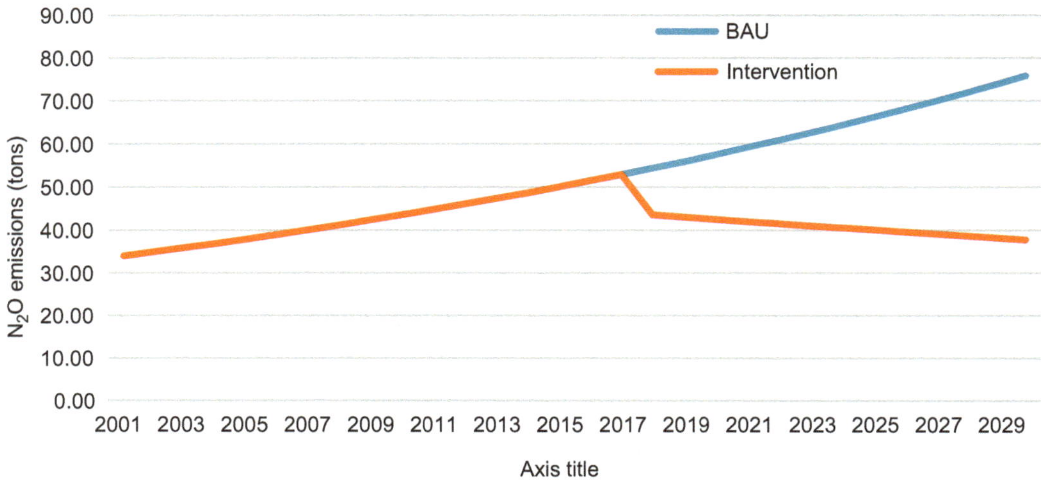


**FIGURE 31.8** BAU and mitigation scenario of $N_2O$ from wastewater.

## Road Transport

$NO_x$ emissions from combustion of diesel and gasoline are targeted through the Bharat Stages I to IV norms for different types of vehicles which are patterned along the Euro emission norms for vehicular pollutants. Table 31.5 below indicates the set of norms in place under Bharat Stages III and IV and timeline of nationwide implementation. Bharat Stage VI norms will be in place next by 2020, and India will directly endorse these stricter norms jumping stage V. Figs. 31.9A and B below show the BAU and intervention scenarios for $N_2O$ and $NO_x$ from diesel vehicles in India.

Another policy initiative taken by the government is to blend 10% ethanol in diesel-driven vehicles. Ethanol contains 35% oxygen and adding oxygen to fuel results in complete fuel combustion, reducing emissions of $NO_x$ along with black carbon and CO (Praptijanto et al., 2015). Further, buses and cars in India are increasingly running on CNG and LPG to combat high levels of urban pollution. As of 2015–16, 2155 TMT CNG was sold in India (MoPNG, 2016). CNG and LPG have clean burning properties. CNG produces 58% less $NO_x$ than diesel (Frailey et al., 2000). In Delhi city all buses run on CNG, so do the three wheelers.

Over and above these initiatives, drastic reduction in emissions is possible if the recommendations on voluntary fleet modernization policy are endorsed. The policy suggestion is to phase out buses, heavy commercial vehicles, light commercial vehicles, and commercial cars and UVs that are more than 15 year old. According to a study carried out by the Petroleum Planning & Analysis Cell (PPAC, 2013), these categories of vehicles together consume 51.1% of the total diesel combusted by the road transport sector. Considering that we are well into 2017, the dictate may be implemented from 2018 onward. A buy-back pricing policy for scrap is also being considered within this

**Table 31.5** Details of Interventions Being Carried Out and Planned in the Road Transport Sector to Reduce Pollutant Emissions From the Road Transport Sector in India

| Bharat Stage (BS) | $NO_x$ Norms (g/km) | $NH_3$ (PPM) | Implemented in India From |
|---|---|---|---|
| **Gasoline Passenger Cars** | | | |
| BSIII | 0.15 | – | 2010 onward |
| BSIV | 0.08 | – | 2017 onward |
| BSVI | 0.06 | – | 2020 onward |
| **Diesel Passenger Cars** | | | |
| BSIII | 0.50 | – | 2010 onward |
| BSIV | 0.25 | – | 2017 onward |
| BSVI | 0.08 | – | 2020 onward |
| **2-Wheeler-Gasoline** | | | |
| BSIV | 0.20–0.39 | – | 2016 onward |
| BSVI | 0.06–0.09 | – | 2020 onward |
| **3 Wheeler-Gasoline** | | | |
| BSIV | 0.40–0.79 | – | 2016 onward |
| BSVI | 0.085–0.10 | – | 2020 onward |
| **Heavy Duty Trucks and Buses-Diesel** | | | |
| BSIII | 5.0 | – | 2010 onward |
| BSIV | 3.5 | – | 2017 onward |
| BSVI | 0.40–0.46 | 10 | 2020 onward |

program as an incentive to do away with the older vehicles. Our estimates indicate that by 2030 the emissions of both $N_2O$ and $NO_x$ will reduce by around 49% with respect to the BAU expected emissions of 18.8 Gg and 3839.5 Gg, respectively, the emission factors of $N_2O$ and $NO_x$ being 3.9 and 800 kg/TJ, respectively (IPCC, 2006; Ramachandra and Shwetmala, 2009).

However, it is to be noted that all-out efforts for mitigating $NO_x$ may lead to increase in $N_2O$ emissions as has been noted in EU (EU, 2016). It states that in EU, by 2030, there will be a rise in $N_2O$ emissions from 12% in 2005 to 15% in 2030 in spite of the various stringent EU norms that are imposed on $NO_x$ and $NH_3$ emission from transportation that India also follows as Bharat norms.

## Energy Sector

In this section we discuss possible mitigation scenarios for abating $NO_x$ emissions from fuel combustion in power generation, mobile towers, gen sets, agriculture pump sets, tractors, in residential and commercial sectors, and from Industries. It is clear that substantial $NO_x$ emissions are emitted from power generation from coal and from

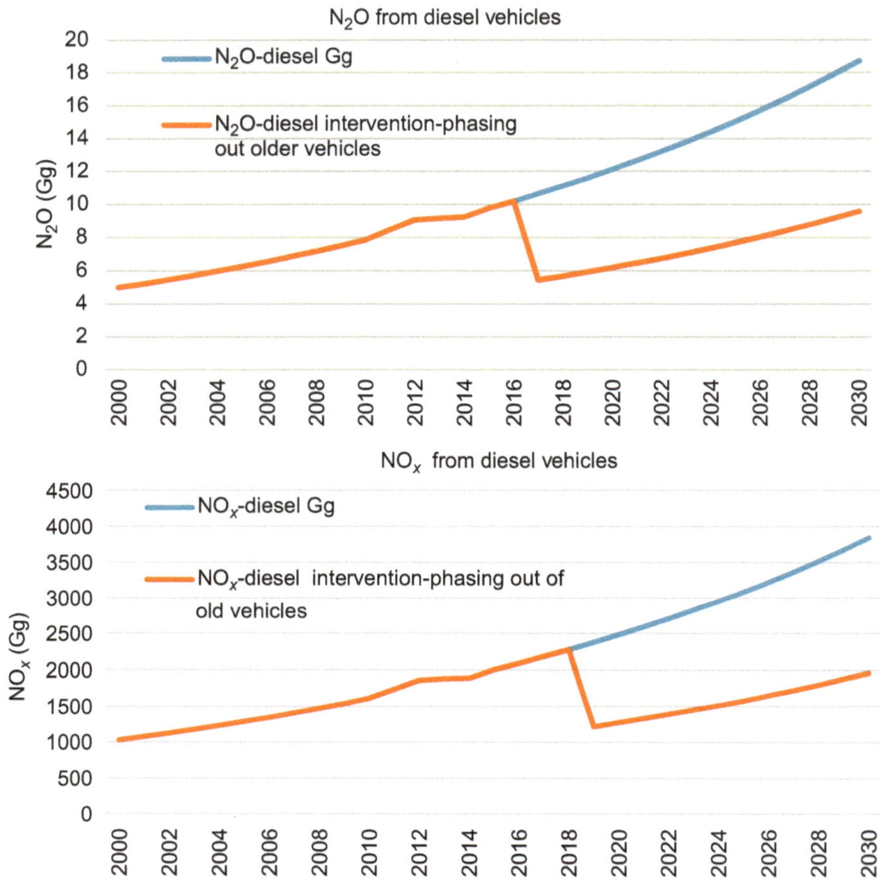


**FIGURE 31.9** Impact of phasing out of diesel vehicles (buses, heavy commercial vehicles, light commercial vehicles, and commercial cars and UVs) on $N_2O$ and $NO_x$ emissions from diesel-run vehicles.

firewood combustion in rural households (see Table 31.6). Mitigation pathways identified in Table 31.6 for reducing $NO_x$ emissions from coal based power generation indicate that about 30-90% reduction is possible using any or a combination of technologies (Bin et al, 2015).

Firewood burning for cooking will be a difficult proposition though other sources of energy including LPG are being propagated in India. This is because firewood is a free commodity for rural households and in far-flung areas without proper connectivity. However, through "Pradhan Mantri Ujjwala Yojana" launched on May 01, 2016, government has introduced an ambitious scheme for providing LPG connectivity to poor rural households without any administrative cost and is expected to benefit 50 million poor households by 2019. Also, improved cooking stoves (*Chulhas*) using biomass can be aggressively marketed, for which the central government has incentives designed.

**Table 31.6** Details of Fuel Use, $NO_x$ Emission and Mitigation Pathways to Reduce Pollutant Emissions From the Energy and Industry Sector in India

| | Fuel | Consumption ('000 tons) 2015–16 | $NO_x$ Emissions (Gg) | Mitigation Pathways | Impact |
|---|---|---|---|---|---|
| Electricity generation | Coal[c] | 566,572 | 3059.49 | • Low $NO_x$ burners<br>• Over fire burners<br>• Fuel re-burning<br>• Fuel biasing<br>• Fuel gas recirculation<br>• Selective catalytic reduction<br>• Selective non-catalytic reduction | 30%–90% reduction |
| | Oil[c] | 197 | 1.53 | All above except for fuel re-burning | 30%–90% reduction |
| | Gas[c] | 10,720 | | **1.** Reducing peak temperatures<br>**a.** Natural gas re-burning (NGR)<br>**b.** Low $NO_x$ burners (LNB)<br>**c.** Inject water or steam<br>**d.** Reduced air preheat<br>**e.** Catalytic combustion<br>**2.** Reducing residence time at peak temperature<br>**a.** Air staging of combustion inject steam<br>**3.** Chemical reduction of $NO_x$<br>**a.** Selective catalytic reduction (SCR)<br>**b.** Selective non-catalytic reduction (SNCR)<br>**c.** Fuel re-burning (FR)<br>**d.** Low $NO_x$ burners (LNB)<br>**4.** Using sorbent in ducts | 60%–90% reduction |
| Mobile towers | Diesel[a] | 1286 | 11.06 | Solar | Zero emission |
| Industry | Diesel[a] | 1533 | 13.19 | Solar | Zero emission |
| Gensets | Diesel[a] | 3154 | 27.12 | Solar | Zero emission |
| Agri-pump sets | Diesel[a] | 2420 | 20.81 | Solar | Zero emission |
| Tractors | Diesel[a] | 5559 | 47.81 | Removing more than 15-year old vehicles | 50% reduction expected |
| Residential | LPG[a] | 10,570 | 5.07 | Solar powered induction stoves | 0 |
| | Wood[b] | 87,600 | 1120.4 | Efficient *chulhas* (burners) | 50% reduction possible |

[a]MoPNG [Ministry of Petroleum and Natural Gas] 2016. Indian Petroleum and Natural Gas Statistics 2015–16. Economics and Statistics Division, Ministry of Petroleum and Natural Gas. Govt. of India, pp. 1–216. http://petroleum.nic.in/docs/pngstat.pdf.

[b]Wood values have been taken as 4 kg/day in 60% of Indian households in rural areas.

[c]CSO [Central Statistics Office] 2016. Energy Statistics India-2016. Ministry of Statistics and Programme Implementation, Govt. of India, New Delhi. Available at:

## Crop Residue Burning

Crop residue burning is convenient to the farmers, because narrow window get a get a between the wet season harvest and the dry season cropping, forcing the farmers to burn the residues to vacate the fields. While several management practices are suggested, some of them involve costly equipment while others involve microbial residue management strategies, none of them were successful on large-scale fields. Some administrative measures such as legal sanctions to biomass burning in Punjab could also have not become successful due to lack of supervision. Making equipment and manpower available to cut the crop residue at a large scale needs to be devised for overcoming this cultural practice prevalent almost all over. India Recently the Government is mulling over systematically buying over the crop residue from farmers and using it for energy production.

# Conclusion

India is faced with multilayered problems at present. On one hand, a sizable section of people in India (similar to several other developing countries and LDCs in South Asia and Africa) still lack improved sanitation facilities. On the other hand, the country increasingly faces a major challenge to secure energy and food security with rising demands of the burgeoning population. Given the level of environmental challenges in the country, N-loading into the atmosphere arising out of various sources often leads to underproductivity and even deaths. A scientific and policy-based socioeconomic approach for wastes will go a long way in solving large parts of such problems. Some of the policy mechanisms through which Nr reaction can be reduced are also considered. It is clear that technologies exist and have to be implemented as policies backed by sufficient funds.

# References

Abrol, Y.P., Raghuram, N., Chanakya, H.N., 2008. Reactive nitrogen in indian agriculture, environment and health. Current Science 94 (11), 1375–1477.

Abrol, Y.P., Pandey, R., Raghuram, N., Ahmad, A., 2012. Nitrogen cycle sustainability and sustainable technologies for nitrogen fertilizer and energy management. Journal of Indian Institute of Science 92, 37–62.

Abrol, Y.P., Raghuram, N., Sachdev, M.S. (Eds.), 2007. Agricultural Nitrogen Use and its Environmental Implications. IK International, New Delhi, p. 552.

Adhya, T.K., Pathak, H., Raghuram, N., Abrol, Y.P., 2016. Reactive nitrogen assessment in South Asia. Current Science 111, 782–783.

Aneja, V.P., Schlesinger, W.H., Erisman, J.W., Sharma, M., Behera, S.N., Battye, W., 2010. Reactive nitrogen emissions from crop and livestock farming in India. In: Pathak, H. (Ed.), ING Bulletins on Regional Assessment of Reactive Nitrogen, No. 3. SCON-ING, New Delhi, 18 pp.

Bin, X., Wilson, D., Broglio, R., 2015. Lower-Cost Alternative De-NOx Solutions for Coal-Fired Power Plants. Power Engineering. Available at: http://www.power-eng.com/articles/print/volume-119/issue-12/features/lower-cost-alternative-de-nox-solutions-for-coal-fired-power-plants.html.

BOBLME [Bay of Bengal Large Marine Ecosystem Project], 2011. Country Report on Pollution – India. BOBLME-2011-Ecology-07, pp. 1–200.

Braun, E., 2007. Reactive Nitrogen in the Environment: Too Much or Too Little of a Good Thing. UNEP/Earthprint, Paris, pp. 1–51.

Census of India 2011, 2011. Provisional Population Totals. Paper 2. In: Series 1. Rural-Urban Distribution. Office of the Registrar General and Census Commissioner, India Ministry of Home Affairs, Govt. of India, vol. 1.

CPCB [Central Pollution Control Board], 2009. Status of Water Supply, Wastewater Generation and Treatment in Class 1 Cities & Class II Towns of India. Control of Urban Pollution Series: CUPS/70/2009 – 10. Central Pollution Control Board (Ministry of Environment and Forests, Govt. of India), Parivesh Bhawan, East Arjun Nagar, Delhi, pp. 1–88, 110032.

CPCB [Central Pollution Control Board], 2015. Inventorization of Sewage Treatment Plants. Control of Urban Pollution Series: CUPS//2015. Central Pollution Control Board (Ministry of Environment and Forests, Govt. of India), Parivesh Bhawan, East Arjun Nagar, Delhi, pp. 1–81, 110032.

CSO [Central Statistics Office], 2016. Energy Statistics India-2016. Ministry of Statistics and Programme Implementation. Govt. of India, New Delhi.

EU [European Union], 2016. EU Reference Scenario 2016 Energy, Transport and GHG Emissions Trends to 2050. Luxembourg. https://ec.europa.eu/energy/sites/ener/files/documents/ref2016_report_final-web.pdf.

FAI [Fertilizer Association of India], 2015. Fertilizer Statistics 2015-15, sixtieth ed. The Fertilizer Associaiton of India, New Delhi.

Frailey, M., Norton, P., Clark, N.N., Lyons, D.W., 2000. An evaluation of natural gas versus diesel in medium duty buses. In: SAE Technical Paper Series 2000-01-2822, Warrendale, Pennsylvania.

Gadde, B., Menke, C., Wassmann, R., 2009. Rice straw as a renewable energy source in India, Thailand, and the Philippines: overall potential and limitations for energy contribution and greenhouse gas mitigation. Biomass and Bioenergy 33, 1532–1546. http://dx.doi.org/10.1016/j.biombioe.2009.07.018.

Galloway, J.N., Cowling, E.B., 2002. Reactive nitrogen and the world: 200 years of change. Ambio 31, 64–71.

Gupta, S., Gupta, R.C., Gupta, A.B., Prakasa Rao, E.V.S., Puttanna, K., Singhvi, A., 2010. Pathophysiology of nitrate toxicity in human and its mitigation measures. In: Raghuram, N. (Ed.), ING Bulletins on Regional Assessment of Reactive Nitrogen, Bulletin No. 20. SCON-ING, New Delhi, 78 pp. http://mospi.nic.in/sites/default/files/publication_reports/Energy_statistics_2016.pdf.

IFA [International Fertilizer Association], 2014. International Fertilizer Association 2014 Database. http://fertilizer.org/ifa/Home-Page/STATISTICS.

INCCA [Indian Network for Climate Change Assessment], 2010. India: Greenhouse Gas Emissions 2007 Ministry of Environment and Forests. Government of India.

India-BUR [India First Biennial Update Report to the United Nations Framework Convention on Climate Change], 2015. Ministry of Environment, Forest and Climate Change. Government of India.

IPCC [Intergovernmental Panel on Climate Change], 2006. In: Eggleston, H.S., Buendia, L., Miwa, K., Ngara, T., Tanabe, K. (Eds.), IPCC Guidelines for National Greenhouse Gas Inventories, Prepared by the National Greenhouse Gas Inventories Programme. Published: IGES, Japan.

Jain, N., Bhatia, A., Pathak, H., 2014. Emission of air pollutants from crop residue burning in India. Aerosol Air Quality Research 14, 422–430. http://dx.doi.org/10.4209/aaqr.2013.01.0031.

Law, Y., lant, P., Yuan, Z., 2012. The effect of pH on $N_2O$ production under aerobic conditions in a partial nitration system. Water Research 45, 5934–5944.

Lok Sabha, 2016. Question No 801. Unstarred Question. Answered on 21.07.2016 Namami Ganga Yojana, Answered by the Minister for Water Resources, River Development and Ganga Rejuvenation. Govt. of India.

MoA [Ministry of Agriculture], 2015. Agriculture Statistics at a Glance- 2014. Published by Directorate of Economics and Statistics. Department of Agriculture and Cooperation, Ministry of Agriculture, Govt. of India and Oxford University Press, New Delhi, India.

MoPNG [Ministry of Petroleum and Natural Gas], 2016. Indian Petroleum and Natural Gas Statistics 2015-16. Economics and Statistics Division, Ministry of Petroleum and Natural Gas. Govt. of India, pp. 1–216. http://petroleum.nic.in/docs/pngstat.pdf.

Mosier, A.R., Bleken, M.A., Chaiwanakupt, P., Ellis, E.C., Freney, J.R., Howarth, R.B., Matson, P.A., Minami, K., Naylor, R., Weeks, K.N., Zhu, Z.-L., 2002. Policy implications of human accelerated nitrogen cycling. Biogeochemistry 57–58, 477–516.

MoSPI, 2016. Ministry of Statistics and Programme Implementation. Statistical Year Book, 2016. Motor Vehicles. Available at: http://www.mospi.gov.in/statistical-year-book-india/2016/189.

NATCOM II [National Communication II], 2012. India Second National Communication to the United Nations Framework Convention on Climate Change. Ministry of Environment and Forests. Government of India.

NSSO [National Sample Survey Office], 2015. Energy sources of indian households for cooking and lighting, 2011-12. In: NSSO 68th Round. (July 2011–June 2012) National Sample Survey Office. Ministry of Statistics and Programme Implementation. Govt. of India. http://mospi.nic.in/sites/default/files/publication_reports/nss_report_567.pdf.

Park, K.Y., Inamori, Y., Mizuochi, M., Ahn, K.H., 2000. Emission and control of nitrous oxide from a biological wastewater treatment system with intermittent aeration. Journal of Bioscience and Bioengineering 90, 247–252.

Pathak, H., 2016. Is nitrogen fertiliser a polluter or producer? The answer lies in its use-efficiency. Indian Journal of Fertility 12, 106–117.

Pellicer-Nàcher, C., Sun, S., Lackner, S., Terada, A., Schreiber, F., Zhou, Q., Smets, B.F., 2010. Sequential aeration of membrane-aerated biofilm reactors for high-rate autotrophic nitrogen removal: experimental demonstration. Environmental Science and Technology 44, 7628–7634. http://dx.doi.org/10.1021/es1013467.

Penman, J., Kruger, D., Galbally, I., Hiraishi, T., Nyenzi, B., Emmanul, S., Buendia, L., Hoppaus, R., Martinsen, T., Meijer, J., Miwa, K., Tanabe, K. (Eds.), 2000. Good Practice Guidance and Uncertainty Management in National Greenhouse Gas Inventories. Institute for Global Environmental Strategies, Japan for IPCC.

PPAC [Petroleum, Planning and Analysis Cell of the Ministry of Oil and Natural Gas], 2013. Report – All India Study on Sectoral Demand of Diesel & Petrol, Ministry of Oil and Natural Gas. Govt. of India, Nielsen, New Delhi, pp. 1–103.

Praptijanto, A., Muharam, A., Nur, A., Putrasari, Y., 2015. Effect of ethanol percentage for diesel engine performance using virtual engine simulation tool. Energy Procedia 68, 345–354.

Ramachandra, T.V., Shwetmala, 2009. Emissions from India's transport sector: statewise synthesis. Atmospheric Environment 43, 5510–5517.

Reis, S., Bekunda, M., Howard, C.M., Karanja, N., Winiwarter, W., Yan, X., Bleeker, A., Sutton, M.A., 2016. Synthesis and review: tackling the nitrogen management challenge: from global to local scales. Environmental Research Letters 11, 120205. http://dx.doi.org/10.1088/1748-9326/11/12/120205.

Seitzinger, S.P., Mayorga, E., Bouwman, A.F., Kroeze, C., Beusen, A.H.W., Billen, G., Van Drecht, G., Dumont, E., Fekete, B.M., Garnier, J., Harrison, J.A., 2010. Global river nutrient export: a scenario analysis of past and future trends. Global Biogeochemical Cycles 24, GB0A08. http://dx.doi.org/10.1029/2009GB003587.

Sharholy, M., Ahmad, K., Mahmood, G., Trivedi, R.C., 2008. Municipal solid waste management in Indian cities – a review. Waste Management 28, 459–467.

Sutton, M.A., Bleeker, A., Howard, C.M., Bekunda, M., Grizzetti, B., de Vries, W., van Grinsven, H.J.M., Abrol, Y.P., Adhya, T.K., Billen, G., Davidson, E.A., Datta, A., Diaz, R., Erisman, J.W., Liu, X.J., Oenema, O., Palm, C., Raghuram, N., Reis, S., Scholz, R.W., Sims, T., Westhoek, H., Zhang, F.S., et al., 2013. Our Nutrient World: The Challenge to Produce More Food and Energy With Less Pollution. Global Overview of Nutrient Management, ISBN 978-1-906698-40-9. CEH, Edinburgh, for GPNM and INI.

Swaney, D.P., Hong, B., Paneer Selvam, A., Howarth, R.W., Ramesh, R., Purvaja, R., 2015. Net anthropogenic nitrogen inputs and nitrogen fluxes from Indian watersheds: an initial assessment. Journal of Marine System 141, 45–58.

Yang, Q., Liu, X., Peng, C., Wang, S., Sun, H., Peng, Y., 2009. $N_2O$ production during nitrogen removal via nitrite from domestic wastewater: main sources and control method. Environmental Science and Technology 43, 9400–9406. http://dx.doi.org/10.1021/es9019113.

Zheng, H., Hanaki, K., Matsuo, T., 1994. Production of nitrous oxide gas during nitrification of wastewater. Water Science Technology 30, 133–141.

# Further Reading

Garg, A., Shukla, P.R., Bhattacharya, S., Dadhwal, V.K., 2001. Sub-region (district) and sector level $SO_2$ and NO($x$) emissions for India: assessment of inventories and mitigation flexibility. Atmospheric Environment 35, 703–713.

Government of India, 2014. Guidelines for Swachh Bharat Mission (Gramin). Available at: http://phed.bih.nic.in/Docs/Guidelines-Swachh-Bharat-Abhiyan.pdf.

Mukherjee, S., Chakraborty, D., 2016. Turning human waste into renewable energy: scope and options for India. In: MPRA Paper No. 73669, Posted 13 September 2016 11:29 UTC. Online at: https://mpra.ub.uni-muenchen.de/73669/.

SACEP [South Asian Co-operative Environment Program], 2015. Nutrient Loading and Eutrophication of Coastal Waters of the South Asian Seas – a Scoping Study. SACEP, Colombo, Sri Lanka, pp. 1–44.

Sarkhel, P., 2012. Municipal solid waste management in India: environmental impacts and institutional reforms. In: Mukherjee, S., Chakraborty, D. (Eds.), Environmental Scenario in India: Successes and Predicaments. Routledge, Abingdon, pp. 171–188.

# Index

'*Note:* Page numbers followed by "f" indicate figures, "t" indicate tables and "b" indicate boxes.'

**B**

## C

**N**

## S

Papier | Fördert gute Waldnutzung
FSC® C083411

Kolibri 360 GmbH
Ferdinand-Jühlke-Straße 7
99095 Erfurt, Deutschland
produktsicherheit@kolibri360.de